Green Politics

Green Politics

An A-to-Z Guide

Dustin Mulvaney, General Editor
University of California, Berkeley

Paul Robbins, Series Editor
University of Arizona

Los Angeles | London | New Delhi
Singapore | Washington DC

Los Angeles | London | New Delhi
Singapore | Washington DC

FOR INFORMATION:

SAGE Publications, Inc.
2455 Teller Road
Thousand Oaks, California 91320
E-mail: order@sagepub.com

SAGE Publications Ltd.
1 Oliver's Yard
55 City Road
London EC1Y 1SP
United Kingdom

SAGE Publications India Pvt. Ltd.
B 1/I 1 Mohan Cooperative Industrial Area
Mathura Road, New Delhi 110 044
India

SAGE Publications Asia-Pacific Pte. Ltd.
33 Pekin Street #02-01
Far East Square
Singapore 048763

SAGE Publications
Publisher: Rolf A. Janke
Assistant to the Publisher: Michele Thompson
Senior Editor: Jim Brace-Thompson
Production Editors: Kate Schroeder, Tracy Buyan
Reference Systems Manager: Leticia Gutierrez
Reference Systems Coordinator: Laura Notton
Typesetter: C&M Digitals (P) Ltd.
Proofreader: Rae-Ann Goodwin
Indexer: Julie Sherman Grayson
Cover Designer: Gail Buschman
Marketing Manager: Kristi Ward

Golson Media
President and Editor: J. Geoffrey Golson
Author Manager: Ellen Ingber
Editors: Jill Coleman, Mary Jo Scibetta
Copy Editors: Anne Hicks, Barbara Paris

Printed in the United States of America

Library of Congress Cataloging-in-Publication Data

Green politics : an A-to-Z guide / general editor, Dustin Mulvaney.

p. cm. — (The Sage references series on green society: toward a sustainable future)
Includes bibliographical references and index.

ISBN 978-1-4129-9679-2 (cloth)—ISBN 978-1-4129-7186-7 (ebk)

1. Environmental policy. 2. Environmental protection—Political aspects. 3. Conservation of natural resources—Political aspects. 4. Sustainability—Political aspects. I. Mulvaney, Dustin.

GE170.G733 2010 320.5′8—dc22 2011002867

11 12 13 14 15 10 9 8 7 6 5 4 3 2 1

Contents

About the Editors

Green Series Editor: Paul Robbins

Paul Robbins is a professor and the director of the University of Arizona School of Geography and Development. He earned his Ph.D. in Geography in 1996 from Clark University. He is General Editor of the *Encyclopedia of Environment and Society* (2007) and author of several books, including *Environment and Society: A Critical Introduction* (2010), *Lawn People: How Grasses, Weeds, and Chemicals Make Us Who We Are* (2007), and *Political Ecology: A Critical Introduction* (2004).

Robbins's research centers on the relationships between individuals (homeowners, hunters, professional foresters), environmental actors (lawns, elk, mesquite trees), and the institutions that connect them. He and his students seek to explain human environmental practices and knowledge, the influence nonhumans have on human behavior and organization, and the implications these interactions hold for ecosystem health, local community, and social justice. Past projects have examined chemical use in the suburban United States, elk management in Montana, forest product collection in New England, and wolf conservation in India.

Green Politics General Editor: Dustin Mulvaney

Dustin Mulvaney is a Science, Technology, and Society postdoctoral scholar at the University of California, Berkeley, in the Department of Environmental Science, Policy, and Management. His current research focuses on the construction metrics that characterize the life cycle impacts of emerging renewable energy technologies. He is interested in how life cycle assessments focus on material and energy flows and exclude people from the analysis, and how these metrics are used to influence investment, policy, and social resistance. Building off his work with the Silicon Valley Toxics Coalition's "just and sustainable solar industry" campaign, he is looking at how risks from the use of nanotechnology are addressed within the solar photovoltaic industry. Mulvaney also draws on his dissertation research on agricultural biotechnology governance to inform how policies to mitigate risks of genetically engineered biofuels are shaped by investors, policymakers, scientists, and social movements.

Mulvaney holds a Ph.D. in Environmental Studies from the University of California, Santa Cruz, and a Master of Science in Environmental Policy, and a Bachelor's Degree in Chemical Engineering, both from the New Jersey Institute of Technology. Mulvaney's previous work experience includes time with a Fortune 500 chemical company working on sulfur dioxide emissions reduction, and for a bioremediation startup that developed technology to clean groundwater pollutants like benzene and MTBE.

Introduction

A hallmark of the past 100 years or so is the greening of political thought and practice. Today there are green political parties, green products, and green political institutions, all of which show how our decisions to organize, consume, and contribute are infused with green politics. Green politics has grown in the popular imagination as well. Everyday there are headlines about climate change, impacts of resource extraction, or chemical pollution in poor neighborhoods. Underlying all of these stories are classic political questions about power, representation, and values.

This collection of entries comprises some of the essential concepts, actors, institutions, and processes in green politics today. It draws on several academic arenas from economics and political science to political ecology and sociology, and focuses on a range of green issues from industrial pollution to indigenous rights. We hope the collection will provide the necessary explanatory frameworks, hypotheses, and case studies to advance important conversations about a just and sustainable future for human civilization.

Much of green politics is, at its root about sustainability, the notion that humans need to protect and conserve Earth's ecological and biogeochemical systems for the long run. Sustainability is a scientific and technical problem at the same time it is a cultural and political one. The difficult challenges we face today require engineers and scientists, but also those with an adept understanding of the political institutions, cultural norms, and social process that shape human civilization's relationship with the planet. It also requires a broader conception of green politics itself. Green politics is not simply about green political parties, environmental law, or deliberating citizen juries, but about everyday activities from deciding to buy organic food to riding a bus.

In many ways green politics is about values. Over the past 100 years, much has been said in political discourse about how to value the nature and the environment. The early debates between conservationists and preservationists, pitted those who wanted to conserve nature for human use against those who sought to preserve nature for its own sake. The seemingly similar ends turn out to have very different outcomes as demonstrated by the political battle over Hetch Hetchy, a large dam placed in the sister valley to Yosemite. John Muir wanted to preserve the valley to inspire future generations, while Gifford Pinchot and others wanted to use the valley for a dam to bring water 300 miles to San Francisco for drinking water and to provide an adequate water supply to fight fires like the one that destroyed the city after the 1906 earthquake.

While this moment illustrates a triumph of conservation over preservation, in reality much of the 20th century saw an exponential growth in areas preserved. This was not simply a U.S. affair either. Vast areas were protected in the tropics and developing countries in the latter part of the 20th century, in many cases pointing to inequitable enclosures of farmers or indigenous lands. The gap between preservation and conservation is being bridged as new forms of ecosystem protection have moved beyond the people versus parks divide and to more integrated approaches to ecosystem protection with multiple levels of human interaction.

The popularization of green politics is often attributed to the *New Yorker*'s serialization of Rachel Carson's *Silent Spring*. This widely read work challenged the chemical corporation's assessments of their own pesticides and attributed high rates of cancer and bird population declines to the use of persistent chemical pesticides. Carson faulted a lack of government regulation for overuse of chemicals. This and several other notable moments in the 1960s—the Cuyahoga River fire, the oil spill in Santa Barbara—led to the passage of major environmental legislation in the United States including the National Environmental Policy, Endangered Species, and Clean Air, and Clean Water acts.

Green politics continued its ascent after another wave of toxics controversies and the emergence of the environmental justice movement. The toxic waste stories out of Love Canal led to the passage of new laws for hazardous waste and provided ammunition for those fighting for environmental justice in some of the country's poorest areas near incinerators, landfills, and petrochemical refineries. These environmental justice groups point to the disproportionate burden of toxic chemicals in their industrialized communities and have helped passed some of the strongest community "right to know" laws in the world.

The greening of politics is evident in the United States in and around Washington, D.C. In the 2008 presidential contest, the green jobs discourse slowly became common tongue by both major political parties. More recently, Congress on both sides of the isles are talking about green jobs as in the midst of an economic recession, showing how the notion that the triple bottom line—protecting jobs, people, and the planet—has gained prominence in the mainstream political discussions, even as adherence to the idea is in and of itself a political act.

Green politics has also gone from the local to the global. From anti-dam movements in India to e-waste activism in Ghana, there is a burgeoning environmental justice movement in the developing world. Despite Lawrence Summers's claim that the third world is vastly underpolluted, these countries are seeking out their own space to make claims about what should and should not be exported to impoverished parts of the planet.

The global character of green politics is also represented in the various global treaties and institutions that have been created over the past century: The Montreal Protocol, the Basel Convention, and the Kyoto Protocol of the United Nation's Framework Convention on Climate Change (UNFCCC). The December 2009 UNFCCC's Conference of the Parties 15 (COP 15) in Copenhagen was one of the most widely covered international environmental negotiations. The Copenhagen Accord—produced in parallel to the UNFCCC process by the United States, India, South Africa, and China—further illustrates the complexity of international environmental negotiations. While getting these countries to commit to emission reductions, thorny questions remain about the extent of emission reductions, cap and trade versus carbon tax, the balance of "common but differentiated responsibilities," and a timetable to achieve emissions reductions. Other outcomes from COP 15 that demonstrate its global significance and complexity include an agreement to reduce emissions from deforestation and a pledge from developed countries of $100 billion annually to help decarbonize developing countries. While the former is seen as a success, concerns that the World Bank will administer the latter fund has left some suspicious because of past projects that have caused environmental harms or exacerbated poverty.

Assigning blame for environmental problems is also inherently political. Of the more contentious ideas in green politics are Malthusianism and "the tragedy of the commons." Malthusian explanations for environmental degradation focus on overpopulation as its root cause at the expense of the high rates of consumption in industrialized countries. Such positions are criticized for placing the burden on the world's poorest, while the developed world continues to consume unabated. Likewise, the tragedy of the commons, the idea that a lack of property rights will cause users to overuse a resource, implies that commodification of everything is the solution to what are otherwise environmental externalities. So where green politics is a struggle between privatization and cooperation, it is important to note that the different policy tools imply different kinds of political assumptions about how the world works. Neoliberal approaches focus on the market as the most efficient means of protecting resources, while the deep ecologists, anarchists, ecosocialists, counterculturalists, pragmatists, and ecofeminists would each have their own say about the best means to achieve those goals.

Green political discourse has become commonplace, but just how we engage in green politics raises some critical questions. Who gets to participate in green politics? Who does green politics represent? This volume seeks to open a dialog about the shape and trajectory of green political thought as it increasingly becomes a part of our everyday lives. As you will find, all shades of green are not created equal.

Dustin Mulvaney
General Editor

Reader's Guide

Politics and Ecology

Biophilia
Biosphere
Deep Ecology
Ecocentrism
Ecological Economics
Ecological Imperialism
Ecological Modernization
Ecologism
Ecology
Ecotax
Industrial Ecology
Political Ecology
Social Ecology

Politics and People

Counterculture
Indigenous Peoples
People, Parks, Poverty
Pinchot, Gifford

Politics Challenges

Acid Rain
Afforestation
Anti-Toxics Movement
Appropriate Technology
Biodiversity
Deforestation
Decentralization
Domination of Nature
Endocrine Disrupters
Environmental Justice
Environmental Management
Equity
Future Generations
Global Climate Change
Globalization
Groundwater
Industrial Revolution
Innovation, Environmental
Kuznets Curve
Limits to Growth
Malthusianism
Megacities
Millennium Development Goals
Nonviolence
North–South Issues
Nuclear Politics
PCBs
Precautionary Principle
Regulatory Approaches
Resource Curse
Revolving Door
Risk Assessment
Risk Society
Silent Spring
Structural Adjustment
Suburban Sprawl
Sustainable Development
Technology
Toxics Release Inventory
Tragedy of the Commons
Transportation
Uncertainty

Politics Laws, Agreements, and Organizations

Politics Parties, Systems, and Economics

List of Articles

List of Contributors

Amar, Shaista Consuelo
University of Houston–Downtown

Arney, Jo A.
University of Wisconsin–La Crosse

Babou, Codandapani
Adhiparasakthi Agriculture College

Baruah, Mitul
State University of New York College of Environmental Science and Forestry

Beder, Sharon
University of Wollongong

Birchler, Susan
Northern Virginia Community College

Böcher, Michael
Georg-August-University

Bohr, Jeremiah
University of Illinois at Urbana-Champaign

Borne, Gregory
University of Plymouth

Boslaugh, Sarah
Washington University in St. Louis

Bremer, Leah
San Diego State University

Bruggeman, James
Independent Scholar

Carr, David L.
University of California, Santa Barbara

Chatterjee, Sudipto
HSG, Sagar University

Chiaviello, Anthony R. S.
University of Houston–Downtown

Coleman, Jill S. M.
Ball State University

Collins, Timothy
Western Illinois University

Darby, Kate
Arizona State University

Davidsen, Conny
University of Calgary

de Souza, Lester
Independent Scholar

Downie, David
Fairfield University

Dudley, Michael Quinn
University of Winnipeg

Duffy, Lawrence
University of Alaska, Fairbanks

Eatmon, Thomas D., Jr.
Allegheny College

Edwards, Ferne
Australian National University

Evans, Tina Lynn
Fort Lewis College

Eysenbach, Derek
University of Arizona

Farley, Kathleen A.
San Diego State University

Finley-Brook, Mary
University of Richmond

Futrell, W. Chad
Cornell University

Gareau, Brian J.
Boston College

Gopakumar, Govind
Rensselaer Polytechnic Institute

Graddy, Garrett
University of Kentucky

Gunter, Michael M., Jr.
Rollins College

Hards, Sarah
University of York

Harper, Gavin D. J.
Cardiff University

Harrington, Jonathan
Troy University

Holden, Madronna
Oregon State University

Holst, Arthur Mathew
Widener University

Howell-Moroney, Michael
University of Alabama at Birmingham

Hurst, Kent
University of Texas at Arlington

Ilangovan, Kumaraswamy
Pondicherry University

Jain, Priyanka
University of Kentucky

Jinnah, Sikina
Brown University

Karlsson, Rasmus
Lund University

Kedzior, Sya Buryn
University of Kentucky

Kelly, Jessica
Millersville University

Kinsella, William J.
North Carolina State University

Kirchhoff, Christine J.
University of Michigan

Köppel, Martin
Independent Scholar

Krishnan, Sinduja
Centre for Environment Education

Kte'pi, Bill
Independent Scholar

Leonard, Liam
Institute of Technology, Sligo

Long, Andrew
Florida Coastal School of Law

Lopes, Paula Duarte
University of Coimbra

Makdisi, Karim
American University of Beirut

Matthews, Todd L.
University of West Georgia

Moran, Sharon
State University of New York College of Environmental Science and Forestry

O'Sullivan, John
Gainesville State College

Patnaik, Rasmi
Pondicherry University

Plec, Emily
Western Oregon University

Pokrant, Bob
Curtin University of Technology

Poyyamoli, Gopalsamy
Pondicherry University

Roka, Krishna
Penn State University

Schmook, Birgit
El Colegio de la Frontera Sur (ECOSUR)

Schnurr, Matthew
Dalhousie University

Siry, Christina
Manhattanville College

Smith, Susan L.
Willamette University College of Law

Snell, Carolyn
University of York

Soria, Carlos Antonio Martin
Universidad Nacional Agraria La Molina / Instituto del Bien Comun

Terrizzi, Alexis
Fairfield University

Vachta, Kerry E.
Wayne State University

Valdivia, Gabriela
University of North Carolina at Chapel Hill

van Bueren, Ellen M.
Delft University of Technology

Van Hooreweghe, Kristen
City University of New York

Waskey, Andrew Jackson
Dalton State College

Woods, Mark
University of San Diego

Yandle, Bruce
Clemson University

Yuhas, Stephanie
University of Denver

Zimmermann, Petra A.
Ball State University

Zivian, Anna Milena
University of California, Santa Cruz

Green Politics Chronology

500: Roman cannon of law, the Justinian Code, establishes "sun rights" to ensure that all buildings have access to the sun's warmth after a long-standing practice of building south-facing windows on buildings to keep them comfortable in the winter.

1306: English King Edward I unsuccessfully tries to ban open coal fires in England, marking an early attempt at national environmental protection. Legend has it that his mother's constant complaining over the thick smoke clouding London leads to the attempted legislation.

1347–1350: In just 3 short years, the Bubonic Plague, a disease transferred from rats to humans, spreads across European trade routes, engulfing the continent and decimating over one-third of the population.

1600s: The Dutch master drainage windmills, moving water out of low-lying lands to make farming available. During the Protestant Reformation, they use windmill positions to communicate to Catholics, indicating safe places for asylum.

1690: Progressive Governor William Penn requires that 1 acre of forest be saved for every 5 that is cut down in the newly formed city of Philadelphia.

1800s: Advances in agricultural and urban technology lead to the First Industrial Revolution. Unprecedented amounts of toxins are released into the air and water. Later in the century, scientists begin to study the possibility of certain chemicals contributing to an increase in Earth's average temperature—a process now referred to as global warming.

1862: U.S. President Abraham Lincoln creates the Department of Agriculture, charged with promoting agriculture production and the protection of natural resources.

1879: The U.S. Geological Survey is established, responsible for examining national geological structure and natural resources.

1912: In one of the nation's earliest efforts to curb industrial and urban waste, the U.S. Congress passes the Public Health Service Act of 1912, which gave significant funding to the U.S. Public Health Service to study problems such as sewage, sanitation, and pollution.

1914–1918: World War I rages throughout Europe. Nations begin to master industrial technologies, and the modern military–industrial complex is born.

1920: The Federal Power Act establishes the Federal Power Commission, an independent agency responsible for coordinating federal hydropower development. The commission is later given authority over natural gas facilities and electricity transmission. It is eventually overtaken by the Department of Energy.

1929–1970: Venezuela is the world's top oil exporter.

1932: The nucleus of an atom is split for the first time in a controlled environment by physicists John Cockcroft and Ernest Walton under the direction of Ernest Rutherford.

1939: In August, a letter is drafted by Hungarian physicist Leo Szilard, signed by American Albert Einstein, and addressed to U.S. President Franklin D. Roosevelt, advising the president to fund nuclear fission research as means to create a weapon in the event that Nazi Germany may already be exploring the possibility. In September, Germany invades Poland, beginning World War II. In October, a secret meeting results in the creation of the Advisory Committee on Uranium for the purpose of securing the element and using it in research to create an atomic weapon.

1940: The British MAUD (Military Application of Uranium Detonation) Committee is established for the purposes of investigating the possibility of using uranium in a bomb. The next year they publish a report detailing specific requirements for its creation.

1942: The Manhattan Engineering District, also known as the Manhattan Project, is established by the Army Corps of Engineers, directed by physicist Robert Oppenheimer. The project makes the development of a nuclear weapon a top army priority and begins to outline methods for construction, testing, and transportation of an atomic bomb.

1945: In July, the world's first nuclear explosion—the "Trinity" test—occurs in the desert of New Mexico. In August, American forces detonate two atomic bombs on Japanese soil. The cities of Hiroshima and Nagasaki were effectively obliterated, losing 140,000 and 80,000 lives, respectively. In the following week, Japan surrenders to Allied powers.

1954: The first Russian nuclear power plant opens.

1955–2003: First proposed in 1955 as the Air Pollution Control Act and later as the Clean Air Act of 1963, the Air Quality Act of 1967, the Clean Air Act Extension of 1970, and the Clean Air Act Amendment of 1977 and 1990 all enact similar legislation regarding hazardous emissions into the atmosphere. Eventually, the Clean Air acts are met with criticism because of their bureaucratic methods. The Clear Skies Act of 2003 amends much of the previous legislation. A significant amount of the new provisions are directed toward energy companies.

1962: Experimental developments into satellite communications between America and Britain, across the Atlantic Ocean, prove successful when Telstar 1 is launched. The world's first working communication satellite delivers transatlantic phone calls, television pictures, and fax messages between the United States and the United Kingdom.

1962: Rachel Carson's *Silent Spring* is a national phenomenon, first published in serial form in the New Yorker and then as a hardcover best seller. This exhaustively researched, carefully reasoned attack on the indiscriminate use of pesticides sparks a revolution in public opinion.

1965: Young lawyer Ralph Nader writes *Unsafe at Any Speed,* an investigation revealing hazards in American automobiles. The book sparks public outrage and influences policy makers to pass laws requiring all cars to have seatbelts. Publicity derived from the book's success gives Mr. Nader a national stage for nonpartisan policy.

1970: Earth Day is made a national holiday to be celebrated on April 22 each year. It is founded by U.S Senator Gaylord Nelson as an environmental teach-in. Today, Earth Day is celebrated by dozens of countries throughout the world.

1970: On the first Earth Day, the Clean Air Extension Act of 1970 is signed into law by the U.S. Congress. Although previous air pollution legislation already existed, the act is immediately recognized as a benchmark for the environmentalist movement by being the first major environment protection law that includes a provision for civil lawsuits. It also sets a standard for all new vehicles to be certified as meeting auto emission standards.

1970: U.S. Congress passes the National Environmental Policy Act, establishing a national policy promoting the enhancement of the environment and creating the President's Council on Environmental Quality as a chief advisor on environmental issues. The most influential part of the act requires all federal agencies to prepare environmental assessments and environmental impact statements documenting the effects of their proposed actions.

1970: As concern with the condition of the physical environment intensifies, U.S. President Nixon creates the Environmental Protection Agency (EPA) to enforce federal environmental regulations. The agency's mission is to regulate chemicals and protect human health by safeguarding air, land, and water. The principle roles of the new agency include the establishment and enforcement of environmental protection standards consistent with national environmental goals; the conduct of research on the adverse effects of pollution and on methods and equipment for controlling it; the gathering of information on pollution and the use of this information in strengthening environmental protection programs and recommending policy changes; assisting others, through grants, technical assistance, and other means, in arresting pollution of the environment; and assisting the Council on Environmental Quality in developing and recommending to the president new policies for the protection of the environment.

1970: Several months after its creation, the EPA opens its doors, with Department of Justice lawyer William Doyle Ruckelshaus as the first administrator. During the EPA's formative years, Ruckelshaus concentrates on developing the new agency's organizational structure, taking enforcement actions against severely polluted cities and industrial polluters, setting health-based standards for air pollutants and standards for automobile emissions, requiring states to submit new air quality plans, and banning of general use of the pesticide dichlorodiphenyltrichloroethane (DDT). Mr. Ruckelshaus later serves as acting director of the Federal Bureau of Investigation.

1970: The National Oceanic and Atmospheric Administration is created for the purpose of developing efficient ways of using the country's marine resources.

1971: The EPA announces the final publication of National Air Quality Standards for six common classes of pollutants—sulfur oxides, particulate matter, carbon monoxide, photochemical oxidants, nitrogen oxides, and hydrocarbons. The emission standards are seen as very strict, and many industries are forced to develop new practices to avoid exceeding the standard. To comply, the city of New York must increase natural gas usage by 300 percent.

1971: U.S. and British scientists begin development of the first wave energy system.

1972: The world's first national environmentalist party, or green party, is formed as the Values Party in New Zealand. Party leaders never get elected to national office but manage to hold local positions.

1972: An outright ban on the use of the pesticide DDT is made by the EPA. In the previous 30 years, the widespread domestic use of the chemical is approximated at 675,000 tons, most of it applied to cotton.

1973: The Arab Oil Embargo begins as a response to the U.S. decision to supply the Israeli military during the Yom Kippur War between Arab nations and Israel.

1973: Aware that vehicles that pass certification standards often fall below those standards after extended use, the EPA regulates required warning systems, including dashboard lights and buzzers, designed to alert drivers that maintenance is required in ordered reduce toxic emissions.

1973: Lead-based gasoline begins to be phased out of the American economy as a result of concern over public health standards.

1974: The U.S. Department of Energy forms a branch dedicated to national research and development of solar energy—the Solar Energy Research Institute.

1974: After sharp increases in the price of oil from the Organization of the Petroleum Exporting Countries leads to a major American energy crisis, the Energy Reorganization Act is signed into law, replacing the Atomic Energy Commission with the Energy Research and Development Administration, responsible for oversight of nuclear weapons, and the Nuclear Regulatory Commission, responsible for commercial nuclear safety. The act also requires the future creation of the Strategic Petroleum Reserve, set to contain 1 million barrels of oil. Because of the same crisis, the U.S. government begins federally funding wind energy research through NASA and the Department of Energy, coordinated by the Lewis Research Center. Under harsh criticism for failing to avoid the crisis, The EPA writes an extensive position paper, identifying the individual household demand of oil as a much larger problem than any single agency can effectively curb.

1974: U.S. Congress passes the Safe Drinking Water Act, putting into motion a new national program to reclaim and ensure the purity of the water consumption. Under the act, each level of government, every local water system, and the individual consumer have well-defined roles and responsibilities.

1975: The Soviet Union and United States discuss joint research opportunities. A multinational agreement, signed in Moscow, leads to joint efforts by the two nations in 11 environmental areas of interest. These include research in the exchange of techniques for building pipelines in areas where the ground is always frozen, measuring the effects of pollutants on life in the ocean, developing methods for protecting such endangered species, and developing methods for control of air and water pollution.

1975: The U.S. government effectively cuts aid to companies in violation of pollution regulation, such as the Clean Air or Water acts. Federal agencies begin withholding contracts, grants, or loans to industrial and manufacturing plants, and various other facilities.

1975: In another response to the energy crisis, Corporate Average Fuel Economy regulations are passed by the U.S. Congress, intending to improve the average fuel economy of consumer vehicles. In 2002, the National Academy of Sciences reviews the regulations and finds they are responsible for a decrease in motor vehicle consumption by 14 percent.

1976: A specially built radiation monitoring van begins touring major metropolitan areas, reporting on the intensity of radiation, primarily emitting from television and radio transmitters.

1976: Passed by U.S. Congress, the Toxic Substance Control Act creates an inventory of regulated chemicals, including asbestos and lead. Any new chemical must be submitted to the EPA before it is manufactured.

1976: The U.S. Congress passes the Resource Conservation and Recovery Act, a landmark bill regulating the disposal of solid and other hazardous waste, after a period of increased municipal and industrial expansion leads to dangerous volumes of waste. Amending the Solid Waste Act of 1965, the new laws set national goals for protecting human health and the environment from the potential hazards of waste disposal, conserving energy and national resources, reducing the amount of waste generated, and ensuring that waste is managed in an environmentally safe manner. The act is expanded and strengthened in 1984.

1977: With the United States still reeling from the oil crisis, the U.S. Department of Energy is created. The new department will coordinate several already established programs, assuming the responsibilities of Energy Research and Development Administration. The Energy Information Administration is responsible for independent energy statistics. The Office of Secure Transportation provides secure transportation for nuclear weapons and materials. The Federal Energy Regulatory Commission is given jurisdiction over commercial energy including electricity and natural gas, as well as managing the Strategic Petroleum Reserve.

1977: The U.S. Congress passes the Clean Water Act 1977, which made major amendments to previous water pollution laws. The new legislation outlaws any discharge of hazardous waste into American waters and is eventually made more stringent with the Water Quality Act of 1987.

1978: The EPA and Department of Transportation join forces to help cities combat transportation pollution and improve air quality standards. The two agencies develop guidelines for urban planning, including increased public transportation and car pooling.

1978: The United States and Canada sign an agreement calling for programs and measures to further abate pollution in the Great Lakes, which are spread across the border.

1978: In a final measure to avoid another energy crisis like the one that began in 1973, the National Energy Act of 1978 is passed by the U.S. Congress. It includes a host of new statues attempting to redefine how the country secures, consumes, and comprehends energy. One of the most influential laws passed as part of the act, the Public Utility Regulatory Policies Act (PURPA), promotes the greater use of renewable energy. The law regulates a market for renewable energy products, forcing electric utility companies to purchase from these suppliers at a fixed price. In maybe the most significant result of PURPA, cogeneration plants become the industry standard. These new plants, encouraged by the act, generate electricity as well as usable steam, which is otherwise wasted. Another law enacted under the National Energy Act, the National Energy Conservation Policy Act, requires utility companies to employ energy management strategies designed to curb the demand for electricity. Another law enacted gives an income tax credit to private residents who use solar, wind, or geothermal sources of energy. Also created is the "gas guzzler tax," which makes the sale of vehicles with a gas mileage below a specified EPA-estimated level liable to fiscal penalty. The Power Plant and Industrial Fuel Use Act and the Natural Gas Policy Act are also passed as part of the National Energy Act.

1979: The Coordination of European Green and Radical Parties forms to coordinate many minority parties throughout the continent. By 1989, green parties would win 26 seats on the European Parliament.

1979: The EPA issues final regulations banning the manufacture and use of polychlorinated biphenyls (PCBs), a highly toxic substance widely used in industrial electrical equipment. PCBs are known to cause cancer and birth defects. The EPA estimates that 150 million pounds of the now outlawed substance is dispersed in domestic air, food, and water supplies, with an additional 290 million pounds stored in landfills.

1979: The German Green Party develops the Four Pillars, which become foundations for green policy groups worldwide. The pillars are ecological wisdom, social justice, grassroots democracy, and nonviolence.

1979: New environmental guidelines, known as the "Bubble Policy," are introduced to entice U.S. companies to explore ways of reducing environmental protection costs. The chemical giant the DuPont Corporation releases a report indicating that they may be able to cut air pollution expenditures from $139 million to $55 million.

1979: Nearly 700 families from the State of New York are displaced from their homes after being exposed to toxic wastes deposited by Hooker Chemical Company.

1980: The U.S. Congress signs the Comprehensive Environmental Response, Compensation, and Liability Act, better known as "Superfund." The new laws are designed to enforce the cleaning of hazardous waste sites. Broad authority is given to the EPA to assess the threat of release of any hazardous substance that may endanger public health or the environment and to remedy the situation using federal resources. In the following year, 114 sites are listed as top priorities for Superfund legislation.

1980: A partial core meltdown occurs at Three Mile Island Nuclear Generating Station, releasing radioactive gases into the Pennsylvania air. An investigation later concludes that no adverse health effects will be perceptible to the community.

1980: U.S. President Jimmy Carter signs the Energy and Security Act of 1980. It consists of six main groups of acts titled U.S. Synthetic Fuels Corporation, Biomass Energy and Alcohol Fuels, Renewable Energy Resources, Solar Energy and Conservation Act, Geothermal Energy, and Ocean Thermal Energy.

1980: As part of the Energy and Security Act, U.S. Synthetic Fuels Corporation is created to provide a viable market for domestically produced, man-made liquid fuel. Their goal is produce 2 million barrels a year within 5 years. In 1985, the corporation is abolished.

1980: The Crude Oil Windfall Profits Act creates what is technically an "excise tax" imposed on the difference between the market price of oil and a base price that is adjusted for inflation. It also increases tax credits for businesses using renewable energy.

1981: The EPA announces that 16 states have the authority to manage their own hazardous waste programs—the first states to manage waste independent of direct federal control.

1980: The American Council for an Energy-Efficient Economy is formed as a nonprofit organization. Their mission is to advance energy efficiency as a fast, cheap, and effective means of meeting energy challenges. The agency works on a state and federal level helping shape energy policy in favor of energy conservation, focusing on the end-use efficiency in industry, utilities, transportation, and human behavior.

1982: All U.S. public, private, and secondary schools are required to test for asbestos in their buildings.

1983: The EPA orders the immediate emergency suspension of ethylene dibromide (EDB) as a soil fumigant for agricultural crops.

1984: The EPA and Department of Defense sign a joint resolution pledging cooperative efforts in safeguarding the quality of Chesapeake Bay waters off the nation's eastern coast.

1984: The Green Committees of Correspondence is the first green political organization in the United States. Its mission is to coordinate many local green groups into a unified party, run for public office, create watchdog groups, and publish green literature.

1984: Disaster occurs in the Indian city of Bhopal as the Union Carbide pesticide plant releases some 42 tons of toxic gas, exposing more than 500,000 people to the hazardous substance. The immediate death toll is tallied at 2,259, and an estimated 8,000–10,000 are killed within 72 hours of the accident. Recognized as the world's worst industrial disaster, later reports approximate the total death toll at approaching 25,000 and anywhere between 100,000 and 200,000 related permanent injuries and adverse health effects. The incident gains worldwide attention and sparks international determination toward safeguarding industry.

1984: U.S. President Reagan signs the Asbestos School Hazard Abatement Act, which authorizes $600 million in federal loans and grants to schools with severe asbestos hazards and demonstrated financial need.

1985: The EPA redefines policy for air toxins, identifying hazards from multiple sources instead of regulating individual pollutants.

1986: The most significant nuclear meltdown in history occurs in Chernobyl, Ukraine. The entire area is subject to nuclear fallout. With thousands of local residents unable to evacuate, generations of families suffer from intense radiation.

1986: Despite concerns over the efficiency of the program, Superfund is amended and reauthorized to incorporate recommendations made in the years following its inception.

1987: With an approaching deadline for regulations enacted by the Clean Air Act that many states do not seem prepared to satisfy, EPA administrator Lee M. Thomas writes an open letter to the 42 state governors. The letter outlines proposed plans for delinquent states, saying that "much long-term planning will need to be done to bring about necessary changes in some of the worst areas." The EPA also estimates that about 70 percent of the nation is not meeting ozone pollution standards.

1987: The National Appliance Energy Conservation Act authorizes the Department of Energy to set minimum efficiency standards for space-conditioning equipment and other larger energy-consuming appliances each year, based on what is "technologically feasible and economically justified." Televisions are originally part of the act's regulations but are later removed.

1988: The EPA issues comprehensive, stringent requirements for the nearly 2 million underground storage tanks, about half of which are used to store gasoline at service stations. Reports conclude that these tanks are susceptible to corrosion and eventual leakage into water supplies.

1989: The Montreal Protocol on Substances That Deplete the Ozone Layer is signed as an international treaty designed to protect the ozone layer by phasing out several substances that are known to be responsible for its harm. Analysis of the agreement claims that if the treaty is adhered to, the ozone layer will successfully recover by 2050.

1990: The U.S. Congress passes the Pollution Prevention Act of 1990, requiring the EPA to establish an Office of Pollution Prevention, develop and coordinate a pollution prevention strategy, and develop source reduction models. The act requires owners and operators of manufacturing facilities to report annually on source reduction and recycling activities.

1991: The U.S. federal government mandates that all of its agencies recycle reusable materials generated from wastes from federal activities. The executive order also encourages an efficient market for items produced using recycled materials.

1991: U.S. President George Bush announces that the Solar Energy Research Institute has been designated the National Renewable Energy Laboratory. Its mission is to develop renewable energy and energy-efficient technologies and practices, advance related science and engineering, and transfer knowledge and innovations to addressing the nation's energy and environmental goals by using scientific discoveries to create market-viable-alternative energy solutions.

1992: Under President Clinton, the Energy Policy Act of 1992 is passed by the U.S. Congress. It is organized under several titles enacting legislation on such subjects as energy efficiency, conservation, and management; electric motor vehicles; coal power and clean coal; renewable energy; alternative fuels; natural gas imports and exports; and various others. Among the new directives is a section that designates Yucca Mountain in Nevada as a permanent disposal site for radioactive materials from nuclear power plants. It also reforms the Public Utility Holding Company Act to help prevent an oligopoly and provides further tax credits for using renewable energy.

1992: Energy Star is established as a unified standard for energy efficient consumer products. The Energy Star logo begins to appear on things such as computers, kitchen appliances, laundry equipment, air conditioners, lighting, and various other energy-saving products. Consumers who own Energy Star–endorsed products can expect a 20 percent to 30 percent reduction in energy usage.

1992: The United Nations Conference on Environment and Development, known as the Earth Summit, is marked as a watershed moment in environmental history. The conference, held in Rio de Janeiro, also represents a diplomatic breakthrough, opening up the possibility for a new era of global economic growth coupled seriously with environmental stewardship. More than 150 countries make commitments to decrease greenhouse gases and address global warming prevention. The most influential part of the agreements, Agenda 21, establishes a global consensus on nearly all forms of pollution in the atmosphere, oceans, and other global resources.

1994: U.S. federal action is taken to address environmental justice in minority populations and low-income populations.

1996: The predecessor to the Green Party of the United States, the Association of State Green Parties, nominates attorney and political activist Ralph Nader and Native American environmentalist Winona LaDuke for president and vice president, respectively, for the upcoming election. The pair makes the ballot in 22 states and wins 0.7 percent of the popular vote.

1997: New comprehensive regulations are issued in the regulation of pesticides. Children are of particular importance in the Food Quality and Protection Act.

2000: The EPA propose a dramatic reduction in the sulfur content in diesel fuels after reports find nearly 15,000 deaths annually associated with soot and smog, mainly from trucks. The EPA also cites diesel fuel pollution as responsible for some 400,000 cases of asthma attacks in the country each year.

2000: The Biomass Research and Development Board is created as part of a U.S. Congress act attempting to coordinate federal research and development of bio-based fuels obtained from living (as opposed to long dead, fossil fuels) biological material, such as wood or vegetable oils. Biofuel industries begin to expand in Europe, Asia, and the Americas.

2000: The Green Party again nominates Ralph Nader and Winona LaDuke for president and vice president, respectively. They draw 2.7 percent of the national vote, one of the strongest showings from a third party in recent American history. The winner of the election, George W. Bush, narrowly edges out incumbent Vice President Al Gore in one of the closest presidential elections in the history of the office. Critics later speculate that had Mr. Nader not been on the ballot, Al Gore would have won. After the success of the election, the Green Party reorganizes and identifies their "10 Key Values"; "Ecological Wisdom," "Global Responsibility," and "Future Sustainability" are all on the list.

2001: The U.S. Congress signs the Convention on Persistent Organic Pollutants (POPs). POPs include already-outlawed—but still used—toxins such as PCBs and DDT.

2001: After the attacks of September 11, the EPA begins to coordinate extensively with the newly created Department of Homeland Security, refocusing their stance on environmental issues as a matter of national security.

2003: In his State of the Union address, U.S. President Bush announces a $1.2 billion Hydrogen Fuel Initiative to reverse the nation's growing dependence on foreign oil by developing the technology needed for commercially viable hydrogen-powered fuel cells. "A single chemical reaction between hydrogen and oxygen," the president tells the nation "generates energy, which can be used to power a car, producing only water, not exhaust fumes. With a new national commitment, our scientists and engineers will overcome obstacles to taking these cars from laboratory to showroom, so that the first car driven by a child born today could be powered by hydrogen, and pollution-free." The initiative will include $720 million in new funding over the next 5 years to develop the technologies and infrastructure to produce, store, and distribute hydrogen for use in fuel cell vehicles and electricity generation.

2003: The Clean School Bus USA program is announced by the EPA, ordered to reduce diesel exhaust exposure to nearly 24 million American children who spend, on average, an hour and a half on school buses each day.

2004: U.S. President Bush's new budget includes $288.2 million for Department of Energy's Weatherization Assistance Program, an increase of $11.2 million above the president's fiscal year 2003 request. By improving the energy efficiency in homes, the program will reduce the energy bills of approximately 126,000 low-income families nationwide in 2003.

2004: The European Green Party is formed from similar separate organizations, refocused on green politics, environmental responsibility, diversity, social justice, inclusive democracy, and nonviolence. The group now includes nearly every member of the European Union, including England, France, Germany, Italy, Russia, the Netherlands, and Spain.

2005: The Energy Policy Act is passed by the U.S. Congress and signed into law by George W. Bush, making sweeping reforms in energy legislation, mostly in the way of tax deductions and subsidies. Loans are guaranteed for innovative technologies that avoid greenhouse gases, and alternative energy resources such as wind, solar, and clean coal production are given multimillion dollar subsides. For the first time, wave and tidal power are included as separately identified renewable technologies. On the local level, individual tax breaks are given to Americans who make energy conservation improvements in their homes. However, total tax reductions greatly favor nuclear power and fossil fuel production, and the bill is later met with criticism. During the 2008 Democratic Primary, candidate Senator Hillary Clinton dubs it the "(Vice President) Dick Cheney lobbyist energy bill."

2007: The Energy Independence and Security Act of 2007 (originally named the Clean Energy Act of 2007) is passed by the U.S. Congress. Its stated purposes are "to move the United States toward greater energy independence and security, to increase the production of clean, renewable fuels, to protect consumers, to increase the efficiency of products, buildings and vehicles, to promote research on and deploy greenhouse gas capture and storage options, and to improve the energy performance of the Federal Government," as well as various other goals. Title I of the original bill is called the "Ending Subsidies for Big Oil Act of 2007." Included in the new provisions is a requirement of government and public institutions to lower fossil fuel use by 80 percent by 2020. Also included is the repeal of much of the legislation included in the Energy Policy Act of 2005.

2009: Amid a global recession, the American Recovery and Reinvestment Act of 2009 is one of the inaugural acts signed by President Barack Obama. Otherwise known as the "stimulus package," it mostly makes provisions for job creation, tax relief, and infrastructure investment, but it is also heavily focused on energy efficiency and science. Multibillion dollar funding is appropriated for energy-efficient building practices, green jobs, electric and hybrid vehicles, and modernizing the nation's electric grid into a smart grid that uses digital technology to save energy. In the official seal of the act, an illustration of a bright green, fertile plant is placed opposite two grinding cogs.

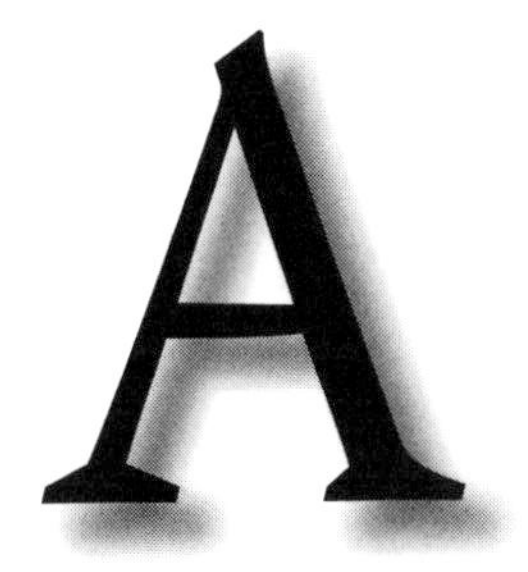

Acid Rain

The average pH of rainwater is 5.6 (pH levels are based on the concentrations of hydrogen ions with respect to pure water). Because water reacts with atmospheric carbon dioxide to form carbonic acid, rain is typically slightly acidic. Acid rain (or acid deposition), however, refers to atmospheric precipitates with pH values below that of rainwater. Acid-causing contaminants may originate in locations far from the acid precipitation, and so this issue crosses boundaries, necessitating political solutions.

Acid deposition is linked to numerous negative environmental impacts, affecting both the natural and built environments. Soils become more acidified, rendering them less hospitable to certain plants. Higher acid levels in lakes lead to a reduction in fish populations. Highly acidic precipitation exacerbates the weathering of many buildings, which is a particular problem with historical edifices. Limestone, for example, is weathered and eroded by acid rain. Although natural causes of acid rain exist, most acid deposition results from the by-products of anthropogenic actions, such as discharges from industrial activity. Natural causes such as volcanic eruptions, thunderstorm activity and lightning, and fire can, however, affect the atmosphere, leading to acidic precipitation.

Trees damaged by acid rain and a bark beetle infestation loom above new growth in Germany's Bavarian Forest near Mount Lusen.

Source: iStockphoto.com

Chemically, sulfur and nitrogen oxides produced by industrial operations enter the atmosphere. Among the largest sources of these oxides is the generation of electrical power, in which coal and oil are burned; vehicular traffic also contributes significantly. The consumption of fossil fuels in both cases sends oxides into the atmosphere. Sulfurous emissions, largely from the production of electricity, combine with atmospheric water and oxygen to produce sulfuric acid (H_2SO_4). Nitrogen oxides, typically from the combustion of petroleum-derived fuels, react similarly to create nitric acid (HNO_3).

Although acid rain may stem from pollution from localized point sources, the ramifications extend regionally. As such, acid rain is truly a regional issue. Atmospheric circulation patterns allow for a relatively efficient transport of pollutants to locales far removed from the original source. For example, acid rain in Scandinavia is often a product of oxides generated in Germany. Similarly, the acid rain falling in the northeastern United States can trace some of its origins to industrial activity in the Midwestern states. In the midlatitudes, prevailing winds are westerly; thus, atmospheric currents can transport these pollutants east of their origins. Mitigation of this problem, then, requires regional understandings and cooperation.

The Discovery of Acid Rain

The term *acid rain* was first used in 1872 by Robert Angus Smith, a Scottish chemist and environmental scientist who developed a simple model of acid deposition in the mid-1800s. His work focused on the British industrial center of Manchester. Despite Smith's initial foray into the problem of acid precipitation, it would be another century before it received the focused attention of scientists. Svante Odén, a Swedish soil scientist, argued that Swedish lakes were suffering environmental degradation that stemmed from acidic compounds generated by industrial activity elsewhere; specifically, (West) Germany and the United Kingdom. Odén's paper was the first to suggest that atmospheric transport could carry industrial contaminants over long distances to cause damage at locations far removed from the original point sources of the pollution. Among Odén's conclusions was that industrial activities and effects were regional, not local, issues.

Although Germany and the United Kingdom disagreed with Odén's results, his work was enough to spur the organization of the United Nations Conference on the Human Environment in 1972, which was the first global environmental conference. Also referred to as the Stockholm Convention (after the city in which it was held), it represented a shift toward a political and scientific awareness of acid deposition as a regional, and even international, phenomenon.

The work of Odén and his colleagues centered on the negative effects of acid deposition on Scandinavian ecosystems and drew the attention of the Norwegian government. Indeed, Odén used earlier Norwegian research to draw some of his conclusions. The natural environment, which is so important to Scandinavian culture, was important enough to warrant the creation (in 1972) of Norway's Acid Precipitation Effects on Forests and Fish Program, with participating scientists from 12 research institutions. Results indicated the widespread acidification of Norwegian forest soils and declines in fish populations.

Negative Effects and Politics

Concern about the negative effects of transported pollutants and acid rain extended beyond Scandinavia. The Organisation for Economic Co-operation and Development (OECD) sponsored the first international project to investigate the acid deposition problem. Eleven countries participated in the venture. In Germany, *Waldsterben* ("forest death"), or forest dieback, received considerable attention in the 1980s. Ostensibly caused by acid rain, this forest decline was said to affect half of all German forests. Bernhard Ulrich of Göttingen University asserted that acid deposition degraded the forest soils, leading to the decline in German woodlands. European policymakers had science to influence them.

Although Germany (along with the United Kingdom) was initially opposed to any policies requiring controls of pollution emission, the issue of *Waldsterben*, particularly in the Black Forest, and the emergence of the Green Party, spurred the German government to embrace and enact procontrol policies. The political turnaround in Germany helped lead to the adoption of the Helsinki Sulfur Dioxide Protocol in 1985.

Europe was not the only region contending with the problem of acid deposition. North America, too, was addressing the issue. Industrial activity in the Midwestern United States was linked to acid rain in the northeastern United States and eastern Canada. Both U.S. and Canadian scientists, following the lead of European scientists, examined acid deposition and its effects in both countries. Publicity yielded by this research created support for continued research. Acid rain was now an environmental issue with considerable public recognition. Because research showed that the transboundary pollutant flow was largely from the United States to Canada, both countries vowed to work together, signing the Memorandum of Intent Concerning Transboundary Air Pollution.

Search for Solutions

Under the administration of President Jimmy Carter, the United States addressed the problem by starting the Initiative on Acid Precipitation in 1979. The following year saw the establishment of the National Acid Precipitation Assessment Program (NAPAP), which functioned to bring together the science and policy sides of the issue. Specifically, the scientific research from the program would be used to inform policymakers. Despite the high aims of the program, it was hampered by the election loss of Carter to the probusiness, antiregulatory philosophy of Ronald Reagan. Reagan's Environmental Protection Agency director Ann Gorsuch Burford weakened NAPAP and the entire Environmental Protection Agency by slashing budgets and, in the case of NAPAP, ceasing important studies. A 1987 report from NAPAP did not find acid deposition to be a problem—a finding that was deemed political and unscientific.

The election of Reagan and the resulting changes in political agendas were setbacks for environmental research and policy, including that of acid deposition. Political tensions between the United States and Canada resulted from Canadian demands that American industrial polluters be regulated or be required to compensate Canada. The Canadians asserted that they were the victims of transboundary pollution, with the resulting acid deposition degrading Canadian soils, waters, and forests. To the consternation of Canadian officials, the Reagan administration took a dim view of scientific research regarding acid precipitation, deeming it scientifically uncertain. However, scientific work highlighting the growing problem of acid deposition continued.

In 1990, under U.S. president George H. W. Bush, the Clean Air Act was updated. This revision specifically addressed the acid rain issue in Title IV of the new act. The Acid Rain Program specifically focused on the reduction of SO_2 and NO_x emissions by using a market-based cap-and-trade approach. The program's initial phase commenced in 1995. Since then, SO_2 emissions have decreased, despite the increase in the amount of electricity generated.

Recognition of the acid deposition problem in East Asia lagged behind that of Europe and North America. Japanese scientists, including Okita Toshiichi, sought the establishment of the Acid Rain Survey (under the auspices of the Environment Agency) in 1983. By 1985, evidence of transboundary air pollution from continental Asia, largely China, was found. Public recognition of acid rain was spurred, expanding political pledges to confront the issue. In 1988, the White Paper for the Environment included discussion about the acid rain problem. In 1997, Japan proposed the creation of the East Asian Acid Deposition Monitoring Network. This network became operational in 2001 (with China's cooperation, as well as that of other East Asian nations). The East Asian Acid Deposition Monitoring Network's purpose is to serve as a collaborative effort to fight the problems of acid deposition, including the monitoring of acid precipitation, gathering of data and statistics, and generation of reports by the involved countries.

See Also: Clean Air Act; Environmental Management; Environmental Movement; Environmental Protection Agency; Global Climate Change.

Further Readings

Kidd, J. S. and Renee A. Kidd. *Air Pollution Problems and Solutions*. New York: Chelsea House, 2006.

Robinson, Peter J. and Ann Henderson-Sellers. *Contemporary Climatology*, 2nd ed. Essex, UK: Longman, 1999.

Schreurs, Miranda A. *Environmental Politics in Japan, Germany, and the United States*. Cambridge, UK: Cambridge University Press, 2002.

Wilkening, Kenneth E. "Localizing Universal Science." In *Science and Politics in the International Environment*, edited by Neil E. Harrison and Gary C. Byner. Lanham, MD: Rowman & Littlefield, 2004.

Petra A. Zimmermann
Ball State University

Afforestation

Globally, forest cover continues to decline as a result of high deforestation rates (13 million hectares [ha]/year), primarily as a result of agricultural expansion in the tropics. At a national and regional level, however, afforestation and reforestation have led to an increase in forest and tree cover in some areas, often despite continued rapid economic and population growth. Such forest resurgence, particularly in Europe and China, lowered the global net forest loss to 7.3 million ha/year between 2000 and 2005. Afforestation refers to the planting of trees in areas that have not been forested in the last 50 years. This includes afforestation of native grassland and shrubland as well as of pasture and croplands that may have been deforested over a half century ago. Afforestation also occurs through the naturalization and invasive spread of trees into nonforested regions. This contrasts with reforestation, which denotes spontaneous forest regeneration or intentional tree planting on previously forested land. As described here, the wide-ranging landscapes that may be described as "afforested" have similarly wide-ranging biophysical and socioeconomic characteristics.

Plantations

Reforestation and afforestation projects, such as these thousands of newly planted pine trees, reduced the annual net loss of the world's forest cover to 7.3 million hectares between 2000 and 2005.

Source: iStockphoto.com

Approximately 45 percent of the mentioned forest resurgence is a result of the expansion of plantation forests established through afforestation, reforestation, and conversion of primary forest. It is estimated that half of the 139.1 million ha of plantations are established through reforestation and afforestation, and half through deforestation and conversion of natural forest in the tropics. Extensive plantation establishment on agricultural land occurred during 1990–2005 in eastern China, southern Europe, and several other industrialized and industrializing nations. The majority (78 percent) of plantations are established for wood and fiber production, with the rest (22 percent) established primarily for soil and water protection purposes. Between 1990 and 2005, productive plantations increased 2.2 million ha/year, whereas protective plantations increased by 380,000 ha/year.

Forest Transition Theory

Drawing on data primarily from developed countries, forest transition theory postulates that economic development leads to a predictable transition from a period of forest contraction to a period of forest expansion (with no distinction made between afforestation and reforestation). Thomas Rudel expanded on forest transition theory and proposed two paths in this forest transition: "the economic development path" and "the forest scarcity path." In the economic development path, increased economic opportunity leads to an increase in nonfarm jobs, leading to an abandonment of agricultural lands followed by spontaneous forest regeneration. Alternatively, the forest scarcity path occurs when landowners plant trees in response to a shortage of forest products. Kathleen Farley proposed a third path, "the reverse economic development path," by which afforestation, funded through national or international forestry programs and, more recently, carbon sequestration initiatives, "seeks" rather than responds to economic development.

Benefits of Afforestation

Afforestation is promoted to support a range of economic, social, and environmental initiatives. Although accounting for approximately 3.5 percent of the total forest area, timber plantations, for example, are approaching 50 percent of global wood production. Tree plantations can additionally provide important revenue for landowners through the extraction of fuel wood and nontimber forest products. Shade coffee production also has led to increased tree cover in Honduras, El Salvador, and Costa Rica, some of which has been established on land deforested more than 50 years ago. Many governments also support forest expansion with the perception that planting trees enhances environmental conditions through decreasing erosion, improving soil and water quality, and increasing water supply. More recently, global climate initiatives, such as the Kyoto Protocol Clean Development Mechanism, have supported afforestation and reforestation projects because of their potential to sequester carbon.

Costs

The assumption that planted forests improve environmental quality has been questioned in recent research that evaluates the ecological outcomes of afforestation compared with previous land covers. The widely used definition of forest as a land cover with at least a 10 percent canopy cover offers considerable latitude for variation in structure and composition of afforested land covers and will accordingly be associated with a wide array of potential ecological services. Likewise, baseline land covers vary substantially among forestation cases, with initial cover in reforestation measured as some type of forest as opposed to a nonforest cover in afforestation cases. As such, despite the tendency to conflate afforestation and reforestation in discussions of forest transitions, both initial and final land use/land cover states determine the appropriate definition when examining the biophysical and socioeconomic implications of forest transitions.

Although plantations have a number of benefits, afforestation of grassland and shrubland generally decreases stream flow and groundwater recharge to varying degrees, depending on forest characteristics and landscape hydrology. Plantation establishment can also cause acidification of stream water and acidification and salinization of soils. Afforestation of agricultural land generally increases soil carbon, whereas afforestation of grassland and shrubland generally causes a loss of soil carbon. Biophysical outcomes of afforestation also depend on the species used, the age of the plantation, and the management practices used. Eucalyptus, for example, results in a greater stream flow reduction relative to other species, and runoff generally decreases with plantation age. Likewise, site preparation and harvesting using heavy machinery will disturb the soil more than less intrusive management strategies.

Although few studies have addressed the effects of afforestation on native biodiversity, research indicates that timber plantations established in native grassland and shrubland generally decrease native plant and bird diversity. The emphasis on forest recovery often assumes that forest cover is the ideal ecological state, marginalizing other ecosystems. Escaped plantation species can also threaten biodiversity outside the established plantation area, as exemplified by the extensive invasion of South Africa's native *fynbos* shrubland system by exotic forestry trees. Conversely, in areas previously deforested, afforestation of converted lands can enhance biodiversity, forming an important part of the wider conservation value of an anthropogenic landscape.

Conclusion

Potential benefits and drawbacks to afforestation depend on the location of forest establishment, on the ecology and cover associated with prior land use, and on the character of the planted forest. In some cases, afforestation may be assumed to improve environmental quality despite prior land uses having superior hydrological services and biodiversity.

Although afforestation can provide both socioeconomic and biophysical benefits under some scenarios, assumptions that increased forest cover will always be beneficial are not supported. Ultimately, the sustainability of afforestation projects may be usefully evaluated not only for their economic value (e.g., the production of goods or carbon sequestration) but also for their effect on water supply, soil quality, and biodiversity protection.

See Also: Biodiversity; Deforestation; Groundwater.

Further Readings

Bass, J. O. J. "Forty Years and More Trees." *Southeastern Geographer*, 46:51–65 (2006).

Bass, J. O. J. "More Trees in the Tropics." *Area*, 36:19–32 (2004).

Carr, D. L. "Proximate Population Factors and Deforestation in Tropical Agricultural Frontiers." *Population and Environment*, 25/6:585–612 (2004).

Carr, D. L., et al. "Agricultural Land Use and Limits to Deforestation in Central America." In *Agriculture and Climate Beyond 2015: A New Perspective on Future Land Use Patterns*, edited by Floor Brouwer and Bruce McCarl. Dordrecht, Netherlands: Springer, 2006.

Duran, S. M. and G. H. Kattan. "A Test of the Utility of Exotic Tree Plantations for Understory Birds and Food Resources in the Colombian Andes." *Biotropica*, 37:129–135 (2005).

Farley, K. A. "Grasslands to Tree Plantations: Forest Transition in the Andes of Ecuador." *Annals of the Association of American Geography*, 97:755–71 (2007).

Farley, K. A., et al. "Effects of Afforestation on Water Yield: A Global Synthesis With Implications for Policy." *Global Change Biology*, 11:1565–1576 (2005).

Farley, K. A., et al. "Soil Organic Carbon and Water Retention Following Conversion of Grasslands to Pine Plantations in the Ecuadorian Andes." *Ecosystems*, 7:729–39 (2004).

Farley, K. A., et al. "Stream Acidification and Base Cation Losses With Grassland Afforestation." *Water Resources Research* 44 (2008).

Food and Agricultural Organization of the United Nations (FAO). "Forest Resources Assessment 2000: Terms and Definitions." Forest Resources Assessment Programme Working Paper 1. Rome: FAO, 1998.

Food and Agricultural Organization of the United Nations (FAO). "Global Forest Resources Assessment 2005: Progress Towards Sustainable Forest Management." FAO Forestry Paper 147. Rome: FAO, 2006.

Guo, L. B. and R. M. Gifford. "Soil Carbon Stocks and Land Use Change: A Meta Analysis." *Global Change Biology*, 8:345–60 (2002).

Hartley, M. J. "Rationale and Methods for Conserving Biodiversity in Plantation Forests." *Forest Ecology and Management*, 155:81–95 (2002).

Hecht, S. B., et al. "Globalization, Forest Resurgence, and Environmental Politics in El Salvador." *World Development*, 34:308–23 (2006).

Hecht, S. B. and S. S. Saatchi. "Globalization and Forest Resurgence: Changes in Forest Cover in El Salvador." *Bioscience*, 57:663–72 (2007).

Jackson, R. B., et al. *Trading Water for Carbon With Biological Carbon Sequestration.* Washington, D.C.: American Association for the Advancement of Science, 2005.

Liu, J. G., et al. "Ecological and Socioeconomic Effects of China's Policies for Ecosystem Services." *Proceedings of the National Academy of Sciences of the United States of America*, 105:9477–82 (2008).

Mather, A. S. "The Forest Transition." *Area*, 24:367–79 (1992).
Pomeroy, D. and C. Dranzoa. "Do Tropical Plantations of Exotic Trees in Uganda and Kenya Have Conservation Value for Birds?" *Bird Populations*, 4:23–36 (1997).
Richardson, D. M. and B. W. van Wilgen. "Invasive Alien Plants in South Africa: How Well Do We Understand the Ecological Impacts?" *South African Journal of Science*, 100:45–52 (2004).
Rotenberg, J. A. "Ecological Role of a Tree (*Gmelina arborea*) Plantation in Guatemala: An Assessment of an Alternative Land Use for Tropical Avian Conservation." *Auk*, 124:316–30 (2007).

Leah Bremer
David L. Carr
Kathleen A. Farley
Independent Scholars

Agenda 21

Agenda 21, the Rio Declaration on Environment and Development, was adopted by more than 178 governments at the United Nations Conference on Environment and Development held in Rio de Janerio, Brazil, June 3–14, 1992. Agenda 21, although not a legally binding program of action for sustainable development, is a document comprising 40 chapters intended to guide the actions of governments, aid agencies, local governments, and other actors on environment and development issues in achieving sustainable development. Agenda 21 covers four broad areas of governance. According to the preamble of Agenda 21, the document addresses the problems that exist today while taking into consideration future generations. The preamble indicates that Agenda 21 represents a political commitment at the highest level to development and environmental cooperation. There is a strong emphasis on the responsibility of governments toward the effective implementation of Agenda 21. However, this assertion of national hegemony is tempered with the recognition that success will require the sustained commitment of other international, regional, and subregional organizations. These groups include nongovernmental organizations and other groups.

Agenda 21 Structure

Agenda 21 is separated into four overarching sections. The first section is social and economic dimensions to development. This includes issues such as production and consumption, health, human settlement, and the promotion of an integrated decision-making process. The second section is the conservation and management of natural resources. This includes the atmosphere, oceans and seas, land, forests, mountains, biological diversity, ecosystems, biotechnology, freshwater resources, toxic chemicals, and hazardous radioactive and solid wastes. Third, Agenda 21 aims to strengthen the role of groups and actors involved in promoting sustainable development. This includes youth groups, women, indigenous populations, nongovernmental organizations, local and regional authorities, trade unions, business groups at all levels, scientific and technical communities, and farms. Fourth, Agenda 21 attempts to outline areas for implementing sustainable development. Areas include finance, technology transfer, information coordination and public awareness, capacity building, education, legal instruments, and institutional frameworks. In essence, Agenda 21 establishes a framework, or "package," of long-term goals. It has been seen as the most comprehensive document negotiated between governments that considers the interactions among economic, social, and environmental trends.

United Nations Conference on Environment and Development

A number of United Nations events led to the development of Agenda 21. The United Nations Conference on the Human Environment held in Stockholm in 1972 was pivotal in raising awareness of humanity's effect on the environment. It was also at this time that the United Nations Environment Programme was created—a significant proponent of sustainable development. In 1987, the World Commission on Environment and Development, or the Brundtland Commission, galvanized the concept of sustainable development on the political stage and began the process of developing a set of principles for sustainable development. It was at the World Commission on Environment and Development that sustainable development was identified as a concept whereby the exploitation of natural resources, combined with the way that financial investment is directed, should influence technical development and institutional change that should apply to both current and future generations. It was these initial principles that ultimately led to the adoption of Agenda 21 at the United Nations Conference on Environment and Development.

The United Nations Conference on Environment and Development was held in Rio de Janeiro in 1992 and is popularly named the Earth Summit. Although the summit was seen as a milestone in environmental politics, a significant number of the planet's leaders converged to discuss possible solutions to today's environmental issues. The summit represented a common recognition by many nations and relevant organizations of the world that issues of development need to be tempered with effective environmental measures. Substantively, a number of agreements emerged from the conference. The Framework Convention on Climate Change set out international targets for reducing the anthropogenic causes of climate change, and 192 countries from around the world joined the treaty. The Biodiversity Convention established broad aims to conserve international biodiversity, with the aim of making use of components of biodiversity in a suitable manner and enabling an equitable distribution of the benefits of using these resources. The Desertification Convention was designed to create localized action frameworks to address the degradation of dry land environments. Most notable was the production of Agenda 21, which mapped out a blueprint for sustainable development by outlining the main issues implicated in global environmental change and how they might be tackled. The United Nations Commission on Sustainable Development was created to monitor and report on the implementation of the agreements made.

Local Agenda 21

There is a strong emphasis in Agenda 21 on achieving sustainable development at the local level in conjunction with further integration of the regional, national, and international levels of governance. Over two-thirds of the 2,500 action items in Agenda 21 relate to the important role of local actors with a focus on local authorities. Each local authority is charged with developing its own Local Agenda 21 strategy, following a consultation period with local residents and interested parties. Ultimately, Local Agenda 21 considers the responsibilities of the local area of governance in achieving sustainable development while maintaining connections with the global and local scales of analysis. The interested parties include the business sector, the voluntary or third sectors, and community groups, as well as underrepresented marginalized groups. In England, Local Agenda 21 has been considered a precursor to identifying and creating mechanisms for achieving sustainable development at the local level. This is particularly evident with the local authority sustainable community.

Agenda 21 set a benchmark for the assessment of progress toward sustainable development. A number of subsequent international meetings have aimed to assess progress toward the goals stipulated in Agenda 21. In 1997, the United Nations General Assembly held a 5-year appraisal of progress toward the implementation of Agenda 21. This meeting was popularly known as Rio Plus Five. In summary, the assessment of the assembly was that progress toward achieving sustainable development was uneven. Processes that were identified as particularly relevant to Agenda 21 and sustainable development were the trends toward increasing globalization, increasing disparities in global wealth, and the persistent deterioration of the global environment.

Criticisms

Agenda 21 has been criticized in two main areas. The first involves implementation. It has been suggested that to date there has been very little progress toward the overall goals established in Agenda 21. There have been suggestions that despite the rhetoric of achieving sustainable development through the framework laid out in Agenda 21, there has been very little institutional support for pursuing these goals and that overall support at the national and international level has been sparse. The second overarching criticism concerns the language of Agenda 21. The document has been criticized for promoting a vision of sustainable development that perpetuates notions of progress through economic growth and industrialization at all costs. There has been particular concern that it perpetuates and extends notions of globalization along the U.S. style of a political pluralist system of democracy.

See Also: Brundtland Commission; Sustainable Development; UN Conference on Environment and Development.

Further Readings

Baker, Susan. *Sustainable Development.* London: Routledge, 2006.

Borne, Gregory. *Sustainable Development: The Reflexive Governance of Risk.* Lampeter: Edwin Mellen, 2009.

United Nations. "Agenda 21." http://www.un.org/esa/sustdev/documents/agenda21/english/agenda21toc.htm (Accessed February 2009).

Gregory Borne
University of Plymouth

AGRICULTURE

Agriculture (*agar + culture*: *agar* meaning land and *culture* meaning cultivation) is the science, art, and business of cultivating soil, producing crops, and raising livestock—simply known as farming, it is a bedrock of human civilization. Thus, by definition, farmers include resource-poor cultivators, pastoralists, fisher-folk, indigenous peoples, women, and agricultural laborers. Domestication of plants and animals was necessary for the evolution of agriculture ca. 10,000–7,000 years B.P. Agricultural practices enabled people to establish permanent settlements and expand urban-based societies. Domestication of plants and animals transformed the profession of the early humans from hunting and gathering to selective hunting, herding, and settled agriculture. Agriculture addresses food/nutrient security and self-sufficiency and is thus very vital to achieving several millennium development goals; it is inseparably linked to several internationally agreed conventions/strategies: the Convention on Biological Diversity, Poverty Reduction Strategy Papers, and Local Agenda 21.

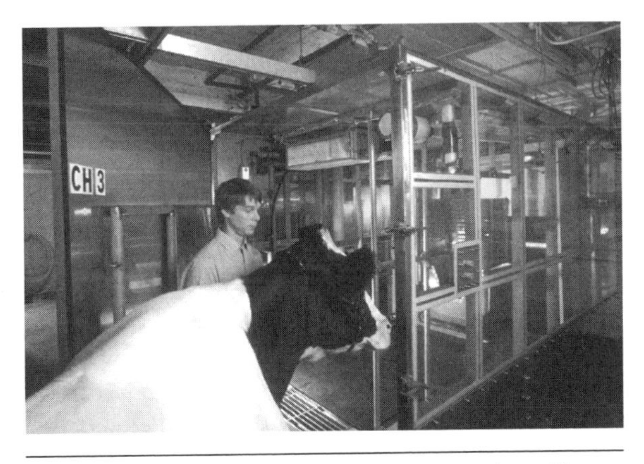

Agriculture may be responsible for as much as 12 percent of greenhouse gas emissions, including methane emissions from cattle. This animal scientist is using a plastic chamber to measure a cow's production of heat and methane.

Source: U.S. Department of Agriculture, Agricultural Research Service/Keith Weller

Until about four decades ago, crop yields in agricultural systems depended on internal resources, recycling of organic matter, built-in biological control mechanisms, and rainfall patterns. Agricultural yields were modest, but stable. Production was safeguarded by growing more than one crop or variety in space and time in a field as insurance against pest outbreaks or severe weather. Inputs of nitrogen were gained by rotating major field crops with legumes. In turn, rotations suppressed insects, weeds, and diseases by effectively breaking the life cycles of these pests. As agricultural modernization progressed, however, the ecology–farming linkage was often broken as ecological principles were ignored and/or overridden. In fact, several agricultural scientists have arrived at a general consensus that modern agriculture confronts an environmental crisis. A growing number of people have become concerned about the long-term sustainability of existing food production systems. Evidence has accumulated showing that although the present capital- and technology-intensive farming systems have been extremely productive and competitive, they also bring a variety of economic, environmental, and social problems.

Of the total land area of about 13 billion hectares globally, crops and pasture occupy almost 5 billion hectares. Thus, agriculture has closer inseparable linkages with ecosystem services. Ecosystem services are the irreplaceable services derived (directly or indirectly) from natural/managed ecosystems on which human welfare depends. Hence ecosystem services must be conserved for sustaining agricultural production. This view is supported by the following facts:

- Three-fourths of the world's poor are rural (will account for 60 percent in 2025) with very limited external support, technological/financial/market/information access
- Two-thirds of the rural poor live on "marginal" farmlands
- Sixteen of 25 biodiversity hot spots are areas with very high malnutrition (one-fifth undernourished)
- Food demand in less developed countries will exceed 50–60 percent by 2030, where 90 percent of the food is grown domestically
- Agriculture is the main employer and creator of wealth in poor, biodiversity-rich countries

Although land and tenurial reforms have been enacted in many countries, their implementation remains fraught with problems. This situation is complicated by the increasing privatization of the commons, which in turn has become a major source of land and resource conflicts that have inseparable linkages for the food security of the poor.

We have to conserve wild biodiversity on farms that are under serious threat because they are important for the following:

- Increasing productivity, food security, and economic returns
- Reducing the pressure of agriculture on fragile areas, forest, and endangered species
- Making farming systems more stable and sustainable
- Contributing to sound pest/disease management
- Conserving soil and increasing natural soil fertility/health
- Reducing/spreading risks
- Optimizing the use of resources and the environment
- Reducing dependency on external inputs
- Improving human nutrition and providing sources of medicines and vitamins
- Conserving ecosystem structure and stability of species diversity

The following facts and figures on the looming threat to agricultural biodiversity are very relevant for ensuring food security on a sustainable basis:

- Since the 1900s, some 75–90 percent of plant genetic diversity has been lost as farmers worldwide have left their multiple local varieties and landraces for genetically uniform, high-yielding varieties.
- Thirty to 50 percent of livestock breeds are at risk of extinction; six breeds are lost each month.
- Today, 75 percent of the world's food is generated from only 12 plants and five animal species.
- Of the 4 percent of the 250,000–300,000 known edible plant species, only 150–200 are used by humans. Only three—rice, maize, and wheat—contribute nearly 60 percent of the calories and proteins obtained by humans from plants.
- Animals provide some 30 percent of human requirements for food and agriculture; 12 percent of the world's population lives almost entirely on products from ruminants.

Emergence of Green Revolution Agriculture

The term *green revolution* refers to the transformation of agriculture that began in 1945 at the request of the Mexican government to establish an agricultural research station to develop more varieties of wheat that could be used to feed the country's rapidly growing population.

In green revolution agriculture (GRA), the major change has been the improvement of irrigation systems, with upstream storage enabling the extension of cultivation into the dry season, enabling intensification and specialization, typified by the introduction in the 1960s of improved high-yielding varieties that require large inputs of chemical fertilizers, pesticides, and intensive irrigation. The resulting expansion of food production has brought Bangladesh, Pakistan, Indonesia, India, China, the Philippines, and others from the brink of starvation to the threshold of national food-grain self-sufficiency.

Effects of GRA

Although GRA has undeniably increased food production, stimulated industrial growth, and fostered political stability, there are several inevitable negative effects:

- Introduction of hybrid monocultures and the consequent loss of traditional germplasm and wild biodiversity in agriculture ecosystems
- Introduction of genetically modified crops and the consequent genetic erosion, unpredictable ecological effects on nontarget organisms, creation of superweeds/new viral genotypes
- Chemicalization/pollution resulting from excessive use of fertilizers/pesticides with irreversible effects on soil quality/biota and water resources as well as health impacts on humans
- Overexploitation/degradation of water/aquifers; salinization, waterlogging, and lowering of water tables
- Heavy reliance on fossil fuels and mechanization; environmental/socioeconomic impacts
- Socioeconomic problems, loss of self-reliance, increasing debt burden, poverty spirals/traps, and suicides

In most developing countries and less developed countries, biosafety/food quality standards to monitor such production systems are absent or are inadequate to predict ecological risks, at least for the local consumers. These trends increase hunger, landlessness, homelessness, despair, and suicides among farmers.

Rediscovery of Sustainable Agriculture

Alternative agriculture practices originated in the 1920s and 1930s, were developed in the 1970s, and mushroomed in the 1980s and 1990s. At this time, organic agriculture has become a global movement, with organized nongovernmental activities developed from the previous individual, separate, and spontaneous activities. In some countries, organic agriculture has even won support from governments, which either gave it legal protection or policy support. Organic food also has found its way into supermarkets. In some countries, there are even management chains for organic food sales. Some of the organic food has entered the urban markets from rural markets and has moved from domestic to international markets.

A large majority of the farmers in developing and less developed countries can be typified as either marginal or small (based on land holdings), and hence

- the majority can afford/use hand tools only;
- they have lower purchasing power for agrochemicals;
- they are confronted with very limited external support and limited access to technological, financial, and market information; and
- they face increasing reductions in subsidies or no subsidies for organic inputs with a stagnating/fluctuating market.

To overcome these serious problems, we have to organically symbiotically link innovations with investments/enterprises by rediscovering the time-tested traditional agricultural strategies such as agroecology, low external input sustainable agriculture, permaculture, biodynamic gardening, and so on. Because of their wider applicability, the first two options need special attention.

Instead of focusing on one particular component of the agroecosystem, agroecology emphasizes the interrelatedness of all agroecosystem components and the complex dynamics of ecological processes. Agroecology must have at least four main characteristics to improve total farm productivity, stability, and resiliency and the sustainability of agroecosystems:

- Socially activating: a high level of self-reliance/participation, challenging the agricultural research and development agenda to forge their own endogenous development path
- Culturally compatible: building on local knowledge and management systems and combining this with the elements of modern science and technology—eco-technologies, geomatics, information and communication technologies, and so on.
- Ecologically sound: optimization and sustainability of the production, sustainable intensification/diversification of products to minimize risks and increase synergism, eco-restoration and integrated farming systems to enhance the nutritional intake/health and ecosystem health
- Economically viable: by minimizing the costs of production through efficiency enhancement and the use of locally available resources, thus decreasing farmers' dependency on the state and industry

Low external input sustainable agriculture integrates the use of low external inputs within the traditional farming systems, preserving the biodiversity of the area and regenerating agriculture. Low external input sustainable agriculture is characterized by

- technical and social options open to farmers who seek to improve productivity and income in an ecologically sound way by optimal use of local resources,
- participatory methodologies to strengthen the capacity of farming communities, and
- combining indigenous and scientific knowledge.

Those strategies that promote the production of healthy, safe, and culturally diverse foods and localization of distribution, trade, and marketing are widely adopted by many farmers throughout the world.

Food Security Issues

Food security facilitates equity in food availability and accessibility at all times; that is, nutritionally adequate food in quantities, quality, and varieties acceptable within the given culture. Food sovereignty refers to the right of peoples, communities, and countries to define their own agricultural, labor, fishing, food, and land policies that are ecologically, socially, economically, and culturally appropriate to their unique circumstances. It includes the true right to food and to produce food, which means that all people have the right to safe, nutritious, and culturally appropriate food and to food producing resources and the ability to sustain themselves and their societies. This entails the support and promotion of local markets/producers over production for export and food imports—thus it supports indigenous communities, especially the rural landless and resource-poor farmers.

Hence, local autonomy, governance, organization, and defense of the commons are at the heart of these concepts. Throughout Latin America and in much of Africa, as well as South and Southeast Asia, farmers, pastoralists, women, indigenous peoples, and migrants are organizing, linking together with their counterparts in the North for working toward these twin goals. They are gaining support from scholars, activists, consumers, and progressive policymakers.

Effects of Climate Change on Agriculture

Although agriculture contributes to climate change, it is also affected by climate change.

- Agriculture releases significant amounts of carbon dioxide, methane, and nitrous oxide—estimated to be 5.1–6.1 billion tonnes carbon dioxide equivalent/year in 2005 (10–12 percent of global anthropogenic greenhouse gas emissions).
- Including indirect contributions (e.g., land conversion to agriculture, fertilizer production and distribution, and farm operations), agriculture releases 17–32 percent of global emissions.
- Agricultural nitrous oxide emissions are projected to increase by 35–60 percent until 2030 as a result of increased nitrogen fertilizer use and animal-manure production.

Food security, at the nexus of a number of issues from energy security to climate change and water scarcity, may be emerging as one of the major risks of the 21st century for the following reasons.

- Moderate local warming (1–3 degrees C) means slightly increased yields in mid to high latitudes, depending on the crops.
- Even small local temperature increases (1–2 degrees C) would decrease yields in low latitudes, especially in tropical and seasonally dry areas, which would increase risk of hunger; further warming of more than 3 degrees C would have increasingly negative effects; global food production is projected to decrease up to 30 percent in Central and South Asia by the mid-21st century.
- Drought/heat waves are very likely to lead to reduced yields and/or crop damage or failure in warmer environments as a result of heat stress.
- Heavy precipitation is linked to damage of crops, soil erosion, and waterlogging of soils.
- Extreme high sea level events are likely to lead to salinization of water bodies.
- Increases in pest and disease infestations as a result of warming temperatures are possible.

Future Perspectives

Increasingly, researchers are showing that it is possible to provide a balanced environment, sustained yields (50 to 180 percent higher yields than GRA), biologically mediated soil fertility, and natural pest regulation through the design of diversified agroecosystems and the use of low-input technologies. With regard to the future strategies for ensuring global agricultural sustainability, these may include, but not be limited to, the following:

- democratize science and technology research (participatory action research);
- de-institutionalize research for autonomous learning and action;
- enable contexts for social learning and action—share information, experiences, and best practice case study documentation through local, regional, national, and international networks;

- undertake genuine agrarian reforms that consolidate ownership, resilience, control, and management of resources by the marginalized groups;
- breed locally adapted seed and livestock and produce on-farm fertilizers, biocides, and so on;
- provide sustainable market chain analysis and direct marketing of value added/quality products;
- develop sustainable agriculture and agricultural support systems for ensuring food/nutrient security with ensured food safety, with a strict process of periodic environmental and socioeconomic impact assessments;
- recognize and reward farmers as the main source of intellectual innovation in agriculture, and as caretakers of biodiversity for food and pharmaceuticals, and recognize all ex situ collections as common global property resources;
- governments and the civil society groups must recognize and build local capacities for participatory risk and hazard mapping to facilitate recovery from and mitigate the effects of disasters with assured emergency food supplies, safety nets, and so on;
- create and ensure enabling policy environments that recognize and build on the rights and capacities of farmers, who should be organized and whose interests should be prioritized in the food production process;
- institute universal recognition of the right to food as the right to life—survival should take precedence over all other imperatives, including the right to profit, corporate opportunism, and advantage; and
- lobby for supportive trade policies to promote sustainable agriculture and ensure that commitments made at multilateral/bilateral levels provide policy for strong sustainability.

See Also: Acid Rain; Afforestation; Agenda 21; Anti-Toxics Movement; Appropriate Technology.

Further Readings

Badgley, C., et al. "Organic Agriculture and the Global Food Supply." *Renewable Agriculture and Food Systems,* 22/2 (2007).

Ching, Lim Li. "Sustainable Agriculture: Meeting Food Security Needs, Addressing Climate Change Challenges." http://www.twnside.org.sg/title2/susagri/susagri055.htm (Accessed April 2011).

IAASTD. "International Assessment of Agricultural Knowledge, Science and Technology for Development." http://www.agassessment.org (Accessed January 2009).

Lal, Rattan, et al. *Food Security and Environmental Quality in the Developing World.* Washington, D.C.: Lewis Publishers, 2003.

Pretty, J., et al. "Resource-Conserving Agriculture Increases Yields in Developing Countries. Environmental Science & Technology Towards a Green Food System: How Food Sovereignty Can Save the Environment and Feed the World." *Grassroots International & Food and Water Watch,* 40/4 (2006). http://www.grassrootsonline.org, http://www.foodandwaterwatch.org (Accessed January 2009).

Gopalsamy Poyyamoli
Pondicherry University

Codandapani Babou
Adhiparasakthi Agriculture College

Anarchism

It is difficult to provide a commonly accepted definition of anarchism, as there are many different strands of thought and practice that are subsumed under the label. However, in its broadest sense, it is derived from the Greek word *anarchia*, which means "no rule." Thus, anarchism is typically taken to refer to stateless societies or social arrangements that lack centralized control of any sort. A green or ecological strand of anarchism exists, putting the focus on a variety of issues related to the environment, technology, and modern social organization. However, ecological or green anarchism is itself fractured into a variety of different viewpoints and agendas.

There are many influential thinkers who argue that anarchism is undergoing a revival, given the decline of Marxian-inspired social arrangements and the ongoing critique of capitalism as a harsh, unyielding social and economic system. Spontaneous uprisings such as those in Seattle at the World Trade Organization meetings in 1999 provide evidence of the continued influence of anarchistic writings and philosophy. However, anarchist thinking has been around for millennia, and most of the current strands of thinking and practice go back into the 19th century. Collectivist-oriented anarchism has predominated over most of this period, inspired by Pierre-Joseph Proudhon and most clearly articulated in the works of Mikhail Bakunin and Peter Kropotkin in the 1800s and by Murray Bookchin in the 20th century. In the United States, more individualistic forms of anarchism have tended to predominate, inspired by Henry David Thoreau, Max Stirner, and many others. Most of these forms of anarchism focus on individual freedom and celebrate capitalism as a social system that can best provide such freedom. Today, individualist and collective forms of anarchism exist alongside anarcha-feminism and green anarchism, among others.

Green anarchism or ecological forms of anarchism not surprisingly emphasize environmental issues. There are several schools of green anarchism, including the anarcho-primitivists and the social ecologists. Green anarchism derives inspiration from the actions of the Luddites in early-19th-century England, the writings of Jean-Jacques Rousseau, and a diverse collection of 20th-century thinkers and social analysts. The anarcho-primitivists include writers such as Derrick Jensen, George Draffan, and John Zerzan, whereas the most notable social ecologist was Murray Bookchin.

The anarcho-primitivists offer a largely negative critique of civilization, particularly focusing attention on the repressive and dominating nature of the social institutions that make up society. These institutions are by nature harmful to humanity and the natural world and are thus not reformable. It is imperative to the anarcho-primitivists that civilization itself must be overcome and that humanity needs to return to a preagrarian state of existence. In effect, anarcho-primitivists are arguing for the spontaneous reemergence of something along the lines of hunter-gatherer living, which was the modus operandi for virtually all of human history. Furthermore, technology is seen as a significant problem by anarcho-primitivists, because it is systemically used to exploit the environment and humanity. Solutions offered (and often practiced) by anarcho-primitivists include the development of ecovillages—small-scale communities that are seen as more ideal for the harmonization of humanity and nature. Other anarcho-primitivists prefer to focus on direct-action activities, usually but not exclusively nonviolent in nature. These activities are directed against what they view as the systems of oppression of humanity and especially of nature.

The other predominant school of thought in green or ecological anarchism is broadly linked to the social ecology approach of Murray Bookchin. In this approach, inspired by the more collectivist writings of both anarchism and Marxism, plus modern ecology and Deep Ecology, the emphasis is placed more squarely on overcoming the social domination of nature by humanity. Concern is centered on exploitative and hierarchical social relationships. Early works by Bookchin expressed an optimistic view about the liberatory potential of technology as a means to overcome social exploitation, though later in his life he seemed less enthusiastic about this possibility. However, the social ecologists remain largely optimistic about the potential for the development of actually existing social relationships that are nonoppressive. Much attention is paid to the development of decentralized, direct democracy, centered in public town meetings. This focus on municipalism and participatory democracy has led critics to argue that this strand of anarchism actually perpetuates existing social inequalities in localities instead of eliminating them, as is the objective, and further that it does not address the underlying domination of nature by humanity.

The different strands of anarchism continue to have a broad influence on certain segments of the population, though they remain almost by inherent design on the margins of political and social activity. The green anarchist critique is one that raises fundamental questions about the nature of social life and its relationship to the natural world, yet it too remains on the margins of environmental discourse and action.

See Also: Capitalism; Decentralization; Deep Ecology; Ecosocialism; Participatory Democracy.

Further Readings

Bookchin, Murray. *Post-Scarcity Anarchism.* Edinburgh, UK: AK, 2004.
Heider, Ulrike. *Anarchism: Left, Right and Green.* San Francisco: City Light Books, 1994.

Todd L. Matthews
University of West Georgia

Anti-Toxics Movement

The Anti-toxics movement's origins can be traced back to Rachel Carson's *Silent Spring* in 1962. The book—initially serialized in the *New Yorker*—highlighted the impact of pesticides such as DDT on plant and wildlife in America in the years following the introduction of scientized methods of agriculture in the United States. In the aftermath of a wider public concern and scientific debate about Carson's work, President John F. Kennedy called on the Science Advisory Committee to investigate issues surrounding the use of pesticides. This inquiry confirmed Carson's position and led to the regulation of the use of chemical pesticides in the United States. Carson has been subject to a number of subsequent criticisms from scientists working for the chemical industry.

The environmental justice movement developed further during the 1960s, when migrant agricultural workers led by Cesar Chavez also challenged the use of pesticides in California. African-American communities mobilized under a racial justice and anti-toxics agenda as part of a number of regional campaigns against toxic plants in Houston, Texas, and Harlem in New York City during the 1960s. In 1978, communities campaigned against the dumping of toxics near their homes in the Love Canal Township near Niagara Falls, New York. The campaign emerged in response to concerns about high rates of cancer and birth defects in the area. Toxic waste had been buried in the region by the Hooker Chemical Company in the 1920s but had begun to seep into the local water supplies.

The Love Canal Controversy

The land surrounding Love Canal was developed for a school and housing despite the warnings of the Hooker Company in the 1950s. Over 50 drums of chemical waste were found at the site during excavation, and the school was built away from the area. In 1957, low-income housing was built on the site. With the construction of a motorway in the 1970s, floods began to occur in the area, often containing toxic residue. In 1978, Lois Gibbs of the Love Canal Homeowners Association led residents in protests over the number of serious illnesses occurring in children in the Love Canal community. The activities of the Homeowners Association led to the discovery of the toxic materials beneath the homes in the area. This information had been withheld from the community when the homes were first built. The Homeowners Association found that over 50 percent of its residents suffered in some way from the effects of the toxic waste beneath their housing.

President Jimmy Carter allocated funds to assist with the Love Canal controversy, which had begun to make national headlines and found coverage in television news broadcasts. An investigation by the U.S. Environmental Protection Agency in 1979 found that the Love Canal area suffered from an abnormally high number of serious illnesses and miscarriages. Pregnant women were evacuated from the Love Canal area as a precaution. The Environmental Protection Agency report also found that up to a third of residents had detectable damage to their chromosomes as a result of exposure to the chemicals at the site. Love Canal was declared a national emergency site in 1980. Over 700 families were evacuated and rehoused, and the Comprehensive Environmental Response, Compensation, and Liability Act, or "Superfund" Act, was signed into law. In 1995, the Environmental Protection Agency won nearly $300 million in compensation for the incident. Legislation against toxic industries continued to be introduced in the aftermath of increased agitation among the public. Many industries relocated to nations that lacked similar legislation but found that local opposition occurred as communities discovered the effects of toxics in emissions over time.

The community-based environmental justice movement developed further in the 1980s in response to the siting of hazardous plants or dumps, often in economically disadvantaged nonwhite neighborhoods. This section of the anti-toxics movement developed in conjunction with the civil rights movement and was organized in response to repeated incidents of siting toxic plants or dumps in minority neighborhoods. Suspicions about toxic industries were heightened by the industrial accidents at Three Mile Island, Chernobyl, and Bhopal (India). Campaigns against incinerators emerged in the 1980s. The emissions from the incineration of waste released dioxins and furans into the atmosphere, and communities near incineration plants began to oppose the incineration industry as a result. Although the concept of "waste to energy" was attractive to many municipalities because of the dual outcomes of waste incineration and energy creation from the same process, concerns raised about emissions by the anti-toxics movement were gradually extended to incineration plants.

Campaigns Against Incineration

The first major campaigns against incineration in the United States occurred at the same time as a "garbage crisis" was being highlighted in the media. This crisis emerged as a result of regional landfills reaching capacity, and in some cases closing down. In addition, scientists such as Barry Commoner and Paul Connett were highlighting the carcinogenic risks posed by incinerator emissions. Another contributory factor was the discovery of a document prepared for the incineration industry in California that suggested ways of overcoming "Political Difficulties Facing Waste to Energy Conversion Plant Siting." The document, sometimes known as the Cerrell Report after the company that had commissioned it, was a handbook on overcoming community resistance to incinerators. It fell into anti-incineration activists' possession and would later be used to demonstrate the technocratic approach taken by the industry when dealing with local communities.

Lois Gibbs of the Love Canal campaign had formed the Citizens Clearinghouse for Hazardous Wastes (CCHW) in the early 1980s. The CCHW would prove to be a major resource for the anti-toxics and anti-incineration movements. The CCHW provided advice for communities opposing municipal and private plants or sites, which came to be known as LULUs—locally undesirable land use cases. The campaigners came to be known as NIMBYs, or "not in my backyard" mobilizations. However, the CCHW and scientific experts such as chemistry professor Paul Connett went beyond the local for the scientific evidence that was provided to local communities; information from Europe, Japan, and Australia was used to provide the data for opponents to toxic plants throughout North America. One significant fact that emerged from this approach was the dramatic increase in dioxin ingestion from the food chain surrounding incinerators. This figure was believed to be as high as 500 times greater than airborne emissions.

The result of the campaigns of the anti-incinerator and anti-toxics campaigns has been increased legislation on emissions and a decline in the construction of incinerators in the United States. The incineration industry has improved its filtering system to reduce emissions to address community concerns. However, the anti-incineration movement also contributed a positive initiative to the toxics debate through its promotion of recycling. The internationally accepted waste hierarchy that places reuse, reduction, and recycling at the top of its pyramid also places incineration and landfilling at the bottom. Paul Connett would also become a lead spokesperson of the zero-waste movement that advocated for the reuse of all materials in production processes.

Campaigns against incineration continue in the United Kingdom, the Republic of Ireland, and throughout Europe. In the United Kingdom, a number of campaigns against incineration have occurred in areas such as Guilford and Brent's Cross and Bristol. Anti-incineration campaigns have also emerged in Russia. Campaigners have the benefit of national and international support networks that evolved from the work of the CCHW and Connett. Anti-toxics campaigners have also been able to link their concerns with the wider issue of climate change by highlighting the negative effect of toxic emissions on the ozone layer. The Internet became a valuable resource for local anti-toxics campaigns that might once have been isolated.

Cultural Accounts of the Anti-Toxics Movement

In 1994, sociologist Andrew Szasz wrote of the methods in which the anti-incineration movement was able to move beyond its NIMBY inceptions in his book *Ecopopulism*. The anti-incineration movement in the northeast of the United States was the subject of a study by Edward Walsh, Rex Warland, and D. Clayton Smith titled *Don't Burn It Here*. The true-life anti-toxics activism of a local mother and legal activist was the subject of a movie titled *Erin Brockovich,* starring Julia Roberts, in 2000. A Movie of the Week about the Love Canal campaign titled *Lois Gibbs and the Love Canal* was made in 1982. The issue of communities opposing toxics plants has become a part of contemporary culture and is regularly featured in television news, documentaries, movies, books, and even comics. Similar to the nuclear industry, the wider toxics and incineration industries have developed a considerable amount of public opposition to their products, and the introduction of innovations or changes is met with a degree of skepticism from concerned communities.

See Also: Environmental Movement; NIMBY; *Silent Spring*; Urban Planning.

Further Readings

Szasz, Andrew. *Ecopopulism: Toxic Waste and the Movement for Environmental Justice.* Minneapolis: University of Minnesota Press, 1994.
Walsh, Edward J., et al. *Don't Burn It Here: Grassroots Challenges to Trash Incinerators.* University Park: Penn State Press, 1997.

Liam Leonard
Institute of Technology, Sligo

Appropriate Technology

Technological innovations have historically driven economic growth and development around the world. Many of these innovations, however, have also led to environmental degradation, natural resource depletion, and social conditions that promote increased economic output at the expense of human rights considerations. Appropriate technology refers to a class of technologies that are characterized as small-scale, affordable, environmentally friendly, locally controlled, and designed to reduce negative consequences associated with traditional technological innovation. Many of the ideas associated with appropriate technology have grown from E. F. Schumacher's seminal work, *Small Is Beautiful*, published in 1973. Then referred to as "intermediate technologies," these development tools were designed to aid poor, rural communities that lacked access to the resources necessary for capital-intensive industrialized development.

Appropriate technologies are used in both industrialized and developing countries. Although industrialized countries have realized the majority of benefits associated with economic growth, many regions within their borders are in need of development. Appropriate technologies have been used to improve the quality of life in these communities. In some instances these technologies are more advanced than those used in developing countries, but they similarly take into account such factors as affordability, sustainability, and the social needs of the community. The following appropriate technologies are a small sample of the many technologies found in communities of the United States, as well as in developing countries around the world.

Solar Greenhouse

Solar greenhouses use energy produced from the sun for food production and plant growth. The trapped heat saves significant energy costs associated with traditional fuel-based heat. The proper selection of heat-absorbing building materials work to warm the interior by releasing stored energy at night when solar energy is not present and indoor temperatures decline. Water may also be used to store heat and can be stored in tanks. Aquariums serve a dual function of storing heat and raising fish.

This barrel is part of a rainwater harvesting system that collects and stores rainwater as it flows down from the roof of a house.

Source: iStockphoto.com

Farmers Market

Appropriate technologies may also be characterized by changes in the activities of members of a community that lead to economic, social, and environmental benefits. Farmers markets are innovations in food production that allow farmers to interact directly with consumers. These markets are usually organized by local governments or citizen groups and require coordination among farmers and consumers. Purchasing food directly from producers allows local consumers to increase the freshness and variety of their produce while decreasing the environmental impacts associated with food distribution across large geographical regions.

Wastewater Treatment

Wastewater management is a critical component of community development. Several technological options are available to meet the ecological, economic, and social needs of developing communities. Among these are natural treatment systems that may be used as an alternative to advanced chemical treatment processes. These systems use both living and nonliving components of the local environment in removing wastewater contaminants through physical, biological, and chemical processes. Conventional technologies include oxidation ponds and activated sludge systems in which waste is decomposed by bacteria under aerobic and anaerobic conditions. Trickling filters are conventional technologies that use rocks or synthetic mediums to remove suspended solids from wastewater. Nonconventional technologies are also used in communities throughout the country. Land treatment applies treated wastewater to agricultural lands and forests, where contaminants are removed naturally. Households may also use on-site treatment systems. Artificial wetlands may also be constructed to mimic natural wetland processes of water treatment. Each of these options varies in the attainable quality of water, construction costs, maintenance requirements, and compatibility with wastewater compositions. Technology adoption may also be limited by available land area, seasonal temperatures, local population, ecosystem characteristics, and end-use considerations.

Solid Waste Recycling

Similar to water treatment, solid waste disposal is a critical and necessary component of local development. Incineration, dumping, and other conventional methods of waste disposal have led to the inefficient use of natural resources, as well as environmental degradation. These disposal sites contain a variety of resources such as paper, glass, metals, and organic materials that may be recovered, recycled, reused, and resold. Source separation and centralized recovery are two methods that may be used to recover solid wastes. The former requires the separation of recyclable products from nonrecyclable products by the consumer. Recyclable materials may then be collected from households or at drop-off sites located in the community. Centralized recovery takes place at local processing stations. At these stations, metallic wastes are separated using magnetic recovery techniques. Combustible waste may also be separated and burned to produce steam that is contracted to building owners in the local community. The ash by-products of the combustion process may also be sold as additives for the production of cement.

The Need for Appropriate Technologies

In developing countries, however, the need for appropriate technologies largely arose from the unresolved issues of growth that were common among developing countries of the 20th century. Unlike the industrialized countries, which were able to take advantage of capital-intensive manufacturing technology to increase employment, developing countries were not able to invest the required resources. Appropriate technologies were meant to be more productive than indigenous technologies but cheaper than the sophisticated technologies of the industrialized world. These types of technologies are often implemented by public–private collaborations with nongovernmental organizations, community-based organizations, firms, local governments, and student groups. They use locally available capital such as indigenous knowledge, natural resources, manpower, and available financial support to raise the standard of living of the community in which they are placed. Appropriate technologies often provide basic human needs. Agriculture, building materials, food processing, renewable energy, safe water, healthcare, and education are the typical focus of appropriate technology projects in the developing world. As in the case of industrialized countries, the social contexts in which appropriate technologies are placed are important factors in the success of these technologies. Technologies that are economically and environmentally feasible may not always meet community-specific needs. For this reason it is often necessary for these technologies to be developed through community-based processes rather than to be transferred from external sources. The following appropriate technologies are a small sample of the many technologies that are found in developing countries and that may also be found in industrialized countries around the world.

Electricity

Energy is an essential element of development and may be provided for in multiple ways using appropriate technologies. Hydroelectric technology harnesses the energy from running water to turn waterwheels that power turbines and generators to produce electricity. The costs of these systems are relatively low, and they require little maintenance. Similarly, wind may be harnessed to generate electricity. Although wind may be more widely available than water, its intermittent nature may require the use of batteries for electricity storage. Photovoltaic systems harness solar energy to produce electricity. The costs of these technologies are high, and they require a high level of technical expertise for manufacturing parts (although assembly is more basic once the components have been made). They may also serve to complement energy sources such as wind power. Animal and plant wastes may be used to produce combustible biogas. Once the bacteria found in these wastes have decomposed the wastes anaerobically, methane is produced. The biogas digester in which fermentation takes place may be made relatively inexpensively.

Natural Pozzolans

Concrete is one of the most significant manufactured materials in the world. The production of Portland cement, a component of concrete, involves burning powdered coal and natural gas, which generates unwanted by-products such as greenhouse gases and particulate matter. Natural pozzolans are an appropriate technology that makes development more sustainable by lowering building costs and reducing pollution. They include diatomaceous earth, volcanic ash, and rice husk ash, which are naturally occurring cementitious materials that extend Portland cement in the concrete mixture. This technology conserves natural capital and simultaneously creates social capital by reducing both environmental degradation and symptoms of poverty. Implementation of natural pozzolan technology has been used for small-scale projects such as gravity-fed water systems for providing safe drinking water to local communities.

Rainwater Harvesting Systems

Rainwater harvesting systems are gravity-fed water systems that collect and store rainwater, usually from the roofs of houses. From the rooftops, water flows through gutters and pipes to a collection tank. The quality of water is dependent on the sophistication of the technology used, as well as the materials that the rooftops, pipes, and collection tanks are made from. Once collected, water is used for domestic and agricultural activities. This technology is especially useful in communities in which water is a scarce resource or where water infrastructure is nonexistent. The costs may vary depending on the system design and materials used for constructing the system, affecting the affordability of the system in some instances.

Cooking Stoves

Cooking stoves fueled by local renewable resources such as biomass are used in communities around the world. The technology is increasingly evolving to accommodate a larger variety of fuel sources, as well as to release cleaner emissions. In the Philippines, for example, considerable developments have been made to the rice husk gasification stove, which allows raw unprocessed rice husks to be used as a fuel for residential cooking. These improvements have provided a cleaner combustion with a higher quality of gas emissions, as most residential stoves are in small kitchens with inadequate ventilation. As the stove is continually improved, it offers an alternative to the commonly used liquefied petroleum gas stoves. Comparing the costs of using a petroleum gas tank to the wasted rice husk biomass left behind by rice mills or even deposited on roadsides highlights the economic advantages to this technology. The technology has significant potential for other agricultural by-products including sawdust, husks from soybeans and cacao, and sugar cane bagasse.

Appropriate technologies have the potential to increase the quality of life in the communities in which they are used while posing negligible costs to the surrounding environment. However, several criticisms have been made regarding this approach to development. Some argue that indigenous technologies have been used throughout history to meet the needs of local populations, discounting the potential for these technologies to steer development in new directions. Others argue that small-scale technologies are not effective for long-term development needs. Proponents counter that appropriate technologies address various areas of quality-of-life needs as an approach to development, rather than serving as narrow remedies to more specific needs. In addition, although some technologies have been successful in meeting development needs, others have faced failure as a result of improper engineering design and specification or incompatibility with the social and cultural patterns.

See Also: Agriculture; Ecological Economics; Industrial Ecology; North–South Issues; Sustainable Development; Technology.

Further Readings

Barrett, Hazeltine and Christopher Bull. *Appropriate Technology: Tools, Choices, and Implications*. San Diego: Academic Press, 1999.

Dunn, P. D. *Appropriate Technology: Technology With a Human Face*. New York: Shocken Books, 1978.

Fritsch, Al and Paul Gallimore. *Healing Appalachia*. Lexington: University Press of Kentucky, 2007.

Schumacher, E. F. *Small Is Beautiful: Economics as If People Mattered*. New York: Harper & Row, 1973.

U.S. Congress. *An Assessment of Technology for Local Development*. Washington, D.C.: Office of Technology Assessment, 1981.

Thomas D. Eatmon, Jr.
Allegheny College

B

Basel Convention

The Basel Convention on the Control of Transboundary Movements of Hazardous Wastes and Their Disposal ("Basel Convention") is a binding international environmental agreement adopted in March 1989 to protect human health and the environment against the adverse effects resulting from the generation, management, transboundary movement, and disposal of hazardous waste. It came into effect in May 1992 and has been ratified by 170 countries (as of January 2009), though not by the United States. Despite notable success in strengthening the Basel regime and constructing requisite norms, critics charge that much work remains, such as ensuring effective compliance, minimizing hazardous waste generation and transboundary movement, enforcing liability on waste exporters, and halting illegal waste traffic.

The Basel Convention prohibits any export of hazardous waste, defined as toxic, poisonous, explosive, corrosive, flammable, ecotoxic, or infectious waste, to Antarctica, to countries that are not signatories to the convention, or to countries that have banned such activity under domestic legislation. Otherwise, it applies the "prior informed consent" principle, whereby waste shipments made without a prior written notification to, and consent of, importing (and transit) states' competent authority are deemed illegal. Basel also requires that parties manage and dispose of their hazardous waste in an "environmentally sound manner" by minimizing waste generation at the source, restricting the waste moved across borders, and treating and disposing of wastes as close as possible to their place of generation. Any transboundary movements of hazardous waste that violate Basel are considered "illegal traffic" and are subject to specific remedies such as the mandatory return of waste to the importing country at the latter's expense.

According to the United Nations Environment Programme, around 400 million tons of hazardous waste is produced annually. Increasing awareness and clout by consumer and environmental groups prompted industrialized countries, which produce roughly 90 percent of the world's hazardous waste, to enact increasingly stringent environmental regulations that made disposing of hazardous waste locally significantly more expensive (100 times more so than in some developing countries) and politically difficult (the "not-in-my-backyard" problem). Some of this waste (10–20 percent) naturally flowed across national boundaries as individual firms searched for cheaper disposal sites in developing countries that lacked the capacity or awareness to dispose of the waste safely.

The hazardous waste trade first gained notoriety in the 1980s following a series of dramatic scandals such as the infamous case of the vessel *Khian Sea*, which departed Philadelphia in 1986 loaded with 14,000 tons of toxic incinerator ash before dumping some of the load in Haiti and then crossing the waters of five continents during an epic voyage lasting 27 months in search of a dump site. Public outrage across the Third World, combined with the mobilization of transnational environmental networks, resulted in the United Nations Environment Programme adopting the "Cairo Guidelines" in 1987, establishing the first set of global procedures to manage and dispose of hazardous waste, based on the prior informed consent principle. However, these guidelines were nonbinding, soft-law instruments. That further, high-profile scandals involving dumped waste were uncovered in countries as far apart as Haiti, Nigeria, and Lebanon during and immediately after the adoption of the "Cairo Guidelines" spoke volumes about the effectiveness of such voluntary provisions. All in all, Greenpeace estimated that over 2.6 million tons of hazardous waste flowed from Organisation for Economic Co-operation and Development countries to non–Organisation for Economic Co-operation and Development countries between 1989 and 1994.

Developing countries viewed this waste trade within the prism of neocolonialism and demanded a comprehensive ban on all hazardous waste exports from rich to poorer countries. For their part, environmental networks considered such migration of the "effluent of the affluent" as violations of environmental justice and potentially a crime against the environment and human rights. Their objective was to stop, or limit, the production of such waste at the source. However, a U.S.-led "veto coalition" of industrialized countries and private business interests strongly resisted this, arguing that restrictions on the waste trade were contrary to the "rational" logic of globalization, free trade, and even development.

The fight was on between these two coalitions, resulting in the compromise adoption of the Basel Convention in 1989, following 2 years of tense negotiations. The convention clearly reflected the power of the "veto-coalition"—it failed to ban any part of the trade, provided no liability provisions to punish violators, and contained only weak enforcement mechanisms and vaguely defined terms. Its effectiveness basically relied on how seriously member states took prior informed consent and environmentally sound and efficient management procedures. Many developing countries refused to sign such a weak agreement, particularly as Basel did little to address the thriving illegal waste trade that operated transnationally and via nefarious criminal networks. Instead, many countries adopted unilateral national bans on hazardous waste imports (88 by 1992) while also concluding regional agreements such as the Organization of African Unity 1991 Bamako Convention that banned the importation of hazardous and nuclear waste into Africa, a favorite target for the "toxic traders."

Since its adoption, the Basel Convention has been strengthened. Fourteen regional centers were established to support capacity building and implementation in developing countries, and environmental networks such as Basel Action Network monitor states' compliance and expose violators. Basel members have also created criteria for environmentally sound and efficient management of hazardous waste, focusing on prevention and minimization procedures and requiring use of cleaner technologies and production methods. In 1995, member states passed the "Basel Ban" amendment to prohibit all transboundary movements of hazardous waste from industrialized countries to developing ones. Although 64 countries have now ratified the Basel Ban and it has been constructed as a morally binding norm, it has not legally entered into force as it requires three-fourths of the original signatory states to ratify it. In 1999, member states adopted the "Basel Protocol on Liability and Compensation" mandating "adequate and prompt" compensation for "damage" caused by transboundary movement of waste, including illegal traffic. With only eight ratifications to date (20 are required), the protocol has floundered.

Industrialists in the North, however, have reacted to the evolution of the Basel regime by creating and then exploiting various loopholes. One such early loophole allowed for export of waste intended for "reuse" or "recycling" purposes, resulting in a sharp rise in "sham recycling" deals in which businessmen simply placed recycling labels on regular waste shipments. Accordingly, Greenpeace noted that over 90 percent of waste was sold for "recycling" in 1992, compared with an average of only 30 percent during 1980–88. Another loophole was in the very definition of "waste." It took a decade of campaigning, for instance, for the Basel Convention to finally decide in 2004 that the global trade in ships destined for breaking (to retrieve scrap metal) in Asia could be considered as toxic waste and thus subject to the convention's rules. The practice, however, continues. Similarly, the recent explosion in electronic waste (e-waste) trade, in which hundreds of thousands of old computers and mobile phones are dumped in developing countries for "reuse" or "recycling" purposes, constitutes a serious threat to the Basel regime. As with ship breaking, workers and communities in target states that do not have the technology or capacity to safely dispose of this waste are thus exposed to a range of health and environmental hazards. Industrialists and free traders argue that this trade creates valuable jobs for poor people.

See Also: Environmental Justice; Globalization; NIMBY; North–South Issues.

Further Readings

Basel Ban Network (BAN). "About the Basel Ban." http://www.ban.org/main/about_Basel_Ban.html (Accessed January 2009).

Basel Convention on the Control of Transboundary Movements of Hazardous Wastes and Their Disposal. http://www.basel.int/convention/about.html (Accessed January 2009).

Greenpeace. "Toxic Trade." http://www.greenpeace.org/international/campaigns/toxics/toxic-trade (Accessed January 2009).

Miller, Marian. *The Third World in Global Environmental Politics*. Boulder, CO: Lynne Rienner, 1995.

Tolba, Mostapha and Iwona Rummel-Bulska. *Global Environmental Diplomacy: Negotiating Environmental Agreements for the World, 1973–1992*, 2nd ed. Cambridge, MA: MIT Press, 2008.

Karim Makdisi
American University of Beirut

Bhopal

Founded in the early 18th century, Bhopal is the capital and second most populous city (with an estimated population in 2001 of 1.4 million) of the central Indian state Madhya Pradesh, located 750 kilometers (460 miles) south of New Delhi. Major industries in the

The photo shows Union Carbide's abandoned methyl isocyanate tanks as they appeared in 2008. These tanks were at the epicenter of the chemical leak that killed as many as 7,000–10,000 people within days and sickened thousands.

Source: Wikipedia

region include textiles, jewelry, jute, cotton, sugar, power-related products (e.g., transformers), and chemicals. However, it was the founding of the Union Carbide India, Limited (UCIL), pesticide plant by the Union Carbide Corporation (UCC) in 1969 and the industrial catastrophe 15 years later that have become synonymous with Bhopal. In the late evening to early morning hours of December 2–3, 1984, a poisonous gas created from methyl isocyanate leaked from the UCIL plant, forming a toxic cloud that enveloped the city, killing and injuring thousands. The Bhopal event highlights the tragic consequences that can result from rapid industrialization in developing countries with substandard safety regulations and from locating hazardous industrial facilities within urban areas. The largest chemical disaster in the world was the product of technological, human, and managerial oversights that led to enduring human health, environmental, and legal ramifications.

The construction of the Bhopal industrial facility was part of larger green revolution efforts begun by the Indian government starting in the 1960s—an agenda designed to improve agricultural productivity. In addition to increasing the usage of genetically modified crops and irrigation techniques, the Indian green revolution included financing agrochemicals in an effort to increase crop yields and become self-sufficient in food production for one of the world's most populous nations. Bhopal was selected as the site for the UCIL pesticide plant for its centralized location within India, available labor force, transportation linkages via railways, and local resource accessibility (e.g., water, electricity). The UCIL facility produced agrochemicals for the local Indian agricultural market such as Sevin, a carbaryl insecticide that was manufactured using the highly volatile methyl isocyanate (MIC). The colorless liquid MIC is an extremely chemically reactive substance, interacting with widespread materials such as common metals (e.g., iron, copper, tin) and water. Although Sevin and other pesticide products can be produced using less dangerous materials, MIC is a more cost-effective chemical intermediate that reduces production time.

A Disaster Waiting to Happen

Initially, MIC was imported from the parent company, but starting in 1979 the Bhopal plant began producing and storing MIC on-site. By 1979, the UCIL facility had been in operation for almost a decade, and the once remote area on the Bhopal outskirts was now a densely populated community. Nearly 100,000 residents lived within a 1-kilometer radius of the facility, primarily in low-income housing and "squatter settlements." Although the corporation was offered a secondary building site in a less-inhabited region, UCIL insisted on building the MIC division next to the existing facility to reduce transport costs. The amount of MIC produced surpassed daily usage requirements by nearly 10-fold, leading to excess MIC being stored in several large holding tanks with capacities four times greater than safely recommended. Safety regulations were further compromised in the overall design and maintenance of the plant. In comparison to similar facilities in North America and Europe, UCIL lacked modern safety systems and precautionary measures for dealing with MIC storage, including preventive maintenance, routine safety inspections, skilled operators, and an emergency plan. Furthermore, the facility was not profitable. Natural calamities (e.g., droughts) limited the ability of Indian farmers to purchase pesticide products, and the Carbide plant, including the MIC storage units, began to fall into disrepair as revenues diminished.

The consequences of such cost-reducing measures initiated at the UCIL plant came to a head on December 2–3, 1984, when at least 27 tons of MIC leaked from its storage tanks. Within a few hours, the deadly MIC cloud had spread nearly 40 kilometers downwind from the plant over a densely settled residential area, exposing nearly a quarter of Bhopal's estimated population of 900,000 to the toxic chemical. MIC gas has a higher density than the chemical composition of surface air, which means it will stay more concentrated at the ground when introduced into the atmosphere. The high toxicity of the MIC gas was further enhanced by prevailing atmospheric conditions—a local temperature inversion and low wind speeds that created stagnant conditions and limited MIC dispersal and dilution. At the height of the gas leak, the MIC concentration was estimated at 27 parts per million (ppm), or over 1,300 times the maximum 0.02 ppm allowable exposure limitations for an eight-hour work shift set forth by the U.S. Occupational Safety and Health Administration; some dispersion models suggest that a few areas were exposed to levels as high as 85 ppm. On the basis of a 1978 publication on MIC usage, the Occupational Safety and Health Administration reported that an MIC exposure of a few ppm will produce marked long-lasting health effects, such as skin and throat irritation, respiratory ailments, and lacrimation, and levels exceeding 21 ppm are almost certainly lethal.

While They Were Sleeping

Bhopal residents were largely asleep, unaware of the toxic cloud that had enveloped their community. The UCIL warning systems failed to sound an alarm until 2 a.m. on December 3, over two hours after the gas leak had begun. Those exposed experienced immediate reactions such as respiratory problems (e.g., breathlessness, choking, chest pain, hypoxia), ophthalmic effects (e.g., burning, blindness), and other poisoning symptoms (e.g., vomiting, convulsions). These symptoms made up most of the direct morbidity, afflicting more than 200,000. The most rapid deaths were caused by acute respiratory distress from severe lung tissue damage, such as bronchial necrosis and pulmonary edema. The UCC put the immediate death toll at 3,800; less conservative assessments suggest as many as 7,000–10,000 people died within the first week. Deaths were not limited to human beings, as 4,000–5,000 large domestic animals, mostly cattle, also perished from the fumes. Physiological and psychological health effects (e.g., birth defects, lung damage, posttraumatic stress disorder, etc.) from the 1984 MIC exposure continue to this day, and toxicity levels near the UCIL plant remain unacceptably high.

What Happened

According to the analysis conducted by the UCC and other independent investigators in the incident aftermath, the chief cause of the toxic gas cloud was the adverse chemical reaction from water being introduced into the number 610 MIC storage tank. This exothermic reaction produced heat and formed carbon dioxide, methylene, and nitrogenous gases. The outcome was a substantial rise in the temperature and pressure within the tank that forced the chemical release valve open and leaked MIC gas. How the water was introduced into the MIC storage tank is a subject of much debate. The UCC maintains that an unnamed "disgruntled employee" sabotaged the plant by deliberately connecting a water line to the tank—a possible terrorist act. Other theories suggest the water was accidentally introduced by workers unfamiliar with the chemical reactivity of MIC during the cleaning of pipes near the tank in the hours before the incident. Workers may have introduced water into the MIC storage tank by not properly closing the vent lines between tanks with slip binds, which, when combined with corrosive metals in the pipes, inadvertently created an MIC catalyst.

Although the water initiated the chemical reaction, the widespread MIC toxic gas expansion occurred from the failure of UCIL to implement safety protocols and containment measures. The UCC investigation acknowledged that several safety systems were not operable during the night of December 2. The hazardous shortcomings of the MIC storage facility included the following: unreliable tank temperature and pressure gauges, ineffective alarm systems, an idle refrigeration unit for cooling the tanks, an unused caustic soda scrubber for gas neutralization, an already-filled reserve tank for MIC potential overflow, and an ineffective, unused flare tower for removing any escaped gases. Conditions were further exacerbated by the absence of hazardous material procedures, an emergency management plan, community safety awareness, and sufficient local healthcare facilities.

The largest chemical disaster in history also gave rise to the largest lawsuit, which spanned the globe, enlisting thousands of plaintiffs and persisting for nearly seven years. Within days of the accident, multibillion-dollar lawsuits were being filed against UCC. In March 1985, the Indian government passed the Bhopal Gas Leak Disaster act that enabled the Indian government to act as the legal representative for all Bhopal disaster victims. Four years later, the Supreme Court of India delivered a final settlement of $470 million dollars for all Bhopal litigants and exempted all UCIL employees from criminal negligence charges. This amounted to an average compensation of only $2,200 per death for the surviving families and a little over $500 for those with long-term injuries. The settlement grossly underestimated the number afflicted and was considered paltry when compared with settlements conferred in U.S. and European courts. The Bhopal disaster and its aftermath underscored the need for enforceable international guidelines regarding environmental safety and establishing industrial disaster preparedness measures in developing nations.

See Also: Corporate Responsibility; Risk Assessment; Urban Planning.

Further Readings

Fortun, Kim. *Advocacy After Bhopal: Environmentalism, Disaster, New Global Orders.* Chicago: University of Chicago, 2001.

Kurzman, Dan. *A Killing Wind: Inside Union Carbide and the Bhopal Catastrophe.* New York: McGraw-Hill, 1987.

Union Carbide Corporation. "Bhopal Information Center." http://www.bhopal.com (Accessed January 2009).

Weir, David. *The Bhopal Syndrome: Pesticides, Environment, and Health.* San Francisco: Sierra Club Books, 1987.

Jill S. M. Coleman
Ball State University

BIODIVERSITY

The term *biodiversity* (biological diversity) is the variation of living organisms in a given ecosystem and is often used to measure an ecosystem's health, conservation value, or degradation. Some have argued that biodiversity conservation is preferable to a species by species approach to conservation that targets protection and preservation of a few flagship, charismatic species. Biodiversity conservation calls for an understanding of an entire ecosystem and highlights the ecological significance of the all the species the ecosystem harbors.

Biodiversity is not uniformly distributed across the globe; conservation experts have identified areas of concentration of biological richness. J. Mc Neeley, the Chief Biodiversity Officer at the International Union for Conservation of Nature (IUCN), identified countries with high biological diversity and termed them *megadiversity* countries. Norman Myers of Conservation International came up with the concept of biodiversity "hotspots" in 1988. He initially identified 10 such hotspots on the basis of anthropogenic pressures and the level of endemism, and through subsequent publications, he raised the number of hotspots to between 18 and 25. There are presently believed to be 33 biodiversity hotspots in the world (Myers, 2000). Biodiversity hotspots became an important tool for directing conservation efforts by global conservation agencies. The World Wildlife Fund (WWF) has identified terrestrial and freshwater "ecoregions," the International Union for Conservation of Nature (IUCN) and the World Conservation Union have identified centers of plant diversity across the world, and BirdLife International demarcated important bird areas.

Identifying Species and Plant Life

The number of species that survives on this Earth runs into the millions. In his 2002 book, *The Future of Life*, Edward Wilson writes that we have only begun to explore life on Earth. Marine biodiversity of our vast oceans, and invertebrates in particular, remains largely unexplored. The bacteria of the genus prochlorococcus—the most abundant organism on the planet—belongs to a group called picoplankton, and are responsible for a large part of organic production in the ocean. The ocean also teems with little known bacteria, aracheans and protozoans. Among the multicellular organisms of Earth in all environments, the smallest species are also the least known. Of fungi, 69,000 have been identified and named, but as many as 1.6 million are thought to exist. Nematode worms make up four out of every five animal species of the world. Fifteen thousand species are known, but millions more await discovery. In the world of flowering plants, 272,000 species have been identified, and the actual number may be 300,000 or more. Every year, 2,000 new species are added to the known list of plant species. In the world of fauna, new species are continually added to the list. Mammals, birds, amphibians, and reptiles continue to be added to the world's biodiversity inventory.

Biodiversity exists at different levels, genes, and species, as well as in ecosystems called alpha, beta, and gama diversity. This encompasses the ecological services that ecosystems provide and caters to the requirements of our food, raw materials, a wide range of our goods and services, genetic material for agriculture, medicines, and industry. Many indexes exist to measure biological diversity, and the Shannon-Wiener index is often used.

Measures to Protect Biodiversity

Biodiversity, however, is also being lost at an unprecedented rate. The rate of loss of biodiversity is not precisely known, but can be correlated with the rate of loss of their habitat. Land use changes, deforestation, degradation, overexploitation, and climate change are some of the known threats causing depletion of biodiversity. While the loss of species over time has been a part of process of the natural course of evolution, recent decades have witnessed significant loss of biodiversity at a rate two to three times faster than has previously occurred in geological history. Efforts are being made by the global community to slow down this loss. Members of the United Nations resolved to address the issue during the Convention on Biological Diversity (CBD) in September 2003. Every year, May 22 is celebrated as International Biodiversity Day. Countries signatory to the convention are mandated to produce status reports to the convention secretariat and at different stages on implementation of their respective National Biodiversity Strategy and Action Plans (NBSAPs).

The 2002 World Summit on Sustainable Development (WSSD) produced a commitment by governments to address sustainable development, and the plan makes a special reference to biodiversity, with a challenging goal of reducing the rate of loss of biodiversity by 2010. Other international processes relevant to conservation are the Ramsar Convention, focusing on conservation of wetlands, Convention of International Trade in Endangered Species (CITES), Convention on Migratory Species (CMS), Convention on Combating Desertification (CCD), and the Inter Governmental Panel on Climate Change (IPCC). Other programs relevant to the conservation of biodiversity are Global Environmental Outlook (involving the United Nations Environment Programme [UNEP], along with collaborating countries), the World Resource Report (UNEP), United Nations Development Programme (UNDP), World Bank, and the World Resource Institute (WRI), Earth Trend (WRI), International Union for Conservation of Nature (IUCN) Red Lists and species survival commission reports, World Development Report (World Bank), Food and Agriculture Organization (FAO) Plant Genetic Resource Assessment, the FAO report on fisheries, forests, and agriculture and the United Nations Educational, Scientific and Cultural Organization (UNESCO) on Man and the Biosphere (MAB) Programme. The Global Strategy for Plant Conservation (GSPC) has specified that by 2010 at least 10 percent each of the world's ecological regions should be effectively conserved, and 50 percent of the most important areas for plant diversity should be assured through effective conservation measures. The Botanical Garden Conservational International in the UK addresses conservation of threatened plant species through a network of botanical gardens across the world. TRAFFIC (Trade Record Analysis of Flora and Fauna in CITES) focuses on trade in floral and faunal species enlisted in the schedules of CITES.

Biodiversity conservation has been a challenging and daunting task to scientists, policymakers, and implementers, as scientific understanding of the intricacies of nature remain inadequate even today. The challenge becomes all the more difficult in a scenario where conservation competes with the pressures of development. Biodiversity hotspots often overlap with areas of human poverty (Fisher and Christopher, 2007), and limited available resources tend to be directed toward addressing the issues of poverty and hunger.

Countries rich in biodiversity have adopted different approaches for conservation (Chatterjee, 1993). The most common approach for an in situ conservation has been to declare areas protected for conservation. As per Article 2 of the CBD, a protected area (PA) is defined as " a geographically defined area which is designated or regulated and managed to achieve specific conservation objectives." The IUCN categorizes six categories of PAs, which are described in the WCPA (World Commission on Protected Areas) framework of the

IUCN. The World Conservation Monitoring Center (WCMC), located in London, maintains a database of PAs declared by various governments. PAs cover around 10 percent of the world's geographical area. While PAs have been largely successful, they are not adequate in protecting the world's full range of biodiversity. Often, elements of biodiversity and ecosystems remain under represented, or not represented at all. Mulongoy and Chape's 2004 study indicates that less than 1 percent of the Earth's marine ecosystem is represented through PAs. Some of the world's protected areas reel with issues such as nonsettlement of rights and unclear boundaries, among others. Assessments made by organizations like the WWF on the management effectiveness of 200 PAs in 34 countries revealed that only 12 percent of the PAs have implemented an approved management plan. Efforts are also being made to include forest areas under community ownership and initiatives under the global protected area network. These areas are called Indigenous Community Conserved Areas (ICCAs).

Biodiversity conservation is thought to be most successful if the developmental aspirations of people are also met. This led to the genesis of Integrated Conservation Development Programmes (ICDPs) that would focus both on conservation and on improving the population's livelihood. CAMPFIRE (Communal Areas Management Programme for Indigenous Resources) in Zimbabwe, the Community Forestry in Nepal, and the Joint Forest Management program in India serve as global examples. A preliminary assessment on such initiatives by McShane and Wells in 2004 concluded that these were reasonably effective in meeting development objectives, but that few had significant positive impacts on conservation.

Some of the newer approaches to biodiversity conservation are forest certification, payment for environmental services (PES), and landscape level conservation. Forest certification involves certifying a forest-based product by a third party on a set of criteria and indicators to ensure that the product has been procured and processed through sustainable means. The Forest Stewardship Council (FSC) and the Pan European Forest Council (PEFC) are examples of such certifying agencies. Examples include forest protection payments in Costa Rica, issuing licenses for hunting trophies in Pakistan (in Chital and Tushi), and contractual protection in the Ben Shen Zan Mountains in the Yunnan Province of China. The International Standards for Sustainable Collection of Medicinal and Aromatic Plants (ISSC-MAP) is attempting to develop criteria and indicators for the sustainable harvest of medicinal plants sourced from the wild.

The seven focal areas selected by the Convention on Biological Diversity (CBD) to assess 2010 biodiversity targets are as follows:

- Reducing the rate of loss of the components of biodiversity, including (i) biomes, habitats and ecosystems; (ii) species and populations; and (iii) genetic diversity.
- Maintaining ecosystem integrity, and the provision of goods and services provided by biodiversity in ecosystems, in support of human well-being.
- Addressing the major threats to biodiversity, including those arising from invasive alien species, climate change, pollution, and habitat change.
- Promoting sustainable use of biodiversity.
- Protecting traditional knowledge, innovations, and practices.
- Ensuring the fair and equitable sharing of benefits arising out of the use of genetic resources.
- Mobilizing financial and technical resources, especially for developing countries, in particular, least developed countries and small island developing states among them, and countries with economies in transition, for implementing the Convention and the Strategic Plan.

2010 is being called the International Year of Biodiversity. The global situation is desperate, but there are encouraging signs that progress can be made. Global population growth has slowed, and with its present trajectory, it is likely to peak at 8–10 billion people by the end of the 21st century. That number of people could be accommodated with a decent standard of living, but just barely. It should also be possible to shelter most of our vulnerable plant and animal species, provided we follow land ethics practices, using the best understanding of ourselves and the world around us that science and technology can provide.

See Also: Convention on Biodiversity; Ecology.

Further Readings:

Chatterjee, S. "Global Hotspots of Diodiversity." *Current Science*, 68/12: 1178–80 (1993).

Chester, Charles. *Conservation Across Borders: Biodiversity in an Interdependent World.* Washington, D.C.: Island Press, 2006.

Fisher, Brendan and Treg Christopher. "Poverty and Biodiversity: Measuring the Overlap of Human Poverty and the Bioversity Hotspots." *Ecological Economics*, 62 (2007).

McNeely, J. A., ed. "Parks for Life: Report of the Fourth World Congress on National Parks and Protected Areas," February 1992. Caracas, Venezuela: World Congress on National Parks and Protected Areas, 1993.

McShane, T. O, et al., eds. *Getting Biodiversity Projects to Work: Toward More Effective Conservation and Development.* New York: Columbia University Press, 2004.

Myers, N., et al. "Biodiversity Hotspots for Conservation Priorities." *Nature*, 403 (2000).

Wilson, Edward O. *The Future of Life.* New York: Time Warner Book Group, 2002.

World Wildlife Fund (WWF). "The Global 200." http://www.worldwildlife.org/science/ecoregions/global200.html (Accessed July 2009).

Sudipto Chatterjee
HSG, Sagar University

Biophilia

Biophilia is the natural tendency to focus on living things and lifelike processes. It is a theory developed by Edward O. Wilson in his book by the same name. Wilson argues that there is an instinctive bond that humans feel for living things, suggesting that biophilia is genetic in origin, a result of evolution, and enables greater survivability. This innate intelligence shapes cognition, emotions, values, and culture. Biophilia fosters a conservation ethic, and elements of modern society that alienate people from the natural world may frustrate innate intelligence and threaten continued evolution.

Relating to life is a biological need—an innate process—and is deeply connected to our mental and physical development. Humans are instinctually drawn to living things, processes, and such, which are aligned with our cognitive development and reason. Humans' ability to understand other organisms is related to our ability to value those organisms, life in general, and ourselves. When we understand other life-forms, we appreciate and understand ourselves more. Humanity is what it is largely because of the unique way we affiliate with and use other organisms. Advances in moral reasoning and development of a deeper conservation ethic are dependent on people's understanding of other organisms and the interconnectivity of life. Greater reporting, observing, and intimacy are required to advance understanding. Personal freedom and a conservation ethic are not at odds with each other; a conservation ethic and affiliation with the biosphere is what secures the stability of our own species.

The biophilia hypothesis proposes that humankind's instinctual interest in other organisms and need to protect and care for them may be the result of evolution.

Source: iStockphoto.com

Humans prefer certain animals because of a kinship feeling: Animals are related life. Humans may relate to animals as if they are people. The observation of the continuity of life with humanity encourages a desire for the protection and continued existence of those other organisms. This desire advances the value of other creatures without diminishing humanity. The insight of relatedness enlarges the circle of ethics beyond our own species to other life. This ethic is not necessarily altruistic. In simple economic measures, species diversity is one of humanity's greatest resources. People completely depend on other living things for their own existence. The more the natural world is explored and cataloged, the more resources people will appreciate and use for their own development. Existing organisms are the result of evolution and are already a very select group. They are our biological inheritance and a natural base for other developments and yet-undiscovered uses. Extinction is a great loss to our own species because we permanently lose that advanced life-form and the gifts it may have offered us.

An Evolutionary Development

The biophilia hypothesis goes beyond just dependence on the natural world for the consumptive physical and material needs people have and extends to the human need for cognitive, intellectual, aesthetic, and even spiritual satisfaction and meaning. The hypothesis suggests that a need for intimate and deep association with living things is born from evolutionary development. The hypothesis compels an examination of why humans protect and cherish life—even life other than their own. It suggests that biophilia is a built-in ethic to protect and care for a broad diversity of life and that it naturally increases the likelihood of personal fulfillment and attaining individual meaning. Innate care for life is part of humanity's evolutionary heritage, and it is inherent and associated with our own genetic fitness and sense of competitive advantage.

The theory sees built-in biophilia as a cognitive set of learning rules and argues that some human values are biologically based and a result of human evolution occurring in an interconnected and complex natural environment. Appreciation for life and natural processes, a form of innate bioculture, evolved as the genes carrying those specific learning propensities were carried through time and privileged by natural selection. Certain genes driving various behavioral responses are a common idea. Certain responses enable better survival, fitness, and greater likelihood to reproduce. Genes carrying such responses subsequently spread through the population. Genes and culture coevolving is a reasonable explanation for various forms of psychological structure, biophilia being one of those forms.

This genetic-based thesis for biophilia is also the basis for the field of "evolutionary psychology." Phobias and biophobias can also be traced to evolution and cast as part of biophilia. Phenomena like the common fear of snakes, a biophobia, can easily be explained in this manner. Snakes, for instance, do create death and illness in primates and mammals all over the world. Primates commonly combine a strong fear of snakes with alarming vocalizations to warn others of the reptiles. Humans, too, are genetically ill-disposed to snake poison, and humans are prone to develop heightened fears and phobias toward snakes with very little education or direct experience. People in cultures all over the world report dreaming of snakes, more than any other animal. Thus, attention to living things and processes appears to have helped humans survive, appears to have shaped inherited cognitive structures, and appears to be cross-cultural.

Biophilia is not a single instinct but a complexity of cognitive and emotional learning tendencies, each of which can be separated and studied individually. Biophilia theory argues that these tendencies that are best suited for learning and living in a complex natural environment are not necessarily capable of adequately adjusting to help us exist in this newer civilization in which we now find ourselves. Human aesthetic, cognitive, and emotional development is also theorized to be a result of conditioning by the natural environment. Our sense of beauty, ethics, and attraction to certain landscapes can all be explained in terms of evolution and biophilia.

The development of biophilia theory also considers culture and its various expressions as being grounded in humans' focus on life and life processes, and researchers use cross-cultural observations to substantiate this hypothesis. Comparisons between indigenous peoples and industrial Western societies raise questions about integrity and wholeness and the consequences of losing biophilia tendencies to aggravated estrangements between people and the natural world.

Biophilia theorists also consider "symbolism" and the role of animals and nature in human communication and cognitive development. It is nature's biotic diversity of kinds and forms that enabled human capacity for metaphor and symbolic communication. The human brain evolved in a biologically diverse natural world, not a modern, manufactured, or bureaucratic world. It is illogical to think that millions of years of evolution have not left a mark on our cognitive structures or can be erased by a few centuries of modernity. The gradual erosion of distinctions between wild and domesticated and natural and artificial is theorized to be potentially problematic.

Theorists see the connections between organisms, understandings of "Gaia," predictable tendencies of organismic symbiosis, and organisms naturally behaving in helpful and predictable manners with one another as evidence of biophilia. A belief in biophilia naturally lends itself to an active defense for a conservation ethic. Biophilia raises questions about modern society, materialism, consumer culture, individualism, the growth ethic, and the need for social and political change. Sensitivity to biophilia calls for action to address current massive species extinction and weakening of ecosystems and theorizes why action may not be taking place.

Biophilia also describes how human consciousness, which developed over the millennia to attend to fast changes and ignore slow changes, no longer fits the modern world. The modern world is characterized by slow changes constantly created by human culture.

Humans are poorly equipped to see how their own culture is changing the world in the course of a few generations. Changes like pollution are happening too slowly for our "fight or flight" nervous systems to become alarmed. The human cognitive structure is designed to become alarmed at sudden changes, not slow changes, so there is an inherent misfit between human consciousness and the world humans have created. Some biophilia theorists make compelling arguments for developing a new consciousness through education to revive an ethic of care and concern for all life and the environment on which all life depends. Although the biophilia hypothesis takes on many forms, scientists delineate the need for more research if it is to become harnessed to create change.

See Also: Biodiversity; Conservation Movement; Deep Ecology; Equity; Intrinsic Value.

Further Readings

Kellert, Stephen R. *The Biophilia Hypothesis*. Washington, D.C.: Island Press, 1993.
Kellert, Stephen R. *Kinship to Mastery: Biophilia in Human Evolution and Development*. Washington, D.C.: Island Press, 1997.
Primack, Richard B. *A Primer of Conservation Biology,* 3rd Ed. Sunderland, MA: Sinauer Associates, Inc., 2004.
Wilson, Edward O. *Biophilia*. Cambridge, MA: Harvard University Press, 1984.
Wilson, Edward O. *The Future of Life*. New York: Knopf, 2002.

John O'Sullivan
Gainesville State College

Biosphere

The biosphere is the highest level of organization of the Earth's biological activity. It processes matter and energy transfer with an efficiency of 10 percent across the Earth's atmosphere, lithosphere, hydrosphere, cryosphere, and anthrosphere. The atmospheric processes involve a vast number of chemical reactions and gas exchanges within the atmosphere. The hydrosphere consists of the Earth's water systems, needed for organisms living on the planet to survive. The geosphere and biosphere are closely connected through soils that consist of an admixture of air, mineral matter, organic matter, and water. The anthrosphere is the dimension of the biosphere that has been—and continues to be—altered by humans for human activities. The human population is now a direct threat to the biosphere through habitat destruction and atmospheric degradation, especially deforestation and greenhouse gas emissions. Human activities are recognized to be the cause of a mass extinction of other species and an ensuing depletion of genetic variation and biological diversity (biodiversity).

The biosphere encompasses all biological activity on Earth, which is of vital importance to the functioning of natural and human-engineered ecosystems, and by extension, the services that nature provides free of charge to human society. The value of these services to the human economy is so huge as to defy quantification, though ecological economists continue the effort to more fully account for costs of production endured by the biosphere as "ecosystem services." The biosphere performs all biological functions, including photosynthesis,

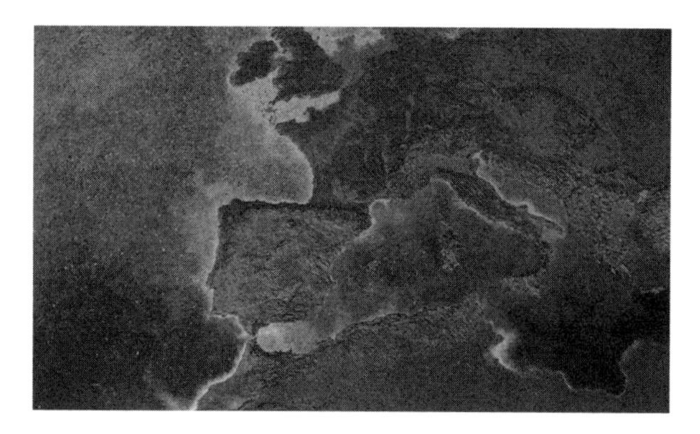

This image of Western Europe created by a NASA remote-sensing satellite is part of an attempt to capture information about the biosphere, such as vegetation cover, that may help scientists better understand changes on a global level.

Source: NASA

decomposition, nitrogen fixation, respiration, and denitrification. The biosphere is structured into a hierarchy of life-forms known as the food chain, in which all life evolves toward increasing complexity. The biosphere makes our planet unique among the planets in the solar system.

The history of the term *biosphere* goes back to 1875, when geologist Eduard Suess originated the term, defining it as "the place on earth's surface where life dwells." Russian scientist Vladimir Vernadsky extended the term in 1929, refining its definition into a form resembling its current ecological sense in his obscure book *The Biosphere*, published in 1926. Thanks to Vernadsky's work, ecology was redefined as the science of the biosphere, and the biosphere concept began to occupy its current central position in Earth systems science.

The biosphere is studied within the scientific fields of biology and ecology. It exists as the highest level of biological organization, beginning with parts of cells and rising to populations, species, ecoregions, and biomes. Biomes describe global patterns of biodiversity within the biosphere's many ecosystems. Living organisms and their remains in the biosphere interact with the other spheres (above) in global biogeochemical cycles and energy budgets, as the biosphere plays it central part in the nurturing of life on Earth.

Researchers directly observe biosphere activity using global remote-sensing platforms, some dispersed on Earth's surface and many launched aboard remote-sensing satellites into near-Earth space over the past few decades. Direct observation has greatly benefited from advanced space-based remote-sensing systems, which are capable of scanning the entire Earth's surface at least once a day. These observations help determine the extent and amount of activity in the biosphere, mainly in terms of vegetation cover and function and characteristic spectrum responses in sensing spectrometers. More remote-sensing efforts in the future will directly observe global patterns of carbon dioxide exchange in the biosphere, caused by respiration, photosynthesis, greenhouse gas emissions, and the combustion of biomass and fossil fuels.

Biosphere Reserves

Humans have established biosphere reserves to help protect natural systems from damage by human activity. These reserves are areas of terrestrial, coastal, and ocean ecosystems that are set aside for the use of other species and to promote the conservation of biodiversity through sustainable use. These reserves can be found around the world and serve as laboratories for testing land, water, and biodiversity. The concept of the biosphere reserves goes back to a 1968 biosphere conference organized by the United Nations Educational, Scientific, and Cultural Organization; as a result of that conference, the Man and the

Biosphere Program was initiated in 1970. One of Man and the Biosphere's original projects was to establish a coordinated world network of sites representing the main ecosystems of the planet, in which genetic resources would be protected and where research on ecosystems as well as monitoring and training could be accomplished.

Each biosphere reserve site has three main functions:

- Conservation: to contribute to the conservation of landscapes, ecosystems, species, and genetic variation;
- Development: to foster economic and human development that is socioculturally and ecologically sustainable; and
- Logistic: to help research, monitoring, education, and information exchange related to local, national, and global issues of conservation and development.

Biosphere reserves are organized into three zones: the core area, which is the only area that requires legal action and that can consist of existing protected areas, such as nature preserves or national parks; the buffer zone, which protects the core area against external pressures; and the transition area, which includes commercial and other human activities, like forest harvesting, sisal and tea plantations, and other similarly nature-oriented uses.

Biosphere II

Biosphere II is an enclosed, transparent geodesic sphere that is designed to contain and support its own atmosphere, mimicking the function of the Earth's biosphere. It was developed in the 1970s in a semidesert ecosystem north of Tucson, Arizona, and several human test subjects were placed into the system, to exist without outside biological support of any kind. This experiment soon resulted in showing scientists how little they understood about the concept of the biosphere. Although Biosphere II has provided highly useful results in experiments with plants and oxygenation, it was ultimately unable to sustain its own atmosphere or replicate the functions of Earth's biosphere. Without resorting to drastic chemical interventions to inject oxygen and diminish toxic levels of carbon dioxide, it was impossible for the complex to support human life. Furthermore, many keystone species, such as pollinators, soon died. However, Biosphere II was a useful laboratory for many years, even if it was unable to demonstrate the actual function of a living, planetary biosphere.

See Also: Biodiversity; Ecology.

Further Readings

Answers.Com. "Eduard Suess: Biography." http://www.answers.com/topic (Accessed April 2009).

Answers.Com. "Vladimir Vernadsky: Information." http://www.answers.com/topic (Accessed April 2009).

Encyclopedia of Earth. "Biodiversity." http://www.eoearth.org/article/Biodiversity (Accessed April 2009).

Encyclopedia of Earth. "Biosphere." http://www.eoearth.org/article/Biosphere (Accessed April 2009).

Hodell, David and Ray G. Thomas. "The Biosphere." http://www.ess.geology.ufl.edu/ess/Introduction (Accessed April 2009).

United Nations Educational, Scientific and Cultural Organization (UNESCO). "Biosphere Reserves." Ecological Sciences: UNESCO/Science. http://www.unesco.org/mab/doc (Accessed April 2009).

Shaista Consuelo Amar
Anthony R. S. Chiaviello
University of Houston–Downtown

Brundtland Commission

The United Nations (UN) established the Brundtland Commission, officially known as the World Commission on Environment and Development, in 1983 to examine the effect of environmental degradation and natural resource depletion on future economic and social development around the world and to propose responses. The commission's 1987 report, *Our Common Future*, is considered a landmark in global environmental politics because it helped to define, legitimize, and popularize the concept of sustainable development.

The World Commission on Environment and Development is commonly referred to as the Brundtland Commission in recognition of its chair, former Norwegian prime minister Dr. Gro Harlem Brundtland. The United Nations Secretary-General's choice of Brundtland to chair the sensitive project reflected the political respect he had earned in many capitals around the world, including both industrialized and developing countries; his experience in environment and development issues (he had also previously served as Norway's Environment Minister); and the belief that he could manage the project to a meaningful and successful conclusion.

In creating the commission, the UN General Assembly requested that it focus its work on proposing long-term strategies for achieving sustainable development; highlighting avenues for more effective cooperation among countries that take account of the interrelationships among people, resources, environment, and development; helping to define shared perceptions of long-term environmental issues; and considering ways that the international community can deal more effectively with environmental concerns, taking into account development needs and other issues. To address these issues, Brundtland and the United Nations gathered leading scientists, lawmakers, diplomats, and ministers to staff or advise the commission. In 1987, after 3 years of research and public hearings on five continents, the Brundtland Commission published *Our Common Future*. The 400-page report was formally submitted to the United Nations General Assembly for consideration and was widely read around the world following its publication by Oxford University Press.

Our Common Future introduced the term *sustainable development* into broad usage, legitimizing the concept in the minds of many observers. The definition of the term used in the report became well known and often cited: "Sustainable development is development that meets the need of the present without compromising the ability of future generations to meet their own needs."

The Brundtland Commission also asserted that the concept of sustainable development contained within it two other intrinsic ideas: "the concept of 'needs,' in particular the essential needs of the world's poor, to which overriding priority should be given; and the idea of limitations imposed by the state of technology and social organization on the environment's ability to meet present and future needs." These elements of detail in the commission's definition drew less attention but influenced thinking on the commission regarding both the priority that it believed should be given to meeting the needs of the world's poor (development) and the reality that certain technological and economic practices were depleting resources at rates that threaten that development.

Our Common Future described a number of key, interconnected environmental and developmental issues—areas in which future developments would be critical to determining whether sustainable development is achievable. These included population and human resources; food security, which also includes issues of distribution and trade; species and ecosystem preservation; sustainable energy paths and consumption patterns; industrial production; and urbanization. The Brundtland Commission concluded that three broad initiatives were necessary to orient these issues on the right path: reexamining crucial environment and development problems and identifying realistic solutions to avoid dire consequences in the future, proposing new international cooperative efforts to create policies to achieve the needed solutions, and developing far greater reservoirs of awareness, knowledge, and responsibility in all political and social communities.

The Brundtland Commission concluded that most environmental, development, economic, and related social problems were fundamentally a singular, interconnected issue. For example, the unequal distribution of economic wealth, political power, and ecological resources around the world contributed significantly (in different ways) to the existence and persistence of poverty and underdevelopment. Poverty, in turn, is both a source and an outcome of environmental degradation. Environmental degradation and uncontrolled resource depletion threaten long-term economic growth for both developed and developing countries, placing limits on the potential to alter this situation and potentially to meet human needs now and into the future. The commission argued that such interconnected cause-and-effect relationships can only be addressed through a new path of global economic growth that is more fairly distributed and based on the principles of sustainable development. Therefore, *Our Common Future* advocated institutional reform that integrated environmental and economic decision making, balancing the terms and patterns of international trade to reduce or eliminate global poverty, limiting pollution, conserving resources, and protecting and enhancing the global commons so that the needs of all humanity can be met.

The Brundtland Commission greatly enhanced global awareness and understanding of sustainable development, the connections between poverty and the environment, and the limits that excessive levels of environmental degradation and resource consumption can place on future prosperity. Published 15 years after the Stockholm Conference (1972), the United Nation's first major conference on international environmental issues, *Our Common Future* provided important conceptual frameworks and political momentum that assisted efforts to convene the 1992 United Nations Conference on Environment and Development (often known as the Rio Earth Summit), adopt Agenda 21 (at the Earth Summit), establish the United Nations Commission on Sustainable Development, and even formulate and adopt the Millennium Development Goals.

See Also: Club of Rome; Future Generations; Limits to Growth; Millennium Development Goals; Precautionary Principle; Sustainable Development.

Further Readings

Hinrichsen, Don. *Our Common Future: A Reader's Guide*. London: Earthscan, 1987.

Richman, Barbara T. "20 Years Into Our Common Future." *Environment*, 49/9 (November 2007).

Tomalty, Ray. "An Enduring Legacy: *Our Common Future* Is as Salient Today as It Was 20 Years Ago." World Commission on Environment and Development. *Alternatives Journal,* 34/1 (January–February 2008).

United Nations. *Our Common Future: Report of the World Commission on Environment & Development.* http://www.un-documents.net/wced-ocf.htm (Accessed February 2009).

David Downie
Alexis Terrizzi
Fairfield University

Bureau of Land Management

The Bureau of Land Management (BLM), an agency under the U.S. Department of the Interior, manages almost 260 million acres of public land in the United States. This amounts to one-eighth of all U.S. land, although 99 percent of BLM land is located in the western United States. BLM lands constitute two-thirds of Nevada, one-half of Utah, and one-third of Wyoming. Although the National Park Service receives a great deal more public attention, the BLM is responsible for three times as much land area as the National Park Service. The agency was established in 1946 to manage public rangelands and public land claims. Today, the BLM manages land for multiple and sometimes competing uses, including grazing, mining, energy production, recreation, and conservation. Unlike the National Park Service, which maintains the reputation of managing some of the country's most spectacular landscapes, the BLM has often been considered the manager of wastelands. The agency's history explains this reputation. The foundations of the BLM lie with two early acts. In 1862, the Homestead Act encouraged westward expansion by providing land to enterprising settlers through the General Land Office. To control some of the degradation caused by cattle in the West, the U.S. government imposed additional grazing regulations via the 1934 Taylor Grazing Act. This act created the U.S. Grazing Service, set aside 80 million acres of public land for grazing, and began charging ranchers a modest fee for grazing on public lands. In 1946, the Grazing Service and the General Land Office merged to create the BLM. Although the Grazing Service had primarily dealt with livestock issues on public lands, the General Land Office had dealt with land management on grazing lands and with minerals management. The merged mandates and priorities of these two organizations contributed to a fairly disorganized and incompetent agency lacking a

The U.S. Bureau of Land Management has been involved in establishing alternative energy facilities on government land, including this geothermal plant in California photographed in June 2008.

Source: U.S. Bureau of Land Management/California

clear mandate. Finally, in 1976, the Federal Land Policy and Management Act provided a clear mandate to BLM: the "management of the public lands and their various resource values so that they are utilized in the combination that will best meet the present and future needs of the American people."

Iconoclast Ed Abbey referred to the BLM as the "Bureau of Livestock and Mining," and many environmentalists have latched onto this conception of the agency. Abbey also called the West "cow-burnt," a condition he may have attributed to BLM management. In 2003, BLM managed over 18,000 grazing permits for about 160 million acres of its land that is used for livestock grazing. There is a spatial mismatch here, however, as over 80 percent of U.S. livestock are raised on private lands in the eastern United States and western public lands produce only 2 percent of total U.S. livestock. Grazing permits are issued for 10 years, and cattle owners pay the BLM a grazing fee of about $1.35 per animal unit month (AUM). An AUM is the amount of food needed for one cow and calf (or one horse, or five sheep or goats) for one month. Many see the low AUM fees as a massive federal subsidy for western ranchers. These grazing permits, developed under the 1934 Taylor Grazing Act, were created to control overgrazing in the West. Grazing on public lands can contribute to environmental degradation, including erosion, disturbance of traditional fire regimes, and destruction of riparian habitat. Increasing environmental concern in the 1960s and 1970s led to the creation of additional regulatory mechanisms to decrease the environmental impact of grazing on public lands. With the advent of the 1969 National Environmental Policy Act, the BLM began conducting Environmental Impact Statements on grazing permits. Later, the Public Rangeland Improvement Act of 1978 led to stock reductions on BLM rangelands. Partly as a result of these regulations, grazing on BLM lands continues to decline: In 1941, BLM issued 22 million AUMs, and in 2008, it issued 12.5 million. New grazing regulations developed in 2006 require ecological data collection and monitoring and open the door for potential stock reductions on BLM lands. Although conflicts still arise among environmentalists, ranchers, and the BLM, laws such as National Environmental Policy Act provide mechanisms for environmental protection.

In addition to grazing, the BLM provides permits and allowances for a suite of extractive industries, including energy, mining, and forestry. Much of the land that BLM manages supports these commercial activities. Mining activities were initially allowed on BLM lands with relatively few regulations or restrictions. Under Interior Secretary Bruce Babbitt, the BLM adopted new requirements and standards to address the environmental impacts of mining. The BLM also leases land for energy resources including oil, gas, and coal: 40 percent of U.S. coal production occurs on BLM lands, as does 11 percent of natural gas and 5 percent of oil production. The BLM manages 700 million acres of underground mineral deposits, many of which are tapped as energy sources. The agency grants permits to companies for all stages of energy production—from exploration through processing—and passes some of the licensing and permitting revenue on to the states. The BLM also manages about 57 million acres of commercial forests. Most of these activities contribute to environmental degradation: Energy and mineral mining contributes to habitat destruction and water contamination, and forestry contributes to an altered fire regime, habitat destruction, and riparian area degradation. Not all activities on BLM land, however, are extractive. Infrastructure on BLM lands contributes to almost half of all U.S. geothermal energy production and about 15 percent of wind power. Also, the BLM manages the wild horse (mustang) population and runs a wildfire management program.

Since the 1976 Federal Land Management and Policy Act, the BLM has managed its land for multiple uses, which has required balancing values of conservation, utilitarianism, and recreation. The Endangered Species Act and other legislation required the BLM to consider conservation values. However, not until some portions of BLM land were designated specifically for conservation did the agency embrace conservation use and value. In 1996, President William J. Clinton designated Utah's Grand Staircase-Escalante region as the first national monument run by the BLM. National monuments are protected areas often created for preservation of a particular cultural or ecological resource. Presidents can designate areas without congressional approval, as Clinton did in 1996. This action was significant in that it provided the first major opportunity for the BLM to manage for conservation. In 2000, the creation of the National Landscape Conservation System provided a framework for more emphasis on conservation and included BLM lands designated as national monuments, national conservation areas, wilderness areas, wilderness study areas, national wild and scenic rivers, and other conservation designations. In addition to providing for conservation, many of these lands also provide for recreational uses.

See Also: Environmental Management; Sagebrush Rebellion.

Further Readings

Muhn, J., et al. *Opportunity and Challenge: The Story of BLM*. Washington, D.C.: U.S. Department of the Interior, Bureau of Land Management, 1988.

U.S. Bureau of Land Management. http://www.blm.gov (Accessed January 2009).

Wilkinson, C. F. *Crossing the Next Meridian: Land, Water, and the Future of the West*. Washington, D.C.: Island, 1992.

Kate Darby
Arizona State University

Capitalism

The nature of the capitalist economic system and its effects on the natural environment is a subject of much debate. This article begins by exploring the basic economic theory that explains why capitalism may result in patterns of production and consumption that ultimately lead to environmental degradation. Next, the article explicates two polar views on capitalism and the environment, first examining the steady state argument, which proposes that the economic freedom that forms the basis for capitalism is at the heart of the environmental crisis. The counterpoint view of the market apologists is then presented; this group of thinkers contends that economic growth and the technological innovation spurred by the market system will produce solutions that will ultimately reduce or even solve environmental problems. After exploring these polar views, this article concludes with thinking in microeconomic theory, which has proposed some ways to encourage environmental protection within the context of the market system.

As awareness of environmental issues grew in the 1970s in the United States, scholars began thinking about the possible systemic causes of environmental problems. One influential group of theorists argued that the very nature of the capitalist system itself was to blame for environmental problems. The basis for their argument was the so-called tragedy of the commons. That is, there are certain common resources (such as air, water, etc.) for which there are no property rights; therefore, in an unregulated market system there is no disincentive to overuse and exploit these common resources because there is little direct cost associated with their use (or abuse) to individual consumers or firms. One of the shortcomings of capitalism is the inability to extend property rights to common resources, which ultimately creates incentives to exploit the environment rather than to protect it.

Using the lens of negative externalities, we are able to get a slightly different view of the same underlying problem. Simply put, an externality is an effect resulting from some market transaction that affects a third party (or parties) who is not involved in the transaction. To take a simple example, consider the production of electricity using coal. In a totally unregulated market, the electric company would produce power by burning coal and sell the power to individual consumers or firms. There would be various costs associated with the electric company's production of power such as the cost of coal, wages for its workers, the cost of plant maintenance, and so on. But this market transaction would also produce externalities in the form of pollution; in the process of producing power, the plant would release emissions into the atmosphere, contributing to air pollution, global warming, and so on. This raises the dilemma of who should pay for this externality. In a completely unregulated market, no one would pay, at least directly, because there is no market for clean air. No one owns the air, so no one can charge the electric company for using (or polluting) it because air is a common resource owned by everyone.

The problem with capitalist production, then, can also be thought of as a problem of uncompensated externalities. Of course, there are countless activities other than power production that also produce negative environmental externalities. Examples include driving, land development, use of recreational and natural areas, and consumption of resource-intensive products; many of our day-to-day activities have environmental side effects that are hard to quantify in exact monetary costs.

In reality, the tragedy of the commons and externality frameworks point to the same underlying problem. Whenever there are resources used in, or affected by, economic activities that are not accounted for in the cost of production or in consumer prices, the net effect will be an overuse and exploitation of those resources because there is no private property interest to protect them.

Working from the premise that the unbridled economic freedom of capitalism was the source of environmental problems, the solution proposed by one early group of theorists was a wholesale institutional change to a regulated steady state economy. This system would involve strong centralized state control and planning, the goal of which would be to conserve natural resources and to protect the environmental commons. This steady state was actually viewed as inevitable by many theorists. They argued that once things got bad enough, there would be no choice but to move to a strong, centralized state to prevent environmental catastrophe.

On the opposite end of the spectrum from the steady state theorists, market apologists held a very different view on capitalism and the environment. Led by economist Julian Simon, the market apologists advanced a strongly optimistic case for capitalism's ability to solve resource scarcity and environmental problems. The main argument of the market apologists was that the technological change spurred through scarcity-induced price change would present new environmentally beneficial products and production processes. In contrast to the steady state argument, the market apologists did not believe that an institutional change from capitalism was necessary but, instead, argued that the adaptive innovative nature of capitalism itself was the key to dealing with resource scarcity and environmental degradation.

The essential problem with capitalism according to steady state theorists was its failure to provide a mechanism for valuing common resources. The main response of market apologists was optimism about the ability of capitalism to produce technologies that would deal with environmental degradation. Since the 1970s, we have seen rapid technological change, some of which has indeed brought improvements to environmentally harmful production processes and has led to the creation of many more environmentally friendly products. Yet despite these advances, there are still many environmental problems in industrialized and developing nations. At the turn of the 21st century, global warming, deforestation, water degradation, air pollution, species extinctions, and loss of land to urban sprawl constitute just a minor sampling of the environmental challenges we are currently facing. Although technology has undeniably aided our progress and led to incremental improvements in environmental quality, it would seem that it has not been the panacea that the market apologists thought it would be.

In the last couple of decades, a new generation of microeconomic theorists has begun to tackle some of the fundamental dilemmas that capitalism presents for the natural environment. Much of that work has employed an externalities framework, focusing on ways in which uncompensated environmental externalities may be better accounted for within market transactions. The idea is that if there can be some reasonably accurate cost assessed for negative environmental externalities, the "true" price of our behaviors will be more accurately reflected. If the costs of certain activities then become too high when those costs are internalized, then the market will function to find less costly, more environmentally friendly ways of doing things so that those environmental harms can be minimized.

There are several avenues for internalizing externalities. Direct regulation, or command and control, is one approach. The government might set pollution targets for industries to meet by a certain time or severely curtail certain production methods (e.g., nuclear power). The advantage of the direct regulatory approach is its apparent administrative simplicity. Command and control does have some shortcomings, though. In particular, command and control usually sets its target objectives without considering the full costs to industry. It also applies fairly uniform standards, not accounting for variation in production efficiencies and processes within industries. This one-size-fits-all approach sometimes may not be the least expensive means of achieving environmental outcomes.

Marketable permits are another approach that has been tried and that has seen some success in polluting industries regulated by the Environmental Protection Agency under the Clean Air Act. The idea behind this approach is to provide each firm with a permit for a certain level of emissions. The permit is marketable, though, and can be purchased by other firms. Economists argue that the added flexibility of buying and selling permits achieves pollution reduction at a lower cost to society than does direct regulation. Consider an industry composed of firms in two groups: those with high levels of pollution and those with low levels of pollution. In a tradable permit system, the low-pollution firms could sell their permits to the high-pollution firms, rewarding the more environmentally friendly practices of the low-pollution firms. At the same time, the cost of purchasing permits for high-pollution forms might be less than the capital costs for upgrading plants, resulting in a lower cost to them. The key to making such a system work is for the regulator to issue a quantity of permits equal to the desired level of pollution. This quantity can be gradually decreased over time, driving up costs gradually and creating incentives for industry to invest in cleaner production processes.

Excise taxes are another way to internalize environmental costs. Perhaps the best example of this is a gasoline tax. In the United States, gas taxes are relatively low compared with nations in Western Europe. Many have argued that low gas taxes in the United States have kept the demand for driving too high, leading to higher levels of auto emissions and air pollution. Some have proposed that if the true social and environmental costs of driving were factored in, the cost of a gallon of gas would be much higher than it is currently. If the gasoline tax were raised to reflect more of these external environmental and social costs, there would likely be at least some decrease in driving or possibly a shift to alternative fuel vehicles. In addition, some have suggested that the revenue generated by a higher excise tax could be used to invest in mass transit and pollution-abatement programs.

In summary, there has been a concerted effort by economists and policymakers to address the commons problems presented by capitalism. Through various regulatory and tax instruments, there has been progress in internalizing many of the external environmental costs associated with market transactions. Yet there is still much room for improvement, as seen in the example of the low excise tax on gasoline in the United States. With our mounting environmental challenges there are clearly areas in which government has failed to fully effect internalization of environmental costs resulting from market activities.

Capitalism is built on the existence of free markets and property rights. Where property rights exist and are enforceable, capitalism functions well as a mechanism for allocating scarce resources. Common resources, such as air and water, for which there are no easily identifiable property rights pose a challenge to unregulated markets because there is a tendency for individuals and firms to exploit those resources with little or no accounting of the associated costs. In light of the dawn of the early environmental movement in the 1970s, thinkers realized these problems inherent in capitalism and proposed a radical institutional shift to an authoritarian steady state. Market apologists responded to this call with a defense of capitalism based on its ability to induce beneficial technological change, which in turn would produce solutions to environmental degradation. Although it does not appear that a transition to a steady state economy is on the horizon, there are still many clearly evident market failures that continue to produce harm to the natural environment. A key 21st-century challenge will be to devise and construct fiscal and regulatory policy that encourages firms and individuals to adequately take stock of the environmental consequences of their activities in capitalist economies.

See Also: Ecological Economics; Limits to Growth; Regulatory Approaches; Steady State Economy; Tragedy of the Commons.

Further Readings

Gilpin, Robert. *Global Political Economy: Understanding the International Economic Order.* Princeton, NJ: Princeton University Press, 2001.

Hackett, Steven C. *Environmental and Natural Resources Economics: Theory, Policy, and the Sustainable Society*. Armonk, NY: M.E. Sharpe, 2006.

Hahn, Robert. "Market Power and Transferrable Property Rights." *The Quarterly Journal of Economics*, 99/4 (1984).

Heilbroner, Robert. *An Inquiry Into the Human Prospect.* New York: W. W. Norton, 1974.

Maurice, S. Charles and Charles W. Smithson. *The Doomsday Myth: 10,000 Years of Economic Crises.* Stanford, CA: Hoover Institution, 1984.

Ophuls, William. *Ecology and the Politics of Scarcity: Prologue to a Political Theory of the Steady State.* San Francisco: W. H. Freeman, 1977.

Pierce, David and R. Kerry Turner. *Economics of Natural Resources and the Environment.* Baltimore, MD: Johns Hopkins University Press, 1990.

Simon, Julian. *The Ultimate Resource.* Princeton, NJ: Princeton University Press, 1983.

Michael Howell-Moroney
University of Alabama at Birmingham

Citizen Juries

Citizen juries are one of a number of recent strategies for increasing public participation in scientific and technological policy decisions. Concerns about a decline in public participation in representative democracy and a "democratic deficit" have led to a focus on more deliberative democracy. This has spurred a search for alternative venues for seeking public input.

Citizen juries, along with consensus conferences, focus groups, polls and referenda, and citizen panels, are one such venue. The idea of creating citizen juries arose in Germany and in the United States as early as the 1970s and has gained traction as concerns about lack of public involvement in traditional representative democracy have increased. Citizen juries have been used mainly in Europe and the United States, but they have also been used in less developed countries as a means of determining public concerns, gauging public sentiment, and gaining public input on complex, often technological, environmental policy issues.

Based on the model of legal trials by a "jury of one's peers," citizen juries bring together a group of citizens to hear information, deliberate, and offer a reasoned decision on policy matters brought before them. Unlike jury trials, however, a citizen jury's findings are not binding on decision makers. Instead, they are taken as a representation of the public-at-large's perception of particular problems facing policymakers, who then respond to the recommendations of the jury but are not bound by them.

Citizen juries generally comprise 12 to 25 citizens selected at random—within certain limits—from a pool of citizens of the institution seeking input. Citizen juries are intended to be representative but also inclusive, so the randomness of the selection is usually combined with a goal of including jurors with as complete ranges of gender, age, and background as possible. The selected citizens then spend several days in a setting that mimics a trial and are presented with a range of information by experts and interested parties. Depending on the model, the jurors may request testimony from specific parties; usually, however, the testimony to be presented is solicited in advance by the final decision makers.

Instead of a judge, a moderator or moderators serve to keep the proceedings fair and topical. Members of the jury have the opportunity to ask questions of the witnesses. Once they have heard all the testimony, the jury members deliberate among themselves and arrive at a decision, which they then report to the convening institution. The report generally includes a description of the underlying reasoning that went into the report, which allows decision makers to gain a sense of the issues and processes of citizen decision making, not just their viewpoints on a predefined issue. Unlike in a jury trial, however, the citizen jury does not make the final determination on what rules will be adopted. Instead, the decision-making institution reviews the deliberations and decision(s) by the citizen jury and takes them under advisement in its decision-making process.

Citizen juries usually serve as only one part of the effort to encourage public participation in decision making on scientific risk issues. They are intended to solicit opinions that might otherwise not be heard through other methods of public participation. They are more intense than, for example, polls or referenda. They are intended to be more representative than public hearings, which self-select for participation by those with existing interests in the issue and/or those with the ability to attend public meetings.

Citizen juries are usually part of a broader scheme to solicit information and public input. Other mechanisms include expert committees (with or without public or lay members), public oversight boards, polls, consensus conferences, focus groups, participatory foresight exercises, and public hearings. Citizen juries differ from these mechanisms in several ways. They are focused on gaining the viewpoint of nonspecialists (as opposed to expert committees). They are short-term, not standing committees or oversight boards. Unlike polls, they provide detailed information to the participants and offer the opportunity for members to pose questions. They are selected through random sampling (with or without a requirement for inclusivity, such as for gender, class, race, age, etc.), whereas public hearings are often dominated by professional groups representing affected parties or specific interest groups, as well as citizens with the time and financial means to attend.

Citizen juries are more similar to focus groups and consensus conferences, both of which aim for inclusivity and representativeness as well as providing mechanisms for information provision, discussion, and debate. Participatory foresight exercises are similar, too, but deal with raising issues for policymakers to consider before actual policy decisions are scheduled. Focus groups are generally moderated discussions without an educational component and are intended to draw out motivations for participants' opinions in a more detailed manner than polls allow. Consensus conferences also include expert testimony and can be very similar to citizen juries in composition, structure, and purpose. One difference is that consensus conferences, as their name implies, are intended to lead to a single group recommendation achieved through consensus. Citizen juries, although usually geared toward reporting a consensus opinion, are not bound to do so. Intended to give policymakers a sense of the will of the citizenry as a whole, they may render majority and minority reports.

Proponents of citizen juries highlight several benefits that they believe citizen juries offer decision makers who are seeking to include the public in policymaking on complex issues of environmental policy. Citizen juries are inclusive, deliberative, and offer citizens a chance to be actively involved in policy-making. Proponents suggest that citizen juries can narrow the gap between "lay" and "expert" knowledge by presenting decisions made by an informed public instead of relying exclusively on advice from predetermined "experts." In addition, citizen juries can provide information about attitude and choice formation, something voting, polls, or other aggregating social choice techniques are less able to do.

Criticisms of citizen juries include that they are insufficiently complete, that they can sway the direction of a public debate before a larger segment of the public has the opportunity to reflect and weigh in on a subject, and that simply including a range of representation on a citizen jury is no guarantee that the jury will, in fact, be representative of the citizenry as a whole. In addition, there remains the question of who determines what question will be put before the citizen jury in the first place—the initial determination of the questions asked and the experts solicited (even if the citizen jury has the ability to change the question and call for additional information) can highly influence the ultimate decision.

Citizen juries have been used to solicit public input in a number of areas of environmental decision making, from park management to water resources to food and agriculture. The technique has extended beyond the United States and Germany to other European countries like Great Britain, Denmark, and the Netherlands, as well as to developing countries like India and Brazil. One of the most recent high-profile examples of citizen juries was as part of a broad-based effort by the British government to involve the public in policy decisions on genetically modified organisms in 2003.

See Also: Institutions; Participatory Democracy.

Further Readings

Dunkerley, D. and P. Glasner. "Empowering the Public? Citizens' Juries and the New Genetic Technologies." *Critical Public Health*, 8/3:181–82 (1998).

Kenyon, Wendy. "A Critical Review of Citizens' Juries: How Useful Are They in Facilitating Public Participation in the EU Water Framework Directive?" *Journal of Environmental Planning and Management*, 48/3:431–43 (May 2005).

PEALS 2004. "The DIY Citizens Jury Project, University of Newcastle, UK." http://www.ncl.ac.uk/peals/research/completedprojects/diyjury.htm (Accessed January 2009).
Smith, G. and C. Wales. "Citizens' Juries and Deliberative Democracy." *Political Studies*, 48/1:51–65 (2000).

Anna Milena Zivian
University of California, Santa Cruz

Clean Air Act

The U.S. Clean Air Act (CAA) is a federal law that attempts to improve air quality by regulating air pollution. The CAA was first enacted in 1970 but underwent major amendments in 1977 and 1990. These amendments added measures to prevent depletion of the stratospheric ozone layer and to control emissions related to acid rain. The act frames the responsibilities of the Environmental Protection Agency (EPA) for establishing national air quality standards and sets deadlines for compliance. Although the CAA seeks to protect the environment from damage caused by air pollutants, the primary goal of the law is to safeguard public health. Many of the programs associated with the CAA are often regarded as successful; however, controversy surrounds the EPA's use of scientific evidence, its lack of conformity with review schedules, the establishment of implementation deadlines, and the costs of emissions compliance.

The content of the CAA (officially Title 42, Chapter 85 of the U.S. Code) includes six titles. Title I addresses the six most common pollutants, referred to as "criteria pollutants" (sulfur dioxide, nitrogen dioxide, carbon monoxide, ozone, lead, and particulate matter). It requires the EPA to determine National Ambient Air Quality Standards, which set permissible levels for each of these pollutants, and that each state to draft State Implementation Plans that describe measures for ensuring compliance. Noncompliant states must implement specified control measures and are subject to possible penalties. Title II regulates the emissions of motor vehicles, aircraft, and other "moving sources." It also supports requirements related to the emissions of vehicle-assembly plants, adoption of low-sulfur diesel fuels, and in certain metropolitan areas, use of reformulated gasoline and vapor recovery nozzles. Title III

Smog above the city of Los Angeles, California, whose severe air pollution helped spur the development of the first Clean Air Act in 1963. In the following decades, parts of the country have enjoyed improved air quality.

Source: iStockphoto.com

controls the emission of hazardous or toxic pollutants connected with serious illness, including lead, mercury, and dichlorodiphenyltrichloroethane (DDT). The CAA Amendment of 1990 expanded the list of included hazardous pollutants from 7 to 187 and called for a shift from a focus on discrete pollutants to industry-wide regulation. Title IV provides oversight of industrial emissions related to acid deposition ("acid rain"). Control measures aim to reduce sulfur dioxide and nitrogen oxides to half of 1980 emission levels. Title V introduces a program for issuing and trading pollution emission permits. The permit-trading program is run by individual states and allows for temporal and spatial distribution of pollutant emissions within and between industries.

Finally, Title VI regulates the production and use of ozone-depleting chemical substances, such as chlorofluorocarbons. It requires the EPA to establish a program to gradually phase out all ozone-destroying chemicals and support the development of "ozone-friendly" substitutes. Although the CAA endows the EPA with the powers to establish and enforce regulations, and in some cases to fund research programs and government initiatives, individual states retain the primary responsibility for implementing plans, issuing permits, and ensuring compliance. Many states have expanded on the CAA by enacting their own legislation detailing implementation procedures or increasing standards beyond those stipulated by federal agencies.

History of Regulation

Municipal laws related to air quality, especially industrial smokestack emissions, have existed in the United States for more than a century. A number of severe smog incidents during the late 1940s and early 1950s in Los Angeles, California; Donora, Pennsylvania; and London, England led to the death and illness of thousands of people. These events served as the catalysts for public awareness campaigns in the United States and abroad and led to the drafting of the first national Air Pollution Control Act in 1955. Although this law was national in scale, it was limited in scope and provided funding only for state-level air pollution control efforts. More comprehensive legislation was passed in 1963 with the first CAA, which provided permanent federal funding for air pollution risk and remediation research and mandated establishment of state pollution control agencies. Once amended in 1965, the first CAA also directed the Department of Health, Education, and Welfare to set motor vehicle emissions standards for the first time. The 1967 Air Quality Act expanded some of these responsibilities but was neither as strong nor as comprehensive as the CAA of 1970. This act represents a complete redrafting from the previous CAA and, most significantly, was approved to coincide with the founding of the U.S. Environmental Protection Agency. Although the law remains largely intact, major amendments in 1977 and 1990 expanded its purview. The 1977 amendments set new air quality standards and adjusted implementation deadlines. The CAA amendments of 1990 legislated emission-permit trading and added provisions for regulating acid deposition and ozone-depleting substances. They also increased standards for the manufacture and sale of gasoline reformulations and increased penalties for noncompliant ("nonattainment") areas.

Although these were the last substantial amendments to the CAA, a number of minor but controversial changes have been issued and adopted in recent years. In 1998, Congress approved the Transportation Equity Act for the 21st Century, which effectively delayed implementation of stricter EPA standards addressing ozone and fine- and coarse-particulate emissions. A subject of much debate in the 105th Congress, these standards had been established in 1997 by the EPA as part of a revision of the National Ambient Air Quality

Standards. Controversy surrounded claims that the new standards would place excessive burdens on industry, transportation, and utilities. The dispute eventually led to a lawsuit (*Browner v. American Trucking Associations*), in which a group of industrial associations argued that the EPA was overstepping its legal authority to pass laws. The EPA maintained that it was restricted from considering compliance costs when imposing new regulations and that the new standards were established to protect Americans from substances that are hazardous to human health. Although the issue was eventually taken before the Supreme Court, it was resolved in Congress with the passage of TEA 21, which included a provision (Title VI) that codified a lengthy implementation schedule devised by President George W. Bush in consultation with the EPA. During the same period, a minor amendment to the CAA allowed limitations on the entry of noncompliant foreign vehicles from Mexico into Southern California (P.L. 105-286). In addition, an amendment to the omnibus appropriations bill delayed the implementation of an EPA ban on the production and use of methyl bromide, a hazardous ozone-depleting chemical employed in agriculture (P.L. 105-277, Section 764). These outcomes led to accusations that EPA officials bowed to pressure from the Bush administration and failed to ensure the standards for air quality recommended by their own scientific advisers. Critics have also pointed to the EPA's failure to review air quality standards every 5 years, as the CAA instructs.

Controversy

More recent controversies surround the need to reduce greenhouse gas emissions (GHGs) and the use of methyl tertiary butyl ether (MTBE) in gasoline. MTBE is an oxygenate used to raise the content of reformulated gasoline to the minimum 2 percent (by weight) required under current EPA standards. Although MTBE gasoline burns cleaner, it has been linked to cases of groundwater contamination and has subsequently been banned in a number of states. Debates continue over the relative benefits of MTBE and ethanol—its only available substitute. The ongoing dispute over the role of the EPA in reducing GHG emissions received renewed attention in 2007, following an executive order and a Supreme Court ruling demanding that the EPA take action under the CAA to reduce motor vehicle–produced GHGs. Efforts to reduce consumption, increase fuel efficiency, and adopt alternative technologies in the effort to reduce GHG emissions will undoubtedly be reshaped by the Obama administration. Other issues that may receive renewed attention in the coming years include efforts to control haze in national parks and wilderness reserves and additional regulation for nitrogen oxides and mercury.

Whether these debates stand as evidence of inherent weaknesses in the CAA or of its lasting robustness and malleability, most of the measures implemented under the CAA are nonetheless widely regarded as successful. In the less than 40 years since its implementation, more areas of the United States have exhibited improved air quality and have CAA-compliant levels of ozone and carbon monoxide than in past decades. Whether the EPA and CAA can be credited with these achievements is another matter for debate: Detractors argue that these developments would have resulted from technological advancements implemented over the past four decades even if Congress had not enacted the CAA, whereas supporters contend that the law fueled these innovations. It is, however, clear that many areas throughout the United States have noncompliant pollutant levels that continue to present a health risk to millions of Americans.

See Also: Acid Rain; Clean Water Act; Environmental Protection Agency; Regulatory Approaches.

Further Readings

Stokstad, Erik. "EPA Adjusts a Smog Standard to White House Preference." *Science, New Series*, 319/5870 (2008).

U.S. Environmental Protection Agency. "Plain English Guide to the Clean Air Act." http://www.epa.gov/air/caa/peg/ (Accessed December 2008).

Sya Buryn Kedzior
University of Kentucky

Clean Water Act

The U.S. Clean Water Act (CWA) is a federal law that attempts to improve surface water quality by regulating the discharge of pollutants into bodies of water. The CWA endows the U.S. Environmental Protection Agency (EPA) with the authority to protect surface water by establishing and enforcing water quality standards and by supporting state and local governments in developing pollution control plans. The CWA does not provide for the protection or regulation of groundwater pollution. The term *Clean Water Act* is commonly used as a blanket term referring to the body of laws included under the 1972 Federal Water Pollution Control Act (Public Law 92-500) and its amendments, including the 1977 Clean Water Act and the 1987 Water Quality Act. These amendments expanded legislation over nonpoint sources of pollution and renewed the timeline and basic principles of the 1972 act. The act is a broad effort to restore and ensure the health of national waterways to protect and promote the propagation of aquatic wildlife and the recreational use of surface water. Under the CWA, the EPA has established use-based criteria for the nation's waterways and a permit-based system for the discharge of pollutants. More than 30 years after it was enacted, the CWA enjoys both a degree of measured success and a share of continuing controversy. Of particular concern are the division of state and federal rights and responsibilities and the congressional failure to significantly revise or revisit the act's shortcomings in reference to the wetlands protection program.

This U.S. Environmental Protection Agency photograph clearly shows waste flowing from an industrial plant into the Calumet River near Lake Michigan before companies were compelled by the Clean Water Act to reduce such emissions.

Source: U.S. Environmental Protection Agency

The content of the CWA (officially Title 33, Chapter 26 of the U.S. Code) includes six titles. Title I establishes grants for research and related pollution control programs that support the act's principal goal of ensuring that the nation's waterways are free from excessive amounts of harmful or toxic substances. Title II was a core provision of the original 1972 Federal Water Pollution Control Act that authorized grants for the construction or expansion of water treatment plants. Under Title II, federal grants covered up to 75 percent of the construction cost for a particular facility or program. This title was superseded by the 1987 Water Quality Act introduction of Title VI, which established the Clean Water State Revolving Fund, a program that requires state governments to match federal loan financing for wastewater treatment projects (see following). Title III mandates the creation of treatment and discharge standards for municipal and industrial point sources of water pollution and frames enforcement standards for noncompliant facilities. One of the most important contributions of Title III is the Water Quality Standards program, which directs the EPA to set allowable pollutant levels for surface water bodies based on their designated use (recreation, agriculture, aquatic life, etc.) and the hazards or risks posed to those activities by various pollutants. Each body of water is assigned a designated use by state authorities, who are responsible for ensuring that it meets the associated water quality criteria. The CWA also employs standards that mandate the adoption of "best practicable technologies" and, later, "best available technologies" for treating wastewater and industrial effluents. States calculate a total maximum daily load—the maximum allowable concentration of a pollutant that can be discharged into a body of water without exceeding water quality standards, and allocate permits to facilities for the discharge of specific pollutants. The 1987 Water Quality Act expanded Title III to include sources of nonpoint pollution. Title IV governs the National Pollution Discharge Elimination System of permit allocation. Permits are allotted according to individual water body total maximum daily load standards. Federal and state permit agencies are required to reissue permits every 5 years and to provide public notice and opportunity for public commentary for pending permits. Although the EPA and state authorities administer the permit program for most bodies of water, permits for the discharge of "dredge and fill" material into national wetlands are managed by the Army Corps of Engineers. Title V addresses the general provisions included in the CWA, including those allowing U.S. citizens to file suit against CWA violators and protecting so-called whistleblowers from negative repercussions of CWA enforcement. Finally, Title VI added to the CWA with the 1987 amendment establishing the Clean Water State Revolving Fund, which replaced the Title II construction grants program. The fund provides long-term, low-interest loans to states and municipalities for the construction or expansion of publicly owned treatment works and other projects to control nonpoint source pollution.

History of Regulation

Water pollution and water quality have been legislated in the United States since passage of the Rivers and Harbors Act in 1899. In the first half of the 20th century, most federal and state regulations were limited to addressing problems related to sewage, sanitation, and refuse that could impair navigation. The 1948 Federal Water Pollution Control Act was the first major U.S. law to directly address water pollution. The act set the standard for future water quality legislation by placing responsibility for establishing and enforcing water quality standards, as well as funding abatement programs, with state authorities. It also articulated federal involvement in water pollution abatement for the "national interest" by providing oversight of interstate water sources and support from the National Public Health Service. The next major law governing water pollution was the 1965 Water Quality Act, which established the responsibilities of the Federal Water Pollution Control

Administration. Although the scope of this agency was limited, it was able to require states to establish water quality standards for interstate bodies of water and to take action when states were not in compliance. Other related laws enacted in the decades before the 1972 CWA included the 1956 Water Pollution Control Act and the 1970 Water Quality Improvement Act. Each of these laws expanded federal participation in water quality protection but maintained state-level authority by restricting federal involvement to interstate bodies of water. By 1970, pressure to revisit this policy had developed in reaction to the growth of the U.S. environmental movement, the establishment of the EPA, and the widely publicized fire on the Cuyahoga River in Cleveland, Ohio, in 1969. These circumstances, in addition to international pressure to renew or expand environmental protection laws before the 1972 United Nations Conference on the Human Environment in Stockholm, Sweden, led to widespread appeals for more comprehensive legal protection for the nation's waterways. The 1972 CWA represents a revision of the 1948 Federal Water Pollution Control Act and the expansion of federal responsibility for pollution abatement.

Shortcomings of the 1972 CWA became apparent soon after enactment of the law. The first of these was the inability to meet the deadlines established for the protection of aquatic life and safety of water-based recreation by 1983 and for the elimination of the discharge of untreated pollutants into navigable waters by 1985. Many of the subsequent amendments to the CWA therefore extended timelines for these tasks and renewed the authority of the act. The second major limitation of the CWA, failure to address sources of nonpoint pollution, became the focus of the 1987 Water Quality Act. Nonpoint pollution, including runoff, siltation, and irrigation return flows, were contributing a considerable amount of pollutants into the nation's waterways and hampering success of the CWA. However, Congress and the EPA were reluctant to legislate nonpoint sources, as they are difficult to identify, and attempts to regulate these sources were resisted by agricultural and industrial interest groups. Legal action forced the EPA to include nonpoint pollutants in the National Pollution Discharge Elimination System permit program, and the Water Quality Act empowered them to direct states to develop nonpoint pollutant treatment programs that would receive partial federal funding. However, nonpoint sources continue to be difficult to address and are identified by the EPA as the top factor in ongoing pollution of the nation's waterways.

Wetlands Controversy

Perhaps the most important ongoing controversy related to the CWA has been wetlands protection. The responsibility of the Army Corps of Engineers to issue permits for dredging and filling wetlands has not prevented discharges of pollutants or the complete drainage or filling of many wetlands. This has led to disputes over the definition of wetlands and lawsuits challenging federal authority over state-sanctioned construction and development programs in wetlands areas. A 2001 Supreme Court decision found that federal agencies did not have authority to prevent the alteration of small-scale wetlands, and another 2006 decision failed to support the CWA's ability to regulate any nonnavigable body of water. These judgments have called the authority of the EPA into question and have added to demands that Congress revise and reaffirm the act once again. In 2007, Senator Russell Feingold (D–Wisc) introduced the Clean Water Restoration Act, a bill that seeks to supersede the recent Supreme Court decisions and to expand federal authority over wetlands and surface water conservation. As of 2009, this bill is still under consideration before Congress (HR-2421).

See Also: Clean Air Act; Environmental Protection Agency; Water Politics; Wetlands.

Further Readings

Ryan, Mark A. *The Clean Water Act Handbook,* 2nd ed. Chicago, IL: American Bar Association, 2004.

Sponberg, Adrienne Froelich. "Supreme Court Ruling Leaves Future of Clean Water Act Murky." *BioScience,* 56/12 (2006).

U.S. Environmental Protection Agency. "Summary of the Clean Water Act." http://www.epa.gov/lawsregs/laws/cwa.html (Accessed January 2009).

Sya Buryn Kedzior
University of Kentucky

Club of Rome

The Club of Rome (CoR) is a private informal group of scholars and businessmen founded by the Italian industrialist Dr. Aurelio Peccei and the scientist Alexander King in 1968 in the city of Rome. The members of CoR at the time of its inception shared concerns about the future of mankind and the ecological well-being of the planet amid growing population, rapid depletion of world resources, and environmental crisis. In 1972, CoR gained the global limelight with the publication of its first commissioned report, called *Limits to Growth*, led by Massachusetts Institute of Technology scholar Dennis Meadows.

Unique circumstances led to the rise of CoR in the 1960s. Some of these were the decline of superpowers; social movements of the 1960s; the end of postwar economic growth; fear of the economic and technological gap between the United States on one hand and Europe and the rest of the world on the other; the rise of the Third World, with its own political voice; Rachel Carson's portrayal of an interconnected biological world in her famous book *Silent Spring*; and National Aeronautics and Space Administration images of a small globe in an infinite universe.

The main founder of CoR, Peccei had experience of many years of traveling and working in Third World countries in Asia, Latin America, and the Middle East, where he witnessed human starvation and environmental problems firsthand. Appalled by the massive gap between the North and the South, he soon realized the need for a new international economic order. He started looking for like-minded people, and in 1968, he founded CoR, a nongovernmental group that did not have a budget or any administrative structure. From the beginning, Peccei decided to limit the membership of the club to not more than 100 people with strong political connections and intellectual affiliations. With a very lean budget, the group networked extensively for funds. Because of the financial independence of the group and the policy that kept elected politicians outside the club, the organization remained politically nonpartisan. Peccei was the main financial backer and intellectual driving force behind the club, and after his demise in 1984, many believed that the club was thrown into a state of disarray.

The purpose and rationale of CoR was best described in a report and proposal called "The Predicament of Mankind," which stated that the dual nature of modern-day technology and scientific advancements had created catastrophic problem scenarios alongside progress. The term *world Problematique* or simply *Problematique* was conceptualized to connote the entanglement of problems, which were held to be incomprehensible as well as inescapable for humans. According to CoR, these problems, such as uncontrolled population growth, disparity of wealth, social injustice, starvation and malnutrition, unemployment, crisis in democracy and civil unrest, decay of the city and depletion of natural resources, and many more, are intimately interrelated. These interconnected problems demand solutions that follow holistic and global approaches as well as a long-term time perspective. CoR further believed that these problems would soon become uncontrollable unless immediate actions were taken to tackle them. The existing political institutions, however, did not have credibility to act, and hence CoR questioned the sovereignty and political will of the nation-state. Also, as the Problematique is experienced universally and is not confined within national borders, the modern nation-state was deemed incapable of taking requisite actions. In fact, the United Nations, which is highly fractured and politicized, as it is based on consensus among different nations, would not be able to act effectively on the Problematique. This task could only be fulfilled by international nongovernmental organizations or an overarching global organization with diverse membership that does not represent any ulterior motives. A four-step agenda to tackle Problematique involved (1) identifying various interrelated problems; (2) a detailed and comprehensive study into the causes and interrelationships between the problems; (3) developing tools and practical steps to tackle the problems; and (4) summoning the political will to implement the solutions.

Peter Moll (1991) divides the work of CoR into five phases: modeling studies represented by *Limits to Growth*, *Goals of Mankind,* and so on; New International Economic Order debates in the United Nations; humanistic and normative studies such as *Goals for Mankind* and *No Limits of Learning*; North-South and developmental studies such as *Tiers Mondes—Trois Quarts du Monde*; and managerial and specific issues reports such as *Road Maps to the Future.*

CoR's most prominent publication was *Limits to Growth* in 1972, research undertaken by Massachusetts Institute of Technology to study long-term global effects of industrialization, rapid population growth, widespread malnutrition, and so on. Scientists Dennis and Donella Meadows, who conducted the study, viewed these problems as interrelated and used sophisticated computer models to project future trends. The study concluded that the exponential increase of population with advances in industrial production will lead to global disaster unless collective political will is applied toward positive changes such as the green revolution and the availability of birth control pills for a steady state economy. In this way, *Limits* brought a utopian perspective to the management of society along the lines of sustainable development. The backdrop of the Middle East oil crisis of 1973 and introduction of sustainable development discourse at the United Nations Environment Conference at Stockholm was ideal for *Limits* to gain mass popularity.

However, various criticisms were mounted at different fronts. For example, the Left criticized *Limits* for ignoring the social and political elements under capitalism that determine the status quo. Further, *Limits* ignored that the working class has faced the ecological crisis at all times, and it is only now that the middle class is facing similar problems. The elite will still find a way to escape such crisis, and hence the changes proposed by *Limits* will never be implemented with an overall consensus. Both Left and Right maintain that the study underrated the positive role of technology. The Right emphasized that in the future market dynamics would take care of resource scarcity automatically by adjusting prices. According to the Third World critics, the zero-growth proposition of *Limits* was only an option for the First World, where affluence made strong economic growth an unnecessary phenomenon. *Limits* was also criticized for its computer modeling and Malthusian doomsday prognosis. Regardless of all the criticism, *Limits* was hugely successful in bringing public attention to resource scarcity and effectively planted environmentalism on the global agenda.

CoR in general has been a catalyst organization in incorporating into the popular culture the notion that the planet Earth is a closed system in which limited natural resources enforce strict limits on the world's population and technology. Because of its flexibility in describing a variety of problems, the term *Problematique* has also become very popular and has been used in development and environmental literature since it first appeared. CoR sought to influence the decision makers instead of participating in the actual policy-making. With the publication and strong promotion of *Limits*, CoR has contributed considerably to this discipline of future studies, environmentalism, and the cultural ecology critique. *Limits* has been a pacesetter in the fields of scarcity environmentalism and global modeling, contributing to an increase in the use of large-scale modeling tools to the interrelated modeling treatments of environment, economics, and development in the 1970s.

See Also: Environmental Movement; Limits to Growth; *Silent Spring*; Sustainable Development.

Further Readings

Moll, Peter. *From Scarcity to Sustainability: Futures Studies and the Environment: The Role of the Club of Rome*. Frankfurt: Peter Lang, 1991.

Neurath, Paul. *From Malthus to the Club of Rome and Back: Problem of Limits to Growth, Population Control, and Migration*. New York: M. E. Sharpe, 1994.

Pauli, Gunter. *Crusader for the Future: A Portrait of Aurelio Peccei, Founder of the Club of Rome*. London: Pergamon Press, 1987.

Priyanka Jain
University of Kentucky

COMMODIFICATION

Commodification constitutes the process of transforming something into a tradable good—a commodity. This process is considered by many as inevitable when the issue is scarcity (e.g., current "water crisis") and the most efficient when the issue is related to negative effects of some activity (e.g., pollution contributing to climate change). Nevertheless, commodification is a political process and requires the definition of property rights, which emanate from a political-legislative act.

Commodification has gained increased relevance since the 1980s, with the shift in the economic paradigm framing political decisions and the government's role in the economy. The neoliberal economic paradigm has (convincingly) presented markets as the most efficient method of allocating resources, goods, and services. Because markets are believed to foster the community's well-being, by creating the best conditions for resource use, conservation, and exchange, resource allocation should unfold under the auspices of the market, based on a price mechanism. As a result, markets will efficiently address issues of wasteful use, overextraction, or negative effects (externalities), such as pollution. This reasoning has also been used in environmental economics to calculate an economic value for environmental services, so that their costs can be internalized and therefore provide the incentives for more efficient use, conservation, and protection of natural resources.

Commodification only works when there is scarcity. Economists classify things into two categories: free and economic goods. The former are abundant, and although they may be useful, they do not have economic value because no one is willing to pay for them. Therefore, they cannot be commodified (e.g., air or sunlight). Economic goods, however, are useful and scarce, and thus they have an economic value. People are willing to pay for them, so they can be traded and can be commodified. Economic goods are further classified, according to each good's consumption, "excludability" and "rivalry." In other words, one has to take into account whether one can be excluded from consuming a good, and whether one's consumption of a good reduces the quantity available for others. Based on these criteria, economists identify four types of goods: private, pure public, toll, and common.

For commodification, the most important characteristic of a good's consumption is its "excludability" nature. The consumption of public goods and common-pool resources is not excludable. No one is willing to pay for seeing the light of a lighthouse (public good) or for entering international waters (common good). The price mechanism does not work in these cases. So, traditionally, the state or some level of government assumes the provision or management of this type of goods. Economic private and club goods are the ones that are most suitable for commodification. To obtain a private or club good, one has to pay a price. If one cannot pay the price, one cannot acquire or gain access to the good. Because a good or service implies the use of labor and of other inputs to produce it and/or deliver it, it is accepted that one should pay to have access to it and, in that way, repay the labor and the inputs (cost recovery) and reward the value-added associated (profit). Thus, there is some consensus that services in general, and manufactured goods in particular, should be paid for.

There is not a consensus, however, when something that was not usually allocated based on price starts being so, or when a service that was charged for at a social price starts being charged for based solely on supply-and-demand dynamics. An example of a commodification process that is currently under way is the one occurring with freshwater resources and water and sanitation services. The acknowledgement that freshwater resources are not abundant and that therefore they have an economic value created the conditions for private actors' participation in this sector and for water to be allocated under the auspices of the market. Historically, water has been under the state's auspices, and although it is an economic private good (consumption is excludable and rival), society, through government, has decided to provide it as if it were an economic public good. This meant that no one was excluded from water provision—universal provision—and that if anything was charged for the water, it was highly subsidized by the state. This type of goods is usually called social goods or services. Because of the vital importance of water, it used to be governed under the state's auspices based on an equity principle. The commodification of water and its associated services implies that they are provided based on a benefit principle (consumers pay), under market rules. Commodification transforms citizens and stakeholders with rights into consumers and clients with economic obligations.

In addition to scarcity, commodification requires a clear definition of property rights. For something scarce to enter the market and become a commodity, either its value or its cost has to be suitable to individual appropriation. For the market to function, it needs a clear and stable allocation of property rights. In fact, economic theory argues that the clearer the property rights system, the more efficiently markets will function. Property rights are important because they reinforce the possibility of exclusion. If something can be individually appropriated, then others can be excluded from its access. This logic underlies the system of property rights over natural resources, including freshwater, and over pollution. In the case of pollution, the theory argues that if one defines property rights associated with pollution, then by controlling the overall level of pollution, one creates scarcity. As a consequence, clearly defined property rights and scarcity create the conditions for pollution to be allocated under the auspices of the market. Again, this is an example of a commodification process, in which pollution, which was a negative effect and a free good, in economic terms, was transformed into an economic good managed by the market.

The commodification process of natural resources and environmental services raises several issues. First, it is not in the market's nature to guarantee universal provision, which in cases such as water resources becomes a matter of survival—a human rights issue. Second, the fact that the price becomes the allocation mechanism will not prevent wasteful use, overextraction, or pollution from occurring, it will only imply that those doing it have to pay for the right to do it. As a consequence, from an environmental point of view, commodification is not a solution per se if thresholds are not defined above which wasteful uses, overextractions, and pollutions are forbidden and not left to market forces. Third, efficiency is always considered from the point of view of the market, just as survival is considered from the point of view of humanity. Efficiency and survival from an ecosystemic point of view are hardly ever taken into account in any governance model.

See Also: Capitalism; Common Property Theory; Steady State Economy.

Further Readings

Bakker, K. J. *An Uncooperative Commodity: Privatizing Water in England and Wales.* Oxford: Oxford University Press, 2003.

Coase, R. H. "The Problem of Social Cost." *Journal of Law and Economics*, 3 (October 1960).

Köllicker, A. "Globalization and National Incentives for Protecting Environmental Goods." MPI Collective Goods Preprint No. 2004/3. Bonn, Germany: Max Planck Institute, 2004.

Rose, C. M. *Property and Persuasion: Essays on the History, Theory, and Rhetoric of Ownership*. Boulder, CO: Westview, 1994.

Paula Duarte Lopes
University of Coimbra

Common Property Theory

Common property theory (CPT) refers to a body of cross-disciplinary literature that deals with the historical and contemporary institutional governance and management of valued resources ranging from fisheries and forests to atmospheric sinks, oceans, and genetic materials. CPT was originally developed to understand the problems of managing what are termed *common-pool resources*. Common-pool resources are valued resources that all can use (principle of the difficulty of exclusion of users) and for which one person's use reduces what is available to others (principle of subtractability or rivalry), thus running the risk of overuse and degradation.

Interest and debate on the special characteristics of common-pool resources and the problems of their management are found throughout history. However, such resources have taken on contemporary significance as a result of the growth of capitalism and its valorization of the idea of private property ownership as the most efficient means to regulate the use of natural and human resources. During the 1950s and 1960s, disciplines as diverse as resource economics, biology, and sociology tended to view so-called communal forms of property ownership as premodern, to be swept away by market or state regulation of extraction, production, and exchange.

Seminal articles by H. S. Gordon on fisheries management and by Garrett Hardin on the tragedy of the commons led to vigorous academic and policy debate over their claims that uncontrolled access to a valued resource resulted in its destruction and that some form of private or state property system would be necessary to ensure proper management of common-pool resources. Underlying their approach was the asocial assumption that humans were motivated solely by self-interest, that they took a largely material and instrumental view of nature, that they could not be expected to act collectively, and that therefore, common property regimes were a hindrance to capitalist (and socialist) forms of progress and should be replaced by private or state control.

Gordon and Hardin's modeling played an important heuristic role in generating considerable economic, biological, and anthropological theoretical and empirical research on the historical and contemporary management of a range of resources, particularly common-pool resources such as fisheries, forests, wetlands, and pastures. Gordon and Hardin were criticized for confusing open access regimes (*res nullius*) with common property regimes (*res communes*) and for failing to consider that controlled community access to common-pool resources could be beneficial for both resource conservation and resources users alike. Critics, drawing on game theory and case studies of resource-based communities around the world, emphasized the cooperative possibilities of human action over atomistic models of human behavior. They argued that under the right social, political, and economic conditions, people were capable of engaging in collective action to achieve increased productivity, greater social equity among resources users, and enhanced environmental and social sustainability.

Early Research

Much CPT research at the time focused on defining the conditions under which small-scale, usually place-based, communities could more effectively manage the extraction and use of common-pool resources such as forests and fisheries. It was particularly influential among governments and international donor agencies concerned about the failures of "top-down" state- and market-based development policies and practices. Emphasis shifted in development circles to management strategies based on greater devolution of governance functions to local communities through the promotion of community management and comanagement schemes. During the 1980s and 1990s, most attention was paid to developing new typologies of property ownership, which went beyond state and private ownership and which were applicable to local communities dependent on resource extraction for their livelihoods. More recently, these typologies have gone through several iterations as a result of the extension of research to wider areas of environmental governance such as climate change.

Common property theorists identified between 30 and 40 conditions that contribute to the establishment and successful management of common-pool resources. These conditions were divided into characteristics of the resource managed, the resource users and their interrelationships, and the relationships among resources, users, and external influences such as the state, markets, and technological innovation. Among the key conditions identified were the number of users, well-defined boundaries of the common-pool resource, spatial concentration of users and of resources to be managed, limited physical mobility of the resources, enforceable measures to control rule breakers and free riders, some history of group cooperation, a high degree of interdependence and mutual support among resources users, a relatively stable population of resource users, limited integration of the community into market relations, and supportive political, legal, and other external institutions.

The sheer number of variables specified as potentially significant in establishing and maintaining common property management systems led to discussion of the statistical and research design problems involved in carrying out comparative analyses across a sufficiently large number of cases to be able to draw any general conclusions about the optimal conditions for the functioning of such systems. Some argued that in the face of such complexity, rather than seeking to identify enabling conditions for successful common property regimes as a basis for establishing universally applicable and context-free generalizations, there should be a shift to understanding how specific common property regimes operate at particular times and places and how such regimes are embedded within specific political, economic, and social contexts.

Criticisms

Apart from such methodological concerns, there were several criticisms of the substantive value of CPT. Some claimed that common property theorists misapplied their concepts and romanticized the idea of common property. For example, common property regimes were said to be confused with customary property systems, which in many instances provide for both private and collective property rights and cannot be considered as exemplars of common property regimes. CP theorists were criticized for a lack of realism in failing to embed their analysis within a broader political ecology framework that focuses on the strong influence "outside" forces have on local community efforts to retain a degree of autonomy in the management of their resources. Many presumed autonomous common property regimes were shown not to be functioning properly or were poorly managed as a result of elite capture, the insufficiency of resources to support livelihoods, and the need for local people to find work away from home or in jobs unrelated to the common-pool resource. Criticisms were leveled about the promotion of common property regimes by some international donor agencies as corporate private property regimes serving wider market agendas rather than promoting local sustainable livelihoods.

Although there is some validity to such criticisms, there is evidence to show that under specific circumstances, common property regimes do work, and the fact that they may be misused or introduced into contexts unfavorable to their success is not a reason for abandoning them altogether. However, one area that is particularly challenging for CPT is its applicability beyond the local or the single scale. The problems of multiscalar management are well illustrated by current international debates and negotiations over the mitigation of greenhouse gas emissions, which involve local people, states, businesses, nongovernmental organizations, international donor agencies, and others. These debates, which seek to provide a new global framework for the governance of the so-called global commons, present particular challenges for CPT. For much of human history, the atmosphere has been considered a public good in the sense that one person's use of it did not interfere with its use by another person. The atmosphere was regarded as an almost infinite sink, which could absorb all manner of pollutants at little or no environmental or human cost. However, the growth of industrial society and the concern over climate change have led to a growing recognition that atmospheric sinks are a type of common-pool resource needed to protect human society and other living things from an atmospheric tragedy of the commons. This requires a form of international environmental governance that goes beyond purely private or state-based property regimes. It remains to be seen what contribution CPT can make to the collective governance of the atmosphere and the biosphere it supports.

See Also: Global Climate Change; Politics of Scale; Tragedy of the Commons.

Further Readings

Ostrom, Elinor, et al., eds. *The Drama of the Commons*. Washington, D.C.: National Academy Press, 2002.

Paavola, Jouni. "Governing Atmospheric Sinks: The Architecture of Entitlements in the Global Commons." *International Journal of the Commons,* 2/2 (2008).

Wagner, John and Malia Talakai. "Customs, Commons, Property, and Ecology: Case Studies From Oceania." *Human Organization,* Special Issue, 66/1 (Spring 2007).

Bob Pokrant
Curtin University of Technology

Conservation Enclosures

Conservation is achieved in two main ways: *in situ* conservation and *ex situ* conservation. *In situ* is the type of conservation in which conservation of endangered species takes place in natural habitats, either by cleaning up the space or by protecting from predators. *Ex situ* conservation may be used as a last resort to an endangered population: the entire population is relocated, usually in the care of humans. This uses some rather questionable methods in laboratory practices. Zoos, seed banks, and botanical gardens are the most common examples of *ex situ* conservation.

Conservation is important to protect biological diversity, including, but not exclusively, of large mammals. Traditionally, the exclusionary process has been used in the creation of protected areas, which exclude the indigenous people from entering or having access to these areas. Neither are they consulted during the conservation process, making them subject to unreasonable pressures to vacate the land and not providing them adequate benefits for relocation. Conservation enclosures provide space for development activities, scientific research, and employment opportunities in tourism and for recreation. The term *park* originates from the Latin *parricus,* meaning enclosure. *Reserve* comes from *reservare,* meaning save. Preservation, protection, and conservation imply that certain areas are kept away from the present demand for any development activities.

The exclusionary process of conservation could have instances of backlash, in which communities that have been ousted may become restricting to conservation. Protected areas could also cause the local communities' traditional beliefs and structure to become questioned. In some cases, areas that were once notified as protected areas have been de-notified when governments realized that valuable resources were available in these protected areas. The Narayan Sarovar Sanctuary in Gujarat, India, was de-notified to allow open-cast mining for a cement company. The area was reduced to one-eighth of its original size, with a comment that the area was in excess of the required amount. However, this was in violation of Section 26A of the Indian Wildlife (Protection) Act, 1972. Similar issues have occurred all over the world.

Legal agreements become binding to the countries that have ratified them. Conservation then becomes included in a country's development plans.

- The Convention on Biological Diversity is one of the most comprehensive treatises, signed and ratified by 165 countries. The convention mainly serves to provide guidelines for countries to perform certain actions for the process of conservation. There is some criticism of the Convention on Biological Diversity being ambiguous about the state of local communities in conservation.
- The Convention on International Trade in Endangered Species of Wild Fauna and Flora is a treaty that prevents trade of endangered species. It also ensures that the trade of specimens does not endanger or threaten their survival.
- The Ramsar Convention on Wetlands, 1971, was specifically designed for the protection and sustainable use of wetlands. There are 158 contracting parties to this convention.
- The Convention to Combat Desertification, 1992, is specific to the problem of desertification of dry arid and semiarid lands in the temperate parts of the Earth and the tropics for many reasons, including climate change.
- The Convention on the Conservation of Migratory Species of Wild Animals deals with the protection of migratory species of flora and fauna to preserve them from extinction.

Countries are measured in terms of their economic development, which directly and indirectly affects conservation and conservation enclosures. In many developing countries, conservation acts become secondary to the countries' growing populations' demands for food, energy, and security.

In some countries like Namibia, conservation efforts have drastically increased numbers of wildlife. To protect wildlife, there is a unique service whereby wildlife is auctioned off to the highest bidder, and in the process, the better stock of wildlife is preserved in private farms. This is legalized by the government and has increased the value of wildlife, literally. The money that is raised from the auction is used for conservation, thereby completing the circle.

The Big Muddy National Fish and Wildlife Refuge on the lower Missouri River, which was established in 1994 to protect the habitat of local wildlife, is an example of an in situ conservation enclosure.

Source: U.S. Fish & Wildlife Service

The exclusionary approach in conservation reserves the right of the government to allow or disallow traditional occupants of forest areas from entering into or benefiting from the forest, especially if the area has been declared as "protected." Some scientists are still adamant about the efficacy of this approach. Some other scientists are now taking into account the increasing power and benefit of the inclusionary process for conservation, which allows occupation of some of the protected lands by indigenous peoples.

An example of this is the Karen people, who live in the highlands of Thailand and practice shifting cultivation. Shifting cultivation connotes destruction of forests; however, as these are a nomadic people, the lands are allowed enough time to rejuvenate before they are occupied again. Although in many countries indigenous communities were persuaded by the government to vacate the area, the Karen are an example of people who have fought the government for their right to occupy the land. Areas of protected lands increasing and land being taken over for development are causes for marginalization of these hill tribes. This is also true in many other parts of the world.

Community forestry includes local communities in efforts to make conservation more decentralized. Forests that occur around the fringes of villages are often included in community forestry. Community forestry has been shown to reap several benefits, many of which are sustainable in the long run.

- Increased access of the local communities to the forest would mean they would gain economically from the forest as well as have a way to live sustainably, enabling the protection of the forests—the main reason for conservation.
- The access itself would mean that local communities are becoming involved in the process of conservation, a factor that would directly affect them.

See Also: Conservation Movement; Indigenous Peoples; Social Ecology.

Further Readings

Adams, William M. and Jon Hutton. "People, Parks and Poverty: Political Ecology and Biodiversity Conservation." *Conservation and Society*, 5/2:147–183 (2007).

Economic Times (July 26, 2008). http://economictimes.indiatimes.com/articleshow/msid-3281849,prtpage-1.cms (Accessed January 2009).

Guha, Ramachandra. "The Authoritarian Biologist and the Arrogance of Anti-Humanism: Wildlife Conservation in the Third World." *The Ecologist*, 27:14–20 (January/February 1997).

Laungaramsri, Pinkaew. *Redefining Nature: Karen Ecological Knowledge and the Challenge to the Modern Conservation Paradigm*. Chennai, India: Earthworm Books, 2002.

Pimbert, Michael P. and Krishna Ghimire. *Social Change and Conservation*. London: Earthscan Publications Ltd., 1997.

Tryzna, Ted. *Global Urbanization and Protected Areas*. Published for the IUCN by InterEnvironment California Institute of Public Affairs. http://data.iucn.org/dbtw-wpd/edocs/2004-047.pdf (Accessed January 2009).

Sinduja Krishnan
Centre for Environment Education

Conservation Movement

Inspiration for the Conservation Movement comes from a long stream of American thought that is itself contradictory. The movement often has been viewed as being at odds with the predominant political economy built on what Morris Udall called the myth of superabundance, alongside deeply held values of individualism, private property, utilitarianism, limitless growth, and constant progress.

The sometimes confusing streams of conservation thought and their relationships with science, ethics, planning, and economics form a complex field of historic action with significant political ramifications, ranging from outright rejection of laissez-faire politics and economics to practicing wise use of resources and amelioration of environmental damage.

Although early-19th-century writers such as Henry David Thoreau and Ralph Waldo Emerson, as well as artists such as George Catlin and John James Audubon, were nurturing a love of nature and the land in our cultural heritage, the country was moving forward with its Industrial Revolution, which built cities, factories, transportation networks, and tremendous wealth for some, while permanently altering both the rural and urban landscapes and the relationship of people to the land. In fact, according to Udall, historic evidence strongly suggests that many early Americans found the land hostile and were themselves hostile toward the land, interested only in using the land to create personal wealth.

On the eve of the Civil War, large-scale commercial timbering spread across the forested hills and valleys of the eastern United States. When George Perkins Marsh published *Man and Nature* in 1864, he issued an early warning about the state of the nation's natural resource base. By the 1870s, scientific consciousness emerged about the need to scientifically manage forests to ensure a future wood supply for the rapidly growing economy; these studies also noted relationships between forests and water and air quality.

Scientific Origins

The Conservation Movement's scientific origins come partly from Europe's forestry practices, which date from the mid-1700s. The founding of the American Forestry Association in 1875 and its merger with the new American Forestry Congress in 1882 signaled a national, well-organized voice for forest management that underpinned the nascent Conservation Movement.

Throughout the 19th century, the federal government recognized the need to deal with natural resources in the face of demands from growing population and industries. The quality of the legislation was often problematic: Enforcement was uneven because of limited political will and the growing political and economic hegemony of major corporations dedicated to large-scale natural resources extraction, especially after the Civil War. Significant federal legislation included the Federal Timber Research Act (1817), Timber Trespass Act (1831), U.S. Dept. of Agriculture and Land Grant colleges (1862), Mineral Land Act (1865), Yellowstone National Park (1872), Timber Culture Act (1873), U.S. Department of Agriculture Division of Forestry (1876), Desert Land Act (1877), Timber and Stone Act and Timber Cutting Act (1878), Yosemite National Park (1890), Forest Reserve Act (1891), and Organic Act (Sundry Civil Appropriations Act), which set out guidelines for a national forest system (1897).

The budding Conservation Movement became more clearly defined and polarized during and after the 1890s. Two major actors, John Muir and Gifford Pinchot, approached conservation from widely different perspectives. Muir, who founded the Sierra Club in 1891, wanted wilderness protection without use based on the intrinsic value of wilderness as a place of refuge and solitude. Pinchot, the nation's first professionally trained forester, gained national prominence when he was named the Department of Agriculture's chief of the Division of Forestry in 1898. He advocated government leadership in the scientific management of natural resources to foster wise use.

Muir countered predominant political and economic values and was closer to the philosophies of Thoreau and Emerson. Because of his opposition to growth, Muir's views were rooted in protest; by seeking to maintain untrammeled wilderness, his approach allowed for species preservation and biological diversity. Pinchot sought a middle ground, with government encouraging efficient use of resources to secure them for future human needs and economic growth in a way said to benefit the greatest number of people in the long run.

The Conservation Movement gained speed and began to address more issues during President Theodore Roosevelt's progressive Republican administration at the beginning of the 20th century. Roosevelt's conservation legacy, which grew out of his experiences as a rancher in the Dakota Territory, is of equal importance to his trust busting and construction of the Panama Canal.

H. W. Brands notes that Roosevelt deemed himself the chief steward of America's resources—a vastly expanded view of the presidency. With help from Pinchot and others, he launched what came to be known as the Conservation Movement, holding seven conferences late in his administration that dealt with various aspects of conservation and resource development: Public Lands Commission, Inland Waterways Commission, National Conservation Commission, Joint Conservation Congress, Conference of Governors (genesis of the National Governors Association), North American Conservation Conference, and the Country Life Commission. Much funding for these meetings came from private foundations because Roosevelt faced opposition from Congress.

Roosevelt's and Pinchot's philosophy of government leadership for conservation did not settle well with Roosevelt's successor. William Howard Taft took a far more pragmatic approach based on private investment to extract natural resources. For example, he dismissed Pinchot, who had charged Interior Secretary Richard A. Ballinger with improprieties, from the government. Ballinger supported private investments in resource development. Harding signed the Withdrawal Act in 1910, which authorized the president to open previously reserved public lands for the development of water-power sites, irrigation, or other public purposes. Meanwhile, the federal government also continued to follow Muir's tradition by adding new national parks, and, after 4 years of discussion, setting up the National Park Service in 1916 under Woodrow Wilson.

Despite conservationists' defeat over damming the Hetch Hetchy Valley near Yosemite to provide water for San Francisco in 1913, the Conservation Movement continued to grow. Through state and national meetings, it came to include women, property owners, sportsmen, and others into conversations about the use and aesthetic values of natural resources. The idea of efficiency—held dear by both Roosevelt and Pinchot—gained appeal as businesses realized that conservation could improve their bottom line. This widening political base fed the movement, although different constituencies did not always agree with each other.

The Movement and the New Deal

The Conservation Movement surged again during President Franklin D. Roosevelt's New Deal. Spurred by the Depression and the withering Dust Bowl, the second Roosevelt, a Democrat, continued the path of his Republican cousin. He vastly expanded the power of the federal government and linked successful agriculture with scientific management and public–private partnerships. Important legislation included the Soil Conservation and Domestic Allotment Act (1935), Civilian Conservation Corps or CCC (1933), and

Tennessee Valley Authority Act or TVA (1933). These acts deeply engaged the federal government with local land owners in using soil conservation to protect watersheds and improve farm efficiency and profits, putting thousands of young men to work on repairing environmentally damaged areas and enhancing natural resources, and planning for resource development in an impoverished area of the country. Roosevelt's brain trust also experimented with planned rural and urban communities that were intended to be environmentally friendly.

As the Conservation Movement evolved, its leaders increasingly recognized the effects of human activities on the environment. For example, in 1910, Theodore Roosevelt connected successful and healthy farm communities with resource conservation, a precursor of sustainability. In 1915, Cornell horticulture professor Liberty Hyde Bailey, a friend of Pinchot's and Roosevelt's and a leader in the nature-study and country-life movements, published *The Holy Earth*, which foreshadowed Aldo Leopold's land ethic. Leopold, a forester and University of Wisconsin professor, published *A Sand County Almanac* in 1948. It culminated his life's work as a high point in conservation thought that underpins notions of sustainability. Leopold extended the definition of the human community to include the surrounding environment—the land community—and said that the environment must be the primary consideration in our activities.

Leopold's thinking emerges in Rachel Carson's *Silent Spring,* published in 1962. Carson's book investigated the effects of the insecticide DDT (dichlorodiphenyltrichloroethane); it created widespread controversy and helped spur the broader-based Environmental Movement that flowered in the 1960s. Pressure from this movement resulted in significant federal and state controls on polluting activities by industries and individuals in the 1960s and 1970s.

See Also: Capitalism; Domination of Nature; Environmental Movement; Forest Service; Intrinsic Value; Land Ethic; Pinchot, Gifford; Political Ecology; Pragmatism; *Silent Spring*; Utilitarianism; Wise Use Movement.

Further Readings

Brands, H. W. *TR: The Last Romantic.* New York: Basic Books, 1997.

Carson, Rachel. *Silent Spring.* New York: Houghton Mifflin, 1962.

Hays, Samuel P. *Conservation and the Gospel of Efficiency: The Progressive Conservation Movement.* New York: Atheneum, 1980.

Leopold, Aldo. *A Sand County Almanac, and Sketches Here and There.* London: Oxford University Press, 1948.

Sparhawk, W. N. "The History of Forestry in America." *Trees: The Yearbook of Agriculture.* Washington, D.C.: U.S. Department of Agriculture, 1949.

Udall, Stewart L. *The Quiet Crisis.* New York: Holt, Rinehart and Winston, 1963.

U.S. Library of Congress. "The Evolution of the Conservation Movement, 1850–1920: Proceedings of a Conference of Governors." http://memory.loc.gov/cgi-bin/query/r?ammem/consrv:@field(DOCID+@lit(amrvgvg16)) (Accessed January 2009).

Timothy Collins
Western Illinois University

Consumer Politics

All economies in the world can be described as being primitive barter economies, command economies, or free market economies. Regardless of the type of economy, they are composed of three basic parts: production, distribution, and consumption. In barter economies, most production is subsistence. Even with limited trading, the choices are usually limited to what is offered for trade. However, archaeological evidence suggests that favorable trading usually occurred because the consumers were pleased with the goods offered. Otherwise, the silk and spice trading of the ancient to medieval world would not have occurred.

Green consumerism, a growing focus of consumer politics, has brought attention to more environmentally friendly transportation, such as bicycling. This has resulted in increasing numbers of bicycle paths and lanes, but safety remains a problem.

Source: iStockphoto.com

Command economies involve totalitarian control. They include economies controlled by Nazis, communists, socialist movements, and others. In the case of the Nazis, private ownership was retained in law, but virtually all economic decisions were directed by the party's commands in production, and often in distribution. Communist and socialist and other totalitarian command economies have offered their workers and consumers an equal share of goods they produce on the principle that equal distribution is economic justice. However, the production in egalitarian command economies creates shoddy goods because production is created by people who have little in the way of incentive to do any work or to do quality work. As a result, consumers in these economies purchase what is available to them simply out of necessity.

Market economies have allowed goods to be produced in large quantities in a competitive producer battle that has often encouraged innovation. Competition occurs in market economies because freedom for many producers to enter into production is allowed. Battles over wages and working conditions have occurred in these economies, but where there is product competition, they play a role that influences production and consumption. The consumer is then the beneficiary of opportunities to "vote" for one producer over another by spending money to purchase one producer's product rather than another's. However, the formation of a monopoly or an oligopoly by one or a few firms can limit consumer choice. At times, a monopoly can offer a price for its products that is cheaper than that for goods made in more competitive markets. This was the case with the kerosene produced by the Rockefeller monopoly. At other times it may be that consumer choice is denied by a manufacturer with a virtual monopoly. This was the situation when Henry Ford made the statement that consumers could buy a Ford automobile in any color they wanted, as long as it was black.

How It Began

The modern consumer rights movement began in response to the oligopoly conditions of the American automobile industry. In the 1950s, U.S. industry was flourishing still as result of its survival and growth during World War II. Much of the industrial plants of Europe and Japan were still being rebuilt. In the rest of the world, industry was still much smaller than in the United States. As a result, the automobile industry could build cars that were responsive to market tastes or to the interests of the companies, but consumer concerns had limited influence in automobile design, safety, or other factors. Into these conditions Ralph Nader entered by first publishing as a study at Harvard Law School articles on consumer safety in the *Harvard Law Record*. In 1959, he published "The Safe Car You Can't Buy"—which was critical of automobile safety—in *The Nation*. In 1965, he published *Unsafe at Any Speed*.

Nader's book was especially critical of General Motors' Chevrolet Corvair because it had been involved in enough accidents in which there were spinouts or rollovers that there were over 100 lawsuits against the company. The book focused public attention on reasons for the growing number of automobile accidents. Eventually, greater attention, which in public affairs almost always translates into spending money on one problem rather than anther one, was applied to automobile safety.

Nader suddenly found that he had attracted a large youth following. The young activists became "Nader's Raiders" and engaged in a number of projects, some of which were of concern to consumers. These and other factors made Nader one of the original advocates for consumer rights.

The consumer rights movement had another source besides Nader—the mental health movement. It had begun in 1868 with the founding of the Anti-Insane Asylum Society by Elizabeth Packard. By the 1950s, modern psychology had emerged from the beginnings made by Sigmund Freud. Its advances had aided psychiatry and the discovery of psychotropic medications. However, patients, their families, friends, and advocates were often at odds with the different models of treatment that were often unsuccessfully applied to the mentally ill. Many of the treatments that are now discredited (e.g., insulin shock therapy and lobotomies) were often administered against the patient's will.

The advent of the civil rights movement in the 1950s and 1960s gave impetus and language to the claims of mental health patients in terms of inequality and civil rights. The fight for patients' rights and against forced treatments led to the organization of groups that engaged in fighting for the mentally ill. These groups included the Insane Liberation Front and the Network Against Psychiatric Assault. The movement for mental health patients' rights often rejected state-run institutions and ranged from psychiatric models to peer-run service providers where there was patient (consumer) choice and participation in their treatment. The patient-centered approach promoted the development of self-help movements in which people who shared similar mental health problems joined together in mutual aid groups and networks.

For those using mental health systems, the focus shifted into a consumer mode in which members (patients and allies) sought to learn as much as possible about the mental health system to gain access to the best services and treatments available. The learning effort was rather like the practical wisdom that when purchasing a piece of real estate, the buyer should seek to know as much as possible about it before making the purchase.

As the mental health movement developed into a mixture of consumer rights and civil rights organizations (patients' rights), activists created nonprofit organizations and sought a "place at the table" to negotiate, defend, or advocate for their cause. The ideologies and practices of the widespread group influenced others in the development of consumerism and consumer politics.

Consumer watchdog groups have continued to lobby, agitate, or advocate on issues, including state and federal funding, that affect the mentally ill. The issues may be over mental health insurance or in opposing moves by physicians to put caps on malpractice insurance claims.

Since the 1980s, consumer advocacy has spread to all manner of products and services. Areas of attention have been in product safety, which led to the creation of the Consumer Product Safety Commission. The Consumer Product Safety Commission has been mandated the mission of protecting the public from unreasonable risks from the thousands of products available in the marketplace. The work of the Consumer Product Safety Commission has been significant because there has been a decline of 30 percent in the number of deaths and injuries associated with consumer products since its inception.

Concerns about product safety have arisen and continue to arise over all manner of products, from lead paint used in toys to adulterated vitamins. The dangers usually capture public attention, which issues a demand for government regulation. Consumer concerns have even extended to animal healthcare after product recalls were issued for products that were adversely affecting animal health.

Consumerism and Politics

The politics of consumerism has as of yet only a limited influence on presidential campaigns. However, the influence of consumer advocates on legislators and regulators has been very important. Issues have ranged from college costs to freedom from Internet taxation. With the growth of the consumer rights movement, there have been organized—in every state and in the federal government—units that are concerned with and regulate consumer issues. The issues from A to Z have grown into a long list covering much of what people consume.

In many states, the state's attorney general usually has an office of consumer protection. These state offices often provide information on a wide range of topics of interest to consumers. Topics can include identity theft, debt management, refund rights, and scams that target elderly individuals, or they may be over interest rates, college costs, or a host of other issues.

In essence, the consumer rights movement has sought to gain information, benefits, and protections for consumers through social, political, and legal actions that would compel everything from honest packaging and labeling to state-of-the-art safety standards, fair advertising, and product guarantees. More broadly, consumer protection also includes protection in the delivery of services so that even college students are now identified as educational consumers or customers.

Green Consumerism

Of significance here is green consumerism, which involves all issues affecting the whole of life as an environmental issue. Consumer environmental issues include not only clear air, water, and land but also the issue of wholesome civic environments. Food and even energy production environmental costs as well as consumption costs are part of the concern of green consumerism.

The broad vision of green consumerism looks to the whole of the environment, both natural and built. Historically, environmentalism has been treated first as a conservation matter, and since the 1970s as an ecological issue that affects the health of humans and nature itself. Having wild spaces or national parks began as a conservation issue, but consumer participation in nature, whether on public or private lands, is of concern. In the case of public lands, it involves traditional environmental issues, and in private ecotourism or other types of tourism or consumption of nature, there is a growing green concern. The fact is that most people today are urbanites or suburbanites who experience nature recreationally by fishing, hunting, birding, hiking, or in some other way as nature consumers. To maintain the purity of the product is in the interest of the public as nature consumers.

Another area of green consumerism is energy production. Concern over oil supplies and prices, as well as the negative effect of automotive and other forms of fossil fuel emissions, has motivated many to seek more efficient forms of energy consumption. In manufacturing, there has been a steady development of more energy-efficient machines from home furnaces to lightbulbs. In the United States, where energy supplies have been much cheaper until recently, there has been a steady resistance to moving rapidly into energy efficiency. The opposition has come from businesses unable or unwilling to engage in research and development of products that would be more energy efficient. However, with the rise of much higher energy prices in the mid-2000s, attention was turned by consumers and manufacturing to energy efficiency.

In effect, the first decade of the third millennium has been a greening decade. Companies like General Electric have opened their research vision to any and all ways to create more efficient energy supplies and systems. With the advent of the Barack Obama administration and a Democrat congress, money was allocated for research in green energy technologies. Creating these various groups will support the "winners and losers." In the case of the fossil fuel industry, it is expected to be a loser. However, the numerous renewable energy sources being investigated may turn out to deliver less than their advocates hope. In any case, the greening of energy consumption will affect the built environment of the world.

The built environment of the United States has often reflected the ideas of remote designers, contractors, urban planners, individuals, or the cost-minded without any regard to the way that building decisions affect civic and/or cultural life. Much of the built environment of the 20th century has revolved around the automobile, with buildings and communities designed to accommodate automobile traffic. This has included filling stations and parking garages. Add to this skyscrapers and concrete jungles in inner cities, and the effect has been one that is unnatural and unhealthy in a variety of ways.

Consumer attention has shifted in recent years to rebuilding the urban environment so that it is amenable to walking, bicycling, light rail, and other more environmentally friendly means of transportation. Bicycle lanes have been built, but in many places bicycling on public streets is still dangerous because of automobile driver indifference. Pedestrian traffic also faces dangers resulting from hazards from automobile and truck traffic.

Green consumerism has been increasingly concerned about food. Concerns about food safety have arisen because there has been a steady occurrence of disease outbreaks caused by contaminated food. These would have gone undetected in earlier years, but advances in technology now make it possible to detect the presence of bacteria or other pathogens adverse to human or animal health. In addition, the use of cobalt radiation or genetic modification of food plants has met such resistance in Europe that the European Union has banned their sale.

European opposition to genetically modified food is based on the observation that, as of yet, it is too new a science to know about its negative possibilities. Another aspect of this concern is the result of an anxiety that genetic changes in foods could affect genetically human biochemistry. In the United States, in contrast to Europe, genetically modified foods and imported (but inspected) foods are sold generally without much opposition, except from those who are virtually a subcult of the organic foods industry. Organic foods are part of green politics. Those in favor of such things as brown eggs and goat milk may not be urban yuppies but may be, often as not, people from a rural background who believe these types of "traditional" and unprocessed farm products are healthier.

The politics of green consumerism is in part the outcome of those opposed to corporatism versus a naturalistic way of life in which there is a tradition of Romanticism or agrarian idealism. At the very least, there is a dose of nostalgia for a simpler time when people were more in tune with the rhythms of nature than of an urban industrial order.

See Also: Green Parties; Organizations; *Silent Spring*; Urban Planning.

Further Readings

Andersen, Robin. *Consumer Culture and TV Programming*. Boulder, CO: Westview, 1995.

Breen, T. H. *The Marketplace of Revolution: How Consumer Politics Shaped American Independence*. New York: Oxford University Press, 2005.

Chatriot, Alan, et al. *The Expert Consumer: Associations and Professionals in Consumer Society*. Burlington, VT: Ashgate, 2006.

Cohen, Lizabeth. *A Consumers' Republic: The Politics of Mass Consumption in Postwar America*. New York: Doubleday, 2003.

Gidlow, Liette. *The Big Vote: Gender, Consumer Culture, and the Politics of Exclusion, 1890s–1920s*. Baltimore, MD: Johns Hopkins University Press, 2004.

Maclachlan, Patricia L. *Consumer Politics in Postwar Japan: The Institutional Boundaries of Citizen Activism*. New York: Columbia University Press, 2002.

Robbins, Paul. *Lawn People: How Grasses, Weeds, and Chemicals Make Us Who We Are*. Philadelphia, PA: Temple University Press, 2007.

Sassatelli, Roberta. *Consumer Culture: History, Theory and Politics*. Thousand Oaks, CA: Sage, 2007.

Andrew Jackson Waskey
Dalton State College

Convention on Biodiversity

Biodiversity (BD) is the genetic/species variation within or between ecosystem(s), biome, or for the entire Earth, and is often used as a measure of the health of biological systems. Biodiversity is a blended word, from biology and diversity, originating from and used interchangeably with biological diversity. This multilevel conception is consistent with the early use of the term *biological diversity* in Washington, D.C., and in international conservation organizations in the late 1960s and 1970s, by Raymond F. Dasmann, who apparently coined the term. Thomas E. Lovejoy later introduced it to the wider conservation and science communities. Subsequently, the 1992 United Nations Earth Summit in Rio de

Janeiro defined *biological diversity* as the variability among living organisms from all sources, including, *inter alia*, terrestrial, marine, and other aquatic ecosystems, and the ecological complexes of which they are part: This includes diversity within species, between species and of ecosystems. This is, in fact, the closest thing to a single legally accepted definition of biodiversity, since it is the definition adopted by the United Nations Convention on Biological Diversity (CBD).

A convention on biodiversity/biological diversity was first proposed by the International Union for the Conservation of Nature (IUCN) in 1981. Negotiated under the auspices of the United Nations Environment Programme (UNEP), the convention opened for signature on June 5, 1992, at the Rio Earth Summit, and entered into force on December 29, 1993.

The new agreements under the CBD commit countries to conserve biodiversity, develop resources for sustainability, and to share the benefits resulting from their use. Under new rules, it is expected that bioprospecting or collection of natural products will be allowed by the biodiversity-rich country in exchange for a share of the benefits. The convention is legally binding, and parties are obliged to implement its provisions. Presently it includes 191 members—190 countries and the European Community. It was ratified by almost every country in the world, with the exception of the United States.

The Conference of the Parties (COP) is the governing body of the convention and advances implementation of the convention through the decisions it makes at its periodic meetings. The COP has established seven thematic programs of work—agricultural biodiversity, dry and sub-humid lands biodiversity, island biodiversity, marine and coastal biodiversity, forest biodiversity, mountain biodiversity, and inland waters biodiversity. Each program establishes a vision for and defines basic principles to guide future work. They also set out key issues for consideration, identify potential outputs, and suggest a timetable and means for achieving these. Implementation of the work programs depends on contributions from parties, the secretariat, and relevant intergovernmental and other organizations. They are periodically reviewed by the COP and the open-ended intergovernmental scientific advisory body known as the Subsidiary Body on Scientific, Technical and Technological Advice (SBSTTA), which conducts assessments of status and trends of, and threats to, biodiversity and provides the COP with scientifically, technically, and technologically sound advice on the conservation of BD and the sustainable use of its components.

The text of the approved convention consists of a preamble, objectives, terminology, principles, and 42 articles, such as jurisdictional scope, cooperation, general measures for conservation and sustainable use, identification and monitoring of activities which have or are likely to have significant adverse impacts on the conservation and sustainable use of BD, *in situ* conservation, *ex situ* conservation, incentive (economical/social) measures, research and training needs, public education and awareness, impact assessment and minimizing adverse impacts, access to genetic resources, access to and transfer of technology, exchange of information, technical and scientific cooperation, handling of biotechnology and distribution of its benefits, financial resources/ mechanisms, relationship with other international conventions, conference of the parties, and the functions of the CBD secretariat and the subsidiary body. It also covers technical and technological advice, reports, settlement of disputes, adoption of protocols, amendment of the convention or protocols, adoption and amendment of annexes, the right to vote, the relationship between this convention and its protocols, signature, ratification, acceptance or approval, accession, entry into force, reservations, withdrawals, financial interim arrangements, secretariat interim arrangements, and depositary and authentic texts. Besides this, there are two annexes on identification and monitoring and arbitration (the latter has two parts and 23 articles).

The objectives of this convention, to be pursued in accordance with its relevant provisions, are the conservation of biological diversity, the sustainable use of its components and the fair and equitable sharing of the benefits arising out of the utilization of genetic resources, including appropriate access to genetic resources and appropriate transfer of relevant technologies, taking into account all rights over those resources and technologies, and appropriate funding.

States have, in accordance with the Charter of the United Nations and the principles of international law, the sovereign right to exploit their own resources pursuant to their own environmental policies, and the responsibility to ensure that activities within their jurisdiction or control do not cause damage to the environment of other states or of areas beyond the limits of their national jurisdiction.

The CBD program of work on Protected Areas (PAs) is a global framework for the establishment of comprehensive, representative and effectively managed national and regional protected area (PA) systems. Parties agreed to close the gaps in the existing systems, enhance management effectiveness, and secure adequate financing.

PAs, the backbone for the stability and functioning of ecosystem processes and the provision of ecosystem services, such as natural carbon storage, water cycles, pollination, control of diseases and flood control, are the foundation for safeguarding ecosystems, species, and genes, in all their abundance and diversity. Properly designed and managed PAs support the livelihoods of local communities and strengthen local and national economies. Protected area networks are our "Safety-Nets for Life on Earth." Thus the establishment and long-term maintenance of PAs is in the interest of the global community.

However, the Life Web is not a fund in itself where applicants can request funding for candidate sites, but it offers a platform where partners can join informally and build support on a bilateral basis. Donors can join in with their existing funding instrument on a co-funding basis. It complements the Global Environment Facility (GEF). While the GEF Focal Strategy for PAS is directed toward strengthening overall PA systems, the initiative adds support to specific sites.

The convention has already accomplished a great deal on the road to sustainable development by transforming the international community's approach to biodiversity. This progress has been driven by the convention's inherent strengths of near universal membership, a comprehensive and science-driven mandate, international financial support for national projects, world-class scientific and technological advice, and the political involvement of governments. It has brought together, for the first time, nations with very different interests.

Continued Challenges

However, many challenges still lie ahead. After a surge of interest in the wake of the Rio Summit, many observers are disappointed by the slow progress toward sustainable development during the 1990s. Attention to environmental problems was distracted by a series of economic crises, budget deficits, and local and regional conflicts. Despite the promise of Rio, economic growth without adequate environmental safeguards is still the rule rather than the exception.

Some of the major challenges to implementing the CBD and promoting sustainable development are as follows:

- Meeting the increasing demand for biological resources caused by population growth and increased consumption, while considering the long-term consequences of our actions.
- Increasing our capacity to document and understand biodiversity, its value, and threats to it.
- Building adequate expertise and experience in biodiversity planning.

- Improving policies, legislation, guidelines, and fiscal measures for regulating the use of biodiversity.
- Adopting incentives to promote more sustainable forms of biodiversity use.
- Promoting trade rules and practices that foster sustainable use of biodiversity.
- Strengthening coordination within governments, and between governments and stakeholders.
- Securing adequate financial resources for conservation and sustainable use, from both national and international sources.
- Making better use of technology.
- Building political support for the changes necessary to ensure biodiversity conservation and sustainable use.
- Improving education and public awareness about the value of biodiversity.

The CBD and its underlying concepts can be difficult to communicate to politicians, administrators, and to the general public. Nearly a decade after the convention first acknowledged the lack of information and knowledge regarding biological diversity, it remains an issue that few people understand. There is little public discussion of how to make sustainable use of biodiversity part of economic development. One of the greatest challenges in sustainable development decisions is the short- versus the long-term time frame, and public versus private profits. Sadly, it often still pays to exploit the environment by harvesting as much as possible as quickly as possible because economic rules do little to protect long-term interests.

The CBD spirit implies a prior informed consent between the source country and the collector to establish which resource will be used and for what, and to settle on a fair agreement on benefit sharing. Bioprospecting can become a type of biopiracy when those principles are not respected. Truly, SD requires countries to redefine their policies on land use, food, water, energy, employment, development, conservation, economics, and trade. BD protection and sustainable use requires the participation of ministries responsible for such areas as agriculture, forestry, fisheries, energy, tourism, trade and finance.

The transition to SD requires a shift in public attitudes as to what is an acceptable use of nature. This can only happen if people have the right information, skills, and organizations for understanding and dealing with BD issues. Governments and the business community need to invest in staff and training, and they need to support organizations, including scientific bodies, that can deal with and advise on biodiversity issues.

A long-term process of public education is also needed to bring about changes in behavior and lifestyles, and to prepare societies for the changes needed for sustainability.

BD underpins SD in many ways; poverty eradication, food/nutrient/livelihood security, provision of fresh water, soil conservation, and human health all depend directly upon maintaining and using the world's BD, and therefore SD cannot be achieved without the conservation and sustainable use of biological diversity. This is especially true for developing countries, where there is higher human population growth as well as accelerating economic growth, triggering climate change, ozone depletion, and the use of hazardous chemicals.

Global Biodiversity Outlook is the flagship publication of the CBD, and preparations are currently underway for the production of its third edition. *Global Biodiversity Outlook 3* will be formally launched in 2010, the year proclaimed as the International Year of BD. Information regarding the status and trends of BD, both at global and regional levels, will be presented, as well as information regarding the progress made in mainstreaming BD issues into the development agenda. There will be an emphasis on case studies that illustrate the positive actions taken to effectively conserve and sustainably use BD. The information contained in *Global Biodiversity Outlook* will provide an assessment of the current status and trends of BD, and will be a vital tool in informing the policy process and shaping international BD commitments post 2010. In particular the *Global Biodiversity Outlook 3* will serve to identify barriers to the further implementation of the CBD and assist in setting implementation priorities.

Conservation and development are no longer seen as conflicting goals, but as mutually interdependent. This change has been in large part due to the process of international consensus building instituted under the convention and its inherent strengths of near universal membership, a comprehensive and science-driven mandate, international financial support for national projects, world-class scientific and technological advice, and the political involvement of governments. By bringing together, for the first time, people with very different interests, the process initiated under the convention offers hope for the future by forging a new deal between governments, economic interests, environmentalists, indigenous peoples and local communities, and the concerned citizen.

See Also: Biodiversity; Political Ecology; Sustainable Development; UN Conference on Environment and Development; UN Framework Convention on Climate Change.

Further Readings

Adams, William M., et al. "Biodiversity Conservation and the Eradication of Poverty." *Science*, 306/1146–1149 (2004).

Cohen, Lizabeth. *A Consumers' Republic: The Politics of Mass Consumption in Postwar America*. New York: Doubleday, 2003.

Johnston, Sam. "The Convention on Biological Diversity: Ten Years on and the Strategic Plan." *Journal of International Wildlife Law & Policy*, 4/2:147–158 (2001).

Le Prestre, Philippe G. *Governing Global Biodiversity: The Evolution and Implementation of the Convention on Biological Diversity (Global Environmental Governance)*. London: Ashgate Publishing, 2003.

United Nations. Report of the UN Conference on Environment and Development. http://www.un.org/documents/ga/conf151/aconf15126-1annex1.htm (Accessed August 2009).

Gopalsamy Poyyamoli
Pondicherry University

Copenhagen Summit

The Copenhagen Summit was an international meeting addressing global climate and climate change. It was the 15th meeting between countries belonging to the United Nations Framework on Climate Change. It represented the most recent attempt to form a global treaty addressing climate change. Government leaders from 192 countries met in Copenhagen, Denmark, from December 7 through December 18, 2009, to discuss cutting global emissions. The goals of the summit included developing a global framework addressing climate change, and garnering international agreement about how to combat climate change. Critics argue that the summit was not successful in achieving either of these goals. Despite hopes that the Copenhagen Summit would produce a legally binding treaty, the conference never overcame negotiating impasses and it is not legally binding. This was largely due to the differing expectations between large developed countries and developing countries.

Goal of the Summit

The main goal of the Copenhagen Summit was to update the Kyoto Protocol. The summit was also the fifth meeting of the parties to the Kyoto Protocol. The Kyoto Protocol was ratified by 187 members of the United Nations Framework on Climate Change and was drafted and adopted in December 1997. The Kyoto Protocol set a goal of reducing greenhouse gases for participating members by 2012. In order to do so, the protocol set emission reduction amounts for each individual participating country and allowed for emissions trading through a cap and trade mechanism. Although a member of United Nations Framework on Climate Change, the United States never formally adopted the Kyoto Protocol. The Kyoto protocol set differing standards for Annex I countries, or developed countries, and Annex II countries, which were defined as developing countries. The reason for the difference was an argument that developed countries had already created a lot of environmental pollution and that limiting developing countries to the same set of reduction standards would be unfair. Although the Clinton Administration did decide to sign the protocol, political leaders in the United States did not agree with dissimilar standards for Annex II countries, especially China. As a result, the treaty was never ratified by Congress.

The Copenhagen Summit provided a forum to update the Kyoto Protocol since it was set to expire in 2012 and presented a new opportunity for world leaders to reach current agreements on climate change and on emission reduction. As noted above, the action-orientated goals included producing a successor to the Kyoto Protocol and to create and international framework for global cooperation for reduction in emissions related to climate change. Each country was expected to develop a goal for reducing absolute emissions.

Product of the Summit

The product of the summit is being called the Copenhagen Accord. It is a nonbinding agreement between five countries including Brazil, China, India, South Africa, and the United States. The other 187 countries in attendance did not sign the accord but did agree to take note of the agreement. The key tenant of the document states that climate change is one of the greatest global challenges facing the world today, and that action should be taken to moderate global temperature increases. Since the document was not signed by the majority of countries in attendance, nor ratified by those that did sign, it is not legally binding. Nor does it contain any legally binding commitments even for the five countries that did agree to a plan for reducing carbon emissions. In other words, neither developing nor developed countries are required to make any formal cuts in their emissions. Without a penalty for noncompliance critics argue that the document is nothing but political jargon.

The Copenhagen Accord also acknowledges a scientific basis for keeping temperature rises below 2 degrees C, but does not contain commitments for reduced emissions that would be necessary to achieve that aim. As an example of those countries that did sign the accord, China has agreed to cut emissions intensity by 40–45 percent below 2005 levels by the year 2020, and the United States has agreed to cut greenhouse gas emissions by 17 percent below 2005 levels by 2020 and 42 percent by 2030. The other three countries listed above have also agreed to reduce emissions by the year 2020. Both the European Union and the United States would like to see developed countries and developing countries comply with similar reductions.

Like climate change agreements in the past, the Copenhagen Summit exposed a conflict between developing countries and industrial nations. As part of the summit Hillary Clinton, Secretary of State of the United States, made an offer of $100 billion dollars in aid to developing countries. This was not successful in gaining their participation. Critics argued the amount was too little too late and that it was not clear how this money would be spent.

U.S. President Barack Obama said he was pleased with the outcome but that there was still a long way to go to build consensus. The Copenhagen Accord does request countries to submit new emissions targets by the end of January 2010. The next round of talks is scheduled for the 2010 United Nations Framework on Climate Change conference in Mexico.

Criticisms

Blame for the lack of consensus was aimed at both developed and developing countries. Those blaming developed countries argued that most UN member states were excluded through the negotiating of the Copenhagen Accord with only a select group of countries. This agreement said that if poorer nations did not sign the accord, they would not be able to access funds to help them adapt to climate change. A second round of criticism, however, was aimed at the unification of developing countries that cooperated at Copenhagen to block attempts at establishing legally binding targets for carbon emissions that would affect their economic growth.

There were also many individuals that protested the meeting of the United Nations Framework on Climate Change. As the summit began, protests simultaneously took place in Copenhagen and other cities throughout the world. It was estimated that there were between 40,000 and 100,000 people in the streets in Copenhagen. Although the protestors had multiple aims, many openly rejected the treatment of developing nations and indigenous rights. Protests also took place in London a week prior to the conference start, demanding British leaders to ask all developed nations to cut their emissions by 40 percent by 2020, and to provide more assistance for the world's poorest countries in adapting to climate change. There were similar protests in Australia.

See Also: Afforestation; Deforestation; Global Climate Change; Kyoto Protocol.

Further Readings

"Copenhagen Climate Talks." *New York Times.* http://topics.nytimes.com/top/reference/timestopics/subjects/u/united_nations_framework_convention_on_climate_change/index.html?scp=1-spot&sq=copenhagen&st=cse (Accessed December 28, 2009).

Cowie, Jonathan. *Climate Change: Biological and Human Aspects.* Cambridge, UK: Cambridge University Press, 2007.

U.S. Congress, "The Kyoto Protocol and Its Economic Implications: Hearing Before the Subcommittee on Energy and Power of the Committee on Commerce." House of Representatives, One Hundred Fifth Congress, Second Session, March 4, 1998. Washington, D.C.: General Printing Office, 1998.

Victor, David G. *The Collapse of the Kyoto Protocol and the Struggle to Slow Global Warming.* Princeton, NJ: Princeton University Press, 2001.

Jo A. Arney
University of Wisconsin–La Crosse

Corporate Responsibility

Corporations are the most common form of business organization and are established as a legal entity separate from its constituent members or shareholders. In a corporation with shares, ultimate responsibility is usually allocated to the board of directors, which is elected by shareholders and accountable to them. All persons, regardless of their position in the corporation, including directors and officers, are evaluated in reference to the fundamental purpose of the corporation. Beyond electing the board of directors, shareholders do not participate in the decision making and operation of the organization. Decision makers in a corporate context, as anywhere else, are liable for the consequences of their decisions. Where the decisions are within the legal authority of the decision maker, liability for the consequences remains with the corporation and not with the individual person. Corporate responsibility then is related to the purpose of the organization as established by law, and where the corporation is a business, the responsibility is to conduct business in compliance with the law. Internally, the board of directors usually bears substantial responsibility for all the corporation's acts or omissions.

All corporations exist as persons only by authority of law; hence the term *legal persons*. As a legal person, the purpose of a corporation is set out in its enabling law. A business corporation therefore is an organization that exists by the authority of business corporation law for business purposes, which is understood to be financial profit. This focus on financial gain has been and remains the responsibility of the business corporation. Where corporations for nonbusiness purposes can and do exist, they are usually mandated with different purposes, and hence different responsibilities. The overriding goal of corporate responsibility is acting in the best interest of the public beyond what constitutes legal obligations; however, the interpretation of this goal in compliance with corporation objectives can vary considerably depending on the business and context.

Since about the mid-20th century, business corporations were demanded by public persons and not by the law to address issues other than financial. With this development, corporate governance involved the exercise of business and social responsibility. What has since come to be known as corporate social responsibility has many definitions, all of which have served to broaden the scope of corporate governance to include contemporary needs and goals of societies locally and globally.

By their actions, for better or worse, corporations have affected the quality of life in the communities within their sphere of influence. For example, employment practices within the company can reinforce or challenge local practices. Such was the case when apartheid was an official policy in South Africa. Business corporations with interests in South Africa, Canada, and the United States were under pressure to operate in South Africa in accord with Canadian and American practices and differently from the local policy of apartheid. At about the same time, a large multinational business corporation selling milk powder to disadvantaged people living under conditions of famine in Africa did respond to international popular pressure. In doing so, this corporation accepted responsibility for the effects of its legitimate business and changed its practice, thereby alleviating the dependencies and health consequences to the population from relying on the powdered milk for babies in the absence of clean water. In this sense, the business corporation's legal obligation to operate for profit was modified with the addition of the responsibility to act contextually in accord with ethical and humanitarian principles or social interests.

As it has developed, corporate responsibility may now include four areas of society: human rights, labor, the environment, and anticorruption. Corporations now have to consider not only more complex regulatory regimes and customer preferences but also concepts of fairness, ethics, industry standards, certification standards of nongovernmental organizations, cradle-to-grave considerations for products, community concerns, and the welfare of the communities locally and internationally.

Role of Stakeholders

Stakeholders in a corporation extend beyond employees, suppliers, and contractors to directors, shareholders, and creditors. Responsibilities of ownership and socially responsible investment principles are advanced by these stakeholders and socially concerned persons.

The manner and participation of interaction between corporations and stakeholders has evolved. Stakeholder influence has been exercised through various forms of shareholder activism, which has resulted in regulatory changes as well as corporate policy amendments. Shareholders voting on resolutions in person or through proxy have affected internal corporate policies to the extent that as early as 1946, the U.S. Securities Exchange Commission adopted a rule requiring companies to include shareholder resolutions with proxy voting forms. The practice of distributing resolutions with proxy voting forms at company expense continues to be a legal requirement, as is the requirement to discuss these resolutions at annual or special shareholder meetings.

Balancing business and other social concerns remains a dynamic exercise that will likely always be an issue. In the 1960s and 1970s, shareholders demanded that corporations address human rights such as equal opportunity and rights for black persons, as well as changes to the business of weapons production. One such proposal to force Dow Chemical to stop producing napalm during the Vietnam War was initially disallowed by the Securities Exchange Commission but later was reversed by a U.S. District Court. Subsequently, other proposals on corporate social responsibility were advanced including consumer interests such as product safety, employment discrimination, and environmental pollution.

The high-profile adversarial approach and public interest issues of the 1960s and 1970s declined during the 1980s for various social and business reasons. Global political and social changes and elevated mergers and acquisitions activity directed attention to market competition. In the 1990s, accidents, unethical business practices, and widening environmental concerns prompted a resurgence in shareholder activism.

Disclosure, transparency, and accountability have been demanded of corporations, and management has been required to integrate social responsibilities throughout the business. Evolving management styles and an interest in quality management systems may have facilitated such considerations. International standards such as those by the International Labour Organization, Responsible Care, and International Organization for Standardization were formally recognized and became prerequisites for businesses in some industries. With a growing interest in continuous business improvement, shareholder activism became an iterative process and more of a dialogue between stakeholders in a continuing engagement.

Corporate Responsibility on a Global Scale

Fair trade practices and nonbusiness certification have heightened awareness of business-related social issues on a global scale. The corporate responsibility to be fair with suppliers and customers must not only be done but must be seen to be done. Forward-looking statements are now part of securities disclosure requirements and are required for some licensed activities. Corporations are expected to be proactive in dealing with social and environmental issues. Some corporations also accept the responsibility for the quality of life of all stakeholders and communities. As some corporations now have operating budgets larger than some of the countries in which they operate, corporate policies and practices can be more significant than some national policies.

Concepts of "sustainable development" have gained credence in some corporations that pride themselves on doing business in an environmentally sustainable manner. What is certain is that corporate governance now includes social obligations. The concept of citizenship for the corporation may apply with greater accuracy, especially as with the inclusion of corporate social responsibility, actions of the business corporation may more closely approximate the actions of human persons.

As with individuals, corporations can and do present a professional face. What substance lies behind that face may be a matter of further inquiry. Compliance with official policies and codes of conduct can be evaded by outsourcing the supply chain to entities that do not subscribe to such codes or cannot be subject to compliance measures. Demands for accountability can therefore be satisfied without effective or material change in business practices and procedures.

To strengthen accountability, nonbusiness sources have taken initiatives to develop multilateral guidelines for transnational corporations and proposals to include them in a binding treaty. Organisation for Economic Co-operation and Development Guidelines for Multinational Enterprises and the Global Compact are two examples of voluntary guidelines that originated outside business or industry. The 2008 crisis of the international credit and financing system and acceptance of corporate "bailouts" from government may indicate a possible pathway to include binding obligations or re-regulation of corporations and to address issues of corporate responsibility—or lack thereof—in a manner that has not been seen for some time. To the extent the conditions on the bailout funds are supported by political will, it may be possible that the nonindustry or corporate rules could evolve to become legally binding globally.

For human persons and, by analogy, legal persons such as business corporations, attitudinal shifts would likely be necessary if responsibility is to be accepted. The prevailing utilitarian approach to the environment, labor, human rights, and ethical behavior has resulted in the current circumstances, in which all are disposable without responsibility in the interests of financial gain. Sustainability as it has developed to date, including principles of liability, polluter pays, and precautionary principles in the Rio declaration, has been shaped in no small measure by corporate business interests. In this sense, including sustainability in corporate social responsibility without the attitude change may be insufficient to achieve a responsibility that includes notions of accountability with acceptance of appropriate enforcement of consequences.

See Also: Bhopal; Environmental Management; Green-Washing; Organizations; Utilitarianism.

Further Readings

Clapp, Jennifer. "Global Environmental Governance for Corporate Responsibility and Accountability." *Global Environmental Politics*, 5/3 (2005).

Linton, April, et al. "A Taste of Trade Justice: Marketing Global Social Responsibility via Fair Trade Coffee." *Globalizations*, 1/2 (2004).

Michael, Bryane. "Corporate Social Responsibility in International Development: An Overview and Critique." *Corporate Social Responsibility & Environmental Management*, 10/3 (2003).
O'Rourke, Anastasia. "A New Politics of Engagement: Shareholder Activism for Corporate Social Responsibility." *Business Strategy & the Environment*, 12/4 (2003).
Pangsapa, Piya and Mark J. Smith. "Political Economy of Southeast Asian Borderlands: Migration, Environment, and Developing Country Firms." *Journal of Contemporary Asia*, 38/4 (2008).

Lester de Souza
Independent Scholar

Cost-Benefit Analysis

One criterion by which public policies are evaluated is the ability to maximize economic efficiency, or the net benefits to society. A policy that achieves this goal provides the greatest amount of benefits to society at the least cost. Therefore, the government, under this standard, should not pursue policies for which the costs to society exceed the benefits or for which the difference between costs and benefits may be increased by alternative policy measures. Cost-benefit analysis (CBA) is an analytical tool that can be used to support such decision making.

Opportunity costs, or the value of resources if allocated for their best alternative use, are used to measure the costs to society of pursuing a public policy proposal. This cost differs from the typical accounting costs associated with expenditures by businesses or government, which may not reveal the true costs to society. For example, the effects of regulation on such economic variables as productivity and unemployment are not captured in accounting costs but may still present costs to society. Opportunity costs may be measured by using data on market trends to determine the value of forgone goods and services.

Benefits to society may be divided into market benefits and nonmarket benefits. Many benefits associated with market interactions may be easily valued in dollar amounts according to the relative gains in the market that can be measured as a consequence of public policy. However, nonmarket benefits—those that cannot be directly valued in terms of dollars—require different methods in assessing their value. An example of nonmarket benefits are those that arise as a result of the preservation of a forest. A forest may have value that stems from use that is not paid for by the consumer, such as recreational activities. Value may also lie in the potential for the forest to provide future benefits that are unknown and that would be destroyed in the absence of the forest. Plant species that have the potential to provide future medical benefits, for example, may be destroyed without the protection of a forest. The existence of the forest also provides benefits to future generations. Whether someone uses or directly benefits from a forest, they may find value in knowing that the forest and its inhabitants may be enjoyed by others. Therefore, the total nonmarket value of a forest is the sum of the values associated with direct use, future options, and existence of the forest.

To quantify these values, analysts using the cost-benefit approach must determine society's willingness to pay for the preservation of the environment or society's willingness to accept recompense for environmental destruction. Analysts may observe this preference in a population through contingent valuation—surveys that measure stated preferences of the population. Travel costs, or the amount spent to use a resource, may also be used to measure nonmarket benefits. This is an indirect approach in which people reveal their preferences through their actions and behavior. Another approach that uses revealed preferences is called hedonic regression. This measure uses the change in prices of related goods to determine people's willingness to pay. For example, if the value of surrounding homes decreases as a result of environmental degradation, this market signal reveals the preference of local homeowners to protect the environment.

The outcome of decisions made in the present has future implications as well. Benefits that occur in the future must be translated into their present values to determine the net benefits to society. Just as compounding can be used to determine the future value of current investments that grow with interest over time, discounting may be used to determine the present value of future benefits. For example, if $500 were deposited into a bank account at 5 percent interest compounded yearly, each year the interest earned on the principal would be added to the principal, which would in turn earn interest. The value of the account would be $525 after the first year, $814 after the 10th year, $5,734 after the 50th year, and $65,751 after 100 years. If the 5 percent rate of yearly compound interest is also used as the discount rate, then discounting an investment that is valued at $5,734 at a 5 percent discount rate would yield a present value of $500. This present value measures the current worth of $5,734, 50 years from now. Thus, the discount rate can be used to make cost-benefit decisions on whether to invest in environmental programs or in other alternatives that would produce greater benefits in the future.

The application of CBA to environmental decision making has been criticized for reasons of moral philosophy, equity, and worth of the environment. Safety from the harmful effects of environmental pollution is viewed by some as part of the fundamental right to liberty. Achieving such a standard based on safety may be "inefficient" from a CBA, as a large amount of resources may be required to acquire low levels of health risks. Many environmental policies have goals of environmental safety as their focus; however, costs underlie all decisions, and at some point the question must be asked, "How much safety is too much?" Should resources needed in other areas be allocated toward maximum safety from all environmental harms? Although safety standards make for sound policy goals, costs and benefits must be considered in achieving the greatest amount of safety with available resources.

Equity arguments against CBA promote distributive justice in the present and in the future. The maximization of net social benefits maximizes the size of the economic pie but may create winners and losers, where some are better off than others as a result of the policy. Thus, distributional equity should be considered alongside efficiency as benchmarks for policymaking. Similarly, it is argued that discounting favors the preferences of the current generation over those of future generations. These arguments favor lower discount rates that place more emphasis on the benefits for future generations.

Others argue that the worth of the environment cannot be measured in monetary units. Attaching a dollar value to environmental benefits lessens the true value of the environment—which could be considered to be priceless. Proponents of CBA, however, argue that dollar values provide a yardstick for comparison. Because so many decisions are made on the basis of dollar value, it is fitting to use this measure for environmental valuation.

These arguments show the normative nature of such societal decision making and struggle to meet multiple demands in the allocation of resources. Despite these arguments,

CBA continues to be used as an analytical tool for providing information to decision makers, who must evaluate the optimal level of resource allocation for the needs of society.

See Also: Ecological Economics; Equity; Intrinsic Value; Risk Assessment; Utilitarianism.

Further Readings

Boardman, Anthony, et al. *Cost-Benefit Analysis: Concepts and Practice.* New Jersey: Prentice Hall, 2006.
Goodstein, Eban. *Economics and the Environment.* New York: John Wiley & Sons, 2007.
Keohane, Nathaniel and Sheila Olmstead. *Markets and the Environment.* Washington, D.C.: Island Press, 2007.

Thomas D. Eatmon, Jr.
Allegheny College

Counterculture

A counterculture represents a cultural group whose lifestyle is opposed to the prevailing culture. Countercultures construct an alternative culture that contests mainstream societal beliefs and values while desiring to influence social change. According to Ken Goffman, aka R. U. Sirius, a U.S. writer, countercultures share the characteristics of the following principles and values: they assign primacy to individuality at the expense of social conventions and governmental constraints, they challenge authoritarianism, and they embrace individual and social change. Countercultures are a form of subculture, with their own shared conventions, values, and rituals, yet they are also systematically opposed to the dominant culture. The term *counterculture* was introduced into public circulation in 1969 by Theodore Roszak in the book *The Making of a Counter Culture*, which discussed the hippie movement of the 1960s—the largest countercultural group in U.S. history. Countercultures occupy both extreme liberal and conservative attitudes and principles, with historic examples including individuals such as Socrates; traditions such as Romanticism, Bohemianism, and the Dandy; and social movements advocating lesbian, gay, and transgender rights and women's liberation. Countercultures play an important role in society, challenging and influencing mainstream societal values while representing an integral element of human nature. This article discusses a brief history of prominent American countercultures—the beats, hippies, diggers, back-to-the-landers, and cyberculture. Each of these countercultures contributed to the development of various principles and practices of the current environmental movement.

The radical questioning of social mores in the United States began most notably in the late 1940s with the Beat Generation: a group of American writers who celebrated nonconformity and spontaneous creativity, later to inspire the Beatnik culture, the yippies, and the hippies of the 1960s. Key Beat Generation writers included Jack Kerouac, Allen Ginsberg, Neal Cassady, and William Burroughs. During this era, countercultural resistance emerged as a result of a combination of youthful rejection of cultural and social norms, the political segregation of the South, the threat of technological bureaucracy, and the event of the

Vietnam War. Couched in discourses of antiwar and nuclear protests, race relations, sexual liberation, women's rights, experimentation with psychedelic drugs, and anticonsumerist themes, the subsequent 1960s encompassed a mass splintering and regrouping of social values and practices. Represented mainly by students, individuals felt that the Corporate State was an oppressive society in which a powerful few monopolized many, controlling the economy, politics, environment, and society. Individuals were taught to be competitive and superficial, with their anonymous, conformist identities easily replaced in a corporate-dominated society. These countercultures retaliated against such a top-down decision-making structure that endorsed economic and material values over all others.

The 1960s are divided into two factions: the New Left, represented by individuals fighting for civil rights with a background in the Free Speech Movement (a student movement at the University of California–Berkeley against technological bureaucracy); and the New Communalists, who were more influenced by the beats, Zen Buddhism, and by psychedelic drugs. Whereas the New Left turned outward toward political action, the New Communalists turned inward toward questions of consciousness. The diggers represented yet another counterculture group from this era, existing from 1966 to 1968 in the district of Haight-Ashbury, San Francisco. The diggers performed free street theater to encourage people to contemplate a new frame of social, political, economic, and cultural mind. They had no apparent organizational structure or leaders and had a strong belief in freedom for the self from societal and legal norms, rules, or expectations and from economic and structural limitations. This freedom was expressed by establishing alternative cashless economies encapsulated by Free Food (scavenging food to give away), the Free Store (providing free essential goods), Free Shelter (an underground network of places to stay), and Free Education (providing survival skills training through socialization).

By the end of the 1960s, many counterculturalists decided to establish alternative egalitarian communities in rural regions, launching the largest wave of communalization in American history. Back-to-the-landers, as they were called, were concerned with their own survival, wishing to break free from the Corporate State to live their beliefs. They aspired to have greater contact with nature, to build their own homes, to produce their own food, to use alternative energies, and to be part of cooperatives that shared similar values. In the early 1970s, approximately 750,000 people in the United States lived in more than 10,000 rural communes, with the movement replicated internationally. However, the Back-to-the-Land Movement did not last long with these numbers, as many individuals soon returned to the cities after discovering that they were unprepared for such reality.

Stewart Brand, author of *The Whole Earth Catalog*, was instrumental in supporting the back-to-the-landers while providing the platform for an ensuing cyberculture and foundation for the World Wide Web. *The Catalog*, published between 1968 and 1998, acted as a tool to empower and inspire individuals by proclaiming a "do-it-yourself" attitude—a vision of technology as a source of individual and collective transformation—and by providing an illuminating media format for future Web development. This publication created a massive shift in attitude toward technology, as computers that had been previously considered as large mainframes locked away and guarded by austere and powerful institutions became decentralized, personalized information systems that empower individuals to network with others, increasing knowledge and participation in world affairs.

These counterculture movements all influenced the development of the environmental movement by including the environment as part of their social concerns, equating "nature" with wild and authentic and in opposition to the conformity of suburbia, the corporate world, artificiality, and the logic of technocracy. Counterculture and environmentalist discourses were also influenced by similar events—the fear of the bomb, the People's Park student protest of 1969, and the Vietnam War. Similar to the counterculture movement, aspects of the environmental movement also called for entire social system change, whereas others challenged single issues against normative behaviors such as the animal liberation movement and ecoterrorism. These principles and actions vary from the quiet and personal, such as a vegan diet, to practicing dramatic and sometimes violent action to deliver positive environmental outcomes in relation to issues such as factory farming, battery hens, whaling, and rainforest logging. This merging of countercultural and environmental values became strongly apparent in 1970 at Earth Day, when more than 20 million people came together from a history of hippies, diggers, beats, the New Left, and the New Communalists and others to demonstrate concern about the environmental crisis. This event was as much a culmination as a beginning, setting the stage for a future of social change.

See Also: Anarchism; Consumer Politics; Environmental Movement; Green Discourse.

Further Readings

Belasco, Warren J. *Appetite for Change: How the Counterculture Took on the Food Industry.* Ithaca, NY: Cornell University Press, 1993.

Cock, Peter. *Alternative Australia: Communities of the Future?* Melbourne, Australia: Quartet, 1979.

Gelder, Ken. *Subcultures: Cultural Histories and Social Practice.* New York: Routledge, 2007.

Goffman, Ken and Dan Joy. *Counterculture Through the Ages: From Abraham to Acid House.* New York: Villard, 2004.

Rome, Adam. "Give Earth a Chance: The Environmental Movement and the Sixties." *Journal of American History* (September 2003).

Roszak, Theodore. *The Making of a Counter Culture: Reflections on the Technocratic Society and Its Youthful Opposition.* London: Faber and Faber, 1970.

Staller, Karen M. *Runaways: How the Sixties Counterculture Shaped Today's Practices and Policies.* New York: Columbia University Press, 2006.

Turner, Fred. *From Counterculture to Cyberculture: Stewart Brand, the Whole Earth Network, and the Rise of Digital Utopianism.* Chicago: University of Chicago Press, 2006.

Ferne Edwards
Australian National University

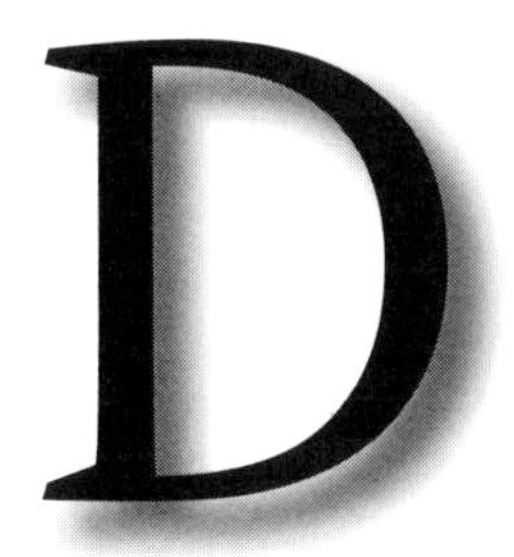

Death of Environmentalism

The "death of environmentalism" refers to a controversial idea within the environmental movement first articulated in a 2004 essay by Michael Shellenberger and Ted Nordhaus. The authors are both veteran activists who have worked professionally for several environmental and labor nongovernmental organizations in California. The pair helped cofound the Apollo Alliance, a network of environmental, private enterprise, organized labor, and civil rights advocates whose primary goal is the creation of three million new green jobs and the weaning of American reliance on foreign oil. Their essay argued that conventional conceptions of environmentalism are outdated and have outlived their usefulness. This is made apparent, they argue, by the unique challenges put forward by global climate change. As such, Shellenberger and Nordhaus are not arguing that environmentalism has died but, rather, that it can no longer deliver on its goals of sustainable society–nature relationships in confronting ecological crises, which will require substantial transformations affecting the lives of billions of people.

While recognizing the significant gains of environmentalism and acknowledging future progress as contingent on that history, they see it as a noble but ultimately inadequate effort today. They particularly focus on its recent weakness over the last few decades, assessing the movement as regressing from its successful status during the 1960s and 1970s (when a myriad of environmental legislation was secured) and critiquing the manner in which environmentalists claim authority to speak for nature by purportedly relaying what the Earth is communicating in an objective framework, resulting in a narrow conceptual divide between what does and does not get defined as environmental. For example, Shellenberger and Nordhaus attack the tactics emphasizing technical policy solutions battling over fuel economy or pollution emission caps, characterizing them as failing to express the graveness of climate change consequences, leaving the movement in a regressed state when compared with earlier decades. This type of environmentalism succeeded in building broad political respect for the environment, but one too shallow to foster the dramatic social, economic, and cultural transformations necessary to answer the challenge of climate change—such as transitioning away from an economy dependent on finite but cheap supplies of fossil fuels, which will require new relations among labor, consumers, and the environment. As a consequence, environmental activists often compensate for shallow support by emphasizing the dire nature of ecological crisis, which Shellenberger and

Nordhaus see as paralyzing, rather than empowering, individuals. The authors referred to "policy literalism" to critique the culture of mainstream environmentalism, which they see as trapped within the realm of technical policy, resulting from the heavy reliance on science to define environmental problems, while ignoring the politics that create the opportunities for such policy. Against the spirit of policy literalism, Shellenberger and Nordhaus attempted to avoid the "death of environmentalism" thesis from having prescriptive components, insisting that solutions will not come from individuals but, rather, from team efforts. Their essay largely rests on placing a politics of limits up against a politics of possibility, where the former may prompt action regulating carbon or other greenhouse gas emissions and the latter focuses on public–private relationships centered around green investment. Defenders of a politics of limits argue that tactics such as carbon taxes and cap-and-trade systems bypass the messy politics of choosing an alternative energy "winner" while simultaneously raising money to pay for such projects. In this, critics see a politics of possibility that strategizes around green investment as an admirable but ultimately impractical implementation of sustainable goals.

The first portion of "The Death of Environmentalism" focused on how environmental movement failure derived from its past policy success, allowing activists to believe they could be an authority on what does and does not count as "environmental." The underlying tendency here is to treat the environment as a thing, isolable in the process of analysis—contradicting the inherent holism they read in Sierra Club founder John Muir. Treating the environment as a thing follows a three-step strategy: define the problem as environmental, create a technical remedy, and promote the technical solution through various tactics such as lobbying, research, advertising, and so on. This type of strategy treats the environment as an end in itself, failing to reach out to new allies, relinquishing aspects of justice, and instead asking citizens what they can do for the environment. Shellenberger and Nordhaus see a fundamental flaw in a thing-based approach in that it implicitly classifies humans as outside the environment, which sends out confusing signals when discussing a man-made phenomenon such as climate change as an "environmental" problem. Movement activists' inability to step outside policy literalism prevents them from considering the figurative truths that shed light on the possibilities opened up by politics; Shellenberger and Nordhaus use this point to critique what they see as an extension of Enlightenment rationality into the environmental movement, where supporters boil their personal politics down to their perception of objective rationality, which frames the problems targeted by environmentalism and how the movement communicates with the public—preventing them from adequately tying problems of global warming to health, labor, culture, or the economy.

This position allowed Shellenberger and Nordhaus to break out of linear rationality to define things such as disaster preparedness as an environmental problem—a point driven home by the aftermath of Hurricane Katrina. Without realizing that potential solutions are immediately structured by the framing of the problem, environmentalism will fail to produce proposals necessary to deal with climate change. At this point, the authors heralded their organizational work with the Apollo Alliance as a guiding example, where new politics are created that wed environmental interests with those of organized labor in the pursuit of creating green jobs, delivering environmental politics into a "postenvironmental" future in which environmental interests are never dealt with as isolated terms.

The second part of "The Death of Environmentalism" focused on what Shellenberger and Nordhaus see as the new politics of possibilities. These politics are contingent on environmentalists, labor, business, and civil rights groups to coordinate goals and action, combating the scope of global warming with equally portioned solutions—a conviction derived from the belief that a movement's success relies more on its vision and values than on technical policy proposals. Shellenberger and Nordhaus use public opinion polling research to argue that vast new opportunities for coalitions exist, as large majorities of working-class citizens who identify with conservative politics under neoliberal regimes support the goals expressed in the Apollo Alliance of creating green jobs and easing the United States off foreign oil. In this, advocates of the death of environmentalism see massive green investment projects as representative of a third wave of environmentalism, following the first wave of conservationism and the second wave of regulation. Sweeping solutions proposed under a single banner enjoy better opportunities for political success, they argue, because they do not use the style of the "laundry list," where a politics of information has the practical effect of heightening the anxiety of individuals who spend the better part of each day attempting to reduce such stress.

In other words, Shellenberger and Nordhaus want environmentalists to stop addressing the public as though they were literate in the jargon of policy. In addition to proposing a more helpful politics that avoids doomsday narratives, supporters of third-wave environmentalism believe they have the strength to confront opposition on the issues of jobs, which has historically been the rhetoric used to divide environmental–labor coalitions. This takes away the ability of industry to frame environmental politics in terms favorable to their interests. By engaging labor in a coalition, environmentalists also gain the opportunity to receive new input, rather than dictate to others what they should be doing to reduce greenhouse gases.

Shellenberger and Nordhaus have been critiqued from the left because of their characterization of third-wave environmentalism as suggesting the need for public–private green projects. For example, political scientist Tim Luke sees transcending second-wave environmentalism as an opportunity to define the environment in terms of public ecology. Luke is in full agreement with Shellenberger and Nordhaus that the practice of mainstream environmentalism has had the practical outcome of turning the environment into one special interest among many others existing in the liberal political order. What he takes issue with is their advocacy for the environmental movement to learn something from right-wing tactics and organization as they pursue green investment projects comparable with building railroads, highways, and the Internet. Luke sees the call to third-wave environmentalism as naive to the corporate interests involved in such public–private investments and is skeptical of whether they will produce the beneficent results that Shellenberger and Nordhaus hope for. The death of environmentalism is then mistaken, he argues, at the point at which its supporters elect markets over the state in solving ecological problems posed by global climate change. Instead, Luke characterizes the proper response as embedded in a democratic public ecology, where networks of organizations and ideas can consider expert knowledge, private property, social inequality, and sustainability when advocating particular social and natural relationships, forming a mixture of state, market, and civil society in pursuit of transforming future ecology. Following the path of public–private green investments will only shift the environment from being a special interest to being an object subjected to the laws of neoclassical economics, unable to establish a political will stable enough to capture significant environmental action, and thus incapable of delivering the movement into a postenvironmental future.

See Also: Conservation Movement; Environmental Movement; Political Ecology.

Further Readings

Apollo Alliance. http://apolloalliance.org/ (Accessed February 2009).
Luke, Timothy W. "The Death of Environmentalism or the Advent of Public Ecology?" *Organization and Environment*, 18/4 (2005).
Shellenberger, Michael and Ted Nordhaus. "The Death of Environmentalism: Global Warming Politics in a Post-Environmental World." http://www.thebreakthrough.org/images/Death_of_Environmentalism.pdf (Accessed December 2008).

Jeremiah Bohr
University of Illinois at Urbana-Champaign

Decentralization

Decentralization is a challenging, complex process of dispersing decision making closer to the point of service or action (i.e., from central government institution to local government) and is the institutional framework in which political, social, and economic decisions are made and carried out. Decentralization means reassigning responsibility, authority, and resources that have a significant degree of autonomy and accountability so that there is system of co-responsibility between institutions of governance at the central, regional, and local levels.

The term attracted attention in the 1950s and 1960s, when British and French colonial administrations prepared colonies for independence by devolving responsibilities for certain programmers to local authorities. In the 1980s, decentralization came to the forefront of the development agenda alongside the renewed global emphasis on governance and human-centered approaches to human development. Today both developed and developing countries are pursuing decentralization policies, as they can address poverty, gender, inequality, environmental concerns, and the improvement of healthcare, education, and access to technology. Central governments are motivated to decentralize for a number of reasons, including meeting the increasing needs of citizens and alleviating the fiscal burdens imposed by expanding demand for infrastructure and services.

For decentralization to be effective, it needs to be driven by common principles of good governance. One of the central pillars of good governance is decentralization. The achievement of good governance (transparency, accountability, collective responsibility, rule of law, legitimacy, minority rights, and absence of corruption—considered pillars of democracy) is a learning process. Good governance and sustainable human development are indivisible. Good governance and decentralization have to be pursued vigorously if the poverty reduction targets and the Millennium Development Goals are to be achieved, as they are dependent on the effective delivery of services at the local level, and it is primarily at the local level that citizens can meaningfully hold their leaders accountable for fulfilling these goals.

Types of Decentralization

Decentralization is a complex and multifaceted concept, embracing a variety of concepts. Different types of decentralization show different characteristics, policy implications, and conditions for success. Three main elements of decentralization are fiscal, political, and administrative decentralization.

Dispersal of financial responsibility is a core component of decentralization. If local governments and private organizations are to carry out decentralized functions effectively, they must have an adequate level of revenues—either raised locally or transferred from the central government—as well as the authority to make decisions about expenditures. Fiscal decentralization can take many forms, including the following:

- Self-financing or cost recovery through user charges
- Cofinancing or coproduction arrangements through which the users participate in providing services and infrastructure through monetary or labor contributions
- Expansion of local revenues through property or sales taxes or through indirect charges
- Intergovernmental transfers that shift general revenues from taxes collected by the central government to local governments for general or specific uses
- Authorization of municipal borrowing and the mobilization of either national or local government resources through loan guarantees

In many developing countries, local governments or administrative units possess the legal authority to impose taxes, but the tax base is so weak and the dependence on central government subsidies so ingrained that no attempt is made to exercise that authority.

Political decentralization aims to give citizens or their elected representatives more power in public decision making, so that the decisions are better informed and more relevant to diverse interests in society than those made only by national political authorities. It is often associated with pluralistic politics and representative government, but it can also support democratization by giving citizens or their representatives more influence in the formulation and implementation of policies. Political decentralization often requires constitutional or statutory reforms, creation of local political units, and the encouragement of effective public interest groups.

Administrative decentralization seeks to redistribute authority, responsibility, and financial resources for providing public services among different levels of governance. It is the transfer of responsibility for the planning, financing, and management of public functions from the central government or regional governments and its agencies to local governments, semiautonomous public authorities or corporations, or area-wide, regional, or functional authorities. Administrative decentralization always underlies most cases of political decentralization. There are three major forms of administrative decentralization:

- Deconcentration is the weakest form of decentralization and is used most frequently in unitary states. It redistributes decision-making authority and financial and management responsibilities among different levels of the national government. It can merely shift responsibilities from central government officials in the capital city to those working in regions, provinces, or districts, or it can create strong field administration or local administrative capacity under the supervision of central government ministries.
- Delegation is a more extensive form of decentralization. Through delegation, central governments transfer responsibility for decision making and administration of public functions to semiautonomous organizations not wholly controlled by the central government, but ultimately accountable to it (e.g., public enterprises or corporations, housing authorities, transportation authorities, special service districts, semiautonomous school districts, regional development corporations, or special project implementation units). Usually these organizations have a great deal of discretion in decision making. They may be exempt from constraints on regular civil service personnel and may be able to charge users directly for services.

- Devolution is an administrative type of decentralization with clear and legally recognized geographical boundaries. When governments devolve functions, they transfer authority for decision making, finance, and management to quasi-autonomous units of local government with corporate status. Devolution usually transfers responsibilities for services to local governments that elect their own functionaries and councils, raise their own revenues, and have independent authority to make investment decisions.

Economic and Market Decentralization

Privatization and deregulation shift responsibility for functions from the public to the private sector. Privatization and deregulation are usually, but not always, accompanied by economic liberalization and market development policies. They allow functions that had been primarily or exclusively the responsibility of government to be carried out by businesses, community groups, cooperatives, private voluntary associations, and other nongovernment organizations.

- Privatization can range in scope from leaving the provision of goods and services entirely to the free operation of the market to public–private partnerships in which government and the private sector cooperate to provide services or infrastructure. Privatization can include allowing private enterprises to perform functions that had previously been monopolized by government; thus, privatization cannot in the real sense be considered equivalent to decentralization.
- Deregulation reduces the legal constraints on private participation in service provision or allows competition among private suppliers for services that in the past had been provided by the government or by regulated monopolies. In recent years, privatization and deregulation have become more attractive alternatives to governments in developing countries. Local governments are also privatizing by contracting out service provision or administration.

Decentralization is one of the preferred green ideological recipes for the reorganization of states and the creation of new sustainable societies. Decentralization, as a specific political proposal, however, has meant different things to different people, varying considerably in its significance according to the different political strategies defended by green thinkers and organizations in the pursuit of their ultimate political goals. Several environmentalists have appealed to the decentralization concept to support the abandonment of contemporary megacities in favor of smaller, theoretically more ecologically sustainable, communes or political communities. Still others have used decentralization to advocate the abolition of nation-states and their refoundation around independent federations of communes called bioregions:

Future Perspectives

The available evidence indicates that we have to ensure that the following conditions are met in order to make decentralization more effective:

- Emphasize local revenue generation and increased financial autonomy to assess taxes or fees and collect/use them responsively.
- Build local government momentum for ecological democracy to guard against bureaucratization/local authoritarianism.

- Use information and communication technologies for facilitating the understanding and support for decentralization and local governance.
- Provide training/capacity development with requisite skills, knowledge, and awareness for mobilizing and empowering communities.
- Link local issues with regional and national issues to take advantage of new developments/schemes.
- Promote transparency, accountability, and active participation and build private–public–civil society partnerships.
- Lobby for strong political will by allying/networking with civil society groups for ensuring "inclusive growth."

See Also: Acid Rain; Agenda 21; Anti-Toxics Movement; Appropriate Technology.

Further Readings

Birkeland, Janis. *Design for Sustainability: A Sourcebook of Integrated, Eco-Logical Solutions*. London: Earthscan, 2002.

Holland, Joe. "The Regeneration of Ecological, Societal, and Spiritual Life: The Holistic Postmodern Mission of Humanity in the Newly Emerging Planetary Civilization." *Journal of Religion & Spirituality in Social Work: Social Thought*, 24/1/2 (2005).

Menon, Sudha Venu. "Grassroot Democracy and Empowerment of People: Evaluation of Panchayati Raj in India, MPRA Paper No. 3839" ICFAI Business School, Ahmadabad. http://mpra.ub.uni-muenchen.de/3839/ (Accessed January 2009).

Sharma, Shalendra D. "Democracy, Good Governance, and Economic Development." *Taiwan Journal of Democracy*, 3/1:29–62 (2007).

United Nations Development Programme (UNDP). "Decentralized Governance: A Synthesis of Nine Case Studies." New York: UNDP, 2000.

Gopalsamy Poyyamoli
Pondicherry University

Deep Ecology

Deep ecology is a radical environmental philosophy and political movement founded upon the holistic belief that all living things have an equal right to life, or have subjective or objective intrinsic values. It has two defining philosophical ideals. First, self-realization, which emphasizes a broadening and deepening of the self toward a sense of personal identity that allows each being's potential to be realized. Second, biological egalitarianism, the principle that humans have no more right to live than any other organism—all living things are equally valuable and deserve the same consideration. In accordance with these principles, deep ecology is considered a biocentric (living-centered), rather than an anthropocentric (human-centered), worldview.

The term *deep ecology* was coined by Norwegian professor of philosophy and accomplished mountaineer Arne Naess in his 1973 article "The Shallow and the Deep, Long-Range Ecology Movements: A Summary." Naess' earlier scholarly pursuits had focused on semantics, unraveling the logic of language and investigating the theoretical reasoning behind positivist science. His later writings on environmental philosophy were influenced heavily by the work of Baruch Spinoza, Aldo Leopold, Rachel Carson, and Mahatma Gandhi, as well as Taoist, Buddhist, and Native American belief systems, particularly their emphasis on organic unity and their rejection of reductive binaries.

Naess articulated an "ecosophy" with two central pillars. The first was a sense of self that went beyond the individual to encompass the entire living world. Naess proposed a conceptual widening of the Self to include the Other, which would allow humans to appreciate the nonhuman world as a part of themselves. This philosophical acknowledgement forces us to recognize that to harm another species is to do harm unto ourselves. The second pillar was the belief that all organisms and entities are equal in intrinsic worth and part of an interrelated whole. Naess rejected the Enlightenment notion that living things can be ranked according to their relative value. He believed that the right to live is a universal right shared by all living things. No one species has more of a right to live or exist than any other.

Naess later expanded these foundational lessons into the eight points of the deep ecology platform:

1. The well-being and flourishing of human and nonhuman life on Earth have value in themselves. These values are independent of the usefulness of the nonhuman world for human purposes.
2. Richness and diversity of life forms contribute to the realization of these values and are also values in themselves.
3. Humans have no right to reduce this richness and diversity except to satisfy vital needs.
4. The flourishing of human life and cultures is compatible with a substantial decrease of the human population. The flourishing of nonhuman life requires such a decrease.
5. Present human interference with the nonhuman world is excessive, and the situation is rapidly worsening.
6. Policies must therefore be changed. These policies affect basic economic, technological, and ideological structures. The resulting state of affairs will be deeply different from the present.
7. The ideological change is mainly that of appreciating life qualities (dwelling in situations of inherent value) rather than adhering to an increasingly higher standard of living. There will be a profound awareness of the difference between big and great.
8. Those who subscribe to the foregoing points have an obligation directly or indirectly to try to implement the necessary changes.

Naess described his approach as "deep" because it was concerned with fundamental questions of humanity's role as part of the biosphere and demanded changes to all facets of human life. He dismissed other environmental worldviews—conservationists, ecosocialists, and environmental justice advocates—as "shallow" for adhering to a dualistic worldview in which the concerns and priorities of humans and nonhumans are considered separately. Deep ecologists Bill Devall and George Sessions contrasted the shallow and deep worldviews according to the following criteria:

Dominant Worldview	*Deep Ecology*
Dominance over nature	Harmony with nature
Natural environment as resource for humans	All nature has intrinsic worth
Economic growth for human populations	Simple material needs
Belief in ample resource reserves	Earth "supplies" are limited
High-tech progress and solutions	Appropriate technology
Consumerism	Doing with enough
Nation-states as governing units	Bioregions as governing units

Within deep ecology, the shallow perspectives contained in the left column are dismissed for prioritizing human management, for their commitment to exploitative practices, and for their materialist and consumer-oriented outlook.

Deep ecology is a radical philosophy in that it suggests that nothing less than a complete transformation of current social and political systems can redress our environmental crises. Deep ecology opposes industrial society and its institutions. It is antihierarchical, favoring leaderless networks with common interests. Decentralization and local autonomy are necessary. Supporters favor organizational units based on ecological rather than political commonalities. One suggestion for reform is to replace nation-states as our basic economic and political units with bioregions—areas with consistent ecological characteristics. This has fueled the secession movement of Cascadia, an amalgam of the Pacific Northwest bioregion, extending from northern California through Oregon and Washington State and extending into western British Columbia and the Yukon. Deep ecologists are committed to creating an ecotopian society in which governing norms and principles are derived from the democratic ideals of the biosphere.

Deep ecology has also influenced the emergence of the environmental direct-action movement, from very radical organizations such as the British Animal Liberation Front (dedicated to liberating animals through violence) to broad-based, more mainstream organizations such as Greenpeace. Many of the founders of Earth First!—a prominent environmental advocacy group that emerged in the southwestern United States in 1979—were followers of deep ecology and sought to undermine the advancement of industrial development by sabotaging projects they deemed destructive to the environment. They engaged in eco-tage, illegal acts of vandalism in defense of ecological goals, targeting those industries deemed particularly harmful and exploitative, especially logging and large-scale dams. It is important to note that Naess never condoned these actions, though he himself did participate in direct action when he tried to block a proposed new dam on a Norwegian fjord. Naess advocated a more nonviolent approach to direct action.

Another important element of deep ecology is its emphasis on emotion and spirituality. Deep ecologists lament the emotional and visceral experience of ecological interconnectedness that our highly anthropocentric culture has socialized out of us. This ecological withdrawal leaves us feeling anger, rage, sorrow, loss, and powerlessness. Deep ecologists have established workshops that provide a forum to recapture these emotional responses. The focus is on bringing emotion back into environmental politics to create a just future for all species. Deep ecologists tend to privilege emotion and experience over rationality and reason.

Deep ecology has been subject to considerable critique. First, critics suggest that deep ecology minimizes the social relations that determine environmental outcomes. By failing to recognize the social roots of our ecological crisis, deep ecology ignores important power categories such as authoritarianism, inequality, gender, and race, which profoundly shape humanity's relationship with the nonhuman world. This critique has found particular resonance among ecofeminists, who have been largely inspired by deep ecology's commitment to holism and its privileging of emotion over reason, but argue that deep ecology fails to sufficiently appreciate the role of patriarchy in oppressing both nature and women.

Second, scholars and activists in the global South have labeled deep ecology a form of elite environmentalism. They contend that elevating the intrinsic value of nonhuman species fails to resonate with those who struggle daily to survive. Third is the accusation that deep ecology's ecocentrism reveals shades of antihumanism. Critics argue that deep ecology adheres to an antihuman ideology in which the needs of the individual and the species are subordinate to the needs of the biosphere. It posits civilization as the enemy of nature, implying that forces detrimental to human populations (war and disease, for instance) might be to the benefit of the planet as a whole. Finally, some suggest that deep ecology is politically naive. It prescribes radical change without a realistic appreciation of the monumental difficulties involved in reorienting human existence toward a vision in which we live as equals among other species. These critics argue that as long as deep ecology refuses to engage with mainstream political processes and debates, the movement will continue to be relegated to fringe status.

See Also: Biosphere; Ecofeminism; Environmental Movement; Intrinsic Value.

Further Readings

Devall, Bill and George Sessions. *Deep Ecology: Living as If Nature Mattered*. Salt Lake City, UT: Peregrine Smith Books, 1985.

Harding, Stephen. "What Is Deep Ecology?" http://www.schumachercollege.org.uk/learning-resources/what-is-deep-ecology (Accessed December 2008).

Naess, Arne. *Ecology, Community and Lifestyle: Outline of an Ecosophy*, trans. D. Rothenberg. Cambridge, UK: Cambridge University Press, 1989.

Naess, Arne. "The Shallow and the Deep, Long-Range Ecology Movement: A Summary" (1979). In *The Deep Ecology Movement: An Introductory Anthology*, edited by Alan Drengson and Yuichi Inoue. Berkeley, CA: North Atlantic Books, 1995.

Matthew Schnurr
Dalhousie University

DEFORESTATION

Deforestation is defined as the complete removal or logging of trees in forested areas. It constitutes a serious threat to human civilization and forest ecosystems because tropical forests especially maintain the structure and function of the Earth system and deliver services from biogeochemical cycling to biotic diversity.

Deforestation contributes to global warming and is one of the major causes of the greenhouse effect. According to the Intergovernmental Panel on Climate Change, deforestation, mainly in tropical areas, accounts for up to 20 percent of the total anthropogenic emissions of carbon, a greenhouse gas that is released during burning and decay. The water cycle is also affected by deforestation. Trees extract groundwater through their roots and release it into the atmosphere; when removed, trees no longer evaporate this water. Deforestation therefore reduces the content of water in the soil, as well as in the atmosphere. Furthermore, soil cohesion is reduced, so that erosion, flooding, and landslides may occur. Deforestation has also, if not most importantly, resulted in reduced biodiversity.

Forests disappear naturally as a result of climate change, fire, hurricanes, or other disturbances; however, most deforestation has been anthropogenic. Although improperly applied logging, fuel wood collection, fire management, and grazing can lead to unintentional deforestation, most anthropogenic deforestation is deliberate. Many developing economies are converting forests and other natural habitat mainly into agriculture. Shifting cultivation is practiced over large areas of land worldwide, raising concern about the efficacy of its practice. However, the rates of deforestation caused by cultivation practices are disputed: Norman Myers attributes 54 percent to shifting cultivation; meanwhile, Edward Barbier finds that intensification of agriculture in shifting cultivation areas makes up 10 percent of tropical deforestation, and expansion of shifting cultivation into undisturbed forests only 5 percent. Gunther Fischer and Gerhard Heilig estimated that cultivated land will increase over 47 percent by 2050, with about 66 percent of it coming from deforestation and wetland conversion. Tropical forest conversion into pasture is another threat for forests in the tropics.

The photograph shows a swath of deforested land in the Amazonian rainforest. Such losses contributed to South America having the second-highest estimated reduction in tropical forests in the 1990s, at 0.4 percent.

Source: iStockphoto.com

Logging, infrastructure, and settlement expansion, and to a minor degree mining, are other important direct drivers. Logging may be a direct source of deforestation or an indirect source resulting from logging roads enabling access for farmers into previously unreachable areas of forest to establish agricultural plots and pasture.

As for whether poverty is an important driver, there is no consensus. One argument is that poor people are more likely to clear forest because of the lack of other economic alternatives; the counterargument states that the poor lack financial ability to clear the forest. The claim that population growth drives deforestation is another disputed topic. Helmut Geist and Eric Lambin showed that population increase caused by high fertility rates is a primary driver of deforestation in only 8 percent of cases. The Food and Agriculture Organization states that global deforestation rates are not directly related to the human population growth rate but, rather, are an indicator of the lack of technological advancement and inefficient governance. Corruption, inequitable distribution of wealth and power, and globalization are also drivers of deforestation.

Obtaining precise figures for deforestation rates has proven difficult. Food and Agriculture Organization data are based largely on numbers provided by forestry departments of individual countries and can therefore be biased. Frédéric Achard and colleagues estimated deforestation based on satellite imagery that exhibits rates of deforestation in the tropics 23 percent lower than the most commonly quoted rates, and for the tropics as a whole, deforestation rates could be in error by as much as ±50 percent.

James Benhin states that despite uncertainty regarding precise deforestation numbers, it is undisputable that the rate of forest loss in the tropics has been on the increase over the last five decades. The Food and Agriculture Organization estimated that about 450 million hectares of tropical forest were lost between 1960 and 1990 and indicated an annual tropical forest loss of around 13 million hectares during the 1990s. That is a mean annual deforestation rate in that period of more than 2 percent. Africa had the highest estimated annual loss of about 0.8 percent, followed by South America at 0.4 percent. However, forest plantations, landscape restoration, and natural expansion of forests have reduced the net loss of forested areas. The net change for 2000–2005 has been estimated at minus 7.3 million hectares annually, and in the 1990s, the net change has been estimated at minus 8.9 million hectares annually.

Including temperate forest, the rate of deforestation globally started to decline in the 1980s, with even more rapid declines from the 1990s to 2005. Based on these trends, reforestation may outstrip deforestation within the next half century and give way to a forest transition. Despite the ongoing reduction in deforestation, the process of deforestation remains a serious global ecological concern and a major social and economic problem in many regions. The decline in the rate of deforestation also does not address the damage already caused by deforestation.

See Also: Afforestation; Agriculture; Biodiversity; Global Climate Change.

Further Readings

Achard, Frédéric, et al. "Determination of Deforestation Rates of the World's Humid Tropical Forests." *Science*, 297/5583 (2002).

Barbier, Edward B. "Explaining Agricultural Land Expansion and Deforestation in Developing Countries." *American Journal of Agricultural Economics*, 86/5 (2004).

Benhin, James K. A. "Agriculture and Deforestation in the Tropics: A Critical Theoretical and Empirical Review." *Ambio*, 35/1 (2006).

Fischer, Gunther and Gerhard K. Heilig. "Population Momentum and the Demand on Land and Water Resources." *Philosophical Transactions of the Royal Society Series B: Biological Sciences*, 352/1356 (1997).

Food and Agriculture Organization of the United Nations (FAO). "Global Forest Resources Assessment 2005." http://www.fao.org/sd/WPdirect/WPan0050.htm (Accessed January 2009).

Geist, Helmut J. and Eric F. Lambin. "Proximate Causes and Underlying Driving Forces of Tropical Deforestation." *BioScience*, 52/2 (2000).

Intergovernmental Panel on Climate Change (IPCC). "Summary for Policymakers." In *Climate Change 2007: The Physical Science Basis. Contribution of Working Group I to*

the Fourth Assessment Report of the Intergovernmental Panel on Climate Change, edited by Susan Solomon et al. Cambridge: Cambridge University Press, 2007. http://www.ipcc.ch/pdf/assessment-report/ar4/wg1/ar4-wg1-spm.pdf (Accessed January 2009).
Myers, Norman. *Conversion of Tropical Moist Forests*. Washington, D.C.: National Academy of Sciences, 1980.

Birgit Schmook
El Colegio de la Frontera Sur (ECOSUR)

Democratic Party

The Democratic Party has been in existence since the late 18th century and has been one of the two main political parties in the United States in the post–Civil War period. As of the inauguration of Barack Obama in 2009, the Democratic Party has had 16 presidents come from its ranks. The Democrats hold a majority of seats in both the Senate and the House of Representatives and control a majority of governorships in states across the country. Democrats have traditionally taken an active protectionist stance toward the environment and environmental issues, favoring governmental policies and programs as a means of ensuring environmental quality.

In 1792, the Democratic-Republican Party was founded by Thomas Jefferson, James Madison, and other opponents of the Federalist Party. The modern Democratic Party generally traces its roots to the election of Andrew Jackson in 1828 and has since occupied a relatively central place as one of two major political parties in the United States, opposite the Republican Party for most of that time period. The Democratic Party has generally positioned itself as more liberal on social and economic issues than the Republican Party, most notably evidenced in the New Deal of Franklin D. Roosevelt and the Great Society of Lyndon B. Johnson. The Democrats were divided until the 1960s, though, largely along regional lines, with Southern Democrats generally being more conservative compared with Democrats from the northeast and the rest of the country. Many of the Democrats from southern states left the Democratic Party following passage of the Civil Rights Act in 1964, culminating with the emergence of the so-called Reagan Democrats in 1980. Since this time, the Democratic Party has increasingly courted and received high levels of support from African Americans, Hispanics, the young, the working class, and highly educated professionals, especially those in academia. In the 21st century, Democrats have increasingly garnered support at the local, state, and national levels, culminating in the November 2008 election of Barack Obama, the first African American president in the history of the United States.

The Democratic Party has generally been seen as a stronger supporter of the environment than the Republican Party, even though many pieces of major environmental legislation have been signed into law by Republican presidents. However, Democrats have championed significant policies and legislation designed to protect the environment, particularly when issues of social justice were involved. The most notable of these efforts was the issuance of Executive Order 12898 by President Bill Clinton in 1994. This executive order called on federal agencies to work toward the achievement of environmental justice whenever possible, particularly in situations that would affect persons of color or low-income populations.

The current agenda of the Democratic Party focuses prominent attention on issues of environmental degradation, particularly in regard to the issue of climate change. Following the lead of former Democratic Vice President Al Gore, a champion in the fight against global warming, the Democratic Party promotes strong initiatives to reverse the rising level of greenhouse gas emissions into the atmosphere. This is in contrast to the 1990s, when members of the Democratic Party in the U.S. Senate joined Republicans to unanimously reject the Kyoto Protocol. Vice President Gore symbolically signed the protocol a year after this vote, but the United States remains one of a few countries worldwide that has not ratified the Kyoto Protocol. As noted, Democrats today are generally strong supporters of fighting climate change, typically linking global warming issues to issues of energy supply and to the development of alternatives to oil and coal. However, this is not a unanimously supported platform within the party, as many Democrats support oil and coal extraction for a variety of reasons. Regardless, the majority agenda of the Democratic Party calls for strong action to counter global warming now and to develop alternative energies that will be less harmful to the climate in the future.

Democrats generally favor strong governmental action to solve problems, including in the arena of environmental protection. There is a preference for the development of strong laws designed to protect environmental quality and natural resources. Democrats also contend that regulatory agencies must be provided with sufficient tools to engage in enforcement activities, or else the laws that are passed are largely rendered meaningless. The Democratic Party believes that the notion of a fundamental dichotomy between economic growth and environmental protection is a false choice and that both can be promoted and ensured through wise action. The policies promoted by candidate Barack Obama during the 2008 presidential election campaign were largely designed around this premise of simultaneously protecting environmental quality while promoting economic growth, most notably through investment in research and development of clean technologies, tax credits for the greening of production processes, and through policies designed to promote the stimulation of environmentally friendly marketplace activity.

The Democratic Party has long championed environmental issues as critical components of its platform, though these issues have become more central in recent decades. Democrats today face a myriad of environmental challenges, most notably the issue of climate change. With the Democratic Party having control of both houses of Congress and the presidency, along with broad public support for their policies regarding environmental protection, it remains to be seen whether significant progress can be made to address these issues.

See Also: Environmental Justice; Equity; Governmentality; Kyoto Protocol; North American Free Trade Agreement.

Further Readings

Democratic Party. "Environment." http://www.democrats.org/a/national/clean_environment/ (Accessed January 2009).

Obama, Barack. "Barack Obama and Joe Biden: Promoting a Healthy Environment." http://www.barackobama.com/pdf/issues/EnvironmentFactSheet.pdf (Accessed February 2009).

Wagner, Heather Lehr. *The History of the Democratic Party (The U.S. Government: How It Works)*. New York: Chelsea House Publications, 2007.

Todd L. Matthews
University of West Georgia

DEPARTMENT OF ENERGY

The U.S. Department of Energy (DOE) is a cabinet-level department charged with national security, research and development, and economic and environmental missions related to energy and energy technologies. Historically the department's missions and budget have emphasized nuclear energy, especially nuclear weapons production and related activities. The department's broader mandate encompasses activities across a wide range of areas including fossil fuels, renewable and alternative energy sources, energy infrastructure, environmental management, and fundamental and applied research. With a budget exceeding $24 billion and more than 100,000 federal and contractor employees, the DOE is a large and complex organization affecting many aspects of environmental and energy politics.

The DOE is an institutional descendant of the Manhattan Project, which developed the first atomic weapons during World War II, and the Atomic Energy Commission, established under the Atomic Energy Act of 1946 to direct activities related to military and civilian uses of nuclear energy. The Atomic Energy Commission managed dual and often conflicting missions, simultaneously promoting and regulating nuclear energy. The Energy Reorganization Act of 1974 sought to separate those missions and to address a changing energy context by abolishing the Atomic Energy Commission and assigning its responsibilities to new agencies. The development of nuclear and non-nuclear energy sources was assigned to the Energy Research and Development Administration, whereas the regulation of commercial nuclear power was assigned to the Nuclear Regulatory Commission. The DOE replaced the Energy Research and Development Administration in 1977, inheriting the development and promotion mission. Following a series of security-related incidents, the National Nuclear Security Administration was established in 2000 as an agency within the DOE housing its nuclear weapons, naval reactors, and nonproliferation programs.

This researcher was working on a device at the Sandia National Laboratory that might someday facilitate reliable production of energy from seawater. The laboratory is part of the U.S. Department of Energy's National Nuclear Security Administration.

Source: U.S. Department of Energy, Sandia National Laboratories

Although the DOE's energy mandate is broad in scope, approximately 60 percent of its budget supports nuclear weapons research, development, production, and maintenance, as well as programs for managing the environmental consequences and waste materials resulting from those activities. From 1942 through the end of the Cold War, the DOE and its predecessor agencies built and operated an extensive nuclear weapons production complex comprising industrial and research facilities throughout the United States. The complex produced more than 70,000 nuclear weapons over that time, along with large quantities of nuclear and chemical waste in solid, liquid, and gaseous forms. Under pressure to maximize production, decision makers compromised in the areas of environmental protection, waste storage and disposal, worker health and safety, and public health and safety. Because of the culture of secrecy surrounding national security programs, those compromises and their consequences were not widely known until after the end of the Cold War. During the 1990s, public activism, investigative journalism, releases of information by insider "whistle-blowers," and congressional inquiries brought a host of problems to light. Those problems included accidental and deliberate releases of radioactive materials into the environment, failures to adequately safeguard workers and neighboring communities, and medical experiments involving radioactive materials performed without subjects' informed consent.

Beginning in 1993, the DOE's Openness Initiative sought to acknowledge those errors and restore public faith in the department. A system of Site-Specific Advisory Boards was created to facilitate dialogue with stakeholders, and the department's Environmental Management Program (which had been created in 1989) undertook a massive effort to address environmental issues across the complex. Some of the weapons production sites have now been closed, and in some cases remediation activities have been declared complete. At the larger sites, remediation activities may need to continue for decades. Maintaining an adequate environmental remediation budget, managing the cleanup effort effectively, and fostering democratic public involvement in cleanup decisions have been persistent challenges for the program.

Permanent disposal of radioactive waste materials from weapons production and commercial reactor operation is another major challenge for the DOE. The Nuclear Waste Policy Act of 1982 mandated the creation of a national waste repository, but that effort has been plagued by technical and political controversies. The Yucca Mountain Repository site in Nevada received congressional and presidential approval in 2002, and the DOE submitted a site license application to the Nuclear Regulatory Commission in 2008. Should it become operational, the Yucca Mountain site would house irradiated reactor fuel (also known as spent nuclear fuel) and high-level waste from civilian and military nuclear fuel reprocessing stored in glass form. The Waste Isolation Pilot Plant near Carlsbad, New Mexico, which became operational in 1999, is designed specifically to house transuranic waste from military programs.

The DOE's system of National Laboratories and Technology Centers includes the three nuclear weapons laboratories, Los Alamos National Laboratory and Sandia National Laboratories in New Mexico and Lawrence Livermore National Laboratory in California, as well as other research and development facilities throughout the United States. Along with their weapons research, development, and design activities, the weapons laboratories support the National Nuclear Security Administration's Stockpile Stewardship Program, intended to maintain the effectiveness of the U.S. nuclear arsenal. The weapons laboratories also conduct research across a broad range of nonweapons areas. Together, the various laboratories and technology centers maintain a set of programs encompassing civilian nuclear energy, nuclear fusion and plasma physics, high-energy physics, advanced computing, genomics, environmental science, nanoscience, fossil fuels, biofuels, wind energy, solar energy, hydropower, electricity storage and delivery, electric grid management, and energy efficiency.

The DOE's Office of Nuclear Energy conducts research and development activities related to new reactor designs, deployment of new nuclear power plants, medical applications, and power generation for outer space applications. A system of domestic and international government-industry partnerships includes the Global Nuclear Energy Partnership, initiated in 2006, and a collaboration agreement reached with India in 2008. Since the passage of the Energy Policy Act of 2005, the DOE has sought to expand the number of operational domestic reactors by providing loan guarantees and risk indemnification for new construction projects. Together with the new international partnerships, these initiatives continue the tradition of promoting nuclear power inherited from the DOE's institutional predecessors.

A number of independent and quasi-independent agencies reside within the DOE. The Federal Energy Regulatory Commission is an independent agency within DOE that regulates interstate transmission of natural gas, oil, and electricity, as well as natural gas and hydropower projects. The Energy Information Administration compiles energy-related statistical data and is charged with preparing policy-independent analyses and reports. Four Federal Power Marketing Administrations sell hydropower produced by federal water projects. The DOE also manages the Strategic Petroleum Reserve, the Northeast Home Heating Oil Reserve, and the Naval Petroleum and Oil Shale Reserves in Wyoming. Together with the U.S. Environmental Protection Agency, the DOE also operates the Energy Star program, promoting commercial and residential energy efficiency.

See Also: Environmental Management; Nuclear Politics; Risk Assessment.

Further Readings

Taylor, B. C., et al., eds. *Nuclear Legacies: Communication, Controversy, and the U.S. Nuclear Weapons Complex*. Lanham, MD: Lexington, 2007, 2008.

U.S. Department of Energy. http://www.energy.gov/ (Accessed February 2009).

Williams, Walter L. *Determining Our Environments: The Role of Department of Energy Citizen Advisory Boards*. Santa Barbara, CA: Praeger, 2002.

William J. Kinsella
North Carolina State University

Domination of Nature

The phrase *domination of nature* implies a neo-Marxist interpretation of the adverse effects that modern society has had on the natural environment. In this interpretation, ecological destruction by society must be understood as being deeply rooted in contemporary social attitudes toward the natural environment. These attitudes see the natural environment as something that must be controlled or dominated by humans through science and technological advancements. Although commonly associated with 17th-century scientific thought, the phrase gained popularity in the mid-1960s, when widespread concern for the environmental impact of industrial society took off in the developed world. In the late 20th and early 21st centuries, concerns about the effect of industrial society on the environment, and its link to attitudes of mastery over nature predominant in the fields of science and technology, have been raised anew as large-scale threats such as genetic modification, global climate change, and ozone depletion have become clearly visible.

The popularization of this attitude toward the environment is often associated with the one-time influential scientific philosopher and Lord Chancellor of England, Sir Francis Bacon (1561–1626), whose works popularized an inductive methodology for inquiry, which indeed is considered a forerunner of the modern scientific method. Bacon held that we must learn the secrets of nature and gain control over "her," and consequently social life would improve because of the knowledge gained. In other words, the driving force behind science, and progress more broadly defined, has been technological control of the natural environment and ourselves, and this was necessary to improve societies and their surroundings. Indeed, the domination of nature was deemed necessary for progress. Bacon believed that "the real business and fortunes of the human race" was the conquering of nature. Nature must be "forced out of her natural state and squeezed and molded" by "the hand of man." Humans must free themselves from nature, which must be "bound into service" and made a "slave." This philosophy went hand-in-hand with mercantilist exploration and the conquest of the Americas and Africa—conquests through which, as Bacon himself reflected, "we have seen what floods of treasure have flowed into Europe by that action [of imperialism], and marveled at how infinite is the access to territory and empire by the same enterprise." Conquests over the mundane world and its uncivilized occupants were viewed as dignified and necessary projects that were justified, in that they would improve nature for the benefits of mankind.

Modern science and technology are seen as encapsulating this desire to overcome the limits set by the natural world. Indeed, although originating with the Enlightenment, the vision of humans dominating nature and the pursuit of material wealth through scientific and technological progress continues to this day. However, critics point out that the domination of nature clearly benefits certain factions of society and supports certain ideologies over others. For example, C. S. Lewis noted that the pursuit of the domination of nature is something that benefits scientists and their sponsors but that may not always benefit the majority of society as the dogma postulates. The domination of nature, in other words, often includes the uneven access to resources, knowledge, and power. The close financial relationship between science and technology and the military, industrial agriculture, and economic development in the industrialized world suggest that technological control is designed to pursue a particular version of progress at the behest of powerful actors in government and the corporate world.

The domination of nature—a view of virulent optimism toward science and technology—can easily be contrasted with the much more cautious (and older) view expressed in ancient Greek myth, that obtaining technological skills (e.g., ability to control fire, to fly) might lead to social destruction just as easily as to social improvements. The Faust legend popularized by Goethe's tragic play of the same name likewise depicts the fears that some held during the early 19th century of tampering with, and the desire to dominate, nature through science. World War I perhaps most clearly marked the modern turn away from unfettered optimism of scientific and technological progress in Europe, whereas caution in the United States occurred closer to World War II, after the atomic bombing of Hiroshima and Nagasaki, Japan. Concerns with genetic modification, global climate change, and ozone-layer depletion are among the new set of global issues that have raised concerns about how power over nature has led to uneven levels of power over people and their surrounding environment and the negative effect this can have on global society.

The issue of morality and responsibility toward nature and society is often raised by scholars when considering the domination of nature. Large-scale ecological degradation, combined with the pursuit of science and technology devoid of a standard of morality, could paradoxically lead—as Lewis and others postulated in the mid-20th century—to the domination of humanity by nature. Science, in its desire to dominate nature in the name of progress, has become subjected to the power of a few people, and it might also become subjected to the same amoral ecological processes that could lead to the demise of the human race, as the Greeks feared millennia ago.

See Also: Ecological Imperialism; Industrial Revolution; Montreal Protocol.

Further Readings

Foster, John Bellamy. *The Vulnerable Planet.* New York: Monthly Review, 1999.
Leiss, William. *The Domination of Nature.* Montreal & Kingston: McGill-Queen's University Press, 1994.
Lewis, C. S. *The Abolition of Man.* New York: HarperCollins, 2001.
Merchant, Carolyn. *The Death of Nature: Women, Ecology, and the Scientific Revolution.* New York: HarperCollins, 1980.

Brian J. Gareau
Boston College

ECOCAPITALISM

The ecocapitalist perspective suggests that sustainability and environmental conservation are entirely compatible within capitalist social relationships. Initial use of the term *ecocapitalism* has been attributed to green political parties. The term has often been employed to show contrast with ecosocialism. The comparison would likely be soundly rejected by ecosocialists, whose central premise is that capitalism is a root cause of much ecological degradation. For some, ecocapitalism is inherently a contradiction in terms. For critics of free market capitalism to accept the unity of the prefix *eco-* with capitalism, the ecological content would need to fundamentally transform capitalism as it is widely practiced today. However, for free market enthusiasts, capitalism would remain largely unchanged in ecocapitalism. The prefix in this case would represent new opportunities to seek profit from ecologically friendly goods and services—a notion that has also been labeled "free market environmentalism" and shares an ideological foundation with green neoliberalism.

A range of ecocapitalist perspectives emerged by the late 1980s: Some radical versions argued for ecological taxes or ecotaxes to replace all other taxes, whereas more conventional positions discussed tradable permits or economic liabilities for polluters. Ecocapitalists advocated improved policy instruments to resolve environmental problems, particularly in situations where public goods, such as the global commons (e.g., oceans, atmosphere), are difficult to protect. Different strains of ecocapitalism were fundamentally the same in that they each called for more accurate valuation of natural capital, which was identified as the base that other wealth and capital built on. Ecocapitalists thus sought to determine a quantifiable value for ecological assets compatible with neoclassical economics. By assigning a value to natural goods and services, a price premium could be calculated accurately for each potential use. Ecocapitalists reject state subsidies that encourage production systems that destroy natural capital; for example, by causing a reduction in the ability of the Earth to perform essential environmental services, such as nutrient cycling or waste filtration.

Ecocapitalists also promote the use of eco-friendly business. This version of ecocapitalism has gained popularity over the past decade. Based in a fundamentally different ideology than earlier applications of the concept, ecocapitalism has come to be used in a nonspecific manner to refer to any merger between environmental consciousness and entrepreneurialism. Ecocapitalism has since been construed to mean many different things. When production focuses on the lowest-cost ways to achieve environmental reductions, large-scale producers may drive out small, local producers, as has sometimes occurred with industrial-scale organic agriculture. The overall sustainability of the production system can be compromised as broader and more holistic—and often more expensive—social and ecological goals are undermined.

Often employed as an example of green business, wetland and biodiversity banking allow for offsets that mitigate negative environmental effects so that development projects can move forward. To do so they must create improvements in another similar and proximate wetland or by protecting a comparable habitat. Offset opponents argue that such policies follow a "polluter pays" principle that permits ecological degradation if the developer can pay the price.

The Kyoto Protocol is often used as a positive example of ecocapitalism, as the sale of carbon credits is supposed to contribute to the reduction of global greenhouse gas emissions. More than 250,500,000 Certified Emission Reductions were offset under Kyoto's Clean Development Mechanism by January 2009. Advocates contend that carbon trade is internalizing costs that were previously external while at the same time incentivizing shifts to better environmental practices. In contrast, critics contend that the market is ill equipped to deal with pollution and may create perverse economic incentives, such with a loophole that allowed companies to earn carbon credits to destroy HFC-23. More HFC-23-emitting plants were then established, even though this path was more expensive overall than not producing the HFC-23 in the first place. A second criticism waged at the Kyoto Protocol is that an agreed-on goal was sustainable development in host countries, but many projects do little more than target emission reductions. They are restricted in what they can achieve because of cost constraints generated by competition to produce low-cost carbon credits for highly competitive global markets.

One of the strongest advocates of the use of the ecocapitalism concept, the company TerraCycle, defines the process essentially as one that internalizes costs that were formerly seen as external (e.g., the waste stream). The company makes profit from what had previously been defined as waste, whether by merely repackaging it (e.g., by selling used wine barrels) or by processing it, as with food waste that worms transform into fertilizer. In addition to reducing the waste stream, people are encouraged to join brigades that have raised more than $10 million for social and education programs in the company's first six years of existence. When brigades send waste to TerraCycle, $.02–0.5 is contributed to a school or charity of the donor's selection.

The chief executive officer of TerraCycle suggests on his blog, titled "The Eco-Capitalist," that the company has created "sponsored waste" out of products not accepted for recycling, such as drink pouches, snack wrappers, or even crushed computers, which the company makes into garbage cans. It also makes handbags from PowerBar wrappers or Capri Sun drink pouches. In some instances, the source product is obscured, but TerraCycle has also created a line of "branded waste," whereby companies like Nabisco promote products like Oreos using handbags made of reused cookie wrappers. If the company's marketing strategy with branded waste is successful, it may have the end result of promoting additional consumption and thus contradict the initial objective of reducing waste.

Although there are companies making major strides to protect the environment, there are examples of corporate green-washing, or using disinformation to suggest an unjustified environmentally friendly public image. Some companies profess to make a more positive environmental impact than exists. This spurs criticism, particularly when vast amounts of money are spent advertising commitment to the environment rather than acting to protect it. Advertising may also be selective or partial, such that positive environmental impacts are highlighted while negatives are ignored. For example, companies that produce genetically modified seeds may stress their use of no-till agriculture, which can reduce erosion and increase the carbon-storage capacity of soils. What they may not mention is the potential for genetic drift, biodiversity loss, or the subsequent need for increased herbicide application in some cases.

Ecocapitalism reflects a contested political discourse in which different visions are at odds. A segment of environmentalists strongly opposes market-based environmental solutions. They question whether schemes such as emissions trading, mitigation banks, and ecosystem markets are about conservation at all or if, in reality, they are a green guise that ensures the ongoing commodification and overexploitation of nature. Taken to the opposite extreme, ecocapitalism is exemplified through probusiness slogans promoted on Libertarian lawyer Rex Curry's website, such as "the color of a healthy environment and the color of money are the same," or "capitalists: the true greens."

See Also: Capitalism; Commodification; Ecological Economics; Ecosocialism; Green Neoliberalism; Green-Washing.

Further Readings

Curry, Rex. "Free Market Environmentalism, Ecocapitalism, Libertarian Environmentalism." http://rexcurry.net/ecoart.html (Accessed January 2009).

Nicols, Dick. *Environment, Capitalism and Socialism*. Sydney, Australia: Resistance Books, 1999.

Sarkar, Saral. *Eco-Socialism or Eco-Capitalism? A Critical Analysis of Humanity's Fundamental Choices*. New York: Zed Books, 1999.

Szaky, Tom. "The Eco-Capitalist." http://blog.inc.com/the-eco-capitalist/ (Accessed January 2009).

Worldwatch Institute. *State of the World 2008: Innovations for a Sustainable Economy*. New York: W. W. Norton, 2008.

Mary Finley-Brook
University of Richmond

Ecocentrism

Sometimes called dark green or deep ecological ethics, ecocentrism is the core of a number of environmental positions focused on protecting holistic natural entities such as species, ecosystems, and landscapes. Ecocentrism uses insights from the science of ecology to locate value within ecological entities, processes, and relationships and represents an alternative to an anthropocentric or human-centered ethic of the environment.

At the end of the classic environmental text *A Sand County Almanac*, Aldo Leopold claims that actions are right insofar as they have a tendency to preserve the integrity, stability, and beauty of biotic communities. Leopold also talks about the value of respecting and protecting species, particular places, wild predation, evolutionary history, ecological energy circuits, wilderness areas, and land health. The land ethic he develops is ecocentric because it focuses on protecting biotic and ecological assemblages, processes, and relationships. This kind of holistic focus is the hallmark of ecocentrism.

The classic conception of ecocentrism is found within the philosophical field of environmental ethics. Environmental ethicists articulate moral norms to govern our actions with nature. There are at least three ways such norms can be justified. First, we can use existing moral norms or ethical theories that are focused on relationships between people. In the context of environmental ethics, such approaches are anthropocentric because our moral duties and obligations only apply directly to people. Second, we can extend existing moral norms or ethical theories that are focused on people to also include nonhuman animals and even plants by arguing that the source of value on which the existing norms or theories are based is also found in more than just humans. For example, if humans have moral value because they are sentient, then animals that are sentient might also have moral value. Moral extentionism that locates value directly in animals yields a zoocentric (centered on zoology) environmental ethic. Moral extentionism that locates value directly in plants as well as animals is biocentric (life-centered). The third way an environmental ethic can be justified is by locating value directly in novel features that classically are not attributed to individual humans. Ecocentrism is found here.

Ecocentrists believe that traditional moral norms and ethical theories that are focused on relationships between humans—including attempts to extend these norms and theories to cover animals and plants—are not sufficiently environmental and thus are inadequate to derive an environmental ethic. They take insights to heart from the science of ecology. Ecologists study the relationships between organisms and their environments, including collections or assemblages of organisms. From an ecological perspective, one cannot fully understand what an organism is without also examining the species of the organism, how the organism interacts within species populations, how the organism is related to ecosystem processes, what the organism eats, what eats the organism, and the like. Ecocentrists claim that we cannot fully understand the value of an individual organism without ascribing or discovering value in these kinds of relationships and at these different levels of organization.

One of the better-known ecocentric environmental philosophers is Holmes Rolston III. Rolston begins with biocentrism and claims that individual animals and plants have intrinsic value because they are teleological centers of life. Species also have intrinsic value because they have biological identities that are reasserted genetically over time, and species lines in part help determine what is and what is to be for species organisms. In terms of niches, species have adaptive fits within nature; this implies they are good right where they are, and humans have duties to let them be and continue to evolve. Ecosystems also have intrinsic value for Rolston. Individual organisms reproduce, species increase their kind, and ecosystems overall increase kinds. Ecosystems have ecological identities that are reasserted over time, and ecosystems in part help determine what is and what is to be for organisms and species. In this sense, ecosystems are prescriptive, selective systems that have systemic, intrinsic value. Rolston believes that if the products of nature are valued—namely, organisms, species, and ecosystems—the processes that produced the products also should be valued. For this reason, he believes that processes such as evolution (natural selection) and relationships such as predation have intrinsic value. All of this makes his environmental ethic ecocentric.

Rolston presents an objectivist account of ecocentrism because intrinsic value for him is objectively found in entities such as species and ecosystems; such value exists in nature independent of human valuers—what philosophers call mind-independent value. Other accounts of ecocentrism are possible. J. Baird Callicott develops Leopold's land ethic into an ecocentric environmental ethic that is decidedly subjectivist. Callicott denies that intrinsic value exists independent of human valuers (subjects) but claims that people can intrinsically value nature over and above using nature as mere resources. Such mind-dependent value is anthropogenic (human-generated) but not anthropocentric; it is ecocentric because it is ascribed to biotic and ecological assemblages, processes, and relationships.

Ecocentrism sits well with people who value biodiversity beyond strict human use and/or appreciation of it. Ecocentrists can locate intrinsic value at different levels of genetic, species, and ecosystem/landscape biodiversity (Rolston) or intrinsically value these levels (Callicott). Other ecological levels of organization such as communities and species populations (or metapopulations) can be intrinsically valuable. Ecocentrism also sits well with the new nonequilibrium paradigm in ecology in which nature is characterized by disturbances and disturbance regimes—all of which can be intrinsically valuable.

It is important to note that ecocentrism is often confused with biocentrism. Biocentrists locate intrinsic value or intrinsically value individual animals and plants, but only as individual organisms. For the biocentrist, species are nothing more than collections of individual organisms, and ecosystems are nothing more than collections of individual organisms located in particular places. In contrast, ecocentrists locate intrinsic value or intrinsically value more than just individual organisms. Confusion between biocentrism and ecocentrism prominently shows up in a number of places, particularly deep ecology. One of the distinguishing marks of deep ecology is supposedly its commitment to biocentrism; most deep ecologists, however, talk about the intrinsic value of holistic entities such as species and ecosystems, thus making deep ecology ecocentric. In addition to deep ecology, many forms of ecofeminism are ecocentric.

Because ecocentrism is rooted in ecology, many scientists who champion the protection of nature—most prominently conservation biologists—are ecocentric. Many people who belong to green political parties are sympathetic to ecocentrism. Rather than reforming economic, legal, and political institutions to better serve the environmental needs of people, many people in green politics seek to restructure or radically recreate these kinds of institutions so that they serve both the needs of people and nature in a holistic sense.

Ecocentrism faces at least three problems. First, ecocentrists might need to worry about getting their metaphysics from biology and ecology. There is controversy over how to define a species and whether species are natural kinds or simply human-imposed categories. The history of ecology is fraught with debates about the existence of entities such as ecosystems, and different models of nature—organismic, community, population, ecosystem, and patch-landscape—have been championed and criticized at different times. Ecocentric value schemes become even more complicated if value is located in the many processes and relationships found in nature. Further, it is not clear that concepts, models, and theories can be taken away from their original scientific meaning and context within ecology to arrive at a theory of value and an environmental ethic. Such a misguided use of ecology is known as "ecologism."

Second, ecocentrists might fall prey to what philosophers call the naturalistic fallacy—a mistake in reasoning when trying to derive values from natural facts. For example, just because species exist (fact), it does not follow that species have value or ought to exist, or just because ecosystems might exhibit properties such as integrity and resilience (fact), it does not follow that ecosystems and these properties have value and ought to exist.

Third, ecocentrists need to address the problem of ecofascism: If holistic ecological entities such as species and ecosystems are the central location of value, then the value of individual organisms seems subservient to these holistic entities. For example, should individual deer be killed to protect an endangered plant species or to reduce deer population pressures on an ecosystem? From the standpoint of environmental justice, should the livelihoods of people who use a particular area be diminished to protect species populations of animals and plants in that area? How can nonhuman nature be protected without harming people and bringing about environmental injustice?

See Also: Biodiversity; Deep Ecology; Ecofascism; Ecologism; Ecology; Intrinsic Value; Land Ethic.

Further Readings

Callicott, J. Baird. *In Defense of the Land Ethic: Essays in Environmental Philosophy.* Albany: State University of New York Press, 1989.
Dobson, Andrew. *Green Political Thought*, 3rd ed. New York: Routledge, 2000.
Leopold, Aldo. *A Sand County Almanac, and Sketches Here and There.* London: Oxford University Press, 1949.
Rolston, Holmes, III. *Environmental Ethics: Duties to and Values in the Natural World.* Philadelphia: Temple University Press, 1988.

Mark Woods
University of San Diego

Ecofascism

The lexicon of environmental politics contains few terms as problematic as *ecofascism*, which, although it can refer to specific historic and contemporary streams in environmental thought, is most often employed as a pejorative accusation aimed at one's ideological opponents.

Depending on the intended rhetorical purpose to which it is put, the term *ecofascism* can refer to the ecological foundations of Germany's National Socialism, radical antihumanist extremism in contemporary environmentalist rhetoric, the use of ecological principles in contemporary movements of the New Right, or as a form of ad hominem attack leveled against environmentalism in general.

The usefulness of ecofascism as a social category is thus rendered problematic, not only by its pejorative usage, but also in the rhetorical use—and abuse—of Nazism as an exemplar of unique evil and the controversial insubstantiality of the term *fascism* itself.

The Environmentalism of National Socialism

Ecology formed a strong undercurrent National Socialist ideology that, when read out of context, sounds quite in keeping with the tenets of latter-day "deep ecology." For example, Adolph Hitler wrote in *Mein Kampf*, "When people attempt to rebel against the iron logic of nature, they come into conflict with the very same principles to which they owe their existence as human beings."

Yet such ecological consciousness was not unique to Hitler or the Nazis; rather, the Nazis were the inheritors of a potent 19th-century German intellectual tradition combining naturalistic and nationalistic conceptions that revered the connections between "blood and soil." The Nazis believed not only in the unity of humans with nature but also in the *volkish* and mystic valorization of the peasantry and the purifying attributes of nature and country living. In contrast, urban life, with its greed, industry, and corruption—of course strongly associated with the Jews—was seen as contrary to traditional German values. Significantly, biological purity was viewed as indistinguishable from racial purity, and prescriptions for the ideal German society adopted a form of ecological determinism in which humans were subject to (ideologically convenient) "natural laws."

Ecology was not merely an ideological matter for the Nazis but contributed to major policies, such as the 1935 Reich Nature Protection Law, agrarian policies, and other efforts at natural conservation. Furthermore, some of the founding fathers of ecological thought such as Ernst Haeckel (who actually coined the term *ecology* in the 1860s) were also purveyors of social Darwinist notions elevating Nordic peoples, as well as a profound antihumanism.

However, there are fundamental differences between Nazi-era conservation and contemporary ecology: the former was romantic, nationalistic, and nostalgic and was not based on a scientific understanding of preserving global ecosystems. Furthermore, the Nazis' conservation ethos did not prevent them from building large-scale industry to support their war efforts, so their commitment to conservation was not powerful enough to supersede their other political ambitions.

Critiques of "Deep Ecology" and the New Right

Humanists and social ecologists such as Murray Bookchin have found odiously misanthropic certain claims, beliefs, and positions adopted by some adherents of "deep ecology," such as the need for aggressive and government-led population control, the pointlessness of foreign aid to famine-stricken nations, and the need for stricter anti-immigration policies. The deep ecologists' call for a radical reduction in our population levels, in particular, are seen by some to be consistent with fascist ideology, particularly when considering the repressive state that would be required to fulfill it.

Bookchin's collaborator Janet Beihl and others have also documented how, over the past few decades, Germany's resurgent "New Right" movements have adapted forms of ecological consciousness into their ideology, often with the intention of widening their appeal. Opposing nuclear power, industrial growth, and American hegemony while espousing mystic New Age beliefs about oneness with nature have helped them win many adherents and, for a while in the early 1980s, saw them in (unsuccessful) competition with the center-left for control of the Green Party. However, others argue that environmental issues came to the attention across the political spectrum in Germany in the 1970s and 1980s, not just on the right, and in any case, those who conflate New Right "green" rhetoric with genuine ecological politics have failed to sufficiently distinguish rhetoric of racist naturalism from genuine ecology.

Pejorative Rhetoric, Ecology, and the Problem with Fascism

Similar to the term *feminazi*, the word *ecofascist* has also been employed—mostly on the political right—by commentators opposed to environmental policies in general, and more particularly at what is seen as the self-righteous and moralistic imposition of "green" rules and regulations on everyone else. At its extreme, this rhetoric is used to accuse environmentalists of wanting to dismantle democratic societies, economic structures, and modernity in general to institute a global state-controlled—and vastly depopulated—green utopia.

In the case of some critiques, there is also a tendency toward a "guilt by association" fallacy, in which ecology advocates and activists are compared with Nazis because the Nazis were conservationists—akin to calling vegetarians Nazis simply because Hitler was a vegetarian.

This is, of course, a common form of rhetorical excess, one in which the term *Nazi* can be casually used to discredit any ideological opponent. As used in the context of environmentalism, however, it is intended to tarnish green political actors with the stain of some of history's most loathsome regimes.

Adolph Hitler's National Socialism, with its institutionalized racism, myth of supremacy, hyperefficient cruelty on the battlefield, and meticulously organized industrial genocide, has become so entrenched in our global cultural memory that it has become our absolute standard of evil. Therefore, attempts to associate environmentalism with not just Nazis but with the ideology of Nazism itself must be taken very seriously.

Historians of ecofascism stress that a belief in the oneness of people and nature does not necessarily lead to or inform fascism. Biehl and Peter Staudenmaier also deplore what they see as the simplistic and misleading appropriation of their work by some conservatives to smear conservationists.

The most significant problem with the term *ecofascism* is that it assumes a shared understanding of what fascism actually is, but this does not exist, either in the scholarly literature or the body politic. Lacking as it does the core readings afforded communism, and exemplified by regimes featuring almost as many differences as similarities, fascism has always proven an enigmatic subject for scholars, who have spent decades attempting to decipher and articulate exactly what fascism represents. One thing that scholars do seem to agree on, however, is that there can be no "generic" fascism: no one list of characteristics can sufficiently describe all the movements and leaders historically referred to as fascistic. Therefore, as it is not really possible to speak of fascism as a unitary movement, it is also essentially meaningless to apply the term to extreme views concerning ecology.

See Also: Deep Ecology; Green Discourse; Green Parties; Political Ideology.

Further Readings

Biehl, Janet and Peter Staudenmaier. *Ecofascism: Lessons From the German Experience.* Oakland, CA: AK, 2001.

Bookchin, Murray. "Social Ecology versus Deep Ecology: A Challenge for the Ecology Movement." *Green Perspectives*, 4–5 (Summer 1987).

Bruggemeier, Franz-Josef, et al. *How Green Were the Nazis? Nature, Environment, and Nation in the Third Reich.* Athens: Ohio University Press, 2006.

Goldberg, Jonah. *Liberal Fascism: The Secret History of the American Left, From Mussolini to the Politics of Meaning.* New York: Doubleday, 2008.

Olsen, Jonathan. *Nature and Nationalism: Right-Wing Ecology and the Politics of Identity in Contemporary Germany.* New York: St. Martin's, 1999.

Michael Quinn Dudley
University of Winnipeg

Ecofeminism

Ecofeminism is a social, political, and academic movement that views the oppression of women and the exploitation of nature as being interconnected. For ecofeminists, feminism is an environmental issue, and the environment is a feminist issue. Many social and environmental problems of the late 20th century are seen as an inevitable outcome of "masculine" behavior. The term *ecofeminism*, or ecological feminism, was coined in 1974 by the French feminist Francoise d'Eaubonne. It is one of the few movements that connect other movements. In addition to the feminist and environmental movements, ecofeminists have extended their analyses to the interconnections among the domination of nature, sexism, racism, and social inequality.

Principles of ecofeminism include the following:

- The domination of women and the domination of nature (as well as other forms of domination such as racism, sexism, and social inequality) are interconnected.
- This domination is justified by a hierarchy that ecofeminists seek to resist on all levels.
- Dualistic thinking, in particular the distinction between culture and nature, supports this domination.
- The central goal is to replace these dominative policies, practices, and philosophies with ones that are not.

The term *ecofeminism* appears to have been published for the first time in 1974 in Françoise d'Eaubonne's book *Le Feminism ou la Mort* (Feminism or death). Yet, its underlying principles had already been stated in Rachel Carson's famous book *Silent Spring* (1962). Other pioneers include Susan Griffin, who made ecofeminism known in the United States; Maria Mies, who did the same for Germany; and Vandana Shiva, who introduced the standpoint to India. In addition to these women, the ecofeminist movement was nurtured by the ideas and writings of Rosemary Radford Ruether, Carolyn Merchant, Irene Diamond, Ynestra King, and Ariel Salleh. Interestingly, ecofeminism departed from spiritual notions with a romantic emphasis on women as protectors of Gaia (mother earth).

Various social movements like the environmental, antinuclear, lesbian-feminist, and peace movements during the late 1970s and 1980s nourished, helped develop, and strengthened ecofeminist ideology, as well as ecofeminist activism. A further inspiration for the growth of ecofeminism—in particular for its academization—was the first ecofeminist conference, titled "Women and Life on Earth: A Conference on Eco-Feminism in the Eighties," held in March 1980 at Amherst College.

Examples of ecofeminist activism include Nobel Peace Prize–winner Wangari Maathai's formation of the Green Belt Movement in Kenya, the main focus of which lies on planting trees to preserve the environment and to improve the quality of life, and India's Chipko (tree-hugging) movement, in which women work together to preserve precious forests for their local communities, for instance.

Defining Ecofeminism

In defining ecofeminism, first it is important to note that there is not one unifying perspective. Just as there is not one feminism, there is not one ecofeminism. Yet, what ecofeminists have in common is the view of seeing interconnectedness between the oppression of women and the exploitation of nature as being interrelated. Ecofeminism challenges the idea that it is possible to solve the destruction of the environment while allowing for gender oppression or vice versa. The range of these women–nature connections explored by ecofeminists is manifold. Some ecofeminists analyze historical, philosophical, conceptual, or symbolic connections. Others discuss epistemological and methodical, economic, scientific, linguistic, or political connections. For instance, ecofeminists analyzing historical connections point to male-centered, culture-defining texts such as the epics of Homer or ancient philosophers outline their early associations of women with nature and their domination by men, whereas ecofeminists investigating linguistic connections point to linguistic links between the oppression of women and nature, such as "taming land" or "reaping nature's bounty," or to instances in which women are placed as being closer to nature by describing them as "wild" and "untamed." Yet, ecofeminism's major project is twofold. In addition to making visible the mentioned women–nature connections, the central goal is to replace these dominative policies, practices, and philosophies with ones that are not.

Second, ecofeminism represents the union of two concerns: ecology and feminism. This already emerges in the term *ecofeminism*, which is a contraction of the phrase "ecological feminism." Thus, ecofeminism combines some assumptions and ideas of a radical ecology perspective with feminism. In a nutshell, ecofeminism is a feminist environmentalism and an environmental feminism.

The "feminist part" of ecofeminism lies in its starting point. Ecofeminism begins with a women and sex/gender analysis focusing on power relations. Humanity is seen as gendered in ways that subordinate and oppress women. Hereby, ecofeminism falls back on various feminisms: liberal feminism, Marxist feminism, socialist feminism, radical feminism, and Third World feminism. Nevertheless, unlike these other feminisms, ecofeminism sees the primary purpose of feminism as a movement that seeks to end all systems of domination, including the suppression of the environment.

In regard to the latter, that is where the "ecological part" of ecofeminism comes in. The mentioned power relations between humans are extended to the relations between humans and the nonhuman nature. The domination of women (as well as the domination of other humans such as poor people, people of color, children, etc.) is seen as being interconnected with the exploitation of nature. Ecofeminists examine how humans use nature, how its domination by men leads to overgrazing and the tragedy of the commons and finally causes the destruction of animal and plant life. Nature is included in the category of the dominated and oppressed.

Third, central to ecofeminism is the view of seeing patriarchy (literally: father rule) as the foundation for oppression, as most other forms of feminism do. Hereby, the patriarchal society is built on four interlocking pillars: class exploitation, environmental destruction, sexism, and racism. The major domination in society stems not from poor by rich, for instance, but from women by men. Men dominate the Earth as they dominate women. Pivotal for the Western patriarchal framework is a pattern of thinking that generates dualisms. These dualisms are created when complementary concepts such as culture/nature, mind/body, male/female, public/private, reason/emotion, or white/black are positioned side by side and seen as mutually exclusive and oppositional. As a result, a value hierarchy is constructed that positions the more "masculine" member of each dualistic pair as more valuable, as superior (e.g., culture is ranked above nature, mind above body, and so on). Ecofeminists argue that this value hierarchy enables the justification for the subordination of some groups or people above others. The "inferior" group lacks the more "valuable" or "superior" characteristics of the dominant group. In regard to ecofeminism, the crucial key here is that men are said to represent the sphere of "humanity and culture," whereas women are part of "nature" (animals, plants, and so on). Men possess "reason," whereas women are at the mercy of "emotion." Yet, an ecofeminist treatment of this insight would be to assess its consequences for women and nature, whereas a feminist treatment would mean assessing its consequences for women only.

Criticisms of Ecofeminism

Yet, ecofeminism has attracted criticism on several points. It has been charged with essentialism because of the strong women–nature connection assumed in some ecofeminist positions. Essentialism claims that cross-culturally and cross-historically, all people of a particular gender or other category share the same traits. In certain ecofeminist writings, women are perceived as having an epistemologically privileged understanding of nature, whereas men are seen as being more related to culture. These writings assume that there is a deep connection between women and nature that either is not attainable for men or that men cannot experience. Despite pushing back gender bias, its polarity is reversed and a new value hierarchy is constructed that will replace the old. Besides being essentialist, this can additionally create new, more subtle forms of oppression for women. Moreover, ecofeminism has been criticized for diverting the focus of the environmental movement into the feminist movement and for running the risk of oversimplification in suggesting that abolishing patriarchy would be the "magic solution" for the social and environmental crisis.

Altogether, despite its critics, ecofeminism outlines important linkages between the domination of women (as well as other forms of oppression like racism, sexism, and social inequality) and the domination of nature. Yet, changing society's value structure certainly is not a quick fix for the pressing environment problems surrounding us like the pollution of rivers and lakes, the depletion of species, or climate change. Nor is it clear whether this approach would be more effective in reducing or stopping environmental destruction than appealing to people's (short- and long-term) interest by giving them incentives for generating or stopping action because of the negotiation of treaties, action programs, and so on.

See Also: Deep Ecology; Domination of Nature; Gaia Hypothesis; Gender; *Silent Spring*; Tragedy of the Commons.

Further Readings

Griffin, Susan. *Woman and Nature: The Roaring Inside Her*. San Francisco: Sierra Club Books, 2000.

Merchant, Carolyn. *The Death of Nature: Women, Ecology, and the Scientific Revolution*. New York: Harper & Row, 1989.

Mies, Maria and Vandana Shiva. *Ecofeminism*. London: Zed Books, 1993.

Sturgeon, Noël. *Ecofeminist Natures: Race, Gender, Feminist Theory, and Political Action*. New York: Routledge, 1997.

Warren, Karen J. *Ecofeminist Philosophy: A Western Perspective on What It Is and Why It Matters*. Lanham, MD: Rowman and Littlefield, 2000.

Martin Köppel
Independent Scholar

Ecological Economics

Despite its name, the field of ecological economics does more than combine "ecology" and "economics." It is a transdisciplinary approach that draws from several fields, including both social sciences (e.g., economics, politics, sociology, ethics, and philosophy) and natural sciences (e.g., biology, physics, and mathematics), and it seeks to synthesize these perspectives to address issues of sustainability comprehensively. In contrast to the neoclassical thinking of an economy as a circular flow of exchange value without any direct link with the biophysical world, ecological economists consider the biophysical elements (energy and matter) as foundations of an economy. Recognizing the Earth's resources as finite and irreplaceable, ecological economists call for a combination of scientific and ethical considerations in human economic activities. Critical to ecological economics is its emphasis on the interconnectedness and complexity of various components of our world and on explaining them with a holistic perspective rather than as fragmented parts. It is because of such characteristics that ecological economics becomes more of a movement than a discipline.

Ecological economics in its current form has its roots in some pathbreaking works from the 1960s and 1970s about economic "growth" and its effects on the environment. Economist Kenneth Boulding's 1966 paper "The Economics of the Coming Spaceship Earth" was significant in shaping the emerging paradigm because it raised serious concerns about the overexploitation of the Earth's resources and its threat to the future. Soon thereafter, ecologist Howard T. Odum completed *Environment, Power, and Society*, which helped create the platform for ecological economics by discussing energy, entropy, human society, and their mutual interactions. Nicholas Georgescu-Roegen's book *The Entropy Law and the Economic Process* further integrated biophysical factors into economics. At this time, computers were just becoming more commonplace, and predictive models were being applied to human problems; the 1972 book *Limits to Growth* by Donella H. Meadows, Dennis L. Meadows, Jørgen Randers, and William W. Behrens III showed the consequences of exponential growth on the Earth's finite resources. This important work alarmed the world about the devastating effects that unchecked growth will have on the planet.

Another important line of thinking that contributed to ecological economics came out of the growing discontent among scholars about the shortcomings of the national accounting systems. Because these systems focused solely on gross domestic product, without any concern for the depletion of natural resources, the approaches only boosted the momentum of environmental destruction. More and more economists and ecologists came into consensus about the need to work together and learn from each other to save the planet. These concerns coalesced into a workshop of ecologists and economists in Barcelona in 1987, where the International Society for Ecological Economics was formed. The society was formally established in the United States in 1988 and has since expanded to include branches in a large number of developed and developing countries around the world. The society's journal, *Ecological Economics*, was first published in 1989, and the field has expanded in several ways since then: Many institutes have been founded in various countries, and hundreds of books on have been written on related topics.

The concept of "entropy" occupies a central place in ecological economists' explanation of the economic system. Borrowing the concept from the Second Law of Thermodynamics in physics, economist Georgescu-Roegen first observed that the economic processes are entropic, which means that "energy" and "matter" are transformed from a state of easy availability to a state of nonavailability in any economic activity. In other words, our economic activities would eventually create a state of unsustainability, as the Earth would run out of its finite stock of energy and materials. Thus, going by the entropy concept, ecological economists observe that there are biophysical constraints to the economy and that it cannot grow indefinitely, as held by neoclassical economists.

Another foundational concept of ecological economics is that of an "open" future, which means we cannot know everything that will happen in the future. Ecological economists make analogies to the concept of evolution in biology. This concept refers to genetic changes in a species through complex processes of mutation and selection that happen over a long period of time. It is an open-ended process because one cannot predict when these mutation processes will occur, when they will stop, and what changes will happen as a result of these processes. Ecological economists introduce this concept of evolution into the economic system, wherein they identify certain elements such as peoples' economic choices and preferences, technologies, and various sociopolitical and economic institutions as "economic genotypes" (i.e., the "potentialities" in an economic system) and show how changes of these genotypes are unpredictable. However, unlike biological evolution, these genotypic changes in an economy occur more rapidly, thereby making the system even more complex and open. Such evolutionary changes in the economic system can be noticed even if we look at the post–Bretton Woods world economic system. Various new institutions (e.g., the World Bank, International Monetary Fund, World Trade Organization, and political entities like the European Union) coupled with unprecedented technologies have changed the economic system at the global as well as the local level. The frequency of economic crises around the globe has also accelerated, thereby contributing further to the openness of future.

The concepts of "ignorance" and "surprise" further elaborate ecological economists' idea of an open future. Ignorance is defined as a state of human inability to predict any probability of a future event, and there is a causal relationship between ignorance and surprise. Although some forms of ignorance may be overcome through knowledge or research, some other forms of ignorance are nonreducible, as they are "phenomenological" and "epistemological" in nature. Phenomenological ignorance occurs from unpredictable changes in a phenomenon or from inherently chaotic dynamics within a system or phenomenon. Epistemological ignorance, in contrast, occurs from the problems in the ways we perceive a phenomenon. In either case, we are unable to predict the future events and, thus, are open to uncertainties. The inability of modern science to have predicted climate change and its various effects on the ecosystem, as well as the necessary remedies, is a good example of nonreducible ignorance and surprise.

The systems approach remains the central tenet of any ecological economist's approach toward sustainability. A systems approach is about viewing the whole world of interacting biotic and abiotic components as a system in which everything is linked with everything and nothing could be understood without understanding the whole system. To explain the systems of human–environment interaction, ecological economists use "complexity theory" and "chaos theory." Both these theories study systems that are highly complex and, therefore, impossible to accurately predict. Chaos theory further argues that there are random emergent properties in a system, which makes the system inherently chaotic and, thus, unpredictable. As a result of such complex and chaotic nature, the system's future cannot be predicted, although we can devise certain coping mechanisms based on the underlying patterns in a system.

Ecological economists, therefore, build "scenarios" and use them as tools to address future issues. Scenarios, which are different from predictions and forecasts, are plausible future images based on identification of the current problem and assumptions about how they can be resolved in future. "Scenario analysis" accepts uncertainty and surprise as integral parts of future coping mechanisms, and in that sense it is different from the "solutions" provided by conventional Western science based on its belief in predictability.

"Transdisciplinarity" facilitates such scenario building by integrating multiple perspectives. Considering the complexity and uncertainties of the systems, ecological economists acknowledge the futility of applying strict disciplinary perspectives; instead they urge multiple conceptual frameworks, a phenomenon that ecological economist Richard Norgaard defines as "conceptual pluralism." It is through such a pluralistic approach that scenario building can be operationalized.

Yet another central theme for ecological economists is the issue of "equity." Although ecological economists accept the paradigm of sustainable development, which seeks to promote intergenerational equity, they go a step forward in raising the issue of intragenerational equity as well. To establish equitable resource distribution, the ecological economists support alternative institutional arrangements, especially with regard to common property resources (e.g., forests, water, lands). Equity is not just a goal in and of itself, they assert, as a system that provides a just and equitable approach toward common properties will also ensure biological and cultural diversity.

So that natural resources are properly measured in economic processes, ecological economists seek to develop new accounting systems in which the ecosystem goods and services (e.g., forests, water, and air quality) are quantified and integrated as parts of the production system. Accordingly, they have proposed changing existing accounting systems and development policies in the entire international development sector (including both nongovernmental organizations and state actors), so that the Earth's finite resources are duly valued.

With regard to its emphasis on distributional issues, the argument for a new ethical discourse still remains one of the more radical elements of ecological economics. Ecological economists assert that economic decisions need to be guided by moral or ethical perspectives if we are to establish a more equitable society. They further argue that this moral aspect should not be limited only to human society but also should be extended with regard to the relationship of humans vis-à-vis environment; that is, intrinsic values in the nonhuman entities irrespective of their use-value in the human world. This call for an ethical discourse, often missing in the scientific discussions, distinguishes ecological economics from other fields in that it does not seek to elevate some professed "neutrality" into a virtue but, rather, calls for frank and open discussion of ethical principles manifested by existing practices and systems.

Ecological economics is often criticized as being embedded within academic circles in the developed world without adequately addressing the historical and geographical specificities of the Third World countries. Another criticism is that ecological economics tends to overplay the "economic" in relation to the "ecological." For example, the environmental valuation system that it prescribes might work against its own interest as the natural resources become more vulnerable to exploitation as a result of the new economic explanation. Nevertheless, the paradigm shift advocated by ecological economics offers a possible route toward achieving a green future. Given the scale and intensity of the world's ecological crisis, with accelerating resource depletion and the prospect of climate change looming, approaches that keep ecological scientists actively engaged in social change are essential. In addition, because of widening discrepancies between rich and poor nations, as well as within them, the social and ethical dimensions of environmental challenges must remain central to achieve real progress. Focusing on growth and related shortsighted interventions risks reinscribing the shortcomings of neoclassical economics, once again, into our built and natural worlds, and continuing to use reductionist problem-solving strategies that keep strict disciplinary boundaries may be fatal for the future of the planet. The novel paradigm that ecological economics offers, through its transdisciplinary approach rooted in a sense of ethics and humility, offers a hopeful path toward a green and sustainable future.

See Also: Capitalism; Ecocapitalism; Ecology; Equity.

Further Readings

Boulding, Kenneth E. "The Economics of the Coming Spaceship Earth." In *Environmental Quality in a Growing Economy: Essays From the Sixth RFF Forum,* edited by H. Jarrett. Baltimore, MD: Johns Hopkins University Press, 1966.

Costanza, Richard, et al. *An Introduction to Ecological Economics.* Boca Raton, FL: St. Lucie, 1997.

Faber, Malte, et al. *Ecological Economics: Concepts and Methods.* Northampton, MA: Edward Elgar, 1996.

Georgescu-Roegen, Nicholas. *The Entropy Law and the Economic Process.* Cambridge, MA: Harvard University Press, 1971.

Norgaard, Richard. *Development Betrayed: The End of Progress and a Coevolutionary Revisioning of the Future.* New York: Routledge, 1994.

Odum, Howard T. *Environment, Power, and Society.* New York: Wiley-Interscience, 1971.

Sharon Moran
Mitul Baruah
State University of New York College of Environmental Science and Forestry

Ecological Imperialism

Similar to "imperialism," ecological imperialism concerns the deliberate act of expanding power, control, and authority by one country over areas located outside its borders—areas that become part of the empire. Whether referring to direct rule, indirect economic control, or ideological or cultural domination, and whether applied to the ancient or modern worlds, imperialism involves the conquest of the powerful over the less fortunate for gains in land, resources, and/or some form of tribute. Ecological imperialism, however, extends understanding of imperialism to include ecological factors for explaining the successful conquest of empires and the demise of the conquered. The phrase is also used to help explain the eventual decline of empires as a result of dismissing the ecological constraints of imperial expansion. Concurrently, ecological imperialism often refers to the act of European powers transferring plant and animal species from one part of the world to another, and subjecting those species and regions to European-style agriculture, which has had immensely negative effects on local ecosystems. In the contemporary period, ecological imperialism manifests itself through the privatization of plant varieties originating primarily in the global South by agro-industry and biotechnology institutions.

Imperialism has existed for thousands of years, as have its ecological consequences. As part of his assessment of ecological imperialism, John Bellamy Foster proposed that societies have overstretched their ecological impact since the dawn of civilization. For example, as the result of a combination of constant warfare, salinization of irrigation water caused by unsustainable increases in agricultural production, and a growing population, the Sumerians experienced ecological collapse around 2,000 B.C.E. The Roman Empire also declined in part because of ecological factors—vast agricultural zones of the empire became desertified as agriculture intensified. Deforestation, overgrazing, and improper irrigation led to soil degradation and disease outbreaks, such as malaria. By 200 C.E., Rome experienced life-threatening food shortages and population decline, contributing to the debilitation of the empire.

Ecological imperialism, however, took off with the emergence of the earliest stage of capitalist development, mercantilism. By the late 15th century, European absolutist states had found new worlds to plunder to feed their people, to expand their markets, and to fund their wars for territorial expansion inside Europe and beyond. Colonialism is the type of imperialism by which European powers expanded the reach of merchant capitalism worldwide, dominating territories that would host their colonies. Justified in part by the European call for the "domination of nature," these colonies extracted natural resources and transformed the land to meet the needs of the colonists and the European markets. By the turn of the 19th century, numerous mammals valuable for their pelts were brought to near-extinction in Europe and North and South America, in some cases leading to severe alterations of the surrounding landscape as a result of their interconnection with the ecosystem. Spanish and Portuguese colonies established in the Caribbean and Brazil to export sugar via slave-based production systems decimated the local environment through deforestation and the de-fertilization of soils commonly associated with sugarcane monocropping, while making these colonies dependent on Europe for food.

Ecological imperialism suggests that European imperialism was victorious because of the successful introduction of European plants and animals and the "successful" dissemination of terminal human diseases as much as for the superior weaponry. For example, Alfred Crosby posits that Europeans transformed the ecology of colonized regions to make it suitable for their needs. For example, European grasses such as rye and oats effectively took over native landscapes in North America while providing food for European horses and ground cover. Europeans tended to settle in temperate climates where their animals and agriculture could flourish, but they also replaced local diversity with monocultures that depleted soils. Diseases such as smallpox, dysentery, and influenza decimated whole native populations in the Americas, clearly contributing to the swift annexation of lands made fertile for European expansion.

By the Industrial Revolution, ecological imperialism acquired a new dimension, the acquisition and transfer of plant species worldwide. World powers such as the Netherlands, France, and Britain established botanical gardens (e.g., Kew Gardens) designed to store plants that would be studied and transferred for large-scale production throughout their respected empires. Coffee, rubber, and other cash crops were extracted from West Africa and Brazil, "improved," and then transferred to foreign regions (often in Asia), where they could be processed with cheaper labor supplies. This global transfer meant that the majority of the world had become subject to commodified agriculture, which replaced complex ecosystems and local agricultural systems with European-style mono-cropping. Here, the land and local ecologies became subject to the capitalist market.

By the mid-20th century, loss of biodiversity became a serious concern among scholars and scientists. Energy-intensive agriculture using artificial fertilizers, pesticides, improved seeds, and heavy machinery was being developed to increase yields and replace traditional agriculture worldwide. As a result, however, plant species used for millennia by indigenous populations were disappearing, wiped out by the few genetic varieties chosen for mono-cropping.

Ecological imperialism points to the exploitative character of this green revolution. For decades, scientists from the developed countries have taken germplasm from less-developed countries to create resistant seeds that are then patented and sold on the market, often to countries in which the initial plant varieties originated. Instead of botanical gardens, today institutions such as the International Agricultural Resource Centers coordinate the extraction of germplasm from the global South for gene banks located in or made available to the industrialized world. For instance, industrialized countries like the United States provide funding to international agricultural improvement centers in the less-developed world and in return receive access to their germplasm for their own agricultural production. In addition, these research centers often are compelled to sell germplasm to breeding companies to fund their operations, which then patent improved varieties from that material. Here, ecological imperialism includes reliance on the industrialized countries for seeds that originated in the global South, but with added dependence on chemical inputs and machinery sold by North-based multinational corporations to successfully use them in agriculture. As a consequence, the global South is robbed of its own ecological diversity and is left instead with economic dependence on the industrialized countries and diminished ecological diversity, with all its negative environmental and social effects.

See Also: Agriculture; Domination of Nature; Green Neoliberalism; Industrial Revolution; North–South Issues.

Further Readings

Crosby, Alfred. *Ecological Imperialism: The Biological Expansion of Europe, 900–1900.* New York: Cambridge University Press, 1986.

Foster, John Bellamy. *The Vulnerable Planet.* New York: Monthly Review, 1999.

Kloppenburg, Jack. *First the Seed: The Political Economy of Plant Biotechnology, 1492–2000.* Madison: University of Wisconsin Press, 2004.

Brian J. Gareau
Boston College

Ecological Modernization

Ecological modernization refers to a particular way of assessing the relationship between development and the environment. Many environmental observers over the last several decades have recognized the increasingly problematic nature of industrial modernization that so often puts the environment and the economy into antagonistic relationships with one another. At the same time, significant progress has been made to relieve this tension—seen in innovations ranging from green chemistry and green accounting to industrial designs that create recyclable products and reduce waste, to environmental political regulations that carefully consider ecological components in moving forward with development. This

Ecological modernization in the United States has been thwarted somewhat by vast areas of urban sprawl that perpetuate the country's car culture, such as this housing development outside of Des Moines, Iowa.

Source: U.S. Department of Agriculture Natural Resources Conservation Service

transformation has been the object of much scrutiny, evidenced in the debates of many social scientists concerned with topics of modernity and the environment as they attempt to identify the actors involved in this change and what it says about the relationship between ideology and discourse regarding observed improvements in nature–society relationships. Ecological modernization then refers to the position held by many social scientists that modern societies possess the tools and insight necessary to deal effectively with environmental crises by using components of the modernizing process, spanning innovations in industrial design, governmental policies, and consumer relationships. Rather than being the source of the problem, ecological modernization conceptualizes industrial modernity as the site of solutions where both environmental and business interests fall under the same rubric, offering an optimistic worldview on the future progress of the states of nature and society.

Originating in Western Europe in the 1980s, ecological modernization theory has remained a dominant perspective in European environmental thought since its inception. Initially, ecological modernization was a reaction to demodernization theories of the late 1970s and early 1980s, which posited that only a radical reorganization of core institutions of modern society could lead to sustainable development. The Dutch National Environmental Policy Plan of 1988 is sometimes identified as a keystone innovation on the grounds that it confirms ecological modernization theory—a reaction to the problems experienced in the Netherlands that personified nature–society antagonisms resulting from very dense populations and pollution levels. The National Environmental Policy Plan called for the familiar drastic reduction in emissions, sometimes referred to as a compartmentalization approach, which divides environmental problems into specific frames such as air or water pollution. Yet its innovation was that it discussed climate change and the need for common interests between the environment and development regarding trade policy, retail consumption, and environmental education. In this sense, the National Environmental Policy Plan is seen as being closely related to the 1987 Brundtland Commission that articulated goals for sustainable development. The German sociologist Joseph Huber is commonly identified as a first scholarly contributor, who throughout the 1980s placed emphasis on technological innovation and market forces as part of a "superindustrialization" that offered the best path to sustainability. Neoliberals engaged with environmental questions have allied themselves with this position, arguing for market reforms and a technological fix as the answers to glaring ecological problems.

However, later ecological modernizationists have rebuffed such alliances by refusing to admit capitalism as a precondition for a clean environment—saying only that it is the most realistic path available. Indeed, scholars such as the environmental sociologist Zsuzsa Gille argue that capitalist relations can regress the state of ecological modernization, pointing to empirical investigations of state socialist programs in Hungary as being ecologically superior to the privatized programs that later took hold under capitalism. The most familiar English-language contributors have been the Dutch sociologists Arthur Mol and Gert Spaargaren (as well as Maurie Cohen, Fred Buttel, and David Sonnenfeld), who define ecological modernization as the "emancipation of ecology," in which environmental crisis prompts a (nonradical) reorganization of rationality, making possible the notion that transcending crisis through an ecological rationality necessitates furthering modern development. This rationality could be manifested in things such as polluter-pay laws or by informing the design process of technology. Ecological modernization recognizes the faults inherent in modern society but challenges any notion that environmental transformation requires a radical social transformation; rather, modernity gains new life with its newfound rationality. The important difference between early and later ecological modernizationists is that although the superindustrialization thesis sees ecological problems as wholly concerning technology and administration, later theorists go beyond this by recognizing a need to consider flexible regulation, cultural education, and new institutional arrangements. Though many of the original scholarly contributors are often European, popular advocates hailing from the United States include Al Gore and columnist Thomas Friedman.

Core Belief

At the core of eco-modernization theory is the belief that environmental concerns actively shape modernity, affecting scholars' engagements with a number of common themes. No longer are science and technology primarily viewed as instigators of ecological degradation, but they are given a new role as potential liberators in preventing and curing future problems. Similarly, economic markets need not be demonized if they can facilitate the exchange of sustainable technology and commodities. This also leads to the expectation that the role of the nation-state will change, bringing it into line with reflexive modernity, where the state focuses on decentralized consensus, bringing nonstate actors into the process of environmental reform. Thus, nongovernmental organizations play a larger part in ecological modernity than they previously did, coming out of the periphery to take a lead role. Underpinning all of these processes is a basic shift in discourse that no longer accepts a relegation of environmental concerns as illegitimate or pet projects. Although not attacking the basics of social relations, ecological modernization does call for a greening of business as usual. Importantly, this perspective does not see capitalism as an inherently unsustainable system; rather, industrialism itself is viewed as the main culprit of environmental degradation. The ecological modernizationist position often assesses capitalism as a dynamic system capable of considering environmental concerns, possessing the capacity to significantly improve production toward sustainable ends, and the most feasible path available toward a sustainable future.

Supporters of the ecological modernization stance sometimes present an optimistic and hopeful view of the improvements made possible through a dematerialization of the economy, where growth is no longer necessarily tied to increased consumption of resources, often citing evidence in Western Europe and Japan. This view finds an ally with the argument for an environmental Kuznets curve, where economists Gene Grossman and

Alan Krueger argue that as a country surpasses a certain threshold of income per capita, economic growth is not only disassociated from environmental deterioration but actually brings improvement. While recognizing that some delinking between materiality and economic growth has occurred, critics note that not all economic growth comes from dematerialized production processes but, rather, from service sectors, which implies that growth in production still degrades ecological interests. Further, they argue that the reason why some countries can make claims to delinking originates in the fact that they have outsourced so much of their production abroad, artificially inflating national figures while causing even greater scarcity of resources in the global South, referencing a world-systems approach.

An alternative explanation for why ecological modernization has not taken root in the United States like it has in Western Europe and Japan focuses on analyzing urban sprawl. Landowning interests—which rely on the availability of cheap fossil fuels—have held major political power throughout American history (the replacement of trolley systems with the establishment of private automobiles serves as an illuminating case study). In addition, an active conservative coalition exists in the United States and is determined to exculpate environmental concerns from policy, manifested, for example, in George W. Bush's response to the Kyoto Protocol. Ecological modernization in the United States would then imply a principle of land management that displaces profits, contracting sprawl while bringing communities of work and residence into closer proximity. This would necessitate creating smaller living spaces and intervening in the American "love affair" with the car in lieu of mass transit, walking, and biking. Maurie Cohen suggested that the history of American environmentalism—marked by an oppositional relationship with industrial corporations—has made it skeptical of ecological modernization as a new form of scientific management, preventing it from integrating ecological modernization into its framework of political goals.

Why the Opposition?

The primary opposition to ecological modernization within social science circles is represented by the "treadmill of production" perspective. Originally formulated by Allan Schnaiberg in 1980, this tradition arose from the observation that technological developments in the postwar West increasingly came with devastating ecological effects, seeing vast expansions in both the extraction of natural resources (withdrawals) as well as the release of pollutants (additions). A basic difference immediately jumps out in that the treadmill perspective offers a much more macroscopic view, looking at how firms operate within society as a whole, whereas ecological modernization focuses more on the effects of production within a single firm or industry. The treadmill of production sees traditional technologies as primarily revolving around withdrawals in terms of ecological effects; it is with technology developed under industrial capitalism that the definitive difference resides—in additions—which are the basis of modern economic growth. Therefore, the treadmill of production sees a basic contradiction existing between capitalism and sustainability. Capitalism relies on economic growth, which relies on the consumption—rather than preservation—of natural resources, which will eventually preclude future growth (hence the centrality of the dematerialization thesis for ecological modernization). The more successful the treadmill of production is in distributing disposable income, the faster the speed the treadmill operates on. This treadmill comes in two different forms: the ecological and the social. The ecological treadmill shows that profits made from increased efficiency are then reinvested into more technology, which then expands the system of production, leading to both greater additions of pollutants and greater withdrawals of natural resources. The social treadmill argues that after each business cycle, profits get reinvested into more efficient technology, gradually displacing workers from the production process (a classic Marxist observation). The treadmill perspective goes beyond the growth/no growth debate by calling into view the very relationship between society and environment, particularly around production. However, instead of focusing solely on productive efficiency (or lack thereof), treadmill theorists also look to social movements to explain how working-class desires for comfortable lifestyles tie them into economic growth coalitions, implying that radical environmental transformation does indeed necessitate radical social transformation, which is what ultimately puts this position at odds with those who confirm ecological modernization.

The treadmill perspective interprets contemporary reactions to environmental problems as a trade-off: States allow private firms unlimited access to natural resources in exchange for the political permit to regulate "end-of-pipe" pollutants. Whereas ecological modernization theorists see this as reflective of an ecological rationality, treadmill theorists see this as nothing new. Gains already experienced in the 1970s were undone through certain political coalitions (e.g., the Reagan or Bush administrations). When positive changes do come, treadmill theorists argue their perspective serves as a better interpretation because (1) firms are often forced to change through state regulation or social activism, (2) ecological improvements are only taken into consideration when it positively affects the economic bottom line, and (3) firms often put up the appearance of improvement through accounting tricks or plain misreporting.

See Also: Brundtland Commission; Ecocapitalism; Industrial Ecology; Industrial Revolution; Kuznets Curve; Kyoto Protocol; Sustainable Development.

Further Readings

Cohen, Maurie J. "Ecological Modernization and Its Discontents." *Futures,* 38 (2006).

Gonzalez, George A. "Urban Sprawl, Global Warming and the Limits of Ecological Modernization." *Environmental Politics,* 14 (2005).

Grossman, Gene M. and Alan B. Krueger. "Economic Growth and the Environment." *Quarterly Journal of Economics,* 110 (1995).

Mol, Arthur P. J. *The Refinement of Production: Ecological Modernization Theory and the Chemical Industry.* Utrecht, Netherlands: Van Arkel, 1995.

Mol, Arthur P. J. and David A. Sonnenfeld, eds. *Ecological Modernization Around the World.* London: Frank Cass, 2000.

Schnaiberg, Allan. *The Environment: From Surplus to Scarcity.* Oxford: Oxford University Press, 1980.

Tellegen, Egbert. "The Dutch National Environmental Policy Plan." *Journal of Housing and the Built Environment,* 4/4 (1989).

Jeremiah Bohr
University of Illinois at Urbana-Champaign

Ecologism

Ecologism is generally considered to be an ideological position that advocates a transformation in human–nature relations, challenges anthropocentric values, emphasizes respect for natural limits, and calls for significant social and economic change. However, the term has a range of divergent definitions and can encompass a spectrum of ideas. Despite this, it has been argued that ecologism is sufficiently comprehensive and systematic to be considered a distinctive new ideology, and this idea is now relatively widely accepted.

Understandings of ecologism can be grouped into two schools, known as "minimalist" and "maximalist." In a minimalist approach, the terms *environmentalism* and *ecologism* are often used interchangeably. Either term can be used as an umbrella term, encompassing a spectrum that runs from "light green" or "ecological modernization" at one extreme to "dark green" or "deep ecology" at the other. It has been suggested by Andrew Vincent that there are several broad themes underlying this kind of ecologism: an emphasis on the interdependence of species and their environment, a skepticism of human dominance, an anxiety about the environmental effects of industrial civilization, and a more positive attitude to nature than is found in other ideologies. Using this broad definition, it is possible to trace the roots of ecologism to the 19th century.

In contrast, the maximalist approach, as propounded by Andrew Dobson, adopts a strict definition of ecologism, clearly differentiating it from environmentalism. In this view, ecologism is seen as more radical than environmentalism, and furthermore, environmentalism is not considered an ideology at all. This is because, unlike ecologism, it does not present a worldview—a vision of a better society and a proposal for how to get there. According to this maximalist conceptualization, environmentalism takes a managerial approach to the environment, assuming that problems can be solved without fundamental changes in current values or patterns of production and consumption. It is rooted in an anthropocentric perspective (prioritizing human needs). This generally results in a pragmatic and reformist political style.

In contrast, ecologism argues that a sustainable society requires radical changes in the human–nature relationship and in social and political systems. It takes an ecocentric perspective. Some suggest that it legitimizes the use of radical strategies to achieve its ultimate goals, as exemplified by direct action groups such as Earth First! Ecologism, under this strict definition, has its origins in the countercultural movements of the 1960s. Some suggest that it is a manifestation of a growing cultural anxiety over environmental risks or link it to increasing levels of postmaterialism.

Proponents of ecologism cannot properly be called *ecologists*, as this term refers to those in the scientific field of ecology. Ecologism is sometimes called "political ecology," and its supporters are called "political ecologists." However, this can also create confusion, as the same term is also used to refer to the study of the relationships between politics and the environment.

Ecologism has a strong philosophical and moral dimension, which is central to its claim of being a distinct ideology. Andrew Dobson suggests that ecocentrism is a core value of ecologism, though its meaning may be interpreted in various ways. In some forms of ecologism, human well-being is still seen as a central concern but is understood in its wider context. Ecologism emphasizes the interconnectedness of people, animals, and environments, and human welfare is seen as inseparable from the health and stability of ecosystems. Brian Baxter argues that the key premise of ecologism concerns the need to take nonhuman creatures into account in any questions of morality. However, this does not necessarily mean that all life-forms are treated as holding equal moral significance. Some forms of ecologism stress the intrinsic value of qualities such as diversity and stability within ecosystems. In general, ecologism can be considered a way of thinking that is not based on the assumption that human beings hold a privileged position in social and political evaluation.

Most proponents of ecologism emphasize the need to respect natural limits, which imply limits to economic growth, population growth, and consumption. Some see this as a defining characteristic of ecologism. However, other proponents of ecologism place less emphasis on this issue. The concept of limits to population is particularly controversial and is seen by some as detrimental to the public acceptance of ecological ideas, as it can lead to a perception of ecologism as an "antihuman" ideology. Another key idea within ecologism is that of drawing lessons from nature and promoting diversity, stability, and cyclical processes within human systems, as well as in natural ones. Finally, a key feature of ecologism is its emphasis on transforming social, cultural, political, and economic systems to create a more sustainable way of life.

Even within ecologism, there is a spectrum of attitudes and beliefs. Recent ecological thought has been profoundly influenced by Arne Naess and his distinction between deep and shallow ecology. This distinction is similar to Andrew Dobson's distinction between environmentalism and ecologism, but deep ecology is not identical to ecologism. Deep ecology is a more clearly defined and radical set of ideas than ecologism, based on the premise that all life has intrinsic value, and that diversity is a value in itself. For deep ecologists, these beliefs create an obligation to act to change the status quo. Other researchers have identified a similar subset of ideas called "radical ecologism." Characteristics of this belief set include an antipathy to modernization, colonialism, and globalization; an ecocentric attachment to native ecosystems; an anticonsumerist ethic, countercultural values, and an interest in decentralized and cooperative lifestyles.

There are several unresolved issues around the topic of ecologism. First, the term can be defined in various ways, though certain themes seem to underpin the disparate definitions. Even when taking a strict definition of ecologism as an ecocentric ideology calling for radical socioeconomic change, the term still encompasses a broad spectrum of ideas. Second, there are still some questions over whether ecologism can be considered a distinctive new ideology. Some claim that it was a short-lived phenomenon during the late 1980s and is now irrelevant, claiming there has been a paradigm shift to "postecologism." The postecologism thesis suggests that ecologism's key components—its diagnoses of problems, values, and strategies for addressing the problems—are now outdated and have largely been abandoned. Others call ecologism merely a cross-cutting set of ideas that can be linked to other ideological categories.

Finally, ecologism's relationships with other ideologies remain an issue of debate. In particular, the question of whether it is naturally oriented to the left or right of the political divide is contentious. Some claim that ecologism is naturally compatible with left-wing positions and radical political causes, and several hybrid ideologies have been developed, including ecosocialism and ecofeminism. Others, however, see elements within ecologism that complement right-wing beliefs. There is also a strong strand of opinion that suggests that ecologism is oriented to neither side but transcends conventional political divides.

See Also: Death of Environmentalism; Deep Ecology; Ecocentrism; Ecological Modernization; Environmental Movement; Green Discourse; Limits to Growth; Political Ecology.

Further Readings

Baxter, Brian. *Ecologism: An Introduction*. Edinburgh, UK: Edinburgh University Press, 1999.
Carter, Neil. *The Politics of the Environment: Ideas, Activism, Policy*, 2nd ed. Cambridge, UK: Cambridge University Press, 2007.
Dobson, Andrew. *Green Political Thought: An Introduction*. London: Routledge, 2007.

Sarah Hards
University of York

Ecology

Ecology is possibly one of the most ancient disciplines. Its evolution has been gradual, and it will continue to evolve with humankind's ability to comprehend and understand his own surroundings and the interaction of its components. Rooted in the Greek word *oikos*, which means home, the subject shares its origin with the study of economics. The term *ecology* is fairly recent, coined by German biologist Ernst Haeckel in 1869. It is pertinent for students of ecology to deeply appreciate its fundamental principles, subdivisions, constraints, and challenges and some of the recent breakthroughs ecologists have made in advancing the subject.

Engagements with the basic tenets of ecology have a long history, with some of the earliest written reflections of the natural world in the writings of Greek philosophers like Hippocrates and Aristotle. In the early 1700s, Anton van Leeuwenhoek, the philosopher and scientist, initiated the study of living beings and their linkages through the food chain and population regulation. There have been several efforts to define "ecology"; possibly the simplest definition offered by E. P. Odum was "the study of the structure and function of nature." Hanns Reiter was probably the first to combine the words *oikos* (house) and *logis* (study of) to form the term *ecology*. French zoologist Isodore Geoffroy St. Hilaire proposed the term *ethology* for the study of the relations of the organisms within the family and society in the aggregate and in the community. Today ethology has become synonymous with animal behavior. St. George Jackson Mivart coined the term *hexicology* in 1894 to describe the study of relations between organisms and their environment as regards the nature of the locality they frequent, the temperatures that suit them, and their relations to other organisms as enemies, rivals, or accidental and involuntary benefactors. Concerned with the sociology and economics of animals, Charles Elton, a British ecologist, defined ecology as "scientific natural history."

Ecologists need to understand both a species and the surrounding ecosystem to address increasing environmental pressures. These scientists are working with Native American tribes to monitor a threatened population of lake sturgeon in Wisconsin.

Source: U.S. Fish & Wildlife Service

The domain and scope of ecology is vast and cuts across many other disciplines that entail human endeavors and nature. Broadly, the subject has two divisions: autecology, which involves study of one organism or an individual species, and synecology, which deals with a group of organisms that are associated together as a unit. A very commonly used term in the science of ecology is *ecosystem*, which is a community of interacting organisms together with the physical environment within which it exists, and with which the species in the community also interact. Ecosystems can be natural, such as terrestrial, including forests and deserts; aquatic (freshwater and marine); or man-made, such as an agricultural, rural, or urban landscape. The boundaries of ecosystems are not very rigid—an ecosystem can be anything from a small pond, to an ocean or a small aquarium, to a vast rainforest; thus, a hierarchy exists from very local to global.

The subject of ecology has gained tremendous relevance in light of the present global efforts to retain healthy ecosystems. Ecosystems are dynamic, but are not randomly moved by uncontrollable forces. Healthy ecosystems maintain certain trajectories that are strongly influenced by species composition, species groups, landscape patterns, soil and atmospheric chemistry, and various other factors. Balances are dynamic rather than static. Understanding the interrelationship between balance and the flux—or between the structure and processes of nature—is the essence of modern ecology. Efforts to understand the ecology of a system are a prerequisite because it is not possible to manipulate an ecosystem without disturbing its elements and constituents. It is the degree of disturbance and the nature and intensity of the effects that ecologists are most often concerned with in the event of manipulation.

Subdisciplines

Ecology has various subdisciplines: Physiological ecology is the study of how environmental factors influence the physiology of organisms; population ecology, the study of dynamics, structure, and distribution of populations; community ecology, the study of interactions among individuals and populations of different species; evolutionary ecology, the study of the physical and biological environment and processes that act as a filter and allow some individuals within a population to reproduce and pass the genetic materials to the next generation and screen others; and ecosystem ecology, the study of the patterns and interactions in an ecosystem. Finally, landscape and global ecology is the study of interactions among different ecosystems. Students of ecology learn about the flow of the energy in the ecosystems from plants (producers), which transform solar energy and carbon dioxide through the process of photosynthesis, through consumers (herbivores), then to carnivores, and finally to decomposers, which are microbes. This happens as a continuous cycle in nature. A very significant aspect of ecology is biogeochemical cycles—the cycling of water—the hydrological cycle, the gaseous nutrient cycle of carbon and nitrogen, and the sedimentary nutrient cycle of sulfur and phosphorus.

The domain of ecology is in spatial (space) and temporal (time) patterns of distributions, and in the abundance of organisms, including their causes and consequences. Several principles operate within this domain, with which ecologists widely agree. Samuel Scheiner and Michael Willig have summarized the principles of ecology.

1. Organisms are distributed in space in heterogeneous distributions.
2. Organisms interact with their abiotic (the nonliving) and biotic (living) environment.
3. The distribution of organisms and their interactions depend on contingencies.

4. Environmental conditions are heterogeneous in space and time.
5. Resources are finite and heterogeneous in space and time.
6. All organisms are mortal.
7. The ecological properties of species are the result of evolution.

Many of the tools and techniques needed for the science of ecology are interdisciplinary in nature. Ecological studies draw and integrate knowledge systems from other disciplines like geography, mathematics, hydrology, physics, chemistry, biology management, social sciences, economics, among others.

Ecology and ecologists are facing constraints and challenges on a global scale. In the face of global changes like deforestation, rapid urbanization, a rise in levels of pollution, climate change, and global warming, the world is experiencing multiple stresses. Central to everyday life is the need to meet the developmental challenge of alleviating poverty for the more than 1.3 billion people who still live on earnings of less than $1 a day; trying to meet the everyday basic needs of food, energy, and water; and the rapid loss of the world's biological diversity. There has been an increase in demand per capita in food and resources (40 percent in grains, 100 percent in fish, and 33 percent in wood). The International Union for Conservation of Nature–The World Conservation Union has reported on the rapid loss of the world's biological diversity. In Europe and Russia, 16 million square kilometers of forests have been reduced to 3.5 million square kilometers. Similarly, only 1 square kilometer of the Asian forests remains of every original 15. Markets have failed to recognize the true value of biodiversity, and institutions have failed to regulate biological resources. In a classic 1997 paper, Robert Costanza valued the world's ecosystem resources at $33 trillion a year. Rise in population, inappropriate use of technology, and inequitable distribution of costs and benefits have exacerbated the problem.

More and more ecologists are involved in managing complex ecological issues; however, their task is made more daunting because they often have to cope with uncertainties. Strong sets of databases often needed to make management decisions are lacking; however, newer tools and techniques in ecology offer many concepts that can be used to address the many issues and to minimize vulnerabilities to natural disasters. Knowledge of a species' functional traits and ecosystem processes can be used to design more productive agroforestry systems and productive ecosystems, and can be applied to improve the lives of millions of people suffering from hunger, lacking clean drinking water and reliable, efficient energy sources, dying from preventable diseases, and suffering disproportionately from natural disasters. Robert Costanza, director of the Gund Institute of Ecological Economics at the University of Vermont, recognizing the coevolution of humans, their culture, and their interaction with the larger ecological system, has emphasized the need for robust analytical and modeling tools. Studying ecology at the landscape level has emerged as a new and effective tool. A landscape is viewed as a mosaic of different land uses, and a holistic view is taken to understand threats and pressures before developing conservation action plans.

As a part of the ecological implications of human impacts of development, Paul Ehrlich and John Holdren developed an IPAT model (Impact = Population × Affluence × Technology), in which the human impact equals the product of population, affluence (quality and quantity of consumption), and technology (efficiency of production and waste assimilation). A very recent approach to assess humankind's impact on the Earth's resources is using an ecological footprint analysis, designed by J. Loh and M. Wackernagel.

An ecological footprint could be used to measure, monitor, and manage consumption of natural resources in pursuit of sustainable development. Ecological footprint analysis seeks to provide a unified comparable measure of human ecological impact.

The World Wide Fund for Nature (known in North America as the World Wildlife Fund) released two Living Planet Indexes, one in 2004, and one in 2008. The indexes showed that the natural ecosystems across all biomes and regions of the world are under severe pressure. The anthropocentric threats identified included habitat loss, fragmentation, expansion of agriculture, overexploitation of species (particularly as a result of fishing and hunting), pollution, and the spread of invasive species. The indicator has been designed to monitor the state of the world's biodiversity. The ecological footprint measures humanity's demand on the biosphere in terms of area of biologically productive land and sea needed to provide the resources we use and to absorb our waste. The Living Planet Index tracked the number of key threatened species of the world using a method similar to the way a stock market is monitored. In 2005, the global ecological footprint was 17.5 billion global hectares (gha) or 2.7 gha per person. (A global hectare is a hectare with world-average ability to produce resources and absorb wastes.) On the supply side, the total productive area or biocapacity was 13.6 gha, or 2.1 gha per person. Humanity's footprint first exceeded the Earth's total biocapacity in the 1980s, and this excess has been increasing ever since. In 2005, demand was 30 percent greater than the supply.

Ecosystems have evolved over large time scales, and even the simplest of ecosystems is highly complex in nature and still not completely understood. We have been through several ecological crises—the widening hole in the ozone layer, acid rain, Minamata disease—caused by severe mercury poisoning, desertification, and global warming, to name a few. The challenge that lies ahead for economists and scientists is immense. The Eco Summit 2007, held in Beijing by the Chinese Academy of Sciences, the International Council for Scientific Union, and Elsevier publications, called for humankind to work together to prevent further ecological deterioration of the Earth. This will require collaboration between civil society, governments, and scientists to apply the goals of ecology to everyday life.

See Also: Biodiversity; Conservation Movement; Environmental Management.

Further Readings

Anitha, K., et al. "Ecological Complexity and Sustainability." *Current Science*, 93/5 (2007).

Begon, M., et al. *Ecology*, 4th ed. Oxford: Blackwell, 2006.

Colinvaux, P. *Ecology*. New York: Wiley, 1966 (1986).

Costanza, R. "The Value of World's Ecosystem Services and Natural Capital." *Nature*, 387:253– 260.

DeClerck, Fabrice. "The Role of Ecological Theory and Practice in Poverty Alleviation and Environmental Conservation." *Frontiers in Ecology and the Environment*, 4/10 (2006).

Dodson, S. I. *Ecology*. Oxford: Oxford University Press, 1998.

Ehrlich, P. R. and J. Roughgarden. *The Science of Ecology*. New York: Macmillan, 1987.

Gross, Matthias. "Restoration and Origins of Ecology." *Restoration of Ecology*, 15/3 (2007).

Kormondy, Edward J. *Concepts of Ecology*. Englewood Cliffs, NJ: Prentice Hall, 1969.

Krebs, C. J. *Ecology*. New York: Benjamin Cummings, 2001.

"Living Planet Report." 2008. WWF International. http://assets.panda.org/downloads/living_planet_report_2008pdf (Accessed February 2009).

Loh, J. and M. Wackernagel, eds. "Living Planet Report." WWF International. http://assets.panda.org/downloads/lpr2004.pdf (2004) (Accessed February 2009).
McIntosh, R. P. *The Background of Ecology*. Cambridge, UK: Cambridge University Press, 1985.
McNaughton, S. J. and L. L. Wolf. *General Ecology*. New York: Holt, Rinehart and Wilson, 1973.
Odum, E. P. *Fundamentals of Ecology*. Philadelphia: W. B. Saunders, 1971.
Rees, W. "Eco-Footprint Analysis: Merits and Brickbats." *Ecological Economics*, 32 /3 (2000).
Scheiner, Samuel M. and Michael R. Willig. "A General Theory of Ecology." *Theoretical Ecology*, 1 (2008).
Stiling, P. *Ecology*. Upper Saddle River, NJ: Prentice Hall, 1992.
Vitousek, Peter M. "Beyond Global Warming: Ecology and Global Change." *Ecology*, 75/7 (1994).

Sudipto Chatterjee
HSG, Sagar University

Ecosocialism

Ecosocialism, sociologist ecology, or green socialism is a theory synthesizing elements of the antiglobalization movement, anarchy, ecology, green politics, socialism, and Marxism. Much of ecosocialism is theoretical work exploring the lines between these other ideologies, how they fit together, and critique of each element. Ecosocialists commonly believe that the capitalist system is unsustainable, filled with contradictions, and the cause of social alienation, vast disparities between the rich and poor, imperialism, destruction of the environment, global hegemonic systems, and repressive states. Ecosocialists support the systematic disassembling of the state and capitalism via nonviolent means and creating a new system that focuses on shared commons, collective ownership, and freely associated producers.

Ecosocialism is a continually developing theory and ideology that is critical of each of the various parts of which the theory consists. Ecosocialists are sometimes described as "Red Greens" because the group is primarily motivated by Marxism, anticapitalist views, and green politics. Red Greens may also be called Watermelons, suggesting they are red on the inside and green on the outside, and the term may be a critique or a compliment, depending on whether or not one thinks one or the other quality should be on the prioritized outside. Some ecosocialists give Marx credit for being one of the first theorists with a comprehensive ecological worldview, and they do not think it's inconsistent to join Marxism with green politics. Other theorists think green politics and ecosocialism are simply a natural extension of Marxism. Modern environmental realities such as population growth, limits to growth and resources, diminished nonrenewable resources, and the limits of technology to deal with climate change all easily lend themselves to discussion of materials, labor, and Marxist analysis.

Some theorists in the green movement take an ecocentric or biocentric approach to environmental issues and argue for the intrinsic worth of all organisms equally. Such theorists do not want to privilege humans and human social justice and critique such as more of the same anthropocentrism that led to current environmental problems. Ecosocialists warn against misanthropy and have a vision that is fundamentally anthropocentric; they accent both social justice and environmental issues. Ecosocialists argue that the most pressing environmental problem is social justice and that the lack of justice leads to other problems like environmental destruction, global warming, and overuse of nonrenewable resources. Ecosocialists argue that the theoretical antecedent of green politics and the environmental movement is Marxist thought.

The Marxist view enables greens to have powerful critique over capitalistic structures. It offers an approach based on material and history and suggests processes for social change. Many ecosocialists think that some form of socialism (human rights) must precede biological egalitarianism (nature's rights). Many greens have said that Marxist theory is too old, rigid, dated, and deterministic to be of use in the current environmental crisis. Ecosocialism is an attempt to make more applicable the theory by applying it to modern society and the existent environmental phenomenon.

What Ecosocialists Question

To make a synthesis between Marxist theory, green politics, and ecology, several fundamental political debates and social questions must be resolved. Customary questions include debates about human nature. Ecosocialists ask whether people are inherently selfish and competitive or generous and caring to others. And they ask whether human nature is shaped more by genetic inheritance or education and the environment.

Ecosocialists inquire about the debate between materialism and idealism, and explore whether people are more influenced by ideas, philosophies, and rational arguments or by material, technology, and economic infrastructure developments.

Ecosocialists analyze the debate between free will and determinism, and query whether or not people are really free to act as agents or are determined by external forces such as God, social class, and other environmental determinants.

Another debate explored by ecosocialists is between collectivism and individualism. They analyze how social change occurs and whether it comes more from organized political action or from individuals changing their own behaviors and thoughts.

Consensuses versus conflict as forces for social change are also explored by ecosocialists. They explore hegemony, class, and powerful leaders who impose their will on people, and they explore political systems and the functions of democratic states and processes leading to agreement and equilibrium.

Relatedly, ecosocialists explore issues of authority and libertarianism, and systematically study whether happy and productive states are more likely to be controlled through rigid hierarchies and elite bodies or self-organized groups lacking centralization and bureaucratization.

They also study advantages and disadvantages between small-scale systems and large-scale systems, and analyze how efficiency changes in various political, economic, industrial, urban, rural, and isolated and integrated systems.

In addition, ecosocialists explore issues of technological determinism and how such influences human agency and social change.

Ecosocialists also explore issues of structuralism and whether individual and group behavior and perception are more a result of cognitive structure or social organization, and ask how it is that different cultural experiences lead one to see the world differently.

Ecosocialism also analyzes the pros and cons of free markets versus intervention, and ask whether or not social needs can be met without planning and external assistance. Ecosocialists explore how intervention stifles creativity and perpetuates inefficiencies.

Finally, ecosocialists conduct analyses between modernist and postmodern schools of thought. They explore the tensions between old world conservative ideologies and new world narratives, and examine the limits of rationalism and the Enlightenment while looking for new rational discourse.

The Emerging Possibility of Ecosocialism

Exploring and synthesizing all of these fundamental issues is much of the work of ecosocialists and the emerging possibility of ecosocialism. Ecosocialists observe that if environmentalism is about ideology, a changed view of the natural world, and changed behavior in relation to the environment, then the majority of educated people in developed nations are now environmentalists to one degree or another. Educated people, arguably led by ecosocialists, are reevaluating economic, social, and political schools of thought and out of necessity bringing in ecological elements. Theorists are exploring questions of political economy, subjective preference theory, costs of production theory, abstract labor theory, and ecocentrism. They are also reevaluating political ideologies in light of environmental concerns. They are systematically analyzing traditional conservatism, market liberalism, welfare liberalism, democratic socialism, revolutionary socialism, mainstream greens, green anarchists, postmodernism, green modernism, and all related arguments in the red-green debate.

A Marxist analysis about the current state of capitalism and related environmental issues is useful because in the 19th century, when Marx was doing his analysis, it was very clear that environmental problems were socially inflicted on the working class. Industrialization and urbanization were related social challenges that came with the revolution of manufacturing, and many of these negative relationships continue to be extensively seen today. Ecosocialists hope that the development of their theory will reform the current damaging capitalistic system and replace it with something positive and sustainable.

Ecosocialists believe there needs to be a balance between economic growth and the prosperity it brings, and environmental protectionism. Ecosocialists argue that ecosocialism can bring both ecological sustainability and a stable economy. They believe that it is possible to create an economy that is ecologically benign through rationally planned development that will serve all of society equally. Ecosocialists believe that the language of sustainability should not be limited to ideas about consumerism but must be applied to issues of economy and society as well. Societies plagued by criminality, chaos, social conflict, and war are unsustainable, and they cause depletion of resources and denigration of the environment. Ecosocialists have also developed responses to questions of unemployment, social security, inequality of talent, motivation problems, state planning, private enterprise, people's wants, human greed, education, distribution, regionalism, globalization, scale, cultural tensions, human rights, the future of science and technology, and future moral, social, and political progress.

Ecosocialists also believe that to attain sustainability, economies that are industrialized need to contract and stabilize into a steady equilibrium. This contraction also requires individuals to accept more basic standards of material living, and ecosocialists point out that this does not mean having less happiness. This more basic standard of living can be acceptable to people if it is done gradually and it is understood that people are becoming more equal. A social environment that stresses equality is vital for people to be content with economic contraction. A guarantee of certain minimal goods will also foster contentedness and restrain social conflicts. To achieve stability in the shrinking economy, planning will be important. A comprehensive plan with price controls will prevent fear and chaos.

Ecosocialists believe that in developing nations, the most important issue is stopping population growth, and they believe the state must take action to facilitate this occurrence. Ecosocialists also believe a socialist framework for the new economy will enable the moral growth and the new moral economy that will be essential for attaining a sustainable system. Ecosocialists are clear that socialism is a moral project. Ecosocialism is dependent on the belief that individuals have the capacity to overcome greed and self-interest and that moral growth is possible.

See Also: Ecocapitalism; Governmentality; Institutions; Limits to Growth; Participatory Democracy; Political Ideology.

Further Readings

Benton, T. *The Greening of Marxism.* New York: Guildford, 1996.

Burkett, Paul. *Marx and Nature.* New York: Monthly Review, 1999.

McBrewster, John, Frederic P. Miller and Agnes F. Vandome. *Eco-Socialism: Socialism, Social Ecology, Green Politics, Green Anarchism, Eco-Communalism, Inclusive Democracy, Environmentalism, Environmental Justice, Anti-Capitalism, Green Party.* Beau Bassin, Mauritius: Alphascript Publishing, 2009.

Pepper, David. *Eco-Socialism: From Deep Ecology to Social Justice.* New York: Routledge, 1993.

Sarkar, Saral. *Eco-Socialism or Eco-Capitalism? A Critical Analysis of Humanity's Fundamental Choices.* London: Zed Books, 1999.

John O'Sullivan
Gainesville State College

Ecotax

An ecotax is a market-based environmental policy instrument that has been implemented in different forms in many countries all over the world. The idea behind ecotaxes is that economic activity leads to environmentally harmful external effects (e.g., carbon dioxide emissions produced by factories or car traffic), and these external effects lead to "external costs" that have to be paid mostly by the entire society rather than the polluter himself. An ecotax, therefore, should help as a price signal to confront polluters with the ecological consequences of their economic action and follows, more or less, the "polluter pays" principle.

Ecotaxes can be levied on emissions like carbon dioxide (carbon taxes), on the consumption of energy (e.g., on fuels, power consumption), or on dangerous goods like batteries or fertilizers like nitrate. In general, ecotaxes are levied with the aim that prices "speak the ecological truth" (Ernst Ulrich von Weizsäcker, former member of the Club of Rome).

Scientifically, ecotaxes have been discussed in economics. In the 1920s, the British welfare economist Cecil Arthur Pigou argued that economic action leads to external effects and external costs. In response to this, he suggested that we "internalize external costs" by raising taxes as an instrument that represents exactly the external costs of economic action. One problem of this so-called Pigouvian tax is that in practice it is impossible to accurately assign external costs to all possible external effects in nature caused by economic action (e.g., which costs do polluted rivers cause, or the health problems of people living near factories?). The Pigouvian approach can be characterized as the theoretical root of all ecotax discussions without the possibility of actually being implemented in political practice.

On the basis of these theoretical ideas, the environmental economists William J. Baumol and Wallace E. Oates formulated the standard price approach in the 1970s. The standard-price approach states that the state should first set environmental standards like specific reduction goals for carbon dioxide emissions. Second, the state should levy taxes on emission-relevant actions that lead to a reduction of emissions as a result of the polluters' prevention reactions to increasing prices. Such a policy could encourage enterprises to invest in newer technologies for emission reduction or consumers to buy smaller cars. Ecotaxes are also seen as drivers for innovations. The standard set by the state can be fulfilled by the appropriate prices manifested in higher taxes. In contradiction to the Pigouvian approach, taxes are not oriented toward the "real external costs," which are very difficult to estimate, but to standards set by a government's environmental policy goals. The standard price approach can be labeled as the scientific background of environmental policy discussions about ecotaxes.

Problems of ecotaxes include that it is very difficult to estimate in advance which tax rate actually leads to the desired reduction of emissions. The implementation of goal-oriented and palpable ecotaxes is therefore a trial-and-error process, and as a result the tax is extremely controversial.

One important discussion asks whether tradable permits might be a better market-based environmental policy instrument in general, since emission goals can definitely be reached by emissions trading. Another, more politically oriented, criticism questions whether ecotaxes are really applied to reach environmental policy goals, or to simply raise revenues for the state. An important study by R. W. Hahn shows that in most countries where ecotaxes have been introduced, they fulfilled a function for raising revenues for the state, whereas the stimulus for changing environmentally related behavior was too small to really gain positive ecological effects. So in political debates, many critics of this instrument accuse the state of just implementing ecotaxes for raising revenues without being really interested in a change of ecological behavior. Another potential problem with ecotaxes in political practice is that it is very difficult for politicians to implement market-based instruments that lead to higher prices for consumers and enterprises: Higher taxes are no support for politicians' interests of earning votes, so in political debates, ecotaxes may not be a popular instrument to be pushed by politicians who want to be elected. This leads from an environmentalist's point of view to the problem that ecotaxes with tax rates high enough to really change the polluters' behavior cannot be enforced in the political process because of the vote-earning interests of politicians and strong opposition by powerful lobbyists from industry. In addition, ecotaxes that can be applied might be too low to gain any substantial ecological effects.

To react to such problems, especially in Europe in the 1980s, the concept of "environmental tax reform" has been developed by ecological scientists, green parties, and environmental nongovernmental organizations. Environmental tax reform consists of combining the collection of ecotaxes with a simultaneous reduction of taxes on labor in a way that no added revenues are gained with ecological taxation. This concept is therefore also known as a revenue-neutral ecological tax reform that may take "double dividend" effects. The first dividend is to set incentives for polluters to reduce pollution by increasing prices. The second dividend is to encourage enterprises to employ workers by decreasing taxes or other costs that emerge when employing people. In Germany, for example, the environmental tax that came into force in 1999 was connected with a lowering of social security costs that have to be paid partly by the companies employing workers. The ecological tax reform therefore is, in the eyes of their advocates, an instrument with which to "kill two birds with one stone." On the one hand it is a market-based instrument to reduce environmental problems; on the other hand, it can be a labor-market instrument as well. Advocates also argue that with the concept of an environmental tax reform, political discussions about the instrument are much less controversial because the prejudice against the state collecting ecotaxes just for raising revenues can be overcome with the idea of revenue neutrality.

In political practice, ecotaxes have been discussed worldwide and implemented since the late 1970s, first in European countries. In the European Union (EU), discussions of implementing a common carbon tax as an important instrument for the EU climate policy began at the end of the 1980s. This ecotax could not be implemented because of conflicts between the different member countries and the former principle of unanimity in the EU regarding tax policy. Nevertheless, different ecotaxes or ecological tax reforms have been implemented in Europe by different forerunner countries such as Denmark, Sweden, the United Kingdom, and Germany. In all of these countries, higher taxes on fuels have been an important part of implementing ecotaxes. Whereas today in nearly every EU country, different ecotaxes exist: in the United States, the instrument only inertly asserts. The Organisation for Economic Co-operation and Development database on economic instruments in environmental policy shows that tax rates for unleaded gasoline in the United States and Canada are the lowest in all Organisation for Economic Co-operation and Development member countries. At this time, under the new Obama administration and possibly changing discussions about U.S. climate policy, there seems to be a possibility for reemerging discussions on environmental policy instruments like ecotaxes. In his current book, Pulitzer Prize winner and *New York Times* columnist Thomas L. Friedman suggests an ecotax as an important instrument for U.S. economic and environmental policy. His claim is that it could help save the world's climate from massive, harmful change and make Western states more independent from nondemocratic, oil-exporting countries. Friedman sees ecotaxes as one of the main policy instruments to react to climate change and regulate pollution within the United States.

See Also: Global Climate Change; Innovation, Environmental.

Further Readings

Baumol, William J. and Wallace E. Oates. "The Use of Standards and Prices for Protection of the Environment." *Swedish Journal of Economics*, 73/1 (1971).

Friedman, Thomas L. *Hot, Flat, and Crowded*. New York: Farrar, Straus & Giroux, 2008.

Hahn, Robert W. *A Primer on Environmental Policy Design*. New York: Harwood, 2002.

Määttä, Kalle. *Environmental Taxes: An Introductory Analysis*. Cheltenham, UK: Edward Elgar, 2006.

Organisation for Economic Co-operation and Development. "The OECD Database on Instruments Used for Environmental Policy and Natural Resources Management." http://www2.oecd.org/ecoinst/queries/index.htm (Accessed February 2009).

Pigou, Cecil A. *The Economics of Welfare*. London: MacMillan, 1920.

Michael Böcher
Georg-August-University

Endangered Species Act

The Endangered Species Act (ESA), enacted in its current form in 1973, is the boldest, most protective U.S. federal environmental law ever written. To forestall massive extinction of species by human activity, it seeks to prevent extinction of any plant or animal species in the United States by protecting both the species and the ecosystems on which they depend. In addition, in restricting the import and export of endangered and threatened species and possession of illegally taken wildlife, the statute implements the protections given to animal and plant species worldwide by the Convention on International Trade in Endangered Species of Wild Fauna and Flora, restricting international commerce in plant and animal species harmed by trade. Although the ESA has been amended four times since 1973, the essential protections provided by the act remain largely undiminished.

The ESA identifies species to be protected by listing them as endangered or threatened (E&T) species. Species in the ESA include species, subspecies, or distinct populations of a species. Endangered species are species that are in danger of extinction throughout all or a significant part of their range. Threatened species are species that are likely to become endangered in the foreseeable future. All plant and animal species, except pest insects, are eligible for listing.

Listing of a species as endangered or threatened may be initiated by the agencies that administer the ESA—the National Atmospheric and Oceanic Administration National Fisheries (NOAA Fisheries), which protects marine species, and the U.S. Fish & Wildlife Service (FWS), which protects all other species—or by a citizen petition. When considering whether to list a species, the agency focuses solely on the biological status of the species and threats to the species' continued existence. The agency considers five factors: (1) damage or destruction of the species' habitat; (2) human overuse of the species; (3) nonhuman dangers to the species, such as disease and predation; (4) the inadequacy of existing protection of the species; and (5) other human and natural factors that affect the continued existence of the species. Listing decisions are made through notice and comment rulemaking, using the best available scientific and commercial information. The benefits and costs of listing a species as endangered or threatened cannot be considered in making the listing decision. Within the agency, listing documents are subject to peer review of other scientists. During the second term of President George W. Bush, the FWS was racked by scandal when a political official in the Interior Department intervened in listing and other decisions that must be based on science. Ultimately, after an internal investigation, the official resigned and FWS was forced to revise dozens of FWS decisions.

Because of habitat loss, the Florida panther has become one of the most endangered animals in the United States, with an estimated population of only 90–100 left in the wild.

Source: U.S. Fish & Wildlife Service

Of the nearly 1,900 species listed worldwide as endangered or threatened, over 1,300 reside in the United States. There are an additional 69 species currently proposed for listing.

Candidate Species

Candidate species are species for which the agencies have sufficient information to propose listing, but that have not been listed because higher-priority listing actions are consuming the agencies' limited resources. FWS and NOAA Fisheries cooperate with state and local governments, tribes, private landowners, environmental groups, and other nongovernmental organizations to encourage voluntary actions to reduce the threats to over 250 species that are currently candidate species. Nonfederal landowners can enter into candidate conservation agreements with FWS and NOAA Fisheries to implement measures that reduce or remove threats to candidate species and other at-risk species so that the need to list those species is avoided. In return, landowners may receive regulatory assurances that if a species covered by the agreement is subsequently listed, the landowners will not be required to take any actions other than those specified in the agreement. In the event of listing, the landowners are also given an enhancement of survival permit, allowing an incidental take for activities authorized by the agreement.

Critical Habitat Designation

FWS and NOAA Fisheries are required to identify critical habitat for a listed species, to the extent prudent and determinable, either when a species is listed or within 1 year of the listing decision. Critical habitat is a geographic area that contains physical or biological features essential to the conservation of the species that may need special management or protection. It can include areas not occupied by the species at the time of listing, such as historical range from which the species has been excluded by human activity, that nonetheless are essential to conservation of the species.

Unlike the listing decision, which is based solely on scientific and commercial information about its biological status, the agencies can consider economic factors in designating critical habitat. The agencies can exclude an area from being designated critical habitat if economic analysis suggests that the benefits of excluding it outweigh the costs of exclusion, unless the failure to designate the area as critical habitat may lead to extinction of the species.

Designation of critical habitat for listed species has become extremely controversial, especially in the West, because federal agencies may not take actions adversely affecting critical habitat. Many large animal species require large amounts of habitat, and many activities in the western United States are dependent on federal action because of the federal government's extensive land holdings, so western resource users tend to vehemently oppose designation of critical habitat. As a result, only 540 of nearly 1,900 listed species have designated critical habitat.

Ramifications of Listing

Once a species is listed, it receives significant protections. First, no federal agency can take any action that jeopardizes the continued existence of the species or adversely affects critical habitat. Second, no person can take any action that kills, harms, or harasses a listed animal without a permit. Third, FWS or NOAA Fisheries must develop a recovery plan for each listed species that identifies actions necessary to recover the species to the point where it can be delisted, that is, removed from the list of endangered and threatened species.

Consultation

The ESA section 7 consultation requirement applies only to federal agencies. The federal agency planning to take an action ("action agency") must consult with FWS or NOAA Fisheries concerning any action that it authorizes, funds, or carries out to ensure that the action will not jeopardize the continued existence of an E&T species. Thus, consultation is required before a federal agency issues a license or permit, such as a Federal Energy Regulatory Commission license to build a hydroelectric facility or a Clean Water Act section 404 permit to fill wetlands for real estate development. Consultation is required before issuing a grant to build an alternative energy pilot project or funding highway construction. Consultation is necessary when carrying out projects such as the U.S. Forest Service selling timber or building roads in a national forest or before the Department of Defense builds a new barracks on an air force base.

Federal agencies avoid actions that might jeopardize listed species by consulting with FWS and NOAA Fisheries to determine whether listed species exist in the area affected by a proposed action. If listed species are present, the federal agencies, in consultation with FWS or NOAA Fisheries, conduct a biological assessment to determine whether the proposed action may adversely affect a listed species.

If there may be an adverse effect, the action agency seeks a Biological Opinion from FWS or NOAA Fisheries on whether the action will jeopardize the continued existence of a listed species. If the proposed action will jeopardize a listed species, it cannot be undertaken, absent an exemption secured through a painstaking, time-consuming, politically dangerous decision by a cabinet-level Endangered Species Review Committee, otherwise known as the God Squad. The committee has met only three times in the more than 30 years it has existed.

If jeopardy will occur unless the proposed action is changed in certain respects, the FWS or NOAA Fisheries issues a "conditional jeopardy" opinion that identifies a "reasonable and prudent alternative" to the proposed action that will not jeopardize the species.

Throughout the process of making these decisions, the action agencies are actually responsible for the accuracy of the factual decisions as to presence, adverse effect, or jeopardy, and those decisions may be subject to judicial review. However, throughout the process, the action agencies consult with the experts at FWS and NOAA Fisheries, whose expert advice also may be subject to judicial review. Action agencies that ignore the advice of the expert agencies are likely to have their decisions overturned by the courts.

The consultation process occurs in conjunction with the environmental impact assessment process under the National Environmental Policy Act. If an endangered species is present, the agency proposing action should prepare an ESA biological assessment in conjunction with preparing an environmental assessment. If the proposed action may adversely affect an E&T species, the agency will initiate formal consultation with FWS or NOAA Fisheries, seeking a biological opinion as it prepares an environmental impact statement to comply with the National Environmental Policy Act.

Action agencies often informally consult with FWS or NOAA Fisheries to minimize or eliminate the effect of a proposed agency action on listed species. Through changes in the proposed agency action, action agencies may be able to avoid a biological assessment or the need for a formal biological opinion. This agency's willingness to change the proposed action to avoid formal consultation is similar to the strategy employed by agencies to mitigate or change agency actions to avoid preparation of a full-scale environmental impact statement under the National Environmental Policy Act. Allowing agencies to use these strategies when they possess good information and act in good faith can ultimately provide enhanced protection for endangered and threatened species. However, allowing agencies to use such strategies to avoid consultation may backfire and injure endangered or threatened species, if either the agency does not intend to implement the changes or conditions necessary to avoid impacts on endangered or threatened species or if the agency has bad information about the effectiveness of the changes or conditions.

At the end of the second term of President George W. Bush, the Interior Department made a controversial change in the consultation process that would have allowed action agencies to decide that an action did not adversely affect any endangered or threatened species without any consultation with FWS or NOAA Fisheries. The effectiveness of that administrative rule, along with the other "midnight rules," was suspended on the first day of the Obama administration, and that rule is expected to be rejected by Congress under the Congressional Review Act.

Section 9 Takings

ESA section 9 makes it unlawful to "take" a threatened or endangered species. The ESA defines the term "take" to include actions that harass, harm, pursue, hunt, shoot, wound, kill, trap, capture, or collect any threatened or endangered species. Under the FWS and NOAA Fisheries regulation, "harm" includes significant habitat modification, where it actually kills or injures a listed species through impairment of essential behavior (e.g., nesting or reproduction). Obviously, a broad variety of activities beyond obvious hunting and poaching qualify as a taking under section 9. For example, even failure to take appropriate regulatory action by a local government may amount to a taking.

Persons who knowingly violate the ESA by unlawfully taking an endangered or threatened species may be criminally prosecuted and are subject to a year in prison and a fine of $50,000 per violation. In addition, they may be subject to civil enforcement action by the government or a citizen suit by concerned citizens seeking injunctive relief and civil penalties of up to $25,000 per violation.

Other Prohibited Activities

In addition to takings prohibited under section 9, the ESA makes a variety of activities related to endangered or threatened wildlife unlawful. It is unlawful to import or export listed species, engage in any commercial activity connected to interstate or foreign commerce with respect to a listed species, offer to sell or sell listed species in interstate or foreign commerce, or take listed species on the high seas. It is also unlawful to possess, ship, deliver, carry, transport, sell, or receive listed species that have been unlawfully taken.

Plants receive somewhat less protection from the ESA. Although they are not protected from taking on nonfederal lands, it is unlawful to remove and reduce to possession an endangered plant or to maliciously damage or destroy an endangered plant on federal lands. However, on nonfederal lands, it is unlawful to remove, cut, dig up, or damage or destroy any endangered plant if those acts are in knowing violation of any state law or regulation or in the course of a violation of a state criminal trespass law.

Incidental Takings

Where, incidental to otherwise legitimate activity, nonfederal activities that will result in the taking of an endangered or threatened species may occur, the person engaging in that activity may seek an incidental take permit from FWS or NOAA Fisheries. Under ESA section 10, the applicant must submit a habitat conservation plan with the permit application. The purpose of the habitat conservation planning process associated with the permit is to ensure that the effects of the authorized incidental take are minimized and adequately mitigated.

If an incidental take is expected as a result of a federal action, the Biological Opinion allowing that action will contain an incidental take statement that protects the government and private parties from enforcement action or citizen suits.

Recovery Planning

The FWS and NOAA Fisheries are responsible for preparing and implementing a recovery plan for each E&T species, unless the plan will not promote conservation of the species. The ESA requires that the agencies give priority to preparing plans for E&T species most likely to benefit from such plans, especially because of construction, development projects, or other economic activity. Recovery plans specify the steps necessary to conserve the species and set recovery goals that, when met, allow delisting.

FWS and NOAA Fisheries have prepared recovery plans for slightly more than 1,100 of the 1,300 species listed in the United States, in roughly 615 separate recovery plans. Some recovery plans cover more than one species, and a few species have separate plans covering different parts of their ranges.

Nonfederal landowners who are willing to voluntarily manage their land to protect listed species may enter into a Safe Harbor Agreement with FWS or NOAA Fisheries. In exchange for restricting the use of their land in ways that benefit listed species, the landowners receive assurances that no additional future regulatory restrictions will be imposed on the use of their property.

Recovery and interstate commerce permits may be issued to allow for incidental take as part of recovery activities. A typical use of a recovery permit is to allow scientific research on a listed species to understand better the species' long-term survival needs. Interstate commerce permits allow transport and sale of listed species across state lines, such as for a breeding program.

See Also: Regulatory Approaches; Environmental Protection Agency; Fish & Wildlife Service.

Further Readings

Baur, Donald C. and W. M. Robert Irvin. *The Endangered Species Act: Law, Policy and Perspectives*. Chicago, IL: American Bar Association, 2001.

Bean, Michael J. and Melanie J. Rowland. *The Evolution of National Wildlife Law: Third Edition (Project of the Environmental Defense Fund and World Wildlife Fund-U.S)*. Santa Barbara, CA: Praeger, 1997.

Endangered Species Act. http://www.fws.gov/endangered/pdfs/ESAall.pdf (Accessed February 2009).

Goble, Dale D., J. Michael Scott and Frank W. Davis. *The Endangered Species Act at Thirty: Vol. 1: Renewing the Conservation Promise.* Washington, D.C.: Island Press, 2005.
Listing Statistics. http://www.fws.gov/endangered/listing/Index.html (Accessed February 2009).
NOAA Fisheries Endangered Species Regulations, 50 CFR Part 222. http://frwebgate6.access.gpo.gov/cgi-bin/TEXTgate.cgi?WAISdocID=319974288391+4+1+0&WAISaction=retrieve (Accessed February 2009).
NOAA Fisheries Protected Resources Program. http://www.nmfs.noaa.gov/pr/ (Accessed February 2009).
U.S. Fish & Wildlife Service Endangered Species Program. http://www.fws.gov/endangered/ (Accessed February 2009).

Susan L. Smith
Willamette University College of Law

Endocrine Disrupters

Endocrine disruptors are substances that act like hormones in a body. Examples include DDT, PCBs, and the controversial Bisphenol A. They are hormonally active agents that are exogenous (originate outside of affected organism). If present and active they may affect either cells or an individual organism. In medicine, exogenous factors may be pathogens or therapeutics. These exogenous factors include DNA transfers, viral infections, and carcinogens. When these substances act like hormones, they affect the endocrine system's functions, which are to release hormones that are chemicals used as an informational system in the body. Hormones are analogous to the signals sent through the nervous system but use blood vessels rather than nerve tissue. In humans, the major endocrine system parts include (from head to toe) the pineal gland, pituitary gland, thyroid gland, thymus, adrenal glands, and pancreas. Reproductive glands in females are the ovaries and in males the testes. Endocrine pathologies number in the hundreds—some are common (diabetes mellitus, thyroid disease, and obesity), and others such as Cushing's syndrome (pituitary gland) or Addison's disease (adrenal glands) are rare. Endocrine pathology classifies endocrine disorders as primary, secondary, and tertiary. Cancer can also affect the endocrine system in a number of ways.

Endocrine disrupters have been linked in numerous studies to negative biological effects. These studies of laboratory animals given low doses of various endocrine disrupters suggest that they can produce adverse effects in humans. Tragically, some endocrine disrupters have been administered as therapies. For example, in the 1970 the drug diethylstilbestrol was prescribed to over five million pregnant women as a way to block spontaneous abortions. However, it was discovered that the children born to women who had received diethylstilbestrol developed significant reproductive problems during puberty. The problems included disruption of reproductive development and vaginal cancer in some cases.

In 2007, the chemical bisphenol A came under scrutiny as an endocrine disrupter. In March 2007, a class action lawsuit was filed in California against manufacturers of plastic baby bottles on the grounds that the chemical was in the plastic bottles and posed a developmental health risk.

DDT (dichlorodiphenyltrichloroethane) was hailed as a magic chemical for killing insect vermin such as lice, roaches, ants, and mosquitoes, as well as other insects. However,

Adult and juvenile brown pelicans at a nesting colony in Florida. Brown pelicans suffered population declines because dichlorodiphenyltrichloroethane (DDT) weakened the shells of their eggs, which then ruptured during incubation.

Source: U.S. Fish & Wildlife Service

after years of often indiscriminate use, negative effects appeared that were described by Rachel Carson in *Silent Spring* (1962). Her argument was that pesticides, herbicides, and other chemicals used in the environment were affecting wildlife, and thus could also affect humans. The most infamous example of DDT affecting wildlife was the effect it had on large raptors' eggshell thickness. The shell must be thick enough to withstand the stresses of incubation but not so thick that the chick cannot peck its way out at the appropriate time. DTT was found to interfere with the shell thickness as an endocrine disrupter.

Negative health effects attributed to chemicals that cause endocrine disruption include reproductive problems such as sterility, abnormalities, spontaneous abortions, premature puberty, interference with the immune system, and other problems. The issue that is a major concern of public health is whether the large number of chemicals that have been put into the environment, into the food chain, or into human-built environments are of sufficient strength to affect human endocrine systems. The studies conducted on laboratory animals use large or massive doses of chemicals to test for their potential as endocrine disrupters. However, human exposure is to low levels of potential endocrine disrupters.

The effect between high-level dosages in laboratory animal studies and the low-level dosages to which humans are exposed is a matter of ongoing investigation. The sources for human exposure to endocrine disrupters are numerous, but food is the most common. Human exposure to DDT, polychlorinated biphenyls, and residues of hormones used in poultry, meats, and fish are prime suspects. In the case of fish, exposure of fish—whether freshwater or saltwater, wild or farm raised—does not seem to matter much because of pollutants in the water that enter into the fish and thus into the food chain.

Environmental exposure in homes to chemicals used for cleaning, finishing, flame retarding, or other uses have also been implicated as potential endocrine disrupters. Chemicals naturally occurring in wood can cause health problems at high levels. Chemicals that are used to treat wood or to manufacture flooring or to treat wood against termites or other insects can also threaten humans. Even if the ingestion is at a low level of limited amounts of dust from the wood products, the effects on infants and children can be significant. Studies have demonstrated much higher levels in very young children compared with their parents. In other cases, the contamination of water supplies by agricultural chemicals poses general health problems including endocrine disruption.

To keep endocrine disrupters from affecting the environment, numerous laws have been passed in the United States and in other countries. In the United States, federal legislation includes (as amended) the Toxic Substances Control Act of 1976; the Federal Insecticide,

Fungicide, and Rodenticides Act (1947); the Federal Food, Drug, and Cosmetic Act of 1938; the Clean Water Act (1985); the Safe Drinking Water Act of 1986; the Clean Air Act of 1963; the Food Quality Protection Act of 1996; and the Safe Drinking Water Act of 1996. Federal legislation has been amended numerous times as new problems arise and new solutions are proposed.

In contrast to the European Union, which created the Registration, Evaluation, and Authorization of Chemicals program (2006), the United States, in the Toxic Substances Control Act, grandfathered most chemicals. The Registration, Evaluation, and Authorization of Chemicals program requires the importation or manufacture of chemicals in excess of 1 ton to be registered.

To protect against endocrine disrupters, the Environmental Protection Agency began development in 1998 of the Endocrine Disrupter Screening Program. The program began an analysis of chemicals in commercial use that at the time numbered over 85,000. The screening was to use in vitro test systems on a variety of animals from tadpoles to mice: If the screening produces any reaction with estrogen receptors or with androgen receptors, then further testing is ordered. However, the program did not begin until 2007, when tests with sufficient sensitivity were adopted.

Fear and suspicion of science and government regulators affects the politics of endocrine disrupters. For some the issue is proof of damage, because it is expensive to protect against the disrupters. Simply banning them may create serious unintended consequences. For example, DDT was banned, but DDT-soaked mosquito nets are very effective against malaria, which kills millions, as well as other deadly mosquito-borne diseases. To simply ban it on the grounds that it is, if mishandled, a potential endocrine disrupter is action likely to evoke political battles. Most other, if not all other, chemicals can evoke similar costly battles. Scientifically based decision making that is open to ongoing modification needed to adjust to the evolving science is the approach that probably will not satisfy environmental or business ideologues, but it is the approach that sound regulations would follow.

See Also: Anti-Toxics Movement; Clean Air Act; Clean Water Act; Environmental Protection Agency; PCBs; *Silent Spring*.

Further Readings

Berkson, Lindsey and John R. Lee. *Hormone Deception: How Everyday Foods and Products Are Disrupting Your Hormones—And How to Protect Yourself and Your Family*. New York: McGraw-Hill, 2000.

Guillette, L. J., Jr. and A. Crain, eds. *Environmental Endocrine Disrupters*. London: Taylor & Francis, 2000.

Metzler, M. *Endocrine Disrupters*. New York: Springer, 2002.

Naz, Rajesh K., ed. *Endocrine Disrupters: Effects on Male and Female Reproductive Systems*. London: Taylor & Francis, 2004.

Shaw, I., ed. *Endocrine Disrupting Chemicals in Food*. London: Taylor & Francis, Inc., 2009.

Andrew Jackson Waskey
Dalton State College

Environmental Justice

Although there is not a standard definition of what constitutes environmental justice, the environmental justice movement grew out of the inequitable distribution of environmental hazards among poor and minority citizens. Their advocacy is based on the premise that all people deserve to live in a clean and safe environment free from industrial waste and pollution that can adversely affect their health and well-being. Environmental justice seeks to bring this ideal state of the world into reality. A society that embraced environmental justice would ensure that all of the costs that accompany living in an industrialized nation are equally and fairly distributed among citizens and that a nation's minority and/or underprivileged populations are not facing inequitable environmental burdens. From a policy perspective, practicing environmental justice entails ensuring that all citizens receive the same degree of protection from environmental hazards by the government.

The environmental justice movement combines traditional environmentalism with the conviction that all individuals have the right to live in a safe environment. It started as a grassroots movement during the early 1980s in pockets of the country where minorities and the underprivileged faced environmental burdens. It has evolved into community initiatives, federal offices, and a presidential executive order. A criticism of the environmental justice movement is sometimes levied from mainstream environmentalists who believe this is not an environmental movement but a social justice movement.

Environmental justice does not have a birthplace or date of conception. It grew out of the many movements by local groups that sought to protect their own health and well-being, and that of others in their community. One such grassroots start-up formed in the early 1980s in Warren County, North Carolina—the home of a polychlorinated biphenyl landfill. The citizens of Warren and surrounding counties had been the victims of midnight dumping, as industrial polluters dumped thousands of gallons of polychlorinated biphenyl–laden oil along their roadways in the middle of the night. The company and contractor accused of dumping explained that the motive for dumping was to avoid paying for the recycling costs of the oil. In an effort to deal with the dumping crisis, the state of North Carolina created a toxic landfill in Warren County. The residents in Warren County were mainly African American, and it was also one of the poorest counties in the state of North Carolina. Residents felt that the county was picked for the landfill based on its demographic characteristics. Citizens rallied together to try to prevent the dump. They wrote letters, held demonstrations, and

Since the early 1980s, the environmental justice movement has been working toward ending the practice of siting waste-storing or waste-producing facilities, like this power plant, in poor or minority communities.

Source: iStockphoto.com

blocked trucks. Although successful in gaining media attention, the landfill was eventually developed.

The events in Warren County eventually led to the coining of the term *environmental racism*. This term refers to the targeting of communities for the placement of waste-generating or waste-storing facilities or to the discrimination in the enforcement of environmental standards in communities or neighborhoods based on the racial characteristics of the residents. In the book *Dumping in Dixie*, Robert Bullard examines the location of waste disposal facilities in five communities. Using demographic data based on the Zip codes in which the facilities are located, Bullard concludes that the prosperity of a community is not as good a predictor of hazardous waste locations as is the race of the residents. He concludes that choosing where to locate a waste facility often includes racism. Racism has often been used to describe an individual's decision to discriminate against a person or group of people based on their race. This would be applicable in cases in which a company decided to site a hazardous waste–producing facility in a minority neighborhood simply because of the number of minorities living there. Environmental racism is often a form of institutional racism, however. Institutional racism is not based on individual discrimination but, rather, on discriminatory practices that become part and parcel of the practices, traditions, and polices of organizations or society.

An example of institutional racism may be linked to the "not-in-my-backyard" (NIMBY) phenomenon. NIMBY involves groups of citizens engaged in blocking unwanted projects in a given community or neighborhood. NIMBY movements are likely to involve members of a community who have political efficacy and, in turn, believe that they can make a difference. Often, disenfranchised individuals in poor and minority neighborhoods do not exhibit much political efficacy. This means that NIMBY groups are less likely to organize in those neighborhoods. A decision maker looking for a site to place a waste-producing or waste-storing facility would want to avoid areas where there is strong citizen opposition and would be more likely to choose a site based on the path of least resistance. The selection of a minority neighborhood based on these criteria would be an example of institutional discrimination.

Environmental justice is about not only environmental burdens faced by minorities but also those related to an individual's income or gender. Sociologists disagree on the extent to which poverty and race are separated from each other in reference to discrimination. In both the Warren County and Bullard examples given, the individuals in the minority neighborhoods and communities facing environmental burdens were also low-income individuals.

One shining example of a community response to an environmental injustice is the Dudley Street Neighborhood Initiative in Boston. The Dudley Street neighborhood is located less than 2 miles from downtown Boston. In the early 1980s, the Dudley Street neighborhood had a large amount of vacant land. The neighborhood population was made up of minority and low-income families. The vacant land had become the unofficial city dumping ground, where individual citizens and companies discarded their trash, and much of the neighborhood became contaminated by hazardous waste. With the help of volunteers and a small amount of grant money, the citizens were empowered to form the Dudley Street Neighborhood Initiative. The Dudley Street Neighborhood Initiative involved more than 3,000 residents, businesses, nonprofits, and religious institutions. They became dedicated to restoring the neighborhood not only into a healthy place to live but also into a thriving community.

The Dudley Street Neighborhood Initiative has been successful in reversing the environmental dangers presented to members of the community. More than half of the vacant parcels have been cleaned up and transitioned into affordable housing, a new school, a community center, a community greenhouse, parks, playgrounds, and other public areas.

President Bill Clinton signed Executive Order 12898 in February 1994. This order required federal agencies to develop strategies to identify and to address the health or environmental effects of government programs, policies, and actions on minority and low-income populations. He also created an Office of Environmental Justice. This office is now part of the Environmental Protection Agency. This office has been scaled down since the Clinton presidency, but the Environmental Protection Agency still has a National Environment Advisory Council and heads the Federal Interagency Working Group, which gives out grants and awards for communities trying to tackle environmental injustices.

The Green Belt Movement focuses on the environmental hardships borne by women as they attempt to take care of their families in Kenya and is an international example of the principles of environmental justice. The Green Belt Movement was founded by Professor Wangari Maathai. She began this movement in the late 1970s, when she organized poor rural women in Kenya to plant trees. Planting trees promised to generate the wood needed for cooking but was also part of a larger effort to combat deforestation in Kenya. Communities in Kenya have organized to prevent further environmental devastation. The movement has also improved both environmental and societal conditions for women in Kenya. Maathai was awarded the Nobel Peace Prize in 2004 for her work with this movement.

Criticisms

There is a small group of traditional environmentalists who try to distinguish themselves from the environmental justice movement. They have criticized this movement as an attempt to shift the focus of the environmental movement away from important environmental issues and toward more anthropocentric concerns such as racism, classism, and sexism. Other environmentalists embrace environmental justice as a bridge between environmentalism and important social movements.

See Also: Environmental Movement; Environmental Protection Agency; Equity; NIMBY; Participatory Democracy.

Further Readings

Bryant, B., ed. *Environmental Justice: Issues, Policies, and Solutions*. Washington, D.C.: Island Press, 1995.

Bullard, R. *Dumping in Dixie: Race, Class, and Environmental Quality*. Boulder, CO: Westview, 1990.

Cuesta Camacho, D. E. *Environmental Injustices, Political Struggles: Race, Class, and the Environment*. Durham, NC: Duke University Press, 2002.

Dudley Street Neighborhood Initiative. http://www.dsni.org/ (Accessed January 2009).

Lester, J., et al. *Environmental Injustice in the United States: Myths and Realities*. Boulder, CO: Westview, 2000.

Maathai, Wangari. *Unbowed: A Memoir*. New York: Alfred A. Knopf, 2006.

Shrader-Frechette, K. *Environmental Justice: Creating Equality, Reclaiming Democracy*. Oxford: Oxford University Press, 2002.

U.S. Environmental Protection Agency. "Environmental Justice." http://www.epa.gov/environmentaljustice/ (Accessed January 2009).
Wenz, P. S. *Environmental Justice*. Albany: State University of New York Press, 1988.

Jo A. Arney
University of Wisconsin–La Crosse

Environmental Management

Human societies have always had environmental impacts; the major differences between early human civilizations and our contemporary problems are the scales at which humans impact the environment (local/regional/global) and rates of environmental degradation. Human impacts have been magnified by various interconnected factors including population growth and mass consumption. This has led to habitat loss/fragmentation, soil erosion and nutrient depletion, biodiversity loss, water deficits and contamination, food shortages and famines, unsustainable food systems, toxic emissions, chemical pollution, and heavy reliance on nonrenewable resources. Thus, human populations, natural resources, technologies, development, and environmental health are closely/inseparably interrelated. The "environment" is increasingly being managed to meet the goals of sustainability, although the effectiveness of environmental management is debatable.

There are several major challenges for managing our environment that involve balancing many political, ecological and cultural concerns, particularly in rapidly developing countries. High population and economic growth rates have resource consumption and degradation rates larger than the renewable carrying and absorbing capacity. This situation threatens long-term sustainability and exacerbates conflicts between the resource use and users. Existing mechanisms for linking science and policy for environmental management are often highly sectored, whereas the major environmental problems today are increasing across several sectors. Political and administrative support for environmental management is lacking and significant environmental issues (e.g., the effects of agricultural chemicals) are not on policy agendas or actively promoted. Although many environmental and cultural effects of development are proven to be irreversible, the critics of environmental management are taking refuge under the uncertainty and

In an example of ecosystem restoration, the U.S. Fish & Wildlife Service conducted this wetlands habitat restoration project in a former orange grove to benefit the birds of Florida's Pelican Island Wildlife Refuge.

Source: U.S. Fish & Wildlife Service

unpredictability of environmental changes and pushing the burden of proof onto the vulnerable poor. Consequently, the lowest economic classes often endure the maximum impact of nonsustainable development as common public resources (e.g., groundwater, grazing areas, community forests, etc.) on which their very survival depends upon are severely degraded. Traditional ecological knowledge, transmitted orally across generations, could play key roles in sustainable environmental management, such as in the areas of ethnobotany, zoology, pedology, and medicine. Unfortunately, the transference of this knowledge is increasingly threatened by globalization processes (e.g., biopiracy) that result in the consequent loss of local identity, collapse of indigenous social structure, gradual decline of traditional ecological knowledge, and a breakdown of traditional social institutions that were responsible for environmental management.

Environmental management is not, as the term suggests, the management of the environment per se but rather the management of inseparable interactions of physical, chemical, and biological (including human) processes and their consequent impacts on the environment. Hence, it necessarily involves the relationships of the human environment, such as the social, cultural, and economic environment with the biophysical environment. Thus, this topic demands inter- and transdisciplinary approaches; for example, to deal with water management problems, one has to deal with a variety of interrelated fields: hydrogeology, eco-hydrology, watershed planning, sociology, cultural anthropology, environmental economics, and so on.

How Can We Manage the Environment?

To ensure more reliable and effective environmental management, we have to sustainably use ecosystem goods (those that have monetary values: food, fuel, fodder, and fertilizer; construction materials; medicinal plants; wild genes; tourism and recreation landscapes) and conserve and protect ecosystem processes (ecosystem functions with nonmonetary values: hydrologic flux and storage, biological productivity, biogeochemical cycling and storage, decomposition, and maintenance of biodiversity) and ecosystem services (maintaining water and nutrient cycles through restoring soil vegetal cover; generating and maintaining soil through regenerating soil biota; pollination; regulating climate, cleaning water and air by detoxification, and providing beauty and inspiration by restoring native plant and animal communities).

Until recently, ecosystem services were assumed to be inexhaustible, indestructible, and because they had not apparent costs, without value. However, growing demands on the natural environment can no longer be met by tapping unexploited resources, because they are becoming short in supply. These services are not only critical to the bare-bones survival of life but are essential to the economic and technological functions that underlie modern civilization. Unfortunately, the millennium ecosystem assessment has indicated that 60 percent of the ecosystem services (15 of 24) are being degraded or used unsustainably, especially when we know that they are irreplaceable.

Whether it is production or consumption, or whether it is exchange, the goods/commodities and services that are involved can be traced to constituents provided by nature. For effective conservation of ecosystem services, we have to aim to restore ecosystem functions. Ecosystem restoration is founded on fundamental ecological and conservation principles (rehabilitation, reclamation, and bioremediation) and involves management actions designed to facilitate the recovery or reestablishment of native ecosystems. A central premise of ecological restoration is that restoration of natural systems to conditions consistent with their evolutionary environments will prevent their further degradation while simultaneously conserving their native plants and animals. Irrespective of what we are dealing with—ecosystem goods, processes, or services—we have to focus on managing five types of capital: human capital (skills, knowledge, information, ability to work, health, and culture), financial capital (cash, investment, savings, credit, remittances, pensions, and monetary instruments), physical capital (transport, shelter, water, energy, and commons), natural capital (land, water, wildlife, biodiversity, environment, resources, living systems, and ecosystem services), and social capital (networks, groups, trust, and access to institutions). Although all are equally important, natural capital has to be given the utmost priority because it cannot be replaced.

Environmental Management—The Guiding Principles

Several overriding principles should guide environmental management practices. First, define the desirable goals and priorities that are not merely utilitarian or biocentric, but are easily measurable, predictable, and monitorable, and take into consideration the risks, uncertainty, and complexity of the environment. These goals should also redefine the administrative boundaries based on ecosystem or watersheds, not political boundaries. The integrity of life-supporting systems (soil, air, and water) should be recognized, respected, and conserved while opting for multiple use options, sustainable livelihoods, and economic prosperity. This can be done by the integration of sectors and policies at various spatial and temporal levels that are well articulated in the Convention on Biological Diversity, Millennium Development Goals, Commission on Sustainable Development, and Agenda 21.

Ecological, social, cultural, and economic indicators need to be employed to enable participatory monitoring of a sustainable management plan. Ecological history can be used as a starting point to examine the way the environment has been used and misused, indicating the current state of the environment and shaping future environmental management decisions. The maintenance of ecological processes can be done only by limiting resource use, degradation, and pollution rates below the assimilative carrying capacity (i.e., accounting for the number of individuals as well as the intensity of use); however, the actual biophysical and social limits will vary according by site. Resource efficiency (e.g., energy, water, minerals, etc.) must be optimized to ensure sustainability and encourage self-sufficiency for food and energy security. Though environmental changes are riddled with unpredictability and uncertainties, a precautionary approach should be adopted to avoid costlier mistakes that may trigger irreversible changes. Proactive and precautionary approaches have to be preferred, employing the principles of adaptive management (for flexibility and fine tuning), accountability, responsibility, and transparency.

To ensure defined goals and objectives are met on a sustainable basis, private and traditional property rights should not be marginalized. The use of collaborative decision making for community-based comanagement of ecosystems wherever possible will help ensure effective stakeholder analysis and conflict-resolution strategies. Organizational and institutional changes need to be made that catalyze active participation and cooperation to conserve the common property resources. Information and communication technologies, including decision support systems, promote participatory monitoring and feedback. Environmental management participation can also increase with environmental education that promotes sustainable development through experiential learning. Active involvement can be maintained by providing adequate legal mechanisms and financial incentives (and disincentives) for ensuring sustainable environmental management.

Future Perspectives

Environmental management, at the nexus of a number of issues from food, energy, water, and health security to climate change, is emerging as one of the major challenges of the 21st century. The recently emerging ecological footprints methodology (a measure of how much biologically productive land and water area an individual, a city, a country, a region, or humanity uses to produce the resources it consumes and to absorb the waste it generates) indicates that the United States needs 10 times greater area (5 hectares/person) than the developing nations (0.5 hectares/person). For everyone to live at today's U.S. footprint would require three planet Earths. We must think globally, but act locally—both to reduce our individual footprints and as an example to others. Sustainable development (development that meets the needs of the present without compromising the ability of future generations to meet their own needs) must be viewed in a global context as a result of emerging globalization. Recent public awareness of resources and their competing and conflicting uses has identified sustainable development as a desired goal in the environmental decision-making process. The concept of sustainable development provides a framework for integrating environmental sustainability with economic, social, cultural, and political sustainability.

Achieving environmental sustainability requires managing and protecting ecosystems to maintain both their economically productive and their ecological functions, maintaining the diversity of life in both human-managed and natural systems, and protecting the environment from pollution to maintain the quality of land, air, and water. Environmental sustainability would be aided by the transfer of best practice, or "leapfrogging" low-carbon technologies from the richer to the poorer regions. However, we should remember that it starts and ends with us. No set of policies, no system of incentives, and no amount of information can substitute for individual responsibility.

Sustainable management issues continually change over time in response to coevolving social, economic, and ecological systems. Under these conditions, a community-based, adaptive comanagement approach offers an opportunity for more proactive and collaborative approaches to resolving the complex environmental problems we are facing today. Growing skepticism regarding the merits of a technoscientific approach to environmental and ecological crises will be a primary motive for the emerging ecological democracy approach. Ecological democracy can be defined as an alternative democratic model that strives to incorporate interested citizens into environmental decision making in a nonhierarchical manner. This is in contrast to the conventional governance systems that systematically concentrate environmental amenities in the hands of the elites while imposing environmental and ecological degradation on the more vulnerable, poorer communities.

See Also: Agenda 21; Agriculture; Appropriate Technology; Millennium Development Goals; Participatory Democracy; Policy Process.

Further Readings

Barrow, C. J. *Environmental Management for Sustainable Development.* New York: Routledge, 2006.

Buchholz, Rogene A. *Principles of Environmental Management: The Greening of Business,* 2nd ed. Upper Saddle River, NJ: Prentice Hall, 1998.

McConnell, Robert L. and Daniel C. Abel. *Environmental Issues: An Introduction to Sustainability*, 3rd ed. Upper Saddle River, NJ: Prentice Hall, 2007.
Miller, G. Tyler. *Living in the Environment: Principles, Connections, and Solutions*, 14th ed. Belmont, CA: Brooks/Cole, 2005.
Miller, G. Tyler. *Sustaining the Earth: An Integrated Approach*, 7th ed. Advantage Series. Belmont, CA: Brooks/Cole, 2005.
O'Leary, Rosemary. *Managing for the Environment: Understanding the Legal, Organizational, and Policy Challenges*. Hoboken, NJ: Jossey-Bass, 1998.
Ricci, Paolo F. *Environmental and Health Risk Assessment and Management: Principles and Practices (Environmental Pollution)*. New York: Springer, 2005.
Robbins, Paul. *Political Ecology: A Critical Introduction (Critical Introductions to Geography)*. New York: Wiley-Blackwell, 2004.
Russo, Michael V. *Environmental Management: Readings and Cases*. Thousand Oaks, CA: Sage, 2008.
Spaargaren, Gert, Arthur P. J. Mol and Frederick H. Buttel. *Environment and Global Modernity*. Thousand Oaks, CA: Sage, 2000.
U.S. Department of Energy. "Environmental Management Home Page." http://www.em.doe.gov/pages/emhome.aspx (Accessed July 2009).
Weiszäcker, Von. *Factor Four, Doubling Wealth, Halving Resource Use*. London: Earthscan, 2000.
Wright, Richard T. *Environmental Science: Toward A Sustainable Future*, 10th ed. Upper Saddle River, NJ: Prentice Hall, 2007.

Gopalsamy Poyyamoli
Pondicherry University

Environmental Movement

The environmental movement is perhaps the most significant contemporary global movement to have emerged in recent decades. The relationship between humankind and nature has been the subject of much debate and inquiry over time. The environmental movement had its cultural origins in literary accounts of humanity's relationship with nature, beginning from the Romantic poets such as William Blake, John Keats, Percy Bysshe Shelley, and Lord Byron, whose works were concerned with the reconciliation of man and nature. This aesthetic could also be found in subsequent Transcendentalist American literature, such as Henry David Thoreau's *Walden*, published in 1854. The Transcendentalists were interested in the spiritual connections that connected humankind and nature with God and could be seen as the forefathers of deep green ecologists. Charles Darwin's *Origin of the Species* was published in 1859, creating further interest in the understanding of nature.

George Perkins Marsh wrote of the destructive effect of agriculture in his book *Man and Nature* in 1864, and President Theodore Roosevelt would develop the national parks with Gifford Pinchot of the Forestry Service in the early 1900s. In the aftermath of the Industrial Revolution, concerns about protecting wildlife led to the emergence of a progressive conservation movement, alongside federal regulation of natural habitats and the establishment of national parks. Influential conservation groups included the National Audubon Society, founded in 1886, and the Sierra Club, founded by John Muir in 1892. Muir and Pinchot would become adversaries in the campaign to prevent the building of a dam in Yosemite National Park in the early decade of the 19th century.

The preservationist and conservationist movements were prominent in the United States during the final decades of the 1800s and the early decades of the last century. Muir succeeded in having Yosemite National Park created during the 1890s. He was also a central figure in the foundation of the Sierra Club in 1892, which focused on the preservation of the wilderness. The Audubon Society was concerned with the protection of birds, and both groups contributed to the founding of the National Park Service in 1916. Franklin D. Roosevelt would also consider the importance of conservation as part of his New Deal during the Depression years of the 1930s. Americans had become concerned about the Dust Bowl phenomenon that characterized that era, and the Civilian Conservation Corps was founded to address both the ecological and economic problems of the day. Aldo Leopold came to prominence as an advocate of forests and wilderness regions in the 1920s. In the post–World War II era, conservationists were active in highlighting the degradation caused by the urbanization of U.S. society, as well as other concerns such as nuclear warfare. Scientists concerned about the threats posed by industrialization began joining with conservationists to present an environmental critique of modernity. The 1950s witnessed a major controversy about the building of the Echo Park Dam in the state of Colorado. This campaign led to the signing of the Wilderness Act of 1964, which protected millions of acres of wilderness and forests throughout the United States.

Rachel Carson's book *Silent Spring* (1962) highlighted concerns about the effect of science on nature. Beat poets such as Gary Snyder were also influential in the early environmental movement. During the 1960s, a widespread social movement emerged around environmental issues; the environmental movement emerged alongside the mobilization of student, feminist, and antiwar movements. The growing threats of a "population explosion" were expressed in Paul R. Ehrlich's book *The Population Bomb* (1968). Ehrlich predicted the onset of famine from overpopulation and a growing scarcity of resources. Concerns about toxicity and pollution inspired an urban-based environmental movement as an educated population began to question the environmental costs of continued industrialization and urban sprawl. Other influential environmental publications include Garrett Hardin's *The Tragedy of the Commons* (1968), which highlighted humankind's propensity to overconsume commonly held resources, and Donella Meadows' *The Limits to Growth* (1972), which measured the effect of such overconsumption on the Earth's finite resources, as many came to realize the fragility of planet Earth in the wider universe in the aftermath of the moon landings in 1969.

A series of environmental organizations emerged from the counterculture; Friends of the Earth was founded in 1969, the Natural Resources Defense Council (NRDC) in 1970, and Greenpeace in 1971. These organizations would provide the platform for a burgeoning interest in environmental issues following the first Earth Day in 1970. The establishment of the U.S. Environmental Protection Agency (also in 1970) was a response to this wider interest in environmental issues, leading to an increase in environmental regulation. The Brundtland Report (titled *Our Common Future*) from the World Commission on Environment and Development in 1987 set out an understanding of sustainable development on a global level. This was followed by further initiatives such as the 1992 Earth Summit (which introduced Agenda 21), the Rio Declaration, and the Commission on Sustainable Development. Many environmental groups advocated for the Kyoto Protocol on Climate Change in 1997, and environmental nongovernmental organizations continue to lead on issues such as climate change.

The increased incidents of environmental catastrophes such as Three Mile Island, Bhopal (India), and Chernobyl led to increased concerns about environmental degradation globally. In 1992, the Earth Summit led to a further growth in green consciousness; environmentalism became a fashionable cause, with celebrity endorsements and an increased focus on environmental issues from a newly globalized media. Major concerns such as the depletion of the ozone layer, acid rain, global warming, and climate change led to a surge in interest in environmental issues and broadened the scope of environmental movement activism. In recent years, the environmental movement has continued to campaign on a range of issues including sustainable development based on increased ecological and global equity, green consumerism, and green politics with a focus on dignity for all species. Green political parties have not met with varied degrees of political success, remaining marginalized in the United States and the United Kingdom. However, green parties have participated in coalition governments in Germany, Finland, France, Belgium, the Czech Republic, Mexico, and the Republic of Ireland. China's environmental movement is expanding, and as China's economic growth continues, the Chinese environmental movement may be one of the most significant of the coming century.

The environmental movement includes a diverse set of philosophies and principles that range from conservation of species and ecosystems, preservation of habitats and natural areas of special interest, promotion of ecological issues, the embracing of "deep green" notions about the primacy of nature over human development, promotion of sustainable development, biocentricism, and green politics. The late Arne Naess wrote of a deep green life based on his concept of "ecosophy" in 1972. James Lovelock's book *Gaia: A New Look at Life on Earth* (1979) argues that the planet is a single living, self-regulating entity. The concept of "social ecology" was put forward by Murray Bookchin in the 1960s. Bookchin argued that because environmental problems were based on human injustice, environmental issues must be linked with issues of social justice. For Bookchin, the impulse to dominate nature or other humans was at the heart of all environmental and social problems. Social ecology's questioning of the impulse toward growth and development has become one of the environmental movement's most significant critiques of overconsumption. The environmental movement incorporates a range of ideologies across a spectrum that includes what Australian writer Timothy O'Riordan terms "deep green ecologists," who value nature over economic growth, through to "light green technocentrics," who favor growth.

This division between "ecocentric" ecologism and "anthropocentric" human-based concerns has characterized the divisions that are inherent in the environmental movement. The environmental movement recognizes that humanity has a role to play in the planet but argues that this creates certain obligations in relation to maintaining a sustainable present so as not to jeopardize the planet's future. The environmental movement's principles are based on the promotion of a balanced approach to ecological and human rights. However, the environmental movement has been criticized for being too focused on Europe and North America, whereas the emergence of corporate green-wash in response to environmental concerns has reduced the environmental movement to the role of a special-interest lobby group, lacking a wider appeal in spite of increased interest in environmental issues among the wider public.

The environmental movement includes many groups with participatory activist roots, such as Earth First!, whereas others have embarked on a more professionally negotiated platform, such as the World Wildlife Fund. However, it was the sinking of the Greenpeace ship *The Rainbow Warrior* in Auckland Harbor in 1985 that gave rise to the environmental movement's global activist appeal. *The Rainbow Warrior* was en route to protests against nuclear testing in Moruroa, in the French Polynesian Islands of the South Pacific. New Zealand became a nuclear-free zone as a result of the outcry of the sinking and loss of life. Between 1981 and 2000, the Greenham Common Women's Peace Camp protestors campaigned against nuclear weapons at the Greenham U.S. Air Force base in the United Kingdom. The camp participants were protesting about both the human and environmental costs of nuclear war; the last Trident missiles were removed from the base in 1991. The Sea Shepherd Society, headed by Paul Watson, has led direct activist campaigns against the whaling industry, sinking two Icelandic whaling vessels in 1986.

The antinuclear movement was central to the development of the wider environmental movement. In the aftermath of the atomic bombs at Hiroshima and Nagasaki in Japan at the end of World War II, public concerns about nuclear power spread to nuclear energy production and transportation. The proliferation of nuclear weapons during the Cold War between the United States and the Soviet Union highlighted the need to oppose nuclear production for many in the environmental movement. The first major antinuclear protests in the United States were at Seabrook nuclear power plant in New Hampshire in 1977 and 1978. In 1978, 500 people were arrested at a protest against the Diablo Canyon nuclear power plant in California. In 1979, 70,000 people attended a "No Nukes" rally against nuclear power in Washington, D.C., and 38,000 people attended a protest rally at Diablo Canyon. The cultural impact of the antinuclear movement was captured in the 1979 album and 1980 documentary titled *No Nukes*, featuring contemporary artists such as Bruce Springsteen and Jackson Browne. That was also the year of the meltdown at the Three Mile Island nuclear power station in Pennsylvania. The accident occurred in the same week that the movie *The China Syndrome* was released. The movie, starring Jane Fonda, documents a nuclear accident. In 1986, a meltdown at the Chernobyl nuclear plant in the Ukraine was the largest of its kind globally. These events increased public opposition to nuclear power. In 2004, the environmental writer James Lovelock argued that nuclear power was an alternative to fossil fuels and needed to be considered as a clean energy for the future, alongside solar, wind, and wave technologies.

The environmental justice movement emerged during the 1960s, when farm workers led by César Chávez challenged the use of pesticides such as DDT (dichlorodiphenyltrichloroethane) in California. African-American communities mobilized against regional toxic plants in Houston and Harlem during the 1960s. In 1978, families were evacuated from the Love Canal township near Niagara Falls, New York, in response to concerns about high rates of cancer and birth defects. Toxic waste had been buried in the region in previous decades. The community-based environmental justice movement developed further in the 1980s in response to the siting of hazardous plants or dumps, often in economically disadvantaged, nonwhite neighborhoods. The environmental justice movement became associated with the civil rights movement as a result. Attempts to dump toxic waste in the primarily African-American community of Afton in Warren County, North Carolina, in 1982 led to protests and arrests. In 1987, the Commission for Racial Justice published the *Toxic Wastes and Race in the United States* report, which confirmed the extent of dumping in minority communities throughout the United States. In 1990, the First National People of Color Environmental Leadership Summit met in Washington, D.C. Robert Bullard (1990) published a further study on the issue, titled *Dumping in Dixie: Race, Class, and Environmental Quality.* Further support for the environmental justice movement was provided by the Clinton administration.

The growth in consumption in Western society has led to a waste management crisis in many countries. The emergence of incineration in response to this crisis has led to a global mobilization against incineration by communities concerned about the health and environmental risks posed by emissions. In the United States, campaigns against incineration and dumps became an important part of the environmental movement. In addition, moves to promote recycling and waste reduction have been a central part of environmentalists' alternatives to incineration. The "waste hierarchy"' that favors reduction and recycling over dumping and incineration has been adopted as best practice by many municipal authorities. Love Canal residents founded the Citizens Clearinghouse for Hazardous Wastes in 1981. The Citizens Clearinghouse for Hazardous Wastes proved to be an important resource for the anti-incineration campaign. This combination of concerns about hazardous waste, incineration, dumps, and environmental justice led to the development of a wider antitoxics movement that extended links globally, allowing community campaigns to present themselves as more than merely NIMBY or "not-in-my-backyard"-based groups. In 1994, Andrew Szasz published a study on community campaigns titled *Ecopopulism: Toxic Waste and the Movement for Environmental Justice*, which charts the development of local community groups that move "beyond NIMBY" and become part of the wider environmental movement.

Dave Forman's Earth First! has been at the center of direct action activism since 1979. Earth First! was influenced by Edward Abbey's book *The Monkey Wrench Gang* (1975), and the group became synonymous with the antilogging campaign in the Northwestern states of the United States in the 1980s. Earth First! activism has been described as "ecotage"; their environmentally minded campaigns of sabotage have included tree sitting, tree spiking, road blockages, and tunneling to protect forests. A more radical offshoot of Earth First! called the Earth Liberation Front is seen as comprising "eco-terrorists" by some critics of direct-action tactics. The Animal Liberation Front is another group whose tactics in defense of animal rights were criticized. The antifur campaign of People for the Ethical Treatment of Animals also attracted considerable support but was also criticized for their "shock tactics." The group's targeting of corporate groups that violated animal rights brought them widespread media coverage and celebrity support during the 1990s.

Greenpeace is perhaps the most recognizable of all environmental groups. From its origins in Vancouver, Canada, in 1971, Greenpeace has been at the center of environmental movement activism globally. Its protests against nuclear testing in the South Pacific brought the group into conflict with French authorities and led to the sinking of the flagship *Rainbow Warrior* in 1985. In 1995, Greenpeace was involved in one of its most significant campaigns. The group prevented Shell Oil from dumping the decommissioned oil platform Brent Spar in the North Sea. Greenpeace occupied the Brent Spar. With the support of public protests and boycotts across Europe, Shell relented and recycled the oil rig. Greenpeace's Brent Spar protest led to an international moratorium on the dumping of oil rigs. Rex Weyler published a book about Greenpeace titled *Greenpeace: How a Group of Ecologists, Journalists, and Visionaries Changed the World* in 2004. Today, Greenpeace is an international organization that highlights global environmental campaigns on many diverse issues such as providing information about climate change and working toward the preservation of the rainforests. Based in Amsterdam, the Netherlands, Greenpeace has 2.8 million supporters worldwide and maintains offices in 41 countries.

The World Wildlife Fund (WWF) was founded in 1961 in Switzerland. The WWF is known as the World Wide Fund for Nature outside North America. With 5 million members, it is the largest environmental group in the world. The WWF is committed to "halting and reversing the destruction of our environment." The group's focus is on biodiversity, pollution, climate change, and the conservation of forests and oceans. The WWF engages with many multinational corporations to promote education about environmental issues and to prevent further environmental destruction. The WWF's panda logo is one of the most recognizable icons in the modern environmental movement. Friends of the Earth was founded by David Brower following his departure from the Sierra Club in 1969. This represented a new departure for the previously conservationist environmental movement, as Friends of the Earth would go on to oppose nuclear power plants and embrace social as well as environmental justice issues in the subsequent decades. Friends of the Earth has confederations in 69 separate countries, with each national group retaining a degree of autonomy. These national groups are supported by Friends of the Earth International, based in the Netherlands. Friends of the Earth groups campaign on issues such as climate change, desertification, corporate accountability for pollution, nuclear power, and biodiversity.

The NRDC was founded in 1970. It developed with a focus on a reform agenda and now has offices in Washington, D.C.; San Francisco; Los Angeles; Chicago; and Beijing. The group has over 1.2 million members worldwide. The NRDC has had success as a lobby group and has good influence among legislators in Washington, D.C., where they argue for a balanced approach to conservation and sustainable development. The NRDC has remained relevant during recent crises, implementing a study on the health effects of the 9/11 attacks in New York and working with communities in the aftermath of Hurricane Katrina in New Orleans. The NRDC has also reached out to a younger demographic and has developed a platform with the band Green Day to raise social and political consciousness among audiences worldwide. The NRDC has also worked with political leaders such as Robert Kennedy, Jr., to maintain challenges to protect coastlines in Alaska; the group continues to advocate for sustainable development and political justice issues and has focused on clean air and protection of the oceans.

The environmental movement influenced antiglobalization protests on development including the 1999 protests in Seattle against the World Trade Organization and in Davos, Switzerland, against the World Economic Forum through to the 2002 Johannesburg, South Africa, World Summit on Sustainable Development and the G8 protests at Edinburgh, Scotland, and Hamburg, Germany. In the United Kingdom, protestors used direct action to prevent the disposal of the Brent Spar oil rig platform in the North Sea in 1995. The image of "ecowarriors" tunneling into the Earth or chaining themselves to trees as part of their campaigns to protect nature has become a part of contemporary iconography. Earth First!'s antiroads movements in the United States, United Kingdom, and Germany in the 1990s have been followed in the 2000s by Climate Action's "Plane Stupid" protests against airport extensions in the United Kingdom as part of a wider campaign for reducing carbon emissions. The environmental movement has also included mainstream media events such as former U.S. Vice President Al Gore's *An Inconvenient Truth* book and documentary in 2006. The Live Earth concerts of 2007 also served to increase awareness and participation in environmental issues. Scientific evidence of the anthropogenic basis for climate change was presented by the Intergovernmental Panel on Climate Change in 2007, providing a new impetus for the environmental movement of the future. With the election of President Barack Obama in 2008, green issues will remain at the forefront of the political agenda in the coming era, as the challenge of climate change becomes one of the priorities of the age.

See Also: Anti-Toxics Movement; Bhopal; Ecocentrism; Ecologism; Gaia Hypothesis; Green Parties; Intergovernmental Panel on Climate Change; NIMBY; Pinchot, Gifford; *Silent Spring*; Tragedy of the Commons.

Further Readings

Abbey, E. *The Monkey Wrench Gang*. Philadelphia: Lippincott Williams & Wilkin, 1975.

Bookchin, Murray. *The Ecology of Freedom*. Stirling, Scotland: AK, 2005.

Bullard, Robert. *Dumping in Dixie: Race, Class, and Environmental Quality*. Boulder, CO: Westview, 1990.

Carson, Rachel. *Silent Spring*. London: Houghton Mifflin, 1962.

Dobson, Andrew. *Green Political Thought*, 3rd ed. London: Routledge, 2000.

Eckersley, Robyn. *Environmentalism and Political Theory: Toward an Ecocentric Approach*. New York: State University of New York/UCL Press, 1992.

Ehrlich, Paul. *The Population Bomb*. New York: Ballantine, 1968.

Gore, Albert. *An Inconvenient Truth*. London: Rodale Books, 2006.

Hardin, Garrett. "*The Tragedy of the Commons.*" *Science*, 162:1243–48 (1968).

Leonard, Liam. *The Environmental Movement in Ireland*. Dordrecht: Springer, 2008.

Lovelock, James. *Gaia: A New Look at Life on Earth*. Oxford: Oxford University Press, 1979.

Meadows, Donella. *The Limits to Growth*. New York: Universe Books, 1972.

O'Riordan, Timothy. *Environmentalism*. London: Pion, 1981.

Report of the World Commission on Environment and Development: Our Common Future. Oxford: Oxford University Press, 1987.

Rootes, Christopher. *Environmental Movements: Local, National and Global*. London: Routledge, 1997.

Shabecoff, Phillip. *A Fierce Green Fire: The American Environmental Movement*. Washington, D.C.: Island, 2003.

Szasz, A. *Ecopopulism: Toxic Waste and the Movement for Environmental Justice*. Minneapolis: University of Minnesota Press, 1994.

Weyler, Rex. *Greenpeace: How a Group of Ecologists, Journalists, and Visionaries Changed the World*. London: Rodale Books, 2004.

Liam Leonard
Institute of Technology, Sligo

Environmental Nongovernmental Organizations

Nongovernmental organizations (NGOs) are legal entities comprised of individuals and groups that work on a range of issues outside of the scope of government of private companies. NGOs are private, meaning that they are not part of any government, nor are they created by governments. NGOs seek to raise awareness of global problems and often receive money from governments to do so. Many NGOs work as consultants for the United Nations (UN) and other international bodies. Environmental NGOs are often given the acronym ENGOs.

Although there certainly were organizations providing international humanitarian aid long before World War II, the term *NGO* came into use after the formation of the UN in 1945. The pressures of globalization have resulted in the growth of NGOs since that time. Today there are over 3,000 NGOs that consult for the UN. ENGOs have grown in number ever since environmental issues gained salience in the 1970s.

Some scholars argue that there is no distinction between an NGO and a nonprofit organization. Both of these types of organizations exist to address economic failures by filling the voids left by individual governments and the free market. In the United States, however, the title NGO is used to describe a transnational organization that, although it may be partially government supported, is not a government representative. Nonprofit organizations are those that are tax exempt within the United States and that have to roll over any profit into the operation for completing the mission for which they were incorporated.

ENGOs receive some money in the form of grants from governments, but they can also collect dues from their members and/or partners, receive money from the sales of goods and services, apply for grants from other private funding entities, and ask for donations from private donors. They use this funding to work for the cause in their mission statements. For most ENGOs, this mission has to do with sustainability or biodiversity. NGOs seek to bring about change by raising public awareness of issues, lobbying government officials, getting stories into the press, consulting for the UN and other international bodies, mobilizing the public around salient issues, raising financial support, enlisting volunteers, contributing aid to important projects, and other activist activities.

One example of an ENGO is Conservation International. This is a very well-known NGO that was formed in Washington, D.C., in 1987. Its mission is to protect biodiversity and to teach humans how to live in harmony with nature. Conservation International has identified a large number of biological hot spots. Hot spots are locations that have a high amount of biodiversity located in a relatively small area. Members of Conservation International work to prevent extinction in these areas. The extinction of one species, often called an indicator species, can lead to the collapse of the whole system. In hot spots, this could lead to a mass loss of biodiversity.

Conservation International runs a number of programs related to its mission including the Center for Applied Biodiversity, the Center for Conservation and Government, the Center for Environmental Leadership in Business, the Global Conservation Fund, the Global Marine Partnership Fund, and the Sea Turtle Flagship Program, among others. To make these programs successful, and in an effort to achieve its mission, Conservation International has partnered with governments, corporations, communities, and other NGOs.

See Also: Federalism; Globalization; Institutions; Organizations; Sustainable Development; Transnational Advocacy Organizations.

Further Readings

Betsill, M., et al. *NGO Diplomacy: The Influence of Nongovernmental Organizations in International Environmental Negotiations*. Cambridge: MIT Press, 2008.

Bryant, R. *Nongovernmental Organizations in Environmental Struggles: Politics and the Making of Moral Capital in the Philippines*. New Haven, CT: Yale University Press, 2005.

Conservation International. http://www.conservation.org/Pages/default.aspx (Accessed January 2009).

Feher, M. *Nongovernmental Politics*. New York and Cambridge: Zone Books, 2007.

Najam, A. and Yale University. Institution for Social and Policy Studies, Program on Non-Profit Organizations. *Nongovernmental Organizations as Policy Entrepreneurs: In Pursuit*

of Sustainable Development. New Haven, CT: Program on Non-Profit Organizations, Institution for Social and Policy Studies, Yale University, 1996.
United Nations Department of Social and Economic Affairs, NGO Section. http://www.un.org/esa/coordination/ngo/ (Accessed January 2009).

Jo A. Arney
University of Wisconsin–La Crosse

Environmental Protection Agency

The Environmental Protection Agency (EPA) is a government agency in the Executive branch that is charged by Congress with administering and enforcing the numerous laws that deal with regulating the environment. Its mandate continues to generate political struggles because it involves issues affecting business, environmental groups, science organizations, citizens groups, local and state governments, units of the federal government—including members of Congress and the White House—and the public as well. Involved in the political issues are concerns over the financial costs for business, local and state governments, and the public. These may be stated in dollar amounts, but the opposition that usually arises is concerned with health (clean air, safe drinking water, and wholesome environments), the quality of the natural environment, the quality of living in areas in which the poor and minorities live, and safe private and public spaces that may be built on restored landfills or toxic waste dumps.

The EPA was authorized by an act of Congress in 1970 and began operating on December 2, 1970. It was created in response to public demands for improvements in the quality of air, water, and the general environment. Much of the impetus for establishing the EPA came from the new environmental movement sparked by the first Earth Day in April 1970. It was a historical transformation of the policies of the previous 60 years that had focused on conservation while ignoring pollution.

The Environmental Protection Agency works with other government agencies to implement Superfund site cleanups. These U.S. Army Corps of Engineers workers were assessing toxic waste at a Superfund site near Seattle, Washington.

Source: U.S. Army Corps of Engineers

By the time the EPA was authorized, pollution had reached such a high level that the Chicago River caught fire. In other rivers, the fish were dead or diseased, species were threatened with extinction, and the air was so polluted with sulfur dioxide that acid rain was killing forests in northern areas and in Canada. At stake was not just the environment, but human health as well. To meet these challenges to clean air, water, and land, as well as to preserve threatened species, the EPA was assigned the responsibility of eliminating pollution at its source and repairing the damage to the environment.

One of the people who inspired supporters of the EPA was Rachel Carson, who published *Silent Spring*, first as a series in *The New Yorker* magazine and later as a best-selling book. It described the effects that industrial chemicals were having on the environment. The chemicals were being used in agriculture as pesticides; however, in many cases, the chemicals were long lasting and had unintended consequences because they acted adversely on both plants and animals. For example, when DDT (dichlorodiphenyltrichloroethane) spread in the environment, it entered the food chain and acted as an endocrine disrupter for birds by interfering with the thickness of their eggs' shells—too thick and the chick cannot peck free at the end of incubation, too thin and the eggshell will break, killing the chick, with negative consequences for that species and for the balance of nature.

Carson's illustrations of environmental damage to bird eggs were easily transferred to the contents of salads or other foods that could be understood to be poisoning humans. The response to Carson's book in 1962 was a spreading realization that the environment was polluted and an emerging environmental movement. The spreading public horror and, at times, hysteria about pollution produced a demand for government regulations to protect the environment—a response similar to that made after the publication of *The Jungle* (by Upton Sinclair) in 1906 that prompted the creation of the Food and Drug Administration.

The EPA was born of the environmental movement that sought to address the fact of synthetic chemicals pollution. It was assigned the mission to protect human health by safeguarding the natural environment so that nature would be a place of clean air, water, and land. It is on these three things that human life in nature depends.

Congress organized the EPA from several scattered government programs and added more responsibilities. In the 40 years since its inception, it has accomplished much toward providing the United States with a much cleaner and healthier environment. This has meant dealing with pollution, from auto emissions to toxic waste dumps and a long list of other pollutants in between.

President Richard M. Nixon, in response to criticism from a wide array of people in the environmental movement, which by 1970 had spread across the political spectrum of U.S. politics, signed an executive order, Reorganization Plan No. 3. Congress adopted into law a broader version of the order intended to reorganize federal response to the dangers of chemical pollution.

By 2009, the EPA had expanded to employ over 18,000 people working in 10 regional offices and in its main office in Washington, D.C. The EPA headquarters is organized by responsibilities. Its Office of Administration and Resources includes the chief financial officer, general counsel, and inspector general. The Office of Enforcement and Compliance Assurance deals with administration and legal issues. The Office of Environmental Information provides the public with information that enables individuals and businesses to understand public environmental issues. The Office of Environmental Justice administers policies that seek to protect the environment and prevent health hazards, as well as to provide equal access to natural resources. The Office of International Affairs engages in policy development, debates, and implementation with the international community.

The Offices of Air and Radiation; of Prevention, Pesticides, and Toxic Substances; of Solid Waste and Emergency Response; and the Office of Water are engaged with specific environmental problems. The Office of Research and Development is engaged in developing tests for pollution as well as seeking to answer research questions. The EPA also has the duty to review the Environmental Impact Statements developed by other federal agencies.

The EPA also has 10 regional offices across the United States. Each regional office is responsible for implementing the agency's policies within designated states. In some cases, environmental programs have been delegated to states.

All units of the federal government have the responsibility to engage in environmentally responsible activities. In addition, the Departments of Interior and Agriculture also enforce some of the numerous laws passed by Congress that address pollution or other environmental issues. The air quality laws enforced by the EPA include the Air Pollution Control Act (1955), the Clean Air Act (1963, 1966 amendments), the Motor Vehicle Air Pollution Control Act (1965), the Air Quality Act (1967), the National Environmental Policy Act (1969), the Clean Air Act Extension (1970), and the Clean Air Act amendments of 1977 and 1990.

Legislation on water quality regulations and pollution prevention enforced by the EPA include the Water Pollution Control Act (1948), the Water Quality Act (1965), the Clean Waters Restoration Act (1966), the National Environmental Policy Act (1969), the Water Quality Improvement Act (1970), the Federal Water Pollution Control Amendments (1972), the Safe Drinking Water Act (1974), and many more.

Acts related to the land that also deal with the protection of wildlife include the Endangered Species Act (1973), the Marine Mammal Protection Act (1972), the Solid Waste Disposal Act (1965), the Comprehensive Environmental Response, Compensation and Liability Act ("Superfund Act") of 1980, and the Small Business Liability Relief and Brownfields Revitalization Act (2002), which deals with cleaning up land, usually in urban areas that have been polluted by chemicals, that often contains old housing occupied by the poor.

The EPA policy goals are to protect the environment, prevent health hazards, and provide all people equal access to both natural resources and to the decision-making process in matters that deal with the environment. To achieve these goals, the EPA has made environmental justice an integral part of its policies and programs. In 1993, it established the National Environmental Justice Advisory Council as a forum for a wide range of groups (academic, community, environmentalists, industry, indigenous peoples, and state and local governments) to join together to find solutions to environmental justice problems through extended dialogue about specific topics.

The EPA sponsors a number of voluntary programs. The Energy Star Program was begun in 1992 by the EPA to promote energy efficiency. The WaterSense Program was begun in 2006 to promote water conservation. Also connected with water usage is the Safer Detergents Stewardship Initiative. These programs are usually not as politically controversial as those that are mandated.

Several major and expensive programs administered by the EPA are the fuel economy testing it does to give mileage ratings. Its air quality standards use mathematical models to study pollutant dispersal rates. The information developed has a number of practical applications and significant policy implications. Oil pollution prevention at land sites is focused on containment and on accidental spillage events.

Two major new programs involve greenhouse gas emissions and the problem of global warming. During the administration of President George W. Bush, the EPA was a center of controversy over several issues. One that was very important was part of the debate over global warming. In *Massachusetts v. Environmental Protection Agency* (2007), the Supreme Court ruled that the EPA has the authority to regulate the emission of greenhouse gases in automobile emissions because these emissions were included in the Clean Air Act's definitions of air pollutants. It then ordered the EPA to begin regulating emission unless it could demonstrate scientific reasons for not doing so.

In December of 2007, Stephen L. Johnson prevented California and handful of other states from issuing carbon emission standards, something permitted under the rules of the Clean Air Act.Other scientists have also reported that their efforts to conduct science have been interfered with by government officials.

The EPA's mission is to "protect human health and the environment." The number of laws that have been adopted by Congress that fit this goal is rather large. All of this wide range of laws are administered or enforced by the EPA. The issues include clean air, clean water, protecting the environment, and safeguarding human health. The political struggle over these issues usually involves industry, business, and their political supporters.

Mining interests may be compelled by EPA regulations to install expensive equipment and processes to prevent water pollution. Industry may be ordered to install equipment that will reduce air pollution or prevent chemical pollution. Agricultural interests may be blocked from using water supplies for irrigation if the water usage will threaten an insignificant species of fish. The costs to businesses, consumers, and the public in general are bones of contention for those who want looser regulation and those who want a clean environment even if it means higher production costs.

See Also: Clean Air Act; Clean Water Act; Conservation Movement; Endangered Species Act; Environmental Justice; Environmental Movement; Marine Mammal Protection Act; Pinchot, Gifford; *Silent Spring*.

Further Readings

Binns, Tristan Boyer. *EPA: Environmental Protection Agency*. Chicago: Heinemann Library, 2002.

Brulle, Robert J. *Agency, Democracy, and Nature: The U.S. Environmental Movement From a Critical Theory Perspective*. Cambridge, MA: MIT Press, 2000.

Collin, Robert W. *Environmental Protection Agency: Cleaning up America's Act*. Westport, CT: Greenwood, 2005.

Cook, Brian J. *Bureaucratic Politics and Regulatory Reform: The EPA and Emissions Trading*. Westport, CT: Greenwood, 1988.

DeLong, James V. *Out of Bounds and Out of Control: Regulatory Enforcement at the EPA*. Washington, D.C.: Cato Institute, 2002.

Environmental Protection Agency. http://www.epa.gov/ (Accessed June 2009).

Fletcher, Susan, et al. *Environmental Laws: Major Statutes Administered by the EPA*. Hauppauge, NY: Nova Science, 2008.

Haskell, Elizabeth H. *The Politics of Clean Air: EPA Standards for Coal-Burning Power Plants*. Westport, CT: Greenwood, 1982.

Lazarus, Richard J. *The Making of Environmental Law*. Chicago: University of Chicago Press, 2006.

McMahon, Robert. *Environmental Protection Agency: Structuring Motivation in a Green Bureaucracy: The Conflict between Regulatory Style and Cultural Identity*. Eastbourne, UK: Sussex Academic Press, 2005.

Andrew Jackson Waskey
Dalton State College

Equity

The concept of equity requires fairness in the distribution of social benefits and burdens and the entitlement of everyone to an acceptable quality and standard of living. With respect to the environment, equity means that there should be a minimum level of environmental quality that everyone is entitled to. It also means that everyone should have equal access to communal environmental resources and that no individuals or groups of people should be asked to carry a greater environmental burden than the rest of the community as a result of government or business actions.

Equity is not the same as equality. There may be good reasons for people to have different rewards and burdens or to be treated differently. However, equity requires that these reasons be morally relevant; that is, that they be just, fair, and impartial. Impartiality means that factors such as race, religion, color, gender, and nationality are not relevant.

Equity has three aspects: (1) people get what they deserve—fairness; (2) people have certain rights that must be respected; and (3) people's needs should be met and their contribution to meeting such needs is based on their ability to do so.

This means that the distribution of rewards and burdens may be deserved on the basis of a person's efforts, choices, and abilities, but those rewards and burdens should not be out of proportion to the actions or qualities of that person. It also means that there should be limits to the burdens that individuals are subject to and that their basic needs should be met no matter what their abilities. Each person has a right to life, health, and the basic conditions of subsistence, as well as certain political and social rights outlined in the Universal Declaration of Human Rights.

Equity as a concept is fundamental to sustainable development, and it can be applied across communities and nations and across generations. The Brundtland Commission's definition of sustainable development is based on intergenerational equity: "development that meets the needs of the present without compromising the ability of future generations to meet their own needs." The commission also claims that sustainable development requires equity within existing generations.

Intergenerational Equity

Intergenerational equity refers to the need for a just distribution of rewards and burdens between generations and fair and impartial treatment toward future generations. Intergenerational equity has been recognized in various international agreements, and today it is a principle of international law. A number of national laws and agreements also include intergenerational equity.

The nature of our obligation to future generations is controversial. Do we merely need to protect those aspects of the environment necessary for survival and health, such as a minimal standard of clean air and water, or should we ensure a healthy diverse environment for future generations? How do we balance our obligations to current generations with our obligations to future generations when these conflict?

Many economists and businesspeople tend to argue that what is important is to maintain welfare and utility over time and that a community can use up natural resources and degrade the natural environment so long as they compensate for the loss with "human capital" (skills, knowledge, and technology) and "human-made capital" (buildings, machinery, etc.). This is referred to as "weak sustainability."

Understandably, environmentalists generally reject the concept of weak sustainability, even if it incorporates the idea of maintaining minimal environmental functions. They claim that human welfare can only be maintained over generations if the environment is not degraded. They point out that we do not know what the safe limits of environmental degradation are; yet if those safe limits are crossed, the options for future generations would be severely limited.

Second, many do not agree that human and natural capital are interchangeable and believe that a loss of environmental quality cannot be substituted with a gain in human or human-made capital without loss of welfare. Therefore they argue that future generations should not inherit a degraded environment, no matter how many extra sources of wealth are available to them. This is referred to as "strong sustainability."

Intragenerational Equity

Intragenerational equity covers justice and the distribution of resources between nations. It also includes considerations of what is fair for people within any one nation. Environmental burdens are often inequitably distributed. For example, people living in industrial areas are more likely to suffer from air or water pollution. People in the inner city are more likely to suffer from urban decay and traffic problems. In some countries, ethnicity, race, and color seem to be a significant factor in determining who is exposed to environmental burdens.

Poorer people tend to suffer the burden of existing environmental problems more than others do. This is because it costs more to live in areas that have not had their environment degraded. In addition, wealthy areas are more likely to have access to environmental amenities such as parks and protected waterways. Also, more affluent people are better able to fight the imposition of a polluting facility in their neighborhood because they have better access to financial resources, education, skills, and decision-making structures.

Some people are more vulnerable to the health effects of environmental problems because of age and health status. For example, people with existing respiratory problems may be affected more by air pollution, and the very young or the very old may be more vulnerable to environmental pollution in general. Others are subject to greater exposure to chemicals because of their occupation. Workers in certain industries—like mining or mineral processing and the chemical industry—are often exposed to higher health risks than the rest of the community.

Inequities are also caused by the export of hazardous products and wastes to developing countries. Imported pesticides contribute to some half a million poisonings and 40,000 deaths each year. In addition, developing countries are often subject to more of the effects of environmental degradation, more vulnerable to them, and less able to respond and protect themselves from them. For example, the populations of poorer countries are more vulnerable to sea level rise and other effects of climate change, even though they are least responsible for causing it and less able to adapt because of poverty, lack of technology, and population pressures.

Inequities can themselves cause environmental problems. Poverty contributes to environmental degradation because it deprives people of the choice about whether or not to be environmentally sound in their activities. Communities need to have a certain level of security before they will turn their attention to solving environmental problems. However, affluence also contributes to environmental degradation because it is associated with high levels of consumption, resource depletion, and waste accumulation.

Measures to improve environmental problems may be inequitable because they affect some sectors of the community more than others. Measures to protect the environment can affect the competitiveness of national industries in the international market. Environmental measures can cost individual jobs, even if they create new jobs in a different sector of the economy. Environmental charges and taxes can be more burdensome for people with low incomes.

The creation of national parks and wilderness areas can also unfairly affect people who are displaced by those parks or whose access to traditional livelihoods is restricted as a result. Environmental measures can also have inequitable effects if environmental problems are shifted from one place to another or concentrated in one place.

See Also: Brundtland Commission; Environmental Justice; Future Generations; North–South Issues.

Further Readings

Beder, Sharon. *Environmental Principles and Policies*. Sydney: UNSW Press and London: Earthscan, 2006.

Shrader-Frechette, Kristin. *Environmental Justice: Creating Equality, Reclaiming Democracy*. Oxford: Oxford University Press, 2002.

UNESCO Declaration on the Responsibilities of the Present Generations Towards Future Generations, 1997. http://portal.unesco.org/en/ev.php-URL_ID=13178&URL_DO=DO_TOPIC&URL_SECTION=201.html (Accessed November 2005).

World Commission on Environment and Development (Brundtland Commission). *Our Common Future*. Melbourne, Australia: Oxford University Press, 1990.

Sharon Beder
University of Wollongong

Federalism

Federalism is the sharing of power between national and local levels of government such as states, provinces, territories, counties, and cities and is thus a system characteristic that creates considerable opportunities for experimentation in environmental politics. The term is derived from the Latin term *foedus*, meaning league or covenant, and allows the development of multiple channels of authority that, in turn, enlarge the policymaking process along a range of environmental issues from acid rain to climate change.

Although applied in countries as diverse as Australia, Brazil, Canada, Germany, India, Malaysia, Mexico, Nigeria, and the United States, federalism is relatively uncommon across the globe today. Unitary systems such as those found in Cameroon, France, Italy, Japan, Kenya, South Korea, Sweden, and the United Kingdom, to name but a few, are considerably more prevalent. Although some delegation of authority may still occur, it is more along the lines of administrative duties, as the central government retains much more authority than in federal systems. A third and even less common democratic system is a confederation, such as the United States under the Articles of Confederation (1783–89). Confederations tend to be weak and unstable, as political governance is dominated by regional governments, and central sovereign authority is severely limited by an inability to gather taxes and the lack of a chief executive.

In contrast, federalism, and the power-sharing system it creates, fosters innovative experimentation in legislation, including environmental politics. The United States, the first to institute a written constitution formalizing federalism, illustrates just how complicated this can become, with over 83,000 subnational jurisdictions (i.e., states, cities, counties, and school districts), all with the power to tax, spend, and make public policy. A confusing myriad certainly, but also one that is often praised as the ideal "laboratories of democracy," as U.S. Supreme Court justice Louis Brandeis famously asserted in his dissenting opinion to *New State Ice Co. v. Leibmann* (1932).

Nothing illustrates this better on the environmental policy front than the progression of state acid rain programs in the 1970s and 1980s that allowed for testing ideas that eventually coalesced into federal law under the U.S. Clean Air Act amendments in 1990. Environmental activists hope California and the California Global Warming Solutions Act of 2006, which mandates a new reporting system, overall cap on carbon dioxide, and production cut back to 1990 levels by 2020, will play the same defining role in limiting greenhouse gases nationwide.

Still others believe municipalities to be the most appropriate level for effective climate change policy to evolve. And although the United States abdicated international environmental leadership during President George W. Bush's administration in the early 21st century, federalism allowed over 900 different mayors, since Kyoto entered force in February 2005, to carve out their own initiative—the U.S. Mayors Climate Protection Agreement. Under this agreement, participating cities commit to pressure state and federal governments to meet Kyoto reduction numbers, lobby the U.S. Congress for a national emission trading system, and meet Kyoto Protocol targets in their own municipalities.

The historical evolution of federalism offers further insight here. Federalism has roots dating back to the Roman Empire, which facilitated governance over large territories by discouraging those who were conquered from putting up further resistance against Rome, as they were allowed to maintain certain local powers. Modern principles of federalism emerged in the United States with the 1787 Constitutional Convention and the ensuing debate among federalists and antifederalists. Of particular note, *The Federalist Papers*, written by Alexander Hamilton, James Madison, and John Jay, was a direct attempt to convince New York state citizens to vote for Constitutional ratification by highlighting weaknesses within the Articles of Confederation.

Antifederalists such as George Mason were also instrumental in codifying basic principles of American federalism as they led efforts to press for the Bill of Rights, particularly the 10th amendment, which asserts "the powers not delegated to the United States of the Constitution, nor prohibited by it to the Sates, are reserved to the State." Constitutionally, it is also important to recognize the "necessary and proper" clause in Article I, Section 8, as well as the supremacy clause in Article VI, as instrumental in defining the parameters of federalism in the nascent United States and its recognition of national law as "the supreme law of the land."

Indeed, from its first founding, the United States has undergone a general shift in power from states toward the national government (e.g., the landmark Supreme Court case *McCulloch v. Maryland* in 1819, recognizing the creation of a national bank and its freedom from state taxes). Several additional seminal events highlight this progression, from the aftermath of the Civil War and its equal rights amendments to the Great Depression and the New Deal spawned by President Franklin Roosevelt to combat it. World Wars I and II, similarly, ushered in enormous growth in the national government, as did the Cold War and the Vietnam War in particular. The civil rights movement and the Great Society of President Lyndon Johnson further cemented this hierarchy. Under President Richard Nixon, however, rhetoric began to swing in the opposite direction, and by 1980, under President Ronald Reagan's leadership, a series of efforts to reduce national control and return powers to the states was fully underway. As just one example, this "devolution revolution" instituted block grants instead of grants-in-aid, thus freeing state governments to spend funds at their own discretion.

In terms of environmental politics, this meant federally encouraged new freedoms—and handicaps. Justice Brandeis' laboratories of democracy once again surfaced, with innovative experimental offerings, but so did a woeful lack of national guidance on issues such as greenhouse gas emissions. Political scientists offer a helpful categorization for considering this dilemma by dividing federalism into two major types: dual and cooperative. Depending on the political issue at stake, state and local governments interact with one another and the national government in different ways. For one, interaction may mimic a dual or layer cake in displaying distinct, autonomous spheres of authority. The feds handle national defense, but municipalities determine their own property taxes. Of course, not all issues are so cut and dry, particularly in the environmental arena. Here is where the second type of federalism, cooperative federalism—often conceptualized as marble or compound cakes—can be more insightful, as it describes interdependent levels of government characterized by ubiquitous bargaining.

Environmental issues like climate change are a case in point, with emissions standards for automobiles becoming ground zero within this debate. In July 2002, in the face of continued federal inaction, the state of California passed Assembly Bill 1493 to limit greenhouse gas emissions from its motor vehicles. Subsequently, over a dozen other states followed California's lead with similar regulations. However, since these restrictions are tighter than federal levels, states must receive a waiver from the U.S. Environmental Protection Agency to move forward. Of course, with roughly 37 million residents, ranking it as the eighth largest economy in the world, California certainly has the political muscle to continue this fight. That might sound more like combative than cooperative federalism, but if Congressional testimony by ConocoPhillips and General Electric executives in early 2007 is any guide, industry forces pushing mandatory greenhouse gas restrictions may well provide the final momentum in favor of California's tighter restrictions.

See Also: Acid Rain; Clean Air Act; Cost-Benefit Analysis; Global Climate Change; Policy Process; Sustainable Development.

Further Readings

Ayres, Richard E. and Jessica L. Olson. "The New Federalism: States Take a Leading Role in Clean Air." *Natural Resources & Environment*, 23/2:29–32 (Fall 2008).

Butler, David. *Using Federalism to Improve Environmental Policy*. Washington, D.C.: AEI, 1996.

Desai, Uday, ed. *Environmental Politics and Policy in Industrialized Countries*. Cambridge, MA: MIT Press, 2002.

Hamilton, Alexander, et al. *The Federalist: A Commentary on the Constitution of the United States*. New York: Modern Library, 1937.

Holland, Kenneth M., et al., eds. *Federalism and the Environment: Environmental Policymaking in Australia, Canada, and the United States*. Westport, CT: Greenwood, 1996.

McCulloch v. Maryland, 17 U.S. 316 (1819).

New State Ice Co. v. Leibmann 285 U.S. 262 (1932).

Rabe, Barry G. *Statehouse and Greenhouse: The Emerging Politics of American Climate Change Policy*. Washington, D.C.: Brookings Institution, 2004.

Scheberle, Denise. *Federalism and Environmental Policy: Trust and the Politics of Implementation*. Washington, D.C.: Georgetown University Press, 2004.

U.S. Constitution. http://www.house.gov/house/Educate.shtml (Accessed May 2009).

Michael M. Gunter, Jr.
Rollins College

Fish & Wildlife Service

The U.S. Fish & Wildlife Service (FWS) facilitates the conservation, development, and management of the country's fish and wildlife resources, as well as conducts public outreach and education pertaining to the wise use of fish and wildlife resources. Housed within the Department of the Interior, FWS works with others to "conserve, protect and enhance fish, wildlife and plants and their habitats for the continuing benefit of the American people." To fulfill this mission, the service seeks to cultivate an ethic of environmental stewardship guided by ecological principles, scientific knowledge, and a sense of moral responsibility. The FWS's primary tasks include enforcement of the Endangered Species Act and other wildlife laws, management of the nation's wildlife refuges including the Arctic National Wildlife Refuge, direction of the nation's fish hatcheries, distribution of federal aid to state fish and wildlife agencies, and assistance in international conservation efforts. The mission, objectives, and functions of the FWS have been crucial for propagating the ideals of the U.S. conservation movement.

These U.S. Fish & Wildlife Service employees were working with a formerly endangered peregrine falcon at the Monomoy National Wildlife Refuge in Chatham, Massachusetts, in April 2008.

Source: U.S. Fish & Wildlife Service

In the late 1800s, scientists, policymakers, and a growing number of citizens began to pay increasing attention to the rapid depletion and degradation of the nation's natural resources. Early conservationists organized groups such as the Sierra Club (1892) and the Audubon Society (1896) to ensure the continued use of natural resources for present and future generations. Following these actions, Congress undertook a number of conservation measures, including the establishment of the Commission on Fish and Fisheries (1871), which was the precursor to the current FWS. The commission's initial purpose was to investigate declining food fish populations and to promote their regeneration. In 1885, Congress established the Division of Economic Ornithology and Mammalogy (later renamed the Bureau of Biological Survey) within the Department of Agriculture to investigate the relationship between birds and agricultural pests and to delineate the spatial distribution of plants and animals throughout the country. In 1939, the Department of the Interior assumed control over the Commission on Fish and Fisheries and the Bureau of Biological Survey and joined the two in 1940 to create the FWS.

A series of evolving legislation and executive orders granted the FWS authority to carry out its mission to conserve the nation's fish and wildlife resources. The first federal legislation to protect wildlife, the Lacey Act (1900), prohibited the interstate shipment of illegally taken wildlife and the importation of species. In 1903, an executive order by President Theodore Roosevelt established the first Federal Bird Reservation on Pelican Island in Florida. The Migratory Bird Treaty Act of 1918 implemented the Convention Between the U.S. and Great Britain (for Canada) and set limits on migratory bird hunting in North

America. In 1934, the Migratory Bird Hunting Stamp Act, known as the "Duck Stamp Act," required waterfowl hunters to purchase a stamp, which served as a license, with the proceeds going to purchase waterfowl habitat—nearly 5 million acres purchased thus far. In 1937, the Pittman-Robertson Act provided funds for the improvement of wildlife research and dissemination, and the 1950 Dingell-Johnson Act created programs for restoring and improving U.S. fishery resources. The 1956 Fish and Wildlife Act created two new bureaus within the FWS, the Bureau of Commercial Fisheries and the Bureau of Sport Fisheries and Wildlife. The Bureau of Commercial Fisheries was transferred to the Department of Commerce in 1970 and renamed the National Marine Fisheries Service.

Despite the implementation of wildlife legislation and the establishment of national wildlife reserves, wildlife management has, at times, fallen short of achieving conservationist ideals. In 1915, the Bureau of Biological Survey engaged in the systematic elimination of large predators (e.g., wolves, mountain lions, bears, and coyotes) throughout the West, resulting in their near extinction. The eradication of predators led to a surge in the western mule-deer population, resulting in fierce competition for food sources. As a consequence, in the early 1920s, thousands of deer perished along the Grand Canyon's northern rim. A similar situation occurred in the mid-1920s in Kern County, California, where the extirpation of owls, hawks, and other predators led to an inundation of the area by hungry field mice. Meanwhile, along the nation's flyways migratory bird populations continued to decline despite protective legislation. By the middle of the 20th century, a combination of misguided management practices and a national policy of unfettered economic growth contingent on unremitting resource use and transformation of the nation's landscape led to a significant decline in wildlife populations.

The rapid loss of wildlife habitat and subsequent declines in wildlife populations prompted wildlife managers to question the efficacy of policies that failed to take a holistic approach to management. Championed largely by Aldo Leopold, wildlife management underwent a dramatic shift to a strategy that prioritized the interconnections between organisms and their relationship to the environment. This ecosystem approach culminated in one of the most significant pieces of wildlife legislation, the 1973 Endangered Species Act (ESA).

The ESA exemplifies the conservationist ideals embedded in the FWS mission by ensuring that the push for economic growth does not undermine the necessary resources for maintaining fish and wildlife. Although there were previous attempts to protect endangered species, such as the Endangered Species Protection Act of 1966, the ESA provides the enforcement agencies—the FWS and National Marine Fisheries Service—with the authority that is necessary to protect domestic and international species that are in danger of becoming extinct, and the ecosystems on which they depend. The ESA prohibits the "taking" of listed plants and animals without federal permit, which includes prohibitions against the degradation or significant modification of species habitat that disrupt essential behavioral patterns, such as breeding, feeding, or sheltering. For example, in the early 1990s, efforts to protect the northern spotted owl ensured not only the preservation of the owl but also their habitat, which includes old growth forests of the Pacific Northwest.

Success Story

The FWS has been quite successful in implementing the ESA, having lost only seven of more than 1,700 endangered U.S. and foreign species to extinction. Furthermore, nearly one-half of the species listed are stabilized or increasing in numbers. Recent success stories include the delisting of the American alligator, grizzly bear, bald eagle, and peregrine falcon.

Although the FWS has been successful in protecting significant fish and wildlife species and habitats, the agency continues to face a number of challenges. For example, although former FWS employee Rachel Carson documented the dire consequences of toxic chemicals used in agriculture and industrial production in her 1962 classic *Silent Spring*, the pervasive use of chemicals continues to have adverse effects on the nation's fish and wildlife populations. The widespread use of pesticides has led to a significant decline in pollinators (e.g., bees, hummingbirds, and bats) necessary for the continued pollination of flowers and agricultural products. In addition, the consequences of global climate change pose significant challenges for the FWS. For example, the FWS listed the polar bear as threatened in 2008, largely as a result of the decreasing amounts of sea ice crucial to their habitat. However, the decision to list the polar bear provoked controversy, as opponents cited concerns that the polar bear listing would open the ESA for the purposes of establishing climate change policies. Conversely, environmentalists argue that the failure to address global climate change undermines the agency's ability to protect polar bear habitat, an element imperative to their preservation and within the purview of the ESA.

In addition to disagreements about the relationship between climate change and species habitat incited by the polar bear decision, other recent incidents have highlighted the political nature of fish and wildlife management. Julie MacDonald, the deputy assistant secretary of the FWS, resigned in May 2007 after evidence surfaced linking her with political interference in a number of endangered species decisions. Following MacDonald's resignation, the FWS determined that it would reverse seven of eight key decisions where scientific integrity appeared to have been compromised. Furthermore, a last-minute move by the administration of George W. Bush eliminated requirements for federal agencies to consult with FWS biologists on projects involving endangered species—a move that environmentalists argue could pull the rug out from under the ESA. Without public and government support for an ecological approach to wildlife management, it is likely that the conservation of fish and wildlife resources will be jeopardized.

See Also: Biodiversity; Conservation Movement; Endangered Species Act; Environmental Management; Regulatory Approaches; *Silent Spring*.

Further Readings

Burgess, Bonnie. *Fate of the Wild: The Endangered Species Act and the Future of Biodiversity*. Athens: University of Georgia Press, 2003.

Carson, Rachel. *Silent Spring*. Boston: Houghton-Mifflin, 1962.

Hays, Samuel P. *Conservation and the Gospel of Efficiency: The Progressive Conservation Movement 1890-1920*. New York: Atheneum, 1969.

Leopold, Aldo. *A Sand County Almanac, and Sketches Here and There*. New York: Oxford University Press, 1949.

Siry, Joseph. *Marshes of the Ocean Shore: Development of an Ecological Ethic*. College Station: Texas A&M University Press, 1984.

U.S. Fish & Wildlife Service. http://www.fws.gov (Accessed February 2009).

Kristen Van Hooreweghe
City University of New York

Forest Service

From the early colonial vision of forests as a wilderness to be tamed to their current status as major recreational lands and touchstones of environmental controversy, the forests of the United States have always captured the American imagination. The national forests, and the Forest Service that manages them, provide a record of the nation's evolving relationship to the natural environment. The U.S. Forest Service is the largest natural resources research organization in the world, managing 193 million acres of national forests and grasslands. The service, and much of its approach to management, grew out of a desire to manage the timber resource for human consumption. Perhaps more than any other agency, however, the Forest Service became deeply embroiled in the contentious development of U.S. environmental law and policy.

The controversy over clear-cutting forests such as this one in Oregon, which had been a habitat for the northern spotted owl, led to the adoption of new priorities for the U.S. Forest Service beginning in 2000.

Source: U.S. Fish & Wildlife Service

In 1799, the young nation's first federal forest protections grew out of a need for timber to build the U.S. Navy's ships in the 18th century. When these ship-building needs were met by iron, however, the need for and enforcement of these protections faded. For much of the 19th century, federal land policy focused on expansion through taming the wilderness, causing rapid depletion of the United States' vast forest expanse.

By the 1870s, however, concern for the loss of forest resources began to rise. Several states enacted laws for the protection of forests. In 1876, Congress appointed a special agent to assess the state of forests and, in 1881, it established the Division of Forestry (subsequently renamed the Bureau of Forestry)—the precursor of today's Forest Service—within the Department of Agriculture. In 1891, Congress passed the Forest Reserve Act, authorizing the president to designate portions of public land as "forest reserves" managed by the Department of the Interior, and thereby laid the groundwork for today's national forests. The act did not provide significant guidelines for managing the forest reserves, however, and two visions of managing the forests competed: preservation and conservation.

Preservation Versus Conservation

Preservationists, including John Muir and later Aldo Leopold, asserted that the forests should be protected for their intrinsic value and enjoyed primarily as they naturally exist.

Conservationists, including Gifford Pinchot, focused on the use of forests as resources to meet human needs. In the late 19th and early 20th centuries, conservationists saw timber extraction as the primary use of the forest and sought to apply scientific management principles to ensure a consistent supply of the resource. Other recognized uses included grazing, hunting, and water supply.

In the 1897 Organic Act, Congress addressed administration of forest reserves, authorizing federal agency oversight to protect water flows and ensure timber supply. The Organic Act, therefore, marks the beginning of modern comprehensive federal forest management. It also supported conservationist management approaches.

In 1905, Congress transferred authority over the forest reserves to the U.S. Forest Service, which evolved directly from Bureau of Forestry within the Department of Agriculture. Pinchot became the first chief of the U.S. Forest Service. Pinchot maintained a close relationship with President Theodore Roosevelt, and the two men shared similar visions for the management of forest lands.

Pinchot guided the forest service toward a utilitarian "wise use" policy of doing the greatest good for the greatest number of participants in the long run. The "wise use" policy became the management model for the U.S. Forest Service and focused on the optimal use of natural resources. Thus, Pinchot established a conservationist vision for the Forest Service, one that sought to ensure scientific management of forest lands to consistently meet the U.S. demand for timber.

From the Organic Act until the 1960s, national forests were managed primarily to yield timber. Following World War II, demand for timber dramatically increased because of a booming market for housing. The Forest Service met this increased demand through expansion of a method of timber harvest known as "clear-cutting." Before long, a simmering controversy grew from conflict between extensive clear-cutting and other uses, especially recreation. In response, Congress passed the Multiple-Use Sustained-Yield Act in 1960, which required the Forest Service to manage public lands to best meet the needs of the American people. The Multiple-Use Sustained-Yield Act expanded the purposes of the forests to include as coequal uses outdoor recreation, range, timber, minerals, and wildlife and fish purposes. The Forest Service had discretion to determine the appropriate mix of uses. However, the Forest Service continued to make the timber industry its top priority and to grant contracts for massive clear-cutting operations, as it had before the Multiple-Use Sustained-Yield Act was passed. The extensive clear-cutting and other timber-focused policies began to interfere with other forest uses and placed the Forest Service at the center of growing public controversy once again.

Lingering Controversy

The controversy came to a head when forest users filed a lawsuit challenging Forest Service clear-cutting policies. In the Monongahela National Forest of West Virginia, turkey hunters discovered a massive clear-cut while wandering through one of their favorite sites. The Izaak Walton League filed a lawsuit on behalf of the hunters. In 1975, the court effectively compelled an immediate cessation of clear-cutting by concluding that the Organic Act forbade that method of harvesting.

In the aftermath of the Monongahela case, Congress passed the National Forest Management Act of 1976 (NFMA), which remains the most important statute governing Forest Service actions. The NFMA created a set of extensive planning requirements based around land resource management plans that have guided Forest Service decision making ever since. Substantively, the act mandated that the Forest Service ensure that its management tactics not substantially or permanently impair the productivity of forest lands.

Preparation of a land resource management plan requires the Forest Service to consider how its multiple use mandate can best be met in each individual national forest. For example, timber harvesting within a national forest must be based on scientific and other expert determinations of its appropriateness in conjunction with natural features of the landscape and relationship to other uses. The determination of whether a timber sale is appropriate includes an NFMA review that often coincides with the review required under the National Environmental Policy Act. Only after engaging in this process, which includes provision for public review and comment, may the agency authorize timber harvest.

Perhaps the most prominent and important dispute to arise under the NFMA structure centered around the northern spotted owl in the Northwest's old growth forests. In the 1980s, the Forest Service's proposed management guidelines for the future of the northern spotted owl drew criticism from both the timber industry and environmentalists.

In 1991, following extensive litigation, a district court effectively closed a huge expanse of national forest to timber extraction until the agency established a management plan that adequately protected spotted owl habitat. Following enormous public controversy, President William J. Clinton convened the 1993 Timber Summit and formed the Forest Ecosystem Management Assessment Team, which ultimately produced the 1994 Northwest Forest Plan that continues to govern forest use in the Northwest.

Throughout the spotted owl controversy, the Forest Service operated under regulations adopted in 1982. Such regulations provide detailed guidance for the Forest Service to comply with statutes such as the NFMA and have the force of law. In 1999, following a comprehensive review of its planning regulations, the Forest Service's Committee of Scientists recommended extensive changes to the regulations that govern planning under the NFMA. In 2000, these recommendations were largely adopted and new regulations began to reorient Forest Service priorities. The overall goal for managing the national forest system under these rules was ecological, social, and economic sustainability. The regulations placed the highest priority on maintenance and restoration of ecological sustainability as a necessary precondition to support sustainability overall. They also placed a greater emphasis on the use of science in planning, as well as monitoring and evaluation for forest resources.

Reversing Course

In 2001, under President George W. Bush, the Forest Service reverted to the 1982 regulations. It then revised its rules several years later, again altering its priorities. Under the ensuing 2005 regulations, ecological sustainability was deemphasized to be merely one element of the social, economic, and ecological sustainability goal. Thus, the Forest Service currently manages the national forests to meet its multiple-use mandate through planning aimed at achieving a balance of economic, social, and ecological sustainability priorities.

The nation's forests hold one of the most exciting and controversial stories in the evolution of U.S. natural resources management. From taming the wilderness to sustaining national resources for broad public benefit, the Forest Service has been the key player in this fascinating story for over 100 years.

See Also: Endangered Species Act; Fish & Wildlife Service; Land Ethic; Pinchot, Gifford; Sustainable Development; Wilderness.

Further Readings

Houck, Oliver A. "On the Law of Biodiversity and Ecosystem Management." *Minnesota Law Review*, 81/869 (1997).

Steen, Harold K. *The U.S. Forest Service: A History*. Seattle: University of Washington Press, 2004.

U.S. Forest Service. *The U.S. Forest Service: An Overview*. http://www.fs.fed.us/documents/USFS_An_Overview_0106MJS.pdf (Accessed February 2009).

Williams, Gerald. *The Forest Service: Fighting for Public Lands*. Westport, CT: Greenwood, 2007.

Andrew Long
Florida Coastal School of Law

Future Generations

In 1972, the landmark Declaration of the United Nations Conference on the Human Environment laid out an ambitious set of principles to effect preservation and protection of the environment. One of the primary ethical principles espoused in the declaration was an obligation to protect and preserve the planet for the enjoyment of future generations. More recently, the Bemidji statement on Seventh Generation Guardianship has proposed that we look well into the future in considering the consequences of our present actions. As the environmental movement has come of age since the 1970s, this idea of our ethical obligation to future generations has occupied a central place in the moral foundation for environmental protection. This article explores key philosophical disagreements that have arisen over thinking about obligations to future generations. It then examines some difficulties in accounting for future generations' interests in decision making and ends with some mechanisms that are being used to guard the rights of future generations.

Philosophical Disagreements Regarding the Rights of Future Generations

To the casual observer, the ethical imperative of preserving the Earth for future generations might not appear to present any major problems, but there is a great deal of philosophical controversy surrounding the idea. Much of the discussion concerns the challenge of thinking about the rights of "potential persons" in the indeterminate future versus actual persons here and now.

The first type of philosophical problem surrounds the rights of future generations as "potential persons." Some say that we cannot speak of rights of persons who do not yet exist. The argument is that because they have no ability to exercise (or relinquish) their rights claims on the present generation, they do not have rights. Opponents of this view point out that there is a distinction between moral and legal rights, the former existing independently of any legal structure. If the only rights that existed were those currently enforceable under the law, then there would be no social change. Oppressive practices such as slavery and segregation were at one time perfectly legal and enforced by the law, but clearly the rights of those subjected to slavery were violated under that system. In the same way, supporters of the rights of future generations argue that our legal system can (and should) evolve to protect the rights of posterity, particularly with respect to protecting the natural environment.

The second class of problems stems from the temporal distance of future generations. Some argue that we need to consider only our obligations to others living at the same time. One version of this argument is espoused by the cornucopians, who claim that technological innovation will create new resources and solutions to environmental problems, which is effectively saying that the future will take care of itself. This is probably the most extreme case, but there are others who take more intermediate positions on this issue precisely because of the temporal distance between present and future generations.

Another conundrum arising from the temporal distance problem is that preservation of resources for future generations creates an ad infinitum set of obligations: If we preserve resources for the next generation, then the subsequent generation would also have the same obligation. This raises the problem of determining which generation would actually get to use those resources. The counterpoint to this position is to distinguish between resource types. There are certain resources, like endangered species, that do not have much use value but are important to preserve because of their intrinsic value. Other resources, such as energy reserves, are arguably expendable, as long as concerted efforts are made to find alternatives that will supply ample power for the needs of subsequent generations.

To recapitulate, thinking about ethical obligations to future generations does indeed generate some philosophical problems. At the same time, although the indeterminacy and temporal distance of future persons challenges our traditional conceptions of rights, there still appear to be defensible philosophical grounds for considering the effects of our present-day activities on future generations.

Decision Making and Intergenerational Equity

Considering obligations to future generations also has implications for decision making. First and foremost, considering costs to future generations becomes important whenever we contemplate present actions. However, it is often very difficult to quantify those costs, which complicates things. For example, pursuing a development project might produce readily quantifiable (and monetarily measurable) short-term benefits in terms of economic growth, job creation, tax revenue, and so on, but it might also destroy a valuable habitat or eliminate a valued green space in a community. These costs are certainly attributable to the new development, but they are also very difficult to quantify. Although environmental economists have devised some fairly ingenious methods for pricing environmental amenities, generally these values are determined empirically using the willingness to pay of current generations. Therefore, even if some estimates of the value of environmental amenities are available, those estimates cannot account for the possibility that future values of environmental goods such as clean air and water might actually be worth more to future generations.

Intergenerational equity also has implications for some of the core decision-making algorithms that are currently in wide use. Consider the practice of discounting in cost-benefit analysis. The basic procedure with cost-benefit analysis is to tally the monetized costs and benefits of a project over time and then discount those costs and benefits using some predetermined discount rate to obtain their present value. Discounting implicitly assumes that a dollar tomorrow is worth less than a dollar today, presumably because the opportunity cost of that dollar is its investment in some other interest-bearing instrument.

However, by discounting costs—particularly environmental costs—an implicit judgment is being rendered that those costs are somehow less expensive for future generations than for those in the present. By virtue of discounting, the needs of present generations are privileged over future generations, literally by the numbers.

Representing the Interests of Future Generations

As we confront the mounting 21st-century environmental challenges, it is clear that consideration of future generations will continue to occupy a central place in our thinking despite the cognitive challenges it presents. The principal problem is that future generations have no person to represent their interests. Given that future generations cannot speak for themselves, policymakers have begun to develop a way to institutionalize the rights of future generations in the present. Several governments have ombudsmen who generally serve as advisory watchdogs, overseeing and evaluating regulations and legislation from the perspective of environmental stewardship and intergenerational equity. The Commissioner of the Environment and Sustainable Development occupies this role in Canada, for example.

Another way in which the rights of future generations may be institutionalized is through a trust arrangement. The national park system in the United States works in this manner, preserving and managing federal park land in trust for future generations, while allowing current generations to use them. Moreover, many private environmental groups have begun to establish their own land trusts to preserve open space and other environmentally sensitive areas. Finally, in addition to its current usage in preserving lands, legal scholars have begun exploring how the concept of the legal trust could be expanded to protect other environmental resources such as the air and water for posterity.

See Also: Cost-Benefit Analysis; Land Ethic; Precautionary Principle; UN Conference on Environment and Development.

Further Readings

International Human Rights Clinic at Harvard Law School. "Models for Protecting the Environment for Future Generations." http://www.scribd.com/doc/8237457/Models-for-Protecting-the-Environment-for-Future-Generations (Accessed January 2009).

Light, Andrew and Avner de-Shalit, eds. *Moral and Political Reasoning in Environmental Practice*. Cambridge, MA: MIT Press, 2003.

Partridge, Ernest, ed. *Responsibilities to Future Generations*. Amherst, NY: Prometheus Books, 1981.

Scherer, Donald, ed. *Upstream/Downstream: Issues in Environmental Ethics*. Philadelphia, PA: Temple University Press, 1990.

Michael Howell-Moroney
University of Alabama at Birmingham

Gaia Hypothesis

The Gaia hypothesis addresses complex system cycles on Earth by using a medical metaphor of homeostasis for the global ecosystem. This concept expresses a process in which life-forms on the Earth grow, change, and die in ways that lead to the persistence of these or other forms. Specifically, the Gaia hypothesis proposes that life on Earth maintains the Earth's climate and atmospheric composition at an optimum for life. James E. Lovelock formulated the Gaia hypothesis, which was later championed by Lynn Margulis, who elaborated on symbiosis with microorganisms.

Gaia is the name of the Greek Earth goddess. In recent decades, as climate change data accumulated about the connectivity of plants and other organisms with geochemical processes, such as the carbon cycle and oxygen cycle, the idea of Earth as an integrated system gained credibility. The human disruption of Earth system components such as rainforests and the ozone layer were hypothesized to contribute to the serious changes being observed around the Earth. Skeptics contend that supporters of the Gaia tend to attach mysticism and spiritual significance to ecosystem processes. Lovelock, espousing his beliefs on planet sciences, admits to pushing the metaphor of an alive, self-regulating system.

International global researchers in the 2001 Amsterdam Declaration interpreted the Gaia concept as an Earth system science. This Earth system behaves as a single, self-regulating system, composed of physical, chemical, biological, and human components. The interactions and feedback between the components are complex and exhibit multiscale temporal and spatial variability. This declaration or definition has the characteristics of a homeostatic system: maintaining constant conditions, key organ systems, negative and positive feedbacks, and temporal variability. For example, using the Gaia metaphor, the human species could be seen as an organ. With these characteristics, Gaia is commonly seen by scientists not as an individual but as an integrated biosystem—that is, an ecosystem. Recent catastrophic changes such as the ozone hole, global climate change, fisheries failures, and sea otter and sea lion declines have highlighted the behavior of complex system cycles in which stability domains may shift when tipping points are reached.

The interdependence seen in ecosystems and global cycles points out a parallel that exists between the Gaia hypothesis, when it is presented as a metaphor, and Native American ways of understanding how the world works. The experience of applied environmental science seen in traditional activities of hunting, fishing, gathering, and gardening is congruent with the Native American concept that all is "alive." The notion of Gaia, with its implication of the Earth as a living, evolving system with new emergent properties arising from the interaction of animate and inanimate parts brings a holistic view of the Earth seen in traditional systems. Lovelock presents a metaphor of an Earth whose goal is to sustain habitability similar to animism used by indigenous people to explain nature. The "spirits" that exert power over weather and other natural phenomena have human-like personalities, in contrast to the technical explanation of ecosystem science, but many concepts of modern science are still invisible theory to nontechnical people. A typical spirit in animism is the spirit who controls the land (Gaia). Spirits serve as contacts that allow people to maintain harmony with nature. The details of spirit religions vary among cultures but possess common themes of attachment to the land on which they live and a world in motion. The constant flux notion results in a web-like network of relationships. The Gaia hypothesis fits well the indigenous concept that since everything is interrelated, all of creation is related. Similar to Gaia, the only constant is change. In contrast to past perspectives, the oscillations in both natural and man-made chemicals are now expected under certain conditions.

The Gaia hypothesis proposes that life on Earth maintains the Earth's climate at an optimum for life. The carbon and oxygen cycles lead to global ecosystem homeostasis, which in turn results from a large number of local but interacting ecosystems. Lovelock created an imaginary planet called "Daisy World" to illustrate how this complex homeostatic system responds by feedback mechanisms to different stresses.

The Gaia hypothesis, from albedo and climate change to biogeochemistry, focused science's attention on the idea that environmental health is a global issue that concerns all nations and peoples of the world. This global awareness has led to studies of how social systems interact with the ecosystem. The Gaia hypothesis implies an environmental ethic. Although Lovelock initially intended a purely scientific model without any metaphysical or religious implications, by pushing the Mother Earth homeostasis metaphor, he brings a holistic perspective and raises the question of environmental justice that includes moral rights and obligations. The Gaia hypothesis also raises the question of obligations to future generations and to consideration of contaminants in a global context.

Because pollution has no boundaries, environmental rights to a healthy ecosystem have no boundaries. However, to sustain world population numbers and a "developed country" lifestyle, the term *sustainable development* needed to be introduced to address the distribution of ecosystem services. Gaia operates within a set of limits or constraints that lead to feedbacks and self-regulation. Although Gaia does not address individual rights, it does address intergenerational justice and society's legacy to the future of human life on planet Earth.

The Gaia concept resists explanation in the linear cause-and-effect sequential model of science. The emergent properties that make Gaia a real self-regulating system cannot be completely explained in classical reductionist terms. Environmental justice implies not only that we must maintain key features of local ecosystems (i.e., keystone species), maintain the productivity of the ecosystem, and maintain the surface temperature in a proper range but also that the Earth's prosperity is fairly distributed among people and other life-forms. Environmental justice recognizes that access to a healthy environment is a fundamental right of all. This sustainable system involves education in decision-making processes that provide for the needs of the current social system without damaging the ability of future generations to provide for themselves. Because we are just beginning to understand how global systems function, intergenerational environmental justice implies the need for a precautionary principle that would minimize human impacts on other species or their support system.

See Also: Biosphere; Deforestation; Environmental Justice; Global Climate Change; Indigenous Peoples; Precautionary Principle; Tragedy of the Commons.

Further Readings

Cajete, G. *Native Science: Natural Laws of Interdependence*. Santa Fe, NM: Clear Light, 2000.
Lovelock, J. *The Practical Science of Planetary Medicine*. Gloucestershire, UK: Gaia Books, 1991.
Lovelock, J. *The Revenge of Gaia: Earth's Climate in Crisis and the Fate of Humanity*. New York: Basic Books, 2006.
Margulis, L. *The Symbiotic Planet*. London: Phoenix, 1998.
Marten, G. *Human Ecology: Basic Concepts for Sustainable Development*. London: Earthscan, 2007.
Phillip, R. *Ecosystems and Human Health: Toxicology and Environmental Hazards*. Boca Raton, FL: Lewis Publishers, 2001.

Lawrence Duffy
University of Alaska, Fairbanks

Gender

The study of gender and environmental politics is the hypothesis that men and women have differing relationships with the nonhuman world. Men and women politically engage with environmental change differently, and environmental change affects men and women differently. These relational differences are far from passive. Human relationships with the environment are defined by rights, access, use, obligations, and control. Changes in these relationships—through abuse and degradation, structured development or natural disaster—tend to produce different outcomes in the lives of men and women.

According to this perspective, the fusion of gender and environmental politics is a field of both theory and practice. On the theoretical front, inquiry stems from the traditions of feminist scholarship. The first wave of scholarship called for attention to inequalities between men and women with respect to the environment and pointed out regimes of patriarchal control over nature. More recent scholarship focuses on the complexity of gender and sexuality and places the study of women in the context of communities and local environments. From a practical standpoint, activists have seen women as vehicles of progressive environmental change and development and seek to use the social roles of women in projects of political resistance or development.

"Gender" refers to the range of difference between men and women. On one level, it refers to biological distinction—men possess a y-chromosome that triggers development of male sex organs. On another level, gender refers to the social or cultural differences between the sexes. Under this definition, gender differences are often associated with expected societal roles of women and men, how individuals identify themselves (and are identified by others) with respect to these roles, and the significance of sexuality in understanding individual identity.

These Rwandan women used their traditional skills to raise geranium plants in 2007. They planned to develop a business exporting geranium oil for use in perfumes with the help of the World Bank.

Source: U.S. Agency for International Development

Although "gender" and "sex" are often used interchangeably, the preferred academic term explaining differences between men and women is *gender*. This use reflects the rise of feminist scholarship that has argued that the social roles of women are a product of politics and are not necessarily determined by biology. This so-called social construction of gender viewpoint has opened critical paths of inquiry into how politics shape gender roles and how the embodiments of these roles reshape society.

Gender in Environmental Activism

The role of gender in environmental activism remains something of a paradox. Numerous studies have shown stark differences in environmental activism that break along gender lines. In wealthy nations, women are generally more concerned with environmental issues than men. This is explained through the well-worn explanations that women are more conscious of environmental degradation because of its effects on their families. However, when activism is measured, men appear to be more politically active in environmental concerns than women. Measures of activism have ranged from queries about participation in protests, to memberships in nongovernmental organizations and activist groups, and to green consumer practices. Depending on the proxy used for activism, the gender paradox becomes either more convoluted or slightly clearer.

Although broad academic studies may point to ambiguous conclusions, many observers have come to see environmental politics as a movement driven by women. Early conservation movements were led primarily by men representing outdoor activities like hunting, fishing, and so on. In the 1970s, however, concerns over pollution and the effects of pollution on society began to take center stage. Out of community concern, many minority and working-class women were drawn to the burgeoning environmental justice movement, bringing their social issue activist know-how to environmentalism. For example, Lois Gibbs, a Niagara Falls housewife and mother, became a household name as she became the face of the Love Canal pollution issue. Gibbs founded a citizens' movement, mostly composed of concerned mothers, to publicize the threats toxic waste presented to communities and to advocate for government action to clean up.

The fruits of Gibbs's labor paid off, both in her community and in the environmentalist movement in general. In reflecting on the decades since her first foray into advocacy, Gibbs noted that many concerned mothers from across the United States eventually progressed and advanced from grassroots activists to leaders of large nongovernmental organizations and conservation groups. Despite the progress of both women's activism and the environmental movement, there remain some critical points of concern. One pervasive idea about gendered environmentalism is that women are more in tune with the "touchy-feely" holism of nature. Their maternal roles, it is argued, place them in a unique advocacy position—one that is often offered as a point of strength in the ecofeminist movement. In practice, this idea has political implications. Although this maternal credibility may drive passionate arguments, politicians and the media tend to turn toward the male-dominated positivist scientists for data and analysis. Many environmental advocates see this as both disempowering to women and an affront to the moral concerns of environmental justice.

Sometimes, the role of gender in environmental activism is a red herring, obfuscating other political issues. The Chipko movement of the Indian Himalayas is often cited as a model of environmental activism that was spearheaded by women. In the 1970s, frustrated villages in the Indian Himalayas began protesting against state-sponsored deforestation that threatened their livelihoods. By resorting to acts like chaining themselves to the trees, Chipko became emblematic of the power of environmental activism. Some accounts drew on the passionate acts of Chipko women and held them up as a prime example of "feminist principles" that preserved nature by counteracting the destructive forces of masculinist economic development. These accounts hinge on an idea that the protesters were trying to preserve an ecological quality in the face of imminent destruction. Subsequent research, however, posits that this region's forests had been developed, logged, and integrated into the economy for centuries and that the ecology Chipko sought to preserve was actually a product from years of local development (it was not virgin land). Empirically driven research suggests that Chipko was more about the activist concerns over who would control profits from forestry—a community labor dispute. The lesson from Chipko is that the claim of gender injustice, environmental injustice, and feminist activism may be simply romance and myth concealing (perhaps inadvertently) true explanations of environmental conflicts.

Gender in Development

The 1980s gave rise to a series of political movements aimed at incorporating women into environmental development programs. The roots of these "women in development" programs (WIDs) began to take hold in the 1980s. The Third United Nations Conference on Women in Nairobi (1985) brought attention to the concern of gender in sustainable development. By the early 1990s, the United Nations Environment Programme (UNEP), several nongovernmental organizations, and many international government agencies were actively trying to incorporate women into their programs.

Although specific goals may differ from agency to agency, most WIDs assert that when gender is considered in development planning, the plan will be more successful at achieving goals of social equity, economic development, and environmental preservation. According to the UNEP, women's roles in developing countries offer them several unique perspectives. Women are said to have intimate knowledge of the environment, based on their traditional roles as subsistence farmers and water harvesters within rural communities. In addition, women are also seen as keepers of traditional knowledge on matters of environmental importance. Studies have documented the work of women in the preservation of local knowledge about plants and medicinal herbs, maintaining diversity within poultry breeds, developing novel irrigation methods, and wise use of natural resources. Women often bear the biggest challenges of environmental catastrophe. In many drought- and famine-stricken regions, men travel with herds or to urban areas in search of work, and women are left to raise the children. All together, these domestic responsibilities place women at the very center of development work. From an ideal standpoint, development efforts should be specifically directed at empowering women and incorporating their assets into long-term plans.

Moving from theory to practice requires WIDs to consciously structure plans to capture the viewpoints of women. Such an effort is called "gender analysis." The U.S. Agency for

International Development, for example, asks two broad questions of every plan or project: (1) How will gender relations affect the achievement of sustainable results? and (2) How will expected results affect the relative status of men and women? To answer these questions, planners interrogate the roles of men and women within the proposed plan's context. The control of resources, the local empowerment of women, the roles of gender within the local economy, and so on, are all aspects that need to be identified and analyzed. Plans and projects can presumably be tailored to exploit gender roles to achieve maximum results. Analyses like these, which ultimately seek to understand the complexity of women's roles (especially in relation to men's roles), represent an evolution from WID to "gender and development" (GAD).

Critics of WID and GAD abound. Although noble, the specific targeting of women as either subjects for or agents of development did little to address the social inequalities that may be inherent in these societies. GAD theories (and practice) bring to bear tools of social theory to these concerns and seek to uncover the complexity of social and environmental relations. However, WID approaches still have supporters. Feminists have decried the rise of GAD, arguing that the feminist/activist spirit of WID theory is "depoliticized" in favor of the social constructionist viewpoint on gender. These critics may have a point, at least in certain contexts of pervasive patriarchy, but different societies present different challenges. The inclusivity of gender studies may allow for a more thorough analysis of social justice by foreseeing how local empowerment of women (either politically or through the material seizure of certain resources) can alter, for better or worse, conditions of the local environment and society. Although a GAD approach may not be as focused on women, it is still very political.

Frontiers of Gender and Nature

As noted earlier, the field of gender studies has grown increasingly complex. Although there have been significant contributions to understanding the roles of men and women in environmental politics, there has been far less attention paid to the roles of gays, lesbians, and queers in environmental politics. To some, this is indicative of leftist politics in general—preaching inclusion, but not at one's own expense. Queer theory, from a post-structuralist orientation, encourages the abolition of gender as an identity defining category. The "queering" of ecofeminism dismantles the notion that women have a procreative kinship with Mother Nature, and thus have a position of privilege.

Perhaps less radical, but still on the frontier, cross-cultural analyses of gender and the environment are proliferating, especially within the environmental justice movement. Although theorists and policymakers have stressed the importance of local context in gender analyses, new research is looking at how class, ethnicity, and other identities are bound up with gender. Scholars point out that African-American women have been particularly effective in calling attention to environmental justice concerns because they are especially sensitive to the constitutive relationships among class, ethnicity, age, and gender. In a similar vein, legacies of colonialism inform and inspire Native American women's claims to environmental justice

See Also: Ecofeminism; Environmental Justice; Environmental Movement; Gaia Hypothesis; *Silent Spring*; Sustainable Development.

Further Readings

Jackson, Cecile and Ruth Pearson, eds. *Feminist Visions of Development: Gender Analysis and Policy.* New York: Routledge, 1998.
Miller, Stuart. "Women's Work: Women in the Environmental Movement." *E: The Environmental Magazine* (January–February 1997).
Mohai, Paul. "Men, Women, and the Environment: An Examination of the Gender Gap in Environmental Concern and Activism." *Society and Natural Resources,* 5/1 (1992).
Robbins, Paul. *Political Ecology: A Critical Introduction.* Oxford: Blackwell, 2004.
Seager, Joni. *Earth Follies: Coming to Feminist Terms With the Global Environmental Crisis.* New York: Routledge, 1994.
Shiva, Vandana. *Staying Alive: Women, Ecology, and Development.* London: Zed Books, 1989.
Stein, Rachel, ed. *New Perspectives on Environmental Justice: Gender, Sexuality, and Activism.* New Brunswick, NJ: Rutgers University Press, 2004.
Toepfer, Klaus. "Editorial." *Our Planet: The Magazine of the United Nations Environment Programme,* 15/2 (2004).
U.S. Agency for International Development. "USAID Women in Development (WID): Gender Analysis Overview." http://www.usaid.gov/our_work/cross-cutting_programs/wid/gender/gender_analysis.html (Accessed May 2009).

Derek Eysenbach
University of Arizona

Global Climate Change

Since the onset of the industrial revolution (c.1750), human activities have altered the atmospheric composition of the Earth, significantly impacting the terrestrial energy balance. The burning of fossil fuels has substantially increased the amount of particulate matter and concentration of greenhouse gases (GHGs), most notably carbon dioxide (CO_2), methane (CH_4), and nitrous oxide (N_2O), into the atmosphere. The overall impact of human industrial activities on global climate has been a pronounced warming effect, as GHG increases have enhanced the atmospheric greenhouse effect and reduced the amount of outgoing radiation from the Earth. Although global climate change occurs with natural process (e.g., volcanic eruptions, insolation variability), the rate of change is much slower, occurring over millennia rather than the rapid anthropogenic induced climate changes observed over the past century with the modern industrial era. Global climate change impact scenarios have sparked an international debate on policy initiatives for balancing global climate change reduction measures with continued industrial development.

Even if human ingenuity and/or behavior resulted in the total cessation of anthropogenic GHG emissions tomorrow, the persistence of these gases in the atmosphere and the inertia of the geophysical, oceanic, and atmospheric processes that feed off global warming will result in a global mean surface temperature rise approaching at least 2 degrees C over the

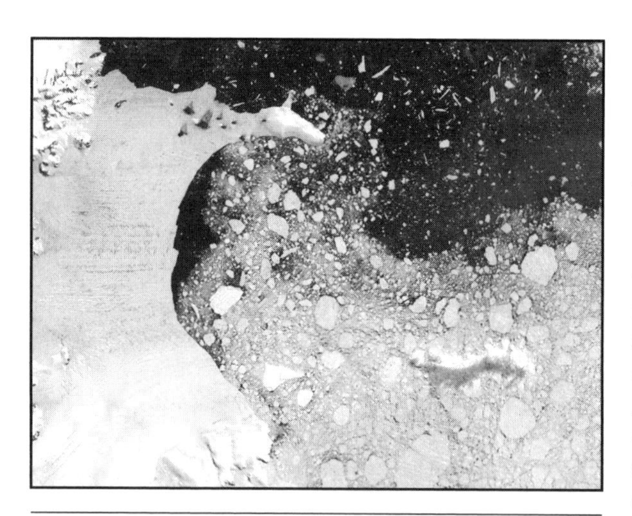

NASA satellites captured this image of the Larsen B ice shelf in Antarctica in 2000 before its rapid collapse in early 2002.

Source: NASA/Goddard Space Flight Center Scientific Visualization Studio

next century. Given the current rate of GHG emissions growth, in spite of any individual and collective reduction efforts, we are almost certain to cause much more warming still.

Unfortunately, the climatic effects that this warming will precipitate are not entirely understood. Computer models used to explain weather and climate phenomena are still limited in their predictive capacities. Aspects of the interactions among atmospheric, land, and ocean systems are still unknown or speculative, and we have no sense of what climatic "tipping points" lurk in the future that might drastically change climate patterns in a time frame that provides us with little or no time to adapt, much less mitigate them.

Background

"Climate change" is closely related to "global warming" but can be thought of as a more complex set of phenomena, many of which are directly dependent on global warming. Discussed elsewhere in this volume, global warming is a phenomenon that results from radiative forcing as a result of various gaseous and particulate substances in Earth's atmosphere. The increase in Earth's mean surface temperature is caused by warming of land, ocean, and/or atmosphere as a result of the amount of the sun's radiant energy that is not reradiated into space but absorbed by the Earth and it components.

Sources of this global warming are natural or anthropogenic. Absent human influence, naturally occurring GHGs (including carbon dioxide) and water vapor (in the form of clouds) provide a relatively stable radiative forcing that maintains the global mean surface temperature at a level conducive to human and other biological life. Abrupt warming may result from variations in solar radiation patterns. Abrupt cooling may result from the reflection back into space of greater amounts of solar radiation by dust resulting from volcanic activity.

A change in global mean surface temperature from global warming results in a variety of physical and chemical changes to Earth systems that often manifest themselves in weather events (such as hurricanes or thunderstorms). The long-term (regional) "average" of these weather events determines what we term our *climate* and defines the weather we expect at any particular time of year at any place on the globe. The change in well-established climate patterns is recognized as climate change.

Although isolated weather incidents cannot be directly linked to a changing climate because of the fundamentally chaotic nature of our climate, there is compelling evidence—much of it provided by National Aeronautics and Space Administration scientist James Hansen—that we are beginning to see the effects of global warming in global climate change. Among the numerous global reports documenting climate change effects are the following:

- The Intergovernmental Panel on Climate Change (IPCC) reported that 11 of the warmest years in the instrumental global surface temperature record occurred during 1995–2006.
- A 2005 Harvard Medical School report concludes that "Climate change and infectious diseases threaten wildlife, livestock, agriculture, forests and marine life, which provide us with essential resources and constitute our life-support systems."
- In its assessment of 2007 global natural and man-made disasters, reinsurance giant Swiss Re highlighted *IPCC Fourth Assessment Report* conclusions that global warming is contributing to an increase in the severity of flood events (1970–2005).
- The U.S. National Oceanic and Atmospheric Administration reported that the combined global land and ocean surface temperature for January 2009 was the seventh warmest since 1880—0.95 degrees F above the 20th-century mean.
- In 2008, the IPCC reported that "Observational records and climate projections provide abundant evidence that freshwater resources are vulnerable and have the potential to be strongly impacted by climate change, with wide-ranging consequences for human societies and ecosystems."
- With respect to climate-related conflict, UNEP states that "[t]he consequences of climate change for water availability, food security, prevalence of disease, coastal boundaries, and population distribution may further aggravate existing tensions and generate new conflicts."

These and many others seem to confirm what James Hansen claimed in his 1988 testimony: Global warming resulting from anthropogenic GHG emissions is resulting in climate change that is already having significant effects on global weather and natural resource patterns. Most recently, though, Susan Solomon and other scientists at the NOAA Earth System Research Laboratory published results indicating that the climate change that takes place as a result of anthropogenic carbon dioxide emissions is "largely irreversible for 1,000 years after the emissions stop."

As our understanding of the Earth's geophysical, oceanic, and atmospheric systems has improved, scientists have created sophisticated global climate models that attempt to reproduce observed weather events and predict future weather and climate trends. Based on a widely accepted set of future GHG emissions scenarios, these models provide an increasingly accurate reflection of extremely complex, interacting systems. Although scientists may have a solid understanding of component processes (e.g., the greenhouse effect, cloud formation, and solar reflectance), the interaction dynamics of land, air, and water continue to pose significant challenges for these models. However, confidence in global climate model forecasts has improved sufficiently since the first IPCC assessment report in 1992. Experts now have "high confidence" that climate change is a reality.

But whether or not specific weather events or regional weather trends (e.g., droughts) can be attributed to climate change, their economic and social impacts provide a glimpse of what lies in store for mankind when climate change actually bites.

Abrupt Climate Change

The Earth's climate is a chaotic system. That is, the regional climate characteristics that we have come to expect reflect one of many possible equilibrium states. This state is generally stable—climate changes come in a continuous fashion, with tomorrow being very much like today. Current global climate models assume that climate change from anthropogenic drivers will occur in a gradual fashion, taking much longer than a human generation to manifest itself. Our adaptation capacity is largely predicated on the assumption that we will have ample time to adapt.

Unfortunately, chaotic systems such as this can be sufficiently perturbed to exhibit unexpected, large-scale shifts in regional weather patterns over decades. Abrupt climate change of this type and magnitude is generally thought to involve sudden changes in oceanic circulation. One of the most prominently discussed disruptions is the thermohaline circulation in the Atlantic Ocean that transports warm surface water northward, tempering the European climate. The *IPCC Fourth Assessment Report* terms such an occurrence "*very unlikely.*" However, it concludes that it is "*very likely* that [the] meridional overturning circulation . . . of the Atlantic Ocean will slow down during the 21st century" (emphasis in the original). This could result in changes in fisheries productivity, ocean carbon dioxide uptake, and unforeseen feedbacks in the climate system.

Scientific evidence holds that abrupt climate change is a remote possibility. However, were some "tipping point" to be crossed, its human and environmental consequences could be so catastrophic that those risks deserve particular care in assessment and response from policymakers. The United States' response to the potential hazards of (abrupt) climate change has been predicated on a traditional requirement that scientists conclusively demonstrate the cause-and-effect relationship in their risk analyses. Given the uncertainties inherent in climate modeling, such certainty is presently unattainable. The economic costs of aggressive mitigation could not be balanced against the quantifiable benefits of those measures. American climate protection policy has emphasized only voluntary action by business to reduce GHG emissions.

The European Union and other nations take a much different approach to their assessment of the risks posed by climate change. These states embrace the precautionary principle in advocating aggressive climate protection measures in spite of not fully understanding the nature or magnitude of the risks from (abrupt) climate change. They understand the consequences of such risks to be too catastrophic to do otherwise. Their support for the Kyoto Protocol and its enforceable emissions targets—widely recognized as being only a first and imperfect step in the international climate protection regulatory process—demonstrates their commitment to sustainable development and environmental stewardship.

What Does the Future Hold?

Although the popular media still—though infrequently—privilege climate change deniers with press time, there is no credible evidence that would lead us to doubt that climate change is real, that it is manifesting itself in contemporary weather events, and that human activity is largely responsible. As the IPCC considers the abrupt melting of the Greenland and/or Antarctic ice sheets as unlikely, though not impossible, gradual climate change dominates current science and policy concerns.

Alarmingly, however, some experts are casting doubt on the accuracy of IPCC "consensus" climate change assessments. Some suggest that the IPCC reference scenarios have seriously underestimated the extent and severity of climate change impacts by underestimating the technological challenges to curbing GHG emissions. The disintegration of the Antarctic Larsen B (2002) and Antarctic Wilkins (2008) ice shelves demonstrates the speed with which global warming–related events can occur. Observations of accelerating rates at which the Greenland ice sheet and Arctic sea ice are melting indicate the extent to which we may have underestimated the current extent of global warming, and recent testimony by James Hansen has warned that global warming has proceeded to such an extent that we are at risk of irreversibly destabilizing the global climate.

Whether or not one agrees with the approaches advocated by the Kyoto Protocol, its successor protocol to the United Nations Framework Convention on Climate Change, and similar regulatory agreements, there is no credible disagreement that we must reduce anthropogenic GHG emissions to avoid many unknown consequences of climate change or reduce their severity. These reductions will take the form of governmental emissions regulations; technological alteration of existing energy production processes to reduce or GHG emissions; transitions to alternative, non-carbon-based forms of energy production; or changes in consumption behaviors to reduce these emissions. To the extent that responses are globally agreed and locally observed, they serve to mitigate future impacts of climate change.

However, to the extent that man has already dumped untold billions of tons of GHGs into the atmosphere, and to the extent that those gases and their warming effects are persistent, the global community must implement measures to adapt to unavoidable climate change. Geographic migration of rainfall and desertification has already been observed and will continue for the indeterminate future. Sea levels have begun to rise and will continue to do so. Daily surface low-to-high temperature ranges are on the rise. Stresses on public healthcare systems resulting from rising average daily temperatures and invasive disease varieties are increasing. The rate of species and ecosystem extinction and migration is increasing. Regional agricultural economies are suffering from the effects of invasive floral species. We are already committed to a wide range of environmental, economic, and social climate change impacts for which we must prepare.

However, although mitigation strategies must be taken uniformly across the globe, adaptation is regionally and locally dependent. Measures taken by coastal communities will need to differ from those necessitated in inland cities. New sources of freshwater must be identified and current consumption habits modified as regional rainfall patterns change and snowfall totals decrease. Urban development patterns and strategies must reflect the present and future reality of rising sea levels and energy and natural resource supply scarcity. In anticipation of less available, and thus more expensive, goods transportation, regional and local economies must further emphasize local supply chains to ensure the welfare of their citizens.

Because of the global nature of the anthropogenic GHG emissions and local (social) impact of climate change effects, there is growing concern that mitigation will be left to the international sphere and localities will only focus on adaptation. However, any international agreement to reduce GHG emissions must ultimately be implemented and enforced at the local level. Therefore, it is essential to see mitigation as an essential form of adaptation. Acting locally, authorities must think globally by embracing climate change mitigation strategies just as enthusiastically as they do adaptation initiatives.

To appropriately balance climate change mitigation and adaptation responses, climate change policy must recognize the fundamentally local nature of any climate protection policy response, and each community must recognize that its welfare is inextricably linked to the well-being of the global community.

See Also: Agenda 21; Environmental Management; Intergovernmental Panel on Climate Change; Kyoto Protocol; Montreal Protocol; Precautionary Principle; Sustainable Development; UN Framework Convention on Climate Change.

Further Readings

Alley, Richard B., et al. "IPCC Fourth Assessment Report: Working Group I Report—Summary for Policy Makers." IPCC Secretariat, 2007.
Bates, B. C., et al., eds. "Climate Change and Water: Technical Paper of the Intergovernmental Panel on Climate Change." IPCC Secretariat, 2008.
Bernstein, Lenny, et al. "IPCC Fourth Assessment Report: Climate Change 2007—Synthesis Report." IPCC Secretariat, 2007.
"Climate of 2008: Annual Report." U.S. National Climatic Data Center. http://www.ncdc.noaa.gov/oa/climate/research/2008/ann/ann08.html (Accessed February 2009).
Enz, Rudolf, et al. "Natural Catastrophes and Man-Made Disasters in 2007: High Losses in Europe." Sigma, 1 (2008). http://www.swissre.com (Accessed February 2009).
Epstein, Paul R. and Evan Mills, eds. "Climate Change Futures: Health, Ecological and Economic Dimensions." (2005) http://www.swissre.com (Accessed February 2009).
Field, Chris. "Decisive Action Needed as Warming Predictions Worsen, Says Carnegie Scientist." Carnegie Institute of Science (2008). http://www.ciw.edu/news/decisive_action_needed_warming_predictions_worsen_says_carnegie_scientist (Accessed February 2009).
Hansen, James. "Global Warming Twenty Years Later: Tipping Points Near." http://www.columbia.edu/~jeh1/2008/TwentyYearsLater_20080623.pdf (Accessed February 2009).
Leoni, Bridgette. "CRED Disaster Figures: Deaths and Economic Losses Jump in 2008." Centre for Research on the Epidemiology of Disasters. http://www.emdat.be/Documents/ConferencePress/pr-2009-01-disaster-figures-2008.pdf (Accessed February 2009).
Matthew, Richard, et al. "From Conflict to Peacebuilding: The Role of Natural Resources and the Environment." United Nations Environment Programme. http://www.unep.org/pdf/pcdmb_policy_01.pdf (Accessed February 2009).
Pielke, Roger, et al. "Dangerous Assumptions." *Nature*, 452/7187 (April 2008).
Solomon, Susan, et al. "Irreversible Climate Change Due to Carbon Dioxide Emissions." *Proceedings of the National Academy of Sciences*, 106/6 (2009).

Kent Hurst
University of Texas at Arlington

Globalization

Although globalization is often seen as an economic phenomenon, it represents multiple processes whereby local populations are influenced by events that take place in different regions, different countries, and even different continents. David Held of the London School of Economics tells us that globalization may be thought of as a process that embodies a transformation in the spatial organization of social relations and transactions, assessed in terms of their extensity, intensity, velocity, and effect, creating transcontinental or interregional flows and networks of activity, interaction, and the exercise of power. It refers to the economic, political, social, environmental, cultural, discursive, and ideological elements of human existence. Each of these domains of globalization are intimately connected and are emphasized differently, depending on the particular perspective. This article highlights the nature of environmental elements in globalization. Humanity's influences on environmental problems are recognized as being linked to the generation and distribution of wealth, knowledge, and power, and to patterns of energy consumption, industrialization, wealth, and poverty. Initially, a brief history of the processes that have led to the rise of globalization with particular reference to environmental issues is outlined. Following this the nature of current global environmental problems is presented. Differing perspectives on how globalization may affect the environment are discussed. Specific global environmental phenomena are then identified, with a particular emphasis on global climate change. Finally, the relationship between global environmental issues and global governance is discussed.

Global environmental problems have led to more climate refugees, who may number as many as 50 million by 2010. The photograph shows a woman in rural Bangladesh, a country that has already suffered environmental displacement.

Source: World Bank

Industrialization and Globalization

There can be little argument that globalization in its current form began in the post–Industrial Revolution period. The latter half of the 20th century has witnessed rapid changes in the global infrastructure, facilitated by technological advancements, notably in transportation and communications, and rapid increases in the global population. These changes have affected all facets of life, from the global to the local scales and in developed and developing countries alike. In fact, it is the continued integration of the global and local arenas that underpins the processes of globalization in contemporary society.

The world's economy has been characterized by growing globalization, which is spurring the increasing integration of the global economy through trade and financial flows and the greater integration of knowledge through the transfer of information, culture, and technology. Globalization raises both fears and expectations. Some suggest that increasing interdependence is good for cooperation, peace, and solving problems. Economic integration may offer dynamic benefits, such as higher productivity. The exchange of goods and services also helps to exchange ideas and knowledge. A relatively open economy is better for learning and adapting foreign, state-of-the-art technology than a closed economy. Others, however, argue that economic interdependence is destabilizing. Rapid flows of money in and out of the economy cause losses, increase inequality, and lower wages, which ultimately results in harm to the environment through a lack of accountability.

Environmental politics must directly consider the effect of globalized trade as it relates to the environment. When considering the environmental impact of trade, there are two overarching interpretations. The first perspective argues that the intensification of international trade through globalization produces an intensification of production and consumption of goods that have been produced in a resource-intensive manner in various countries.

Ultimately, this results in a situation in which the consumers of these goods are so detached from the geographical, not to mention the cultural, site of production that it is extremely difficult to assess the environmental effect of the production of these goods. Moreover, international trade and the processes of globalization are said to enable the importing developed countries to obtain a disproportionate share of the world's natural resources. Such resources include the ownership of genetic information, intellectual property, and information. It is argued that as the world's economic market does not truly reflect the cost of the Earth's resources, through the globalization of trade these resources are exploited at an ever-increasing rate. Moreover, it is argued that the negative implications of globalization of trade are evident through the increased centralization of resources and wealth at the expense of large amounts of the world's populations. This article focuses on the environmental problems that may be considered under the heading of globalization.

Global Environmental Problems

The past three decades have seen a proliferation of evidence of the rise in environmental problems on a global scale. Although environmental problems of the past have always been present wherever complex societies have developed, until the onset of the Industrial Revolution they remained relatively localized. Today there is an overwhelming recognition of the interconnected nature of the planet's environment and the effect that human populations are having on it. As already outlined, changing drivers such as population growth, economic activities, and consumption patterns have placed increasing pressure on the environment. The volume and intensity of human interaction with the environment have produced environmental problems that are global in nature. These global environmental phenomena have produced a situation in which global and local populations are united through the recognition of environmental "crisis." It has become impossible to ignore the fact that people are intrinsically linked on every level on a disparate geographical basis, from the air we breathe to the climate we depend on, to food, water, and every other resource that is used by humanity in a finite global environment. An environmental globalization removes the possibility of any form of opt-out. What is recognized, then, is that the environment itself has an effect on globalization. A number of global environmental problems exist today—transboundary pollution of land, sea, and air; extensive loss of natural habitat and biodiversity; extensive deforestation and desertification; depletion of the ozone layer. Perhaps the most salient example of a global environmental phenomenon is humanity's effect on the global climate. The following briefly elaborates on some of these global environmental phenomena.

Transboundary pollution is an example of the global environment problem. The release of large amounts of synthetic chemicals into the atmosphere and water have lead to a continued buildup of toxic chemicals in the environment that has created conditions of human and biological life that are beyond the tolerable limit of biological organisms. The publication of Rachel Carson's book *Silent Spring* in 1962 is seen as one of the catalysts of the environmental movement and a precursor to more mainstream environmental politics. The book highlighted the detrimental effects that pesticides have on the environment. For example, chlorofluorocarbons have been used in the latter half of the 20th century as industrial solvents, foaming agents, and aerosol propellants. In the 1970s, it was recognized that the unregulated release of chlorofluorocarbons into the atmosphere appeared to be having an adverse effect on the Earth's ozone layer. In the 1980s, a large hole in the ozone layer was discovered over Tasmania, New Zealand, and large parts of the Antarctic.

According to Manfred Steger, 7 of 10 biologists believe that the world is in the midst of the fastest mass extinction of living species in the 4.5-billion-year history of the planet. Three-quarters of the worldwide genetic diversity in agricultural crop and animal breeds has been lost since 1900. Although many environmental problems are global in nature and are considered forces for and products of globalization, there is little dispute that global climate change embodies debates surrounding globalization.

Global Climate Change and Globalization

The following discussion focuses on a single environmental phenomenon to highlight the processes of globalization and the effect this has on environmental politics. As a proviso to the previous statement, to highlight global warming as a single environmental phenomena is profoundly misleading. In actuality, it is a multitude of interacting phenomena that converge at the human environment point of interaction, which in itself is representative of broader discussions on globalization from multiple perspectives.

Global climate change has been described as nothing less than one of the most profound and devastating challenges to face mankind in human history. It is already fundamentally altering the environmental, social, and political makeup of the planet. Through processes of population growth, technological advancement, and altered consumption patterns, humanity has increased the release of carbon dioxide (and other greenhouse gases) into the environment to such a degree that global temperatures are increasing at an accelerated speed compared with available records from previous periods in history. The precise effects of global climate change are difficult to predict, as the factors involved are complex, ambiguous, and uncertain. However, it is generally accepted that through the greenhouse effect, global warming is occurring—global average temperatures increased from 13.5 degrees Celsius in 1880 to 14.5 degrees Celsius in 2000. Further increases in global temperatures will lead to the melting of the polar ice caps, leading to significant sea level rises that will affect large swathes of the planet's low-lying areas. Increased temperatures in the atmosphere will lead to increased evaporation from the world's oceans and to an increase in the intensity, severity, and frequency of global atmospheric phenomena such as tropical storms. Creation of environmental problems through processes of globalization themselves then create new global forces.

Environmental Refugees

Altered global climate has a direct effect on human populations. The idea of environmental refugees draws on recognized refugee patterns that exist as a consequence of economic and political factors. Environmental refugees already represent an increasing problem for the international and national communities as whole populations are displaced by environmental factors such as global warming. In particular, rising sea levels pose a significant problem. According to the United Nations Environment Programme, certain models have forecast that climate change will increase the number of environmental refugees by a factor of 6, to 150 million in the next 50 years. On a shorter timescale, the United Nations University Institute for Environment and Human Security predicts that by 2010 there will be 50 million environmentally displaced people, most of whom will be women and children. In 2005, half of Bhola Island in Bangladesh became permanently flooded, displacing 500,000 people. Some have described the Bhola islanders as some of the world's first climate refugees.

Governing Globalization

These globalized environmental phenomena, products themselves of other facets of globalization of human existence already outlined, in turn create a further evolution of the globalization process by necessitating the restructuring of governance structures on a global basis. Global environmental problems and risks have forced a reordering of the way that environmental issues are dealt with and solutions found. Governance has also become globalized, with increasingly complex interstate interactions and with a growing role for interstate actors in a global governance context traditionally dominated by nations. The response to global environmental issues can no longer exist at the national, regional, or local scale but must be dealt with both above and below the nation-state system. Although nation-states remain significant agents for change on the world stage, the realization of the inadequacies of existing governance frameworks has resulted in the adjustment of these constructs with the development of new forms of responses and programs. In essence, global environmental problems have affected the sovereignty of the nation-state. Sovereignty amounts to a claim that the state has supreme authority over the people within its jurisdiction and, of course, the events. Although states rule, the world corporations have publicly sought the global political stage at gatherings such as the World Economic Forum and at multilateral negotiations such as the multilateral agreement on investment.

The United Nations has played a pivotal role in coordinating responses to the globalization of environmental problems. Following the United Nations Conference on the Human Environment in 1972, the United Nations Environment Programme was created to coordinate the responses to rapidly expanding global environmental problems. The creation of the commission for sustainable development was also a direct result of the need for coordination of the world's response to global environmental issues. *Sustainable development* is a term that has come to define human and environmental interactions in the 21st century. Sustainable development is the promotion of developmental patterns that will enable current human generations to meet their needs without compromising the ability of future generations to meet their own needs. This commonly used definition of the term was coined by the United Nations' World Commission on Environment and Development in 1987. The report of the commission, *Our Common Future,* laid the groundwork for future understandings of sustainable development. The reordering of global governance processes as a response to the globalization of environmental risks has progressed through the medium of a number of conferences and treaties that have been produced over the past three decades. The following text outlines some of these events.

Major Global Environmental Treaties

As has already been mentioned, the political response to these issues exists both above and below the nation-state structure. These have included the Ramsar Convention in Iran in 1972 with regard to the preservation of wetlands; the United Nations Conference on the Human Environment held in Stockholm in 1972; the Convention on International Trade in Endangered Species of Wild Fauna and Flora in Washington, D.C., in 1973; the Marine Pollution Treaty in London in 1978; the United Nations Convention on the Law of the Sea for marine species and pollution regulation in 1982: The 25-year review of this resolution focused on advancing technologies and the exploitation of previously inaccessible resources such as minerals on the ocean floor; the Vienna Protocol on the ozone layer in 1985; the Montreal Protocol on the ozone layer in 1987; the Basel Convention on the regulation of hazardous waste in 1989; the United Nations Conference on Environment and Development in 1992, on the advancement of sustainable development—which, most notably, ratified Agenda 21, which has had a profound effect on global environmental politics; the Jakarta Mandate in marine and coastal diversity in 1995; the Kyoto Protocol on global warming in 1997; the Rotterdam Convention on Industrial Pollution in 1998; and the world Summit on Sustainable Development in 2002. These events are proliferating as the complexity of environmental globalization is recognized. Overarching legal principles have also emerged as a result of environmental globalization; for example, the precautionary principle, the principle of common but differentiated responsibility, and the polluter-pays principle.

Conclusion

This discussion has focused on some of the elements of globalization that relate to the environment and to environmental politics more broadly. In sum, what is important to recognize is that a cyclical relationship exists between the conventional and recognized processes of globalization at the economic and social levels and the creation of global environmental phenomena. In other words, environment issues are now a significant driver of globalization in their own right.

See Also: Biodiversity; Deforestation; Global Climate Change; Industrial Revolution; Precautionary Principle; *Silent Spring*; Sustainable Development; Uncertainty.

Further Readings

Baylis, J. and S. Smith. *The Globalisation of World Politics*. Oxford: Oxford University Press, 2004.

Borne, Gregory. *Sustainable Development: The Reflexive Governance of Risk*. Lampeter, UK: Edwin Mellen Press, 2009.

Gregory Borne
University of Plymouth

GOVERNMENTALITY

Governmentality is a concept coined by Michel Foucault in the late 1970s to describe and analyze the "art of government" or the "conduct of conduct." Governmentality, along with related Foucauldian concepts such as power/knowledge, the gaze, biopower, and technologies of the self, has become influential in social sciences such as sociology, geography, political science, and anthropology. Foucault used these and other concepts to show the coevolution of forms of government, political rationalities, and individuals as political subjects through history. He was especially concerned with how modernity and modern institutions enable novel relationships between government and policy and subject formation and subjectivity. Social scientists have increasingly applied Foucault's concept of governmentality to the creation of environmental subjects through regulation of social interactions with the biophysical world in what has become known as eco-governmentality or green governmentality. As many of these investigations have taken place in the developing world, eco-governmentality is often affiliated with the study of the spread of green neoliberalism through structural adjustment programs and their concomitant environmental conditionalities.

Foucault first presented his views on governmentality in a series of lectures at the College De France in the late 1970s. Rather than being based on a narrow sense of government as state politics, Foucault's understanding of governmentality emphasized the myriad ways in which governments began to increasingly apply the theories and practices of political economy to the disciplining of populations for specific ends, producing specific kinds of subjects in the process. In his lectures, Foucault presented historical reconstructions of governments and political subjects from Ancient Greece to modern forms of government, paying particular attention to the ethical norms of each age. In particular, Foucault showed how the European state of the Middle Ages was a state of justice, maintaining its territory by simply imposing harsh laws on its subjects. During the 15th and 16th centuries, however, Renaissance-era states gradually became administrative states, ordering both people and things to create a more stable society. Rather than a mere defense of territory, then, the state became increasingly concerned with ordering its inhabitants in such a way as to produce specific ends. The emergence of certain forms of knowledge and technical means to achieve these ends constitutes the emergence of governmentality. Foucault then argued that governmentality as a form of power eventually superseded other forms of power such as sovereignty and discipline.

Parallel to his conceptualization of power/knowledge, Foucault's coinage of governmentality involved combining the term for governing with the term for mentality, emphasizing that it is impossible to analyze the technologies and techniques of power without also analyzing the political rationality underlying them. Whereas contemporary usage of the term *government* refers primarily to the political realm, Foucault showed through his lectures how government was used in medical, religious, pedagogical, and other texts up until the 18th century. Thus, along with referring to the administration of the state and its populations, government also signified the management of the body, the family and household, and even the soul. One of the keys to this novel form of power was the gaze of the school, hospital, and prison. That is, an individual who is both constantly within the field of visibility of the school principal, doctor, or prison warden and is also aware of being visible begins to police himself or herself spontaneously. Thus, the gaze and other technologies of power determine the conduct of individuals and objectify them through the application of certain forms of domination. Foucault's goal, however, is not only to describe how apparatuses of power increasingly disciplined individuals, creating the modern subject. Rather, Foucault also traces technologies of the self through which individuals transform themselves, often through the internalized logics of the state. In this sense, the emergence of the sovereign state is codetermined by the emergence of the modern, apparently autonomous individual.

Eco-Governmentality or Green Governmentality

Beginning in the 1990s, political scientists and other social scientists began applying Foucault's insights into the technologies of power and self to analyze the emergence of new understandings of the environment. In particular, scholars such as Timothy Luke (1997) examined the emergence of the "environment" as a conceptual category distinct from previous understandings of the "wilderness" or "nature." The environment is constructed through the expansion of scientific and technological expertise that also reveals novel environmental problems. These environmental problems are then used as justification for specific forms of intervention and management analogous to the construction of madness or disease and resulting interventions analyzed by Foucault. Michael Goldman (2005) in his discussion of the "greening" of the World Bank shows how the logic of environmentally sustainable development is being used to introduce new cultural and scientific logics of interpreting the biophysical realm of a state's territory. Certain areas are thus defined as ecologically important and/or degraded, whereas others are seen as expendable for the sake of improving the populations and natures of other places or the state as a whole. In what is known as green neoliberalism, the World Bank requires mandatory environmental impact assessments and green cost-benefit analyses through environmental conditionalities associated with its loans. These increasingly market-based instruments require forms of environmental monitoring that privilege certain perspectives and forms of knowledge while simultaneously extending the environmental gaze to rural peoples in specific environments. These peoples and their practices are made visible, thus making them accountable for the environmental effects of their actions. These rural peoples then in turn begin to incorporate these new understandings of the biophysical world into their understandings of themselves. For example, traditional modes of farming that included cutting down trees and burning fields to plant crops are suddenly seen to be harmful to biodiversity. Eco-governmentality thus describes how technologies of power and self have been extended to the biophysical world, creating new subjects concerned about the environment. Critics of the green neoliberalism and eco-governmentality point out that although rural peoples in places deemed environmentally important are increasingly constrained in their actions, urban elites throughout the world are allowed to continue consuming much larger shares of the natural world.

Foucault's understanding of governmentality as the "conduct of conduct" has contributed to our understanding of green politics by emphasizing the need to examine the political rationality underlining specific environmental regulations and practices. Moreover, governmentality as an analytical lens decenters the idea of power, exposing the ways in which individuals police both themselves and other individuals. The emergence of the environment and environmentally sustainable development as conceptual categories reorders our relationships not only with the biophysical world but also with the state, enabling the state to extend its influence even further.

See Also: Ecological Imperialism; Green Discourse; Green Neoliberalism; Power.

Further Readings

Agrawal, Arun. *Environmentality: Technologies of Government and the Making of Subjects.* Durham, NC: Duke University Press, 2005.

Dean, Mitchell. *Governmentality: Power and Rule in Modern Society.* Thousand Oaks, CA: Sage, 1999.

Foucault, Michel. *Security, Territory, Population: Lectures at the College de France, 1977–1978*, edited by Michel Senellart. New York: Palgrave, 2007.

Goldman, Michael. *Imperial Nature: The World Bank and Struggles for Social Justice in the Age of Globalization.* New Haven, CT: Yale University Press, 2005.

Luke, Timothy. *Eco-Critique: Contesting the Politics of Nature, Economy, and Culture.* Minneapolis: University of Minnesota Press, 1997.

W. Chad Futrell
Cornell University

GREEN DISCOURSE

"Of our environment, what we say is what we see," argue communication scholars James Cantrill and Christine Oravec in the introduction to their landmark volume *The Symbolic Earth: Discourse and our Creation of the Environment.* This statement captures the fundamental principle underlying the concept of green discourse: Language shapes the human experience of, and effect on, the environment. Although ideas about the environment are based in part on material experiences, language and other forms of symbolic expression become the primary means through which experience is interpreted and meaning conveyed to others. When individuals and groups use language and other symbols to influence public thinking and behavior regarding environmental issues, they are producing green discourse. Green discourse refers to the ways in which symbols are mobilized to convey environmental perspectives. The term has been used in a variety of ways to characterize a wide range of environmental discourses, including image events, toxic tours of contaminated communities, farmers' and ranchers' public responses to environmental management plans, and of course, the rhetorical production of those purporting to conserve, protect, or defend the natural world.

Most commonly, the concept of "green discourse" is used to characterize the language of the environmental movement. Its terminology, however, has proliferated in recent years, reflecting the extent to which perceptions of environmental realities are related to the symbols a society uses. Green discourse often challenges anthropocentric and utilitarian, usufruct views of the environment by positing the intrinsic value of all life. In many cases, green discourses emphasize principles of ecological holism and interconnection among all life forms. An emphasis on wildness and the sublime is also featured prominently in much green discourse. More recently, sustainability and differentiated responsibilities have become part of the dominant discourse of environmentalism.

The history of green discourse is difficult, if not impossible, to trace because it is embedded in literature, art, and philosophy, as well as the discursive production of the contemporary environmental movement. The blossoming of a tradition of green philosophical and literary discourse in the United States is commonly linked with writers such as Henry David Thoreau, Rachel Carson, and Aldo Leopold, though more radical figures such as Edward Abbey (who popularized the term *monkeywrenching*) can also be said to proffer a green discourse. The general emphasis of green discourse in these and other writers' works is on preservation, conservation, protection, and recognition of humans' implication in the more-than-human world. In the realm of art, green discourse is frequently associated with work that articulates a sublime experience of nature, exemplified by early landscape painters of the West such as Thomas Moran. Green discourse might also characterize the symbolic expression contained within online projects such as the Green Museum.

A wide variety of philosophical and religious traditions have profoundly influenced green discourse during the last century. The deep roots of green ideology extend from pre-Socratic Greek to Romantic to Marxist philosophy, while also drawing from indigenous, Christian, Buddhist, Taoist, and other spiritual traditions. Judeo-Christian religious ideology is pervasive as an underpinning of much resource management and wise use discourse oriented toward stewardship and conservation. The discourses of deep ecology and ecological holism are frequently linked to Eastern mystical and spiritual traditions, as well as pan-psychic views of nature as ecologically sacred. Contemporary green discourse also takes many forms. Ecofeminist discourses, for example, analogize the degradation of women and nature under patriarchy. Such discourse resists objectivity and replaces it with a subjective view in which humans love, care, and nurture the environment while recognizing the holism of the deep ecological view. Radical green discourses and Green Party manifestos share principles of nonviolence, social justice, demilitarization, community-based economics, global responsibility, and consideration of the rights of future generations.

The discourse of sustainability, a variation of green discourse that has come to dominate much contemporary organizational and governmental rhetoric about the environment, tends to frame environmental issues as scientific problems with technical solutions. Many environmental questions lack scientific certainty, however, and green discourse is used to highlight the risks of environmental harm resulting from mismanagement or hasty decision making. Advocating for increased regulation and an ethic of prudence in the evaluation of products and processes, environmentalists, scientists, and health researchers alike have called for the implementation of a "precautionary principle" when evaluating and addressing potential environmental risks. Other green discourses stress a human rights perspective that includes intergenerational equity, state responsibility, differentiated responsibilities based on respective capacities of countries or communities, and the right to development for the purposes of eradicating poverty. Such green discourse appears less ecocentric than the radical and philosophical discourse that places human interests alongside (and sometimes beneath), rather than above, the interests of other life forms.

Finally, green discourse can manifest in more pernicious forms, often linked to advertising or public relations campaigns. *Green-washing* is the term used to describe the efforts made by organizations to cover up a lack of environmentally responsible behavior by promoting an environmentally friendly image. Green-washing involves the deliberate dissemination of disinformation that makes an organization or product appear environmentally friendly when, in fact, it is not. Green-washing strategies are frequently used to deflect attention from an organization's poor performance on environmental matters by highlighting a product element or organizational effort undertaken to improve the environment. In contrast with the other examples of green discourse above, green-washing is intentionally disingenuous, as the environmental harms hidden by the discourse far outweigh the benefits being promoted.

See Also: Ecofeminism; Green-Washing; Land Ethic; Political Ideology; Precautionary Principle; *Silent Spring*; Sustainable Development; Uncertainty; Utilitarianism.

Further Readings

Cantrill, James and Christine Oravec, eds. *The Symbolic Earth: Discourse and Our Creation of the Environment.* Lexington: University of Kentucky Press, 1996.

Corbett, Julia B. *Communicating Nature: How We Create and Understand Environmental Messages.* Washington, D.C.: Island Press, 2006.

Cox, Robert J. *Environmental Communication and the Public Sphere.* Thousand Oaks, CA: Sage, 2006.

DeLuca, Kevin. *Image Politics: The New Rhetoric of Environmental Activism.* New York: Guilford, 1999.

Herndl, Craig and Stephen C. Brown, eds. *Green Culture: Environmental Rhetoric in Contemporary America.* Madison: University of Wisconsin Press, 1996.

Pettenger, Mary, ed. *The Social Construction of Climate Change: Power, Knowledge, Norms, Discourses*. Surrey, UK: Ashgate, 2007.
Pezzullo, Phaedra C. *Toxic Tourism: Rhetorics of Pollution, Travel, and Environmental Justice*. Tuscaloosa: University of Alabama Press, 2007.
Torgerson, Douglas. *The Promise of Green Politics: Environmentalism and the Public Sphere*. Durham, NC: Duke University Press, 1999.

Emily Plec
Western Oregon University

Green Neoliberalism

Green neoliberalism is, in the simplest of terms, the idea of a convergence of the liberal desire to expand and intensify market forces and environmentally sustainable development for the global South. The phrase was coined by sociologist Michael Goldman to help explain how these two seemingly conflicting projects/ideologies have become institutionalized by global governing institutions, particularly—but not limited to—the World Bank. The implementation of green neoliberal strategies has led to "environmental states" that open themselves to market forces and the (self-)governance of subjects along neoliberal lines. Although still relatively new, the phrase has been applied to new areas of global environmental governance, including investigations of the Montreal Protocol.

To understand green neoliberalism, we must first understand neoliberalism. Neoliberalism and globalization are often used interchangeably, but this is only the case when the rhetoric of neoliberals matches the policies implemented. Neoliberalism is, in a word, the belief that economically free markets are ends in themselves, and market transactions will lead to the most effective, efficient solutions regardless of the application. Although neoliberalism undoubtedly takes on varying characteristics when applied to real life, we can identify its basic tenets:

- Privatization via the creation of property rights for previously state-owned phenomena
- Maximization of contracts for services rendered, emphasizing the shortening of those contracts, especially for labor agreements
- Opening of markets where they previously did not exist and the intensification of markets, such as through the use of electronic instruments to track worker progress, to automate stock trading, and to assess production procedures
- Deregulation, or the "rollback" of government involvement in social and environmental aspects of daily life, often to establish market proxies for social services previously provided by the state
- Reregulation, or the "rollout" of government policies to facilitate the implementation of the abovementioned aspects of neoliberal agendas
- Increased reliance on civil society groups and private institutions to provide social and environmental needs that governments previously provided

Geographers Neil Brenner and Nik Theodore argue that neoliberalism is a process, taking on various forms depending on the specific spatial and temporal configurations of social, political, and environmental life, which they describe as "actually existing neoliberalism." In other words, neoliberalism never exists as an ideal type but, rather, is embedded in specific articulations of social life that make neoliberal policy impossible to fully realize. It is from this tradition that Michael Goldman is able to provide a sociological, yet uneven, spatial, and temporal character to green neoliberalism.

Green neoliberalism emphasizes the global dimension of neoliberalism, in which globalization—regulated by global governing institutions—is seen as encouraging or coercing nation-states into conducting themselves like business firms. Nation-states are encouraged to "sell themselves" to the global market as prime sites for private investment, not simply for selling goods. Green neoliberalism also includes the idea that governments must attract business to efficiently manage their natural resources. Nation-states are seen as being much less adept at managing the natural environment because of a lack of competition. This ideology is disseminated by global financial organizations, like the International Monetary Fund, through its macroeconomic policies, and by global development banks, like the World Bank, through its financial and technical assistance programs to less-developed countries. Scholars have criticized both the International Monetary Fund and the World Bank for being undemocratic and lacking transparency (e.g., the World Bank president is always an American appointed by the president of the United States, the International Monetary Fund managing director always a European). These global organizations, especially the World Bank, are seen as spearheading the implementation of green neoliberalism in the less-developed world. Because of its enormous operations involving private consultants, engineers, and nongovernmental organizations, and the ability to disseminate massive amounts of data and other forms of knowledge, the World Bank is seen as a key institution in the creation of business-friendly "environmental states." The World Bank has effectively established a dominant, long-standing discursive understanding of environmentalism and a dominant way of implementing that discourse in development practices.

The history and genealogy of the World Bank's transition to neoliberalism is immensely complex. Let us simplify it by noting that the World Bank implemented policies that were supported by Margaret Thatcher, Ronald Reagan, Milton Friedman, and various laissez-faire economists of fiscal discipline as a reaction to the financial crises of the 1970s. Since that time, environmental projects supported by the World Bank were increasingly subjected to rigid cost-benefit analyses that often pitted local environmental costs against national economic benefits. However, because of its hierarchical structure, and the need to implement projects fast, the World Bank's scientific data are not rigorous by any standards. World Bank data, however, shape the environmental assessments that determine whether a project's environmental costs outweigh the economic benefits.

However, green neoliberalism is about more than simply subjecting the local environment to market parameters. It is also about creating actors and nation-states that govern themselves and their environmental projects along neoliberal lines—a green neoliberal governmentality. Trained by the World Bank, and connected to the bank for funding and contracts, professionals working on development projects in government, the private sector, and international civil society learn to approach environmental problems through a neoliberal lens, which helps maintain the World Bank's green neoliberal agenda.

Because international civil society groups are so thoroughly immersed in the World Bank's version of environmentalism, they increasingly offer little resistance to it. Indeed, the inclusion of global civil society networks more centrally into global environmental governance has modified the very doctrines and belief systems under which those networks operate. Far more damaging than simply green-washing the image of global governing institutions, or appeasing civil society to World Bank–style development projects through inclusive reforms, civil society participation has served to discipline these traditionally antagonizing forces to fit the global neoliberal context. Many civil society groups are provoked to take on a new form of environmentalism: Food security, development, social justice, and ecological sustainability will all be provided through the provisions of the free market.

Other studies of global environmental governance that have followed Goldman's analysis of the World Bank report findings similar to those explained here. Recent alterations in the language of what constitutes a suitable exemption to the phase-out of ozone-depleting substances under the Montreal Protocol, for example, mark the green neoliberalization of this global treaty, which has had a negative effect on the ability of civil society groups to contest the persistence of these harmful chemicals.

See Also: Globalization; Governmentality; Green-Washing; Montreal Protocol.

Further Readings

Brenner, Neil and Nik Theodore. "Cities and the Geographies of 'Actually Existing Neoliberalism.'" *Antipode*, 34/3 (2002).

Castree, Noel. "Neoliberalising Nature: The Logics of Deregulation and Reregulation." *Environment and Planning*, 40/1 (2008).

Gareau, Brian J. "Dangerous Holes in Global Environmental Governance: The Roles of Neoliberal Discourse, Science, and California Agriculture in the Montreal Protocol." *Antipode*, 40/1 (2008).

Goldman, Michael. *Imperial Nature: The World Bank and Struggles for Social Justice in the Age of Globalization*. New Haven, CT: Yale University Press, 2005.

Harvey, David. *A Brief History of Neoliberalism*. New York: Oxford University Press, 2005.

Brian J. Gareau
Boston College

Green Parties

The first green party was established in Tasmania, Australia, in 1972, emanating from a clean water conservation campaign. Green parties were founded in New Zealand (the Values Party) in 1972 and in the United Kingdom (the Ecology Party) in 1973. The German Greens or *Die Grünen* were also established in 1979 and ran candidates in the 1980 national election. By the mid-1980s, a Green Party had been established in the United States in the aftermath of wider campaigns by environmental movement participants around issues such as conservation and preservation, opposition to nuclear power, and campaigns against toxic industries.

The first green group in the United States was known as The Green Committees of Correspondence, and they organized in local confederations until 1991. These groups organized under the principle of decentralization and maintained local structures. An interesting facet of the Green Committees is their organizational structure, which was based on bioregional rather than political considerations. The group was focused on local politics and was initially opposed to involvement in national politics.

The Green Committees of Correspondence was renamed the Greens/Green Party in 1991. This group continued to organize at the substate level; national politics were seen as outside the local ethos of green politics. The Association of Autonomous State Green Parties emerged with an agenda to participate in national elections in the 1990s. The Association of Autonomous State Green Parties put forward consumer activist Ralph Nader as a candidate in the 1996 presidential campaign. This move led to a nationally based party called the Green Party of the United States (GPUS) being formed in 2001. The GPUS became the third party in U.S. politics at this time and organized on a political platform similar to the established green parties operating in Europe. GPUS set out "10 key values" as guiding principles for their activists:

1. Grassroots democracy
2. Social justice and equal opportunity
3. Ecological wisdom
4. Nonviolence
5. Decentralization
6. Community-based economics and economic justice
7. Feminism and gender equity
8. Respect for diversity
9. Personal and global responsibility
10. Future focus and sustainability

This Green Party of Canada campaign sign appeared in Montreal in October 2008. Green parties have spread throughout the world and have gathered small but growing numbers of votes in North America.

Source: Flickr/Sweet One

Another group known as the Association of State Green Parties worked to develop the electoral capacity of local green organizations, often in tandem with GPUS. In the 2000 presidential election, the Green Party again nominated Ralph Nader and Native American activist Winona LaDuke for president and vice president. The Green Party was on the ballot in 44 of the 50 U.S. states and received 2.7 percent of the vote. The candidacy of Nader was criticized in the aftermath of George W. Bush's marginal success over green-minded Democratic candidate Al Gore. The Green's response was that it represented a platform for change from the policies of both mainstream parties. The Association of State Green Parties took the name "Green Party of the United States" in 2001 and successfully applied for recognition from the Federal Electoral Commission.

In 2002, John Eder became the first Green Party candidate to win a seat in a state legislature, winning in the Maine House of Representatives—a seat that Eder successfully defended in 2004. The 2004 presidential campaign was marked by Ralph Nader seeking an endorsement rather than a nomination from the Green Party, although prominent party member Peter Camejo would run as Nader's vice presidential candidate. The Green's nominated candidate was David Cobb. The controversy between supporters over the endorsement and the nomination approaches again divided the Greens. The greens experienced electoral successes in 2006 for Green Party candidates at the local level, although the removal of candidate Carl Romanelli from the Pennsylvania ballot created further problems for the party, including the loss of their third-party status in that state. The Green's presidential candidate in 2008 was former congresswoman Cynthia McKinney, with Rosa Clemente nominated for vice president. They received 0.5 percent of the national vote. Ralph Nader declined the Green Party endorsement as part of his Independent campaign.

The first green party in Europe was the Ecology Party in the United Kingdom in 1972. The party has prominent spokespersons such as their chair, Jonathon Porritt, author of environmental books such as *Seeing Green: The Politics of Ecology Explained (*1984) and *Capitalism as If the World Matters* (2007). The Ecology Party received 40,000 votes in the 1979 general election and 54,000 votes in the 1983 general election, increasing its profile and membership considerably in the process. Under the title Green Party, nearly 90,000 votes were won in the 1987 general election. However, the United Kingdom Green's most successful campaign came in 1989, when the party won 2 million votes, representing 15 percent of the overall vote in the European elections. This result pushed the Liberal Democrats into fourth place, but the Greens were unable to gain any seats because of the United Kingdom's majoritarian electoral system. The United Kingdom Green Party was divided into the Green Party of England and Wales in 1990, after the Green parties of Scotland and Northern Ireland organized independently. The groups still cooperate, although the Northern Ireland Greens merged with their counterparts in the Republic of Ireland in 2007. The Green Party in the Republic of Ireland has been a participant in the coalition government of that state since 2007.

The German Greens, or *Die Grünen*, was founded in 1979. Leading figures such as Petra Kelly united the party around political issues such as opposing nuclear power and weapons, as well as highlighting concerns about increased industrialization. The party split in 1982 with a faction forming the Ecological Democratic Party. The remaining members of *Die Grünen* organized the party around campaigns of political protest. After the reunification of Germany in 1990, *Die Grünen* achieved 7.3 percent of the national vote, gaining 49 seats. In 1998, the party went into coalition with the Social Democrats for the first time, and the Greens had some policy successes, including the phasing out of nuclear weapons. By 2002, *Die Grünen* was up to 8.6 percent of the vote and continues to gain support in opposition.

The Australian Green Party emerged from its origins in Tasmania to form from a coalition of conservationists and antinuclear protestors. Regional green groups organized under a national banner in 1984 after encouragement from German green activist Petra Kelly. The Australian Greens are organized through federal structures and have a consensus-based National Council. The Australian Green Party was formed from that alliance in 1992. The Australian Green's charter contains the following principles:

- Social justice
- Sustainability
- Grassroots democracy
- Peace and nonviolence

The Australian Greens won an increase in support for their campaign of support for asylum seekers in 2001. The party won 7.2 percent of the vote in the federal elections of 2004. This percentage increased to 9 percent of the federal vote in 2007, representing 1.38 percent of the overall national vote. The party continues to poll well in regional elections across Australia.

The Green Party of Aotearoa New Zealand is committed to environmental principles and preservation of the indigenous Maori culture. The New Zealand Greens emerged in 1972 from the Values Party, the first nationally organized green party in the world. The Greens became associated with the left-wing Alliance during the 1990s but set out on an independent path to raise their profile from 1997 on. The party achieved 5.16 percent of the vote in the 1999 general elections, rising to 5.3 percent in the 2005 elections and 6.72 percent in the general election of 2008.

Green parties have also participated in coalition governments in Finland, France, Belgium, the Czech Republic, and Mexico. In France *Les Verts* formed in 1982. In 1989, the party received 10.6 percent of the vote in elections to the European Parliament. Green parties in the European Parliament formed the European Green Party in 2004—the first political grouping to do so. The European Green grouping has been Eurosceptic in the past, as the European Union was seen to be the opposite of the decentralization ethic favored by most greens. The European Greens also oppose increased links between the European Union and the North Atlantic Treaty Organization. The Greens have since become more proactive on European Union affairs and were supportive of wider European integration.

In Africa, the Federation of Green Parties of Africa links the green parties of the continent, and in the Americas the Federation of the Green Parties of the Americas unites greens from both North and South America. The Asia-Pacific Green Network unites green parties from Asia and the Pacific Ocean. In the contemporary era, green parties continue to work for environmental politics and justice under their slogan of "think globally, act locally."

See Also: Democratic Party; Environmental Movement; Policy Process; Political Ideology.

Further Readings

Dobson, Andrew. *Green Political Thoughts*. New York: Routledge, 2007.

Frankland, E. Gene, et al., eds. *Green Parties in Transition: The End of Grass-Roots Democracy?* Aldershot, UK: Ashgate, 2008.

Hawkins, Howard. *Independent Politics: The Green Party Strategy Debate*. New York: Haymarket, 2006.

Leonard, Liam. *The Environmental Movement in Ireland*. Dordrecht, Netherlands: Springer, 2008.

O'Neill, Michael. *Green Parties and Political Change in Contemporary Europe*. Aldershot, UK: Ashgate, 1997.

Porritt, Jonathon. *Capitalism as If the World Matters*. London: Earthscan, 2007.

Porritt, Jonathon. *Seeing Green: The Politics of Ecology Explained*. London: Blackwell, 1984.

Talshir, Gayil. *The Political Ideology of Green Parties: From the Politics of Nature to Redefining the Nature of Politics*. New York: Palgrave Macmillan, 2002.

Liam Leonard
Institute of Technology, Sligo

Green-Washing

Green-washing refers to disinformation disseminated by an organization so as to present an environmentally responsible public image. One important characteristic of green-washing is the suggestive and manipulative use of information. A poor record of environmental performance itself is not described as green-washing, only a false representation of a poor performance as environmentally friendly is. Green-washing is not just failing to report negative information, but a selective process of omitting information to display a different public image. This highlights another key factor of green-washing: intention. Any form of intentional misguidance, misrepresentation, or selective disclosure of information is called a *green-wash* if this practice intends to improve the overall perception of environmental performance. Green-wash practices can occur on multiple levels. One type of green-washing targets the immediate experience of the end consumer through misleading facts on a product. This is arguably the most obvious to consumers and to green-wash critics. Environmentally suggestive language has experienced an immense growth in marketing and product labeling, with many legally uncontrolled uses of terms such as *green*, *natural*, *organic*, *environmentally friendly*, *eco-friendly*, *recycled*, and *biodegradable*. Green-wash refers to uses of these and other terms in a misguiding fashion with levels of truth ranging from lacking external validation of claims to suggestive or incorrect facts.

The environmental marketing and activist organization Terrachoice identifies six major sins of green-washing concerning product claims. They range from falsely claiming or implying non-existent product certifications (sin of lying) by choice of words or graphic elements on the label, to intentionally unclear information (sin of vagueness), and disputable claims that lack external validity mechanisms (sin of no proof). Three additional types of green-washing are concerned with falsely selective proportionalities of claims, such as advertising a single environmentally positive aspect while omitting all other negative information (sin of tradeoff); claims that may be true, but distract from greater environmental concerns (sin of the lesser of two evils); and claims that may be substantiated but suggest a positive magnitude that ignores their actual irrelevance or small contribution to the overall product (sin of irrelevance). The latter includes cases of million-dollar marketing campaigns to advertise a much smaller amount spent on an environmental project, or a marketing campaign where the status quo is claimed as a new achievement, for instance, "phosphate free." Further types of green-washing in this category concern strategic influences on the company or entity's public image beyond the information on a particular product. Examples are rebranding practices to avoid past associations with negative public images; using suggestive imagery in advertising to allure to new "green" values; or explicitly stating exaggerated environmental claims on the entity's performance. Other key green-washing practices encompass buttressing tactics that seek to weaken the environmental criticism as a measure to defend the status quo, for example by casting doubt on the severity of the problem or risk; by emphasizing scientific uncertainties around the environmental accusations; or by acknowledging the existence of a problem but questioning the suggested environmental alternative.

The cost of being caught green-washing is a relative unknown for the organization, usually depending on nonbinding regulations, media reaction, and public interest. Where green-washing is revealed and publicly criticized, it may erode public trust in the organization involved or possibly the entire sector associated with similar economic practices. Environmental advances offer many economic advantages for individual organizations, such as brand loyalty of consumers, higher investor confidence, and an overall gain in the organization's reputation and public image that translates into a stronger economic performance.

Market-led sustainability is a term often used to describe neoliberal self-regulatory mechanisms where the consumers reward or penalize a company for its environmental performance by choosing which brand to buy. However, this system is only as good as the consumer information that the market self-regulating behavior is based on. Misinformation is often how green-washing occurs in the first place, and it also influences how identified cases of green-washing are penalized by the public.

Misinformation and lack of information is at the core of the problem in green-washing. The discovery, prosecution, and public punishment of green-washing cases are similarly influenced by imbalances of the information that is available and interpreted by the public. In particular, green-wash criticism and public reactions against green-washing are often based on the perceived magnitude of the case. The dramatic effect of a single spectacular act of green-wash causes a lot of media attention and criticism, which is why organizations may strategically pursue smaller breaches of conduct.

Multiple, subtle acts of green-wash, however, may evade public attention, legal investigation, and erode the public perception more effectively in the long run. Cases with the most media coverage are not necessarily the most severe cases of green-washing, and even the biggest green-wash offender may not be the most severe offender to the environment. Companies that refer to environmental information in their advertising are more vulnerable than companies that have a worse environmental performance, but do not include environmental information in their advertising. The public becomes more upset with companies that do not live up to their claims than if they had not made any claims in the first place.

Numerous voluntary environmental standards, such as certifications, eco-standards, eco-labeling, and corporate voluntary self-regulation, have been introduced by corporate industries for consumer-oriented marketing purposes. There are also an increasing number of socially screened mutual funds and indexes that seek to attract more environmentally inclined investors, for example the Dow Jones Sustainability Indexes, Domini 400, and the FTSE4Good Index. Industry-led environmental programs, for example the World Business Council for Sustainable Development, Chicago Climate Exchange, or the U.S. Environmental Protection Agency's partnership programs, seek to promote large-scale collaboration platforms for voluntary self-regulation and corporate environmental standards. There is a widely acknowledged lack of legally binding regulations to control green-washing in national legislation, including in the United States, as well as a lack of effective international control mechanisms against global green-washing practices. Mandatory public disclosure of social and environmental information, as well as external audit concepts, have been widely suggested to create better control mechanisms of the gap between corporate image and substance of claims.

The ISO 14000 environmental management standards are among the most widely adopted global voluntary standards, established by the International Organization for Standardization (ISO) in the early 1990s. ISO 14000 standards are, however, not at all sufficient, as some environmental groups argue. ISO 14001, for example, only requires regulatory compliance with the country of its international headquarters, making its environmental standards only as effective as individual national regulations, which are often affected by the corporate-driven interests of an industry dependent country. Moreover, other standards, such as those suggested in Agenda 21, call for a stricter compliance within all countries where the entity operates.

Addressing the lack of systematic regulatory mechanisms, numerous nongovernmental organizations (NGOs) have launched anti-green-wash campaigns for public awareness and action. They investigate cases of green-washing and bring them to the attention of the media and the public. They also act as a form of nongovernmental enforcement and sanctioning by embarrassing green-washing firms through the media and damaging their public image. Key anti-green-wash organizations and campaigns include the Greenwash

Academy Awards, a green credibility index for businesses, Greenpeace's Stop Greenwash campaign, the U.S.-based CorpWatch, and the U.K.-based Corporate Watch, and numerous other NGOs that participate in anti-green-wash activism.

Ultimately, the damage of green-washing and the causes of corporate environmental impact need to be understood as part of a deeper structural critique. While corporate entities need to be held responsible for sound environmental practices, environmental criticism on this level alone is not sufficient because it would fail to address the driving forces behind resource demand levels in the first place. Blaming an oil company for producing oil only criticizes its means of production, but fails to acknowledge that it is our society that causes the existence and quantity of oil demands with increasingly unsustainable consumption levels overall. "Eco-friendly" products, even when the product claim is just, are often marginally better than conventional products in the resources and energy needed for production and transport, waste production, and overall environmental impact. This may be one of the most dangerous effects of green-washing on our society. It creates the illusion that "green" change is occurring across corporate industries and governance, while actually suppressing the momentum for real social and economic change in global environmental issues.

See Also: Agenda 21; Industrial Ecology.

Further Readings

Futerra Sustainability Communications. "The Greenwashing Guide." http://www.futerra.co.uk/downloads/Greenwash_Guide.pdf (Accessed January 2009).

Laufer, William S. "Social Accountability and Corporate Greenwashing." *Journal of Business Ethics*, 43:253–261 (2003).

Ramus, Catherine. "When Are Corporate Environmental Policies a Form of Greenwashing?" *Business and Society*, 44/4 (2005).

Sarkis, Joseph. *Greening the Supply Chain.* New York: Springer, 2006.

TerraChoice Environmental Marketing, Inc. "The Six Sins of Greenwashing: A Study of Environmental Claims in North American Consumer Markets." (2007) http://www.terrachoice.com/files/6_sins.pdf (Accessed January 2009).

United Kingdom Department for Environment, Food and Rural Affairs. "Green Claims—Practical Guidance: How to Make a Good Environmental Claim." (2003) http://www.defra.gov.uk/environment/consumerprod/pdf/genericguide.pdf (Accessed January 2009).

Conny Davidsen
University of Calgary

Groundwater

Water exists in many forms including water vapor, liquid, and ice, all of which are found on Earth including atmospheric moisture, precipitation, soil moisture, groundwater, ice, snow, oceans, and seas; within plants and animals; in lakes, streams, and rivers; and in man-made reservoirs. Water cycles between these forms both visibly, as when atmospheric moisture falls as rain, and invisibly, as when precipitation seeps into the soil to become groundwater. Water makes up 70 percent of the Earth's surface. Most of the water is contained in the oceans, with only a small percentage available as freshwater, most of which is frozen in glaciers.

Groundwater reaches the surface through springs like this one in the Westcave Preserve in Texas. Groundwater is the biggest source of freshwater for human consumption.

Source: iStockphoto.com

Freshwater constitutes approximately 2.5 percent of all terrestrial water sources, with only 0.75 percent being groundwater. This amount of water sounds minuscule, but a comparison between available surface and groundwater resources reveals quite a difference. At any given time, the Earth contains roughly 30,300 cubic miles of surface water resources contained in lakes and streams compared with millions of cubic miles of groundwater available within a half-mile of the land surface. With that comparison in mind, it comes as no surprise that groundwater is the largest single supply of freshwater available for human consumption. Groundwater resides below the surface of the Earth in porous layers of rock and sand called aquifers. Soil and rock strata differ in porosity, in the amount of water that can be held in the soil, and in permeability—the rate at which water flows through the soil. An aquifer that holds groundwater between layers of relatively impermeable rock or soil (e.g., clay or shale) is called a confined aquifer. Some confined aquifers are under pressure, such that when a natural opening, such as an artesian spring, or a man-made opening, such as a well, taps into the confined aquifer, water emerges from the surface under pressure. Unconfined aquifers are characterized by relative permeability between the ground surface and the water table below the surface where groundwater is stored.

Groundwater can be found almost everywhere on Earth at varying depths, some very near the surface, as in the case of wetlands, or at some depth below the surface, as in some dry areas in the western United States. Groundwater close to the surface may be just a few hours old, whereas groundwater located far beneath the surface of the Earth might be thousands of years old. Often, the older the groundwater, the more difficult it is to replenish by natural means.

Recharging Groundwater

Groundwater is most often recharged through percolation from rainfall and snowmelt or from nearby surface water. Surface water under the influence of groundwater describes locations where groundwater and surface water are interconnected. In these cases, either water from lakes, rivers, and streams can flow into the groundwater, in essence recharging the aquifer, or groundwater can flow into the surface water, diminishing the aquifer. Rivers and streams can receive as much as half of their flow from groundwater. Groundwater also plays a large role in the health of wetlands.

Water percolates through the Earth's surface at varying rates and flows down by gravity until the soil or rock is no longer permeable. The groundwater then flows laterally until it finally reaches the surface again and is discharged from the aquifer in springs, lakes, rivers, or the ocean. A few special cases exist in which groundwater movement is restricted. One such case is when the aquifer is perched above an impermeable layer of rock or sediment and above the water table. Another exception is ancient, or fossil, groundwater usually found in deserts. Fossil aquifers are essentially nonrenewable. Groundwater is predominantly used for drinking water supply and irrigation. Worldwide, approximately one and a half billion people depend on groundwater for their drinking water supply. In the United States, the U.S. Geological Survey estimates total groundwater withdrawals for 2000 were 80 billion gallons per day. This amount is more than twice the amount of groundwater withdrawn daily in the United States in 1950. The majority of groundwater withdrawals in the United States are used for irrigation, such as the 174,000 square-mile Ogallala Aquifer in the High Plains region; however, a third of the population relies on groundwater for drinking water.

Groundwater Issues

Groundwater is susceptible to two categories of issues: quality impairment and unsustainable use. Quality impairments can be natural or man-made. Because groundwater is held beneath the Earth's surface in aquifers, it is exposed to naturally occurring chemicals and minerals that dissolve or dissociate into the water by natural chemical processes. One such naturally occurring contaminant is arsenic. A famous example of the effects of this naturally occurring contaminant is the widespread arsenic poisoning that occurred in Bangladesh after the public water supply was changed from surface water to groundwater. The switch was made to improve the water supply because surface water supplies were contaminated with pathogens and other waterborne disease vectors. Unfortunately, once the population switched to consuming groundwater, it became evident that the aquifer was contaminated with arsenic, causing another public health issue. Groundwater can also be affected by anthropogenic contaminants. Anthropogenic contaminants are transported into the groundwater from agricultural runoff, urban runoff, municipal and industrial discharges, and seepage of directly injected wastes. Once contaminated, groundwater is notoriously difficult and expensive to treat and often must be abandoned altogether. The United States has moved to enact policies to reduce the likelihood of anthropogenic contamination of groundwater. One such policy is the 1974 Safe Drinking Water Act and its controls on underground injection of wastes.

Groundwater mining is the second serious issue facing groundwater resources. When groundwater is withdrawn at a rate that exceeds the rate of natural replenishment, the result is depletion of the aquifer—termed *groundwater mining*. In addition resulting in an unsustainable use of the resource, groundwater mining can cause significant effects on surface waters and on land overlying the aquifer, causing subsidence. Excessive pumping can rob nearby surface waters of their flow because, as the water table in the aquifer lowers, surface waters flow into the groundwater by gravity. Eventually, the surface water can cease to flow altogether, such as the drying of the Santa Cruz River near Tucson, Arizona.

Groundwater mining is a serious concern in rapidly growing areas of the southwestern United States, including Arizona. Recent efforts to reduce the Arizonans' reliance on groundwater resources and to recharge the aquifers have focused on the creation of Groundwater Management Areas to restrict groundwater withdrawals. These efforts have resulted in some aquifer recovery. Over time, officials in Arizona hope to achieve sustainable groundwater use such that they meet the needs of the present without compromising future generations' use of the resource.

See Also: Brundtland Commission; Innovation, Environmental; Urban Planning; Water Politics.

Further Readings

Alley, W. M., et al. "Sustainability of Groundwater Resources." U.S. Geological Survey Circular 1186, Denver, Colorado, 1999.

Bagla, Pallava. "Arsenic-Laced Well Water Poisons Bangladeshis." *National Geographic* (June 5, 2003).

Gilliom, R. J. "Pesticides in U.S. Streams and Groundwater." *Environmental Science & Technology* (May 15, 2007).

Glennon, Robert. *Water Follies.* Washington, D.C.: Island, 2002.

Opie, John. *Ogallala: Water for a Dry Land.* Lincoln, NE: University of Nebraska Press, 2000.

U.S. Geological Survey. "Groundwater Use in the United States." http://ga.water.usgs.gov/edu/wugw.html (Accessed February 2009).

World Commission on Environment and Development (Brundtland Commission). *Our Common Future.* New York: United Nations, 1987.

Christine J. Kirchhoff
University of Michigan

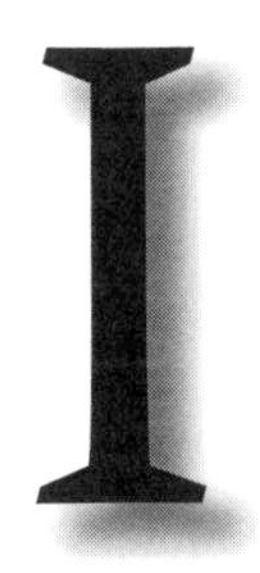

Indigenous Peoples

This article presents an overview of the environmental values, knowledge, and subsistence strategies of indigenous peoples both in their traditional contexts and in the contexts of colonialism and globalization. It discusses the current status of indigenous peoples in line with the 2007 United Nations Declaration on the Rights of Indigenous Peoples. It outlines indigenous environmental activism and precedent-setting legal cases. Finally, it discusses the model of biocultural diversity illustrated in Indigenous Conservation Areas that indigenous peoples manage and protect in line with traditional human–nature partnerships.

Historical Placement of Indigenous Peoples in the Human Timeline

The Industrial Revolution took place approximately seven generations ago. In contrast, the tenure on Earth of nonindustrialized peoples represented by today's indigenous cultures is an estimated 36,000 generations. Indigenous hunting-and-gathering peoples represent 99 percent of historic human cultures; in 2009, indigenous peoples still represent 90 percent of global cultural diversity. Indigenous cultures are intimately tied to their geographic homelands. Whatever the archeological evidence for the length of continuous residence for some on their lands—up to 100,000 years—indigenous peoples characteristically see themselves, their ways of life, and their lands as created in concert with one another.

This indigenous political group called the Confederation of Indigenous Nationalities of Ecuador (CONAIE) protested the 2002 summit of the Free Trade Area of the Americas in Quito, Ecuador, on October 31, 2002.

Source: Wikipedia/Donovan & Scott

Many of the indigenous peoples of the Americas, Australia, and Africa were hunters and gatherers. Others were farmers, like the dryland farmers of the American Southwest, who still live in villages thousands of years old, or the rice farmers of Asia. Others were fisherman, like the Coast Salish of Puget Sound, who lived in multifamily cedar longhouses that might cover an acre of ground. Still others practiced shifting horticulture, like the Kayapó of Brazil, who took more than one generation to move over their traditional territory, seeding wild gardens as they went. And some lived as island fisherman and yam gardeners in the Pacific or nomadic herders in the Middle East and Africa. Altogether, subsistence strategies of indigenous peoples entail some combination of hunting, gathering, fishing, herding, agriculture, and shifting horticulture in a context of ongoing flexibility and adaptation.

Today, many indigenous peoples are bicultural, as they both adhere to their ancient values and ways of life and respond to the pressures of the modern world.

Colonialism, Development, and Indigenous Subsistence Strategies

The 2007 United Nations Declaration on the Rights of Indigenous Peoples defines indigenous peoples as cultural communities having prior residence on lands within modern nation-states. It affirms the rights of these peoples both to cultural self-determination and to control their traditional lands and natural resources. It also points out the current marginalization of these peoples, resulting from a colonial legacy of disease, poverty, violence, starvation, slavery, and forced sterilization. Though indigenous peoples traditionally stabilized their populations by various child-spacing methods, many have a higher birth rate today, as they attempt to replace populations stressed to the brink of extinction. To set the repopulation of indigenous communities in the context of environmental impact, on a daily basis, a person living in a modern industrialized nation uses as much as two dozen times the natural resources as does an individual living in an indigenous community.

Colonialism propagated stereotypes that continue to hamper the entry of indigenous peoples into the modern global arena as equal partners. According to the colonial paradigm, higher placement on the ladder of progress gave nation-states the right to take over the lands of others and homogenize the cultures they encountered to their own. However, the predominant distinction between industrial and indigenous technologies lies not in their comparative placement on an evolutionary scale but in the fact that indigenous technologies adapt human activities to particular lands in a framework of dynamic mutualism, whereas industrial technologies adapt lands to human activities in a hierarchical framework of human control.

Botanist M. Kat Anderson's research with diverse native peoples of California reveals the sophisticated botanical methods used to increase particular plants and catalyze the diversity and abundance of traditional landscapes. A recent United Nations Educational, Scientific and Cultural Organization report on indigenous knowledge uses the parallel codevelopment of human cultures and ecological systems in the Amazon to counter the misperception that indigenous peoples either ravaged their lands or had no effect on them whatsoever.

Both of these misperceptions are linked to the faulty assumption that indigenous peoples lack the technical knowledge with which to make key subsistence choices. However, hunting-and-gathering populations depend for their survival on detailed environmental knowledge that gives them the option of practicing agriculture—though they may choose not to do so. Danish agroeconomist Ester Boserup's research into the institution of intensive agriculture exposes a reason for the choice not to take up this subsistence strategy: although it may increase the carrying capacity of the land, it is also likely to increase the workload of local populations.

Further, intensive agriculture is simply not suitable to all landscapes, as indicated by the failure of the green revolution that imposed modern farming techniques on Third World lands in a generalized fashion. Tens of thousands of families in the New Agricultural Movement in Bangladesh are currently reclaiming their lands in the wake of the devastation caused by the green revolution, using traditional methods combining diverse wild cropping, multicropping, and animal husbandry. In Oaxaca, Mexico, indigenous Mixteca farmer Jesus Leon Santos won the Goldman Prize for his use of pre-Hispanic farming techniques to restore farmlands left barren by industrial farming. As in the case of the New Agricultural Movement, Santos' methods have not only restored the fertility of local lands but also have spurred the recovery of local water tables.

Vandana Shiva documents the mismatch between generalized technologies and particular lands—and the impoverishment of local populations through the extraction of material and cultural resources in development projects that used them. Along with Veronika Bennholdt-Thomsen and Maria Mies, Shiva proposes a model of development that follows the "subsistence perspective" of indigenous peoples to counter what she terms "mal-development."

Climate change is another serious threat to indigenous peoples that is posed by the activities of industrialized nations. This especially affects island nations (some of whom have already begun to evacuate their homelands as ocean waters rise and salinate their water tables); those living on polluted mountaintops and in circumpolar regions, where permafrost is melting and game animals and reindeer herds are under stress; and those in drought-ravaged areas around the globe.

Indigenous Knowledge and Modern Science

Though indigenous populations are among the poorest of the world's poor, this is not an artifact of their traditional lifestyles. Richard Lee quantifies the results of the subsistence strategies of the San people of the Kalahari Desert in a seminal study that indicates their longevity, health, and nutritional well-being, as well as the minimal labor required to sustain their way of life. Such well-being is reliant on environmental knowledge kept in oral tradition in narrative form. Modern studies have compared indigenous knowledge with that of modern science, indicating their comparable validity—even though their methods and emphases differ.

The predictive efficacy of indigenous knowledge is illustrated by the case of the Moken in the Surin Islands, Thailand; the Ong and Jarawa in the Andaman Islands, India; and the Simeulue Island peoples in Indonesia, who forecast the massive 2004 tsunami and prepared accordingly. They survived unscathed as that tsunami took hundreds of thousands of lives in neighboring developed nations. Though they went unheeded, fishermen in south India tried to warn local authorities of the coming tsunami a few days before it struck.

Problems arise in the patenting of aspects of traditional knowledge—pharmaceuticals, for instance. Patents are problematic not only when they fail to fairly compensate communities of origin but also when they assume private ownership of community knowledge, which by cultural tradition belongs to future as well as current generations.

Indigenous Worldviews and Values

Indigenous cultures exhibit striking diversity not only between peoples living in distinct landscapes but also between peoples who share the same lands—as in the case of the Mapuche, who resisted incorporation into the Incan Empire, and the San peoples of the Kalahari, who shared their lands with Bantu pastoralists while maintaining a hunting-and-gathering lifestyle. Yet for all their diversity, communities living in intimate connection with the natural world share similar worldviews along the following dimensions:

- The sense of the intrinsic value and spiritual authority of natural life and the systems that sustain it
- The sense of nature as teacher: the idea that humans become human by means of their embeddedness in the natural world, as expressed in the quote with which Deborah Rose titles her book on Australian Aborigines: "Dingo makes us human"
- The sense of the unique value of individual persons, other natural life, and particular lands such that none of these can be replaced by or exchanged for any another
- The sense of the fundamental interdependence of natural life in the earthly family in which all life is kin
- The sense of responsibility for one's actions flowing from honoring the unique value of others in an interdependent world

Connected with this worldview, indigenous cultures characteristically maintain these values:

- Gratitude and reverence
- Sharing
- Cooperation
- Reciprocity
- Humility
- Balance

These worldviews and values have substantial environmental impact. The fishing practices of the Columbia River peoples in the U.S. Pacific Northwest derived from their respect for salmon nations as equal spiritual partners with humans. During their thousands of years on the Columbia River, those with this stance harvested an annual salmon take seven times the size of the modern one without harming the sustainability of the runs. Their worldview also led them to careful preservation of salmon habitat, as expressed to an Indian Agent in an 1851 protest of the denigration of salmon habitat created by sawmills at the mouth of the Columbia.

The Bemidji Statement of Seventh Generation Guardianship, drawn up jointly by the Indigenous Environmental Network and the Science and Environmental Health Network, is a model of the responsibility entailed in values that make those who hold them "Guardians of the Future."

Indigenous Environmental Activism

Though the United Nations Declaration on the Rights of Indigenous Peoples has no provisions for enforcement, it lends moral weight to struggles to protect indigenous lands from encroachment and destruction, such as the Igorot (Philippines) battle against giant dams;

the battle against drilling in the Arctic National Wildlife Refuge, to which the Gwich'in have contributed substantial leadership; the indigenous campaigns in Ecuador and Nigeria against oil drilling by Chevron; and the struggle that reversed the forced removal of the San from their ancestral Kalahari Desert, where they opposed diamond mining.

In South America, indigenous peoples have fought the ravages of globalization and protected their biotically rich lands in Bolivia, Brazil, Ecuador, Peru, Uruguay, and Venezuela. In Mesoamerica, the Mayangna and Mistiko have protected their rainforest against illegal logging through successful land claims and peaceful patrol of the boundaries of the Bosawas Biosphere Reserve.

Native American legal battles have set important precedents in protecting the environmental commons, as in water rights cases pressed by several tribes in the American West beginning in the 1970s, which resulted in the adoption of cumulative assessment in setting water quality standards for the Colorado River. A suit pressed by the Onondaga Nation of central New York excludes repossession of traditional property now held by non-Indians. Instead, it pursues the Onondaga Nation's right to assume environmental trusteeship over traditional lands and thus set standards for the cleanup of toxins deposited by industrial activity.

In Ecuador, *Pachamama* is an indigenous term for the sacred life-giving qualities of nature. This is also a key term in the Ecuadorian constitution, which asserts the legal rights of Pachamama and gives legal standing to those who sue on her behalf. The "Pachamama" organization formed at the request of indigenous elders, who lobbied for the inclusion of the rights of Pachamama in the Ecuadoran constitution.

In that indigenous stewardship of traditional lands challenges corrupt governmental regimes—as well as multinational timber and oil companies; uranium, gold, coal, and diamond mining companies; endangered species black markets; and drug cartels—indigenous environmental leaders have been subject to considerable violence. In the early stages of founding the Greenbelt Movement that won her the Nobel Peace Prize, Wangari Maathai suffered persecution by the Kenyan government that included being arrested and beaten. In the Sierra de Petatlán of Mexico, Felipe Arreaga Sanchez, his wife Celsa Valdovinos Rios, and Albertano Peñaloza, whose work earned them the Chico Mendes award for environmental heroism, have endured imprisonment, family murders, and continual threats in retaliation for their work protecting local forests. In the Amazon, violence inherited from the colonial period prevails against indigenous peoples like the Suri, who work to protect their traditional forests against illegal logging. Organizations such as Survival International and Amnesty International seek to protect indigenous environmentalists through global publicity on their behalf.

Biocultural Diversity

Over 95 percent of the world's high-biodiversity areas overlap with lands claimed by indigenous peoples, partly because biodiversity is central to indigenous subsistence and ecological management strategies, and partly because indigenous lands have not been subject to the intensive development and industrialization that has destroyed biodiversity elsewhere. As a result, today indigenous peoples are traditional stewards of 80 percent of the Earth's remaining biodiversity, even as they make up 90 percent of its cultural diversity.

In this context, the United Nations Programme on the Environment stresses the importance of biocultural diversity, recognizing the coevolution of human cultures and ecological systems. The Willamette Valley in Oregon, which European explorers nicknamed the "gourmand's paradise" for the diversity and abundance of its natural food resources, was a prime example of biocultural diversity. It evolved in concert with thousands of years of indigenous activity geared to increasing local plant and animal habitat.

Given the coevolution of human cultures and natural landscapes, the Convention on Biological Diversity has a stated goal of preserving indigenous knowledge to preserve global biodiversity. This model is an alternative to contention between indigenous populations and conservation agencies that prioritized their goals over human rights, or wilderness and national park set-asides that cut off indigenous access to traditional areas. The cooperative model is taken up today by agencies such as the World Wildlife Fund, which work to protect biodiversity on indigenous lands by enlisting indigenous leadership and designing ways to relieve the crushing poverty that subjects local peoples to pressure to participate in black markets for endangered species. In Northern California and Washington, indigenous peoples who initiated or joined legal battles to protect roadless areas retain access to these areas for the purposes of ceremony and the collection of materials necessary to cultural practices. This is a distinct change from policies that formerly excluded peoples from areas that coevolved in concert with the activities of their ancestors.

Indigenous Community Conserved Areas

In their recent *Resurgence* article, Ashish Kothari and Neema Pathak detail the importance of Indigenous Community Conservation Areas that replicate traditional human–nature partnerships in the Amazon and Australia; in the lands of the Qashqai in Iran and the Borana in Ethiopia and Kenya; in thousands of sacred groves in India and sacred crocodile ponds in Mali; in managed community forests in Africa, South Asia, and North America; in critical habitats of wild animals in southern India; and in ecosystems overseen by farming or mixed rural–urban communities, such as the Potato Park in the Andean highlands of Peru and the rice terraces in the Philippines.

Indigenous Community Conservation Areas cover an estimated 12 percent of the Earth's surface and provide connectivity across areas crucial for the migration of people, wildlife, and gene pools of both plants and animals. In an era of climate change, Indigenous Community Conservation Areas protect natural areas such as rainforests that sequester carbon and harbor reservoirs of biological and cultural diversity to help sustain the resilience of living systems that face the challenge of climate change.

See Also: Biodiversity; Ecofeminism; Global Climate Change; Intrinsic Value; Social Ecology.

Further Readings

Alcorn, Janis. "Beauty and the Beast." *Resurgence*, 250 (2008).

Anderson, M. Kat. *Tending the Wild*. Berkeley: University of California Press, 2005.

Bennholdt-Thomsen, Veronika and Maria Mies. *The Subsistence Perspective*. London: Zed Books, 1999.

Boserup, Ester. *The Conditions of Agricultural Growth*. Chicago, Aldine: 1965.

Boserup, Ester. *Economic and Demographic Relationships in Development*, edited by T. Paul Schultz. Baltimore, MD: Johns Hopkins University Press, 1990.

Davis, Shelton H. *Victims of the Miracle, Development and the Indians of Brazil*. London: Cambridge University Press, 1977.

Diamond, Stanley. *In Search of the Primitive*. New Brunswick, NJ.: Transaction Books: 1974.
"Guardians of the Future," http://guardiansofthefuture.org (Accessed February 2009).
Hove, Chenjerai and Ilija Trojanow. *Guardians of the Soil*. Munich, Germany: Frederking & Thaler, 1996.
Keller, Robert H. and Michael F. Turek. *American Indians and National Parks*. Tucson: University of Arizona Press, 1998.
Kimmerer, Robin. "The Rights of the Land." *Orion* (November–December 2008).
Kothari, Ashish and Neema Pathak. "Defenders of Diversity." *Resurgence*, 250 (2008).
Lee, Richard B. and Irven DeVore, eds. *Man the Hunter*. Chicago: Aldine: 1968.
Maffi, Luisa. "Cultural Vitality."*Resurgence*, 250 (2008).
Mander, Jerry and Victoria Tauli-Corpuz, eds. *Paradigm Wars, Indigenous Peoples' Resistance to Globalization*. San Francisco: Sierra Club Books, 2006.
Martin, Gary. "Restoring Resilience." *Resurgence*, 250 (2008).
Mathaai, Wangari. *Unbowed*. New York: Alfred A. Knopf, 2006.
Rose, Deborah Bird. *Dingo Makes Us Human*. Cambridge: Cambridge University Press, 1992.
Salick, Jan and Anja Byg, eds. *Indigenous Peoples and Climate Change*. Oxford: Tyndall Centre for Climate Change Research, 2007.
Schaefer, Carol. *Grandmothers Counsel the World*. Boston: Trumpeter Books, 2006.
Shiva, Vandana. *Biopiracy: The Plunder of Nature and Knowledge*. Boston: South End Press, 1997.
Shiva, Vandana, et al. *Biodiversity: Social & Ecological Perspectives*. Atlantic Highlands, NJ: Zed Books, 1991.
Suzuki, David and Peter Knudtson, eds. *Wisdom of the Elders*. New York: Bantam, 1992.
Turner, Nancy J. *The Earth's Blanket: Traditional Teachings for Sustainable Living*. Seattle: University of Washington Press, 2005.
UNESCO LINKS Programme. "Indigenous Knowledge and Changing Environments." International Experts Meeting, Cairns, Australia, 2007.
Wilson, Ken. "Guides and Gatherers." *Resurgence*, 250 (2008).

Madronna Holden
Oregon State University

Industrial Ecology

Public opinion and various legal cases against industrial accidents and/or pollution have led to greater awareness among industrialists, authorities, and communities of the urgent need for industrial sustainability. Sustainable industrialization has emerged as a strategic, balanced, and developmental trend for facilitating socioeconomic benefits for the present generation without sacrificing the net industrial profits or compromising the needs of future generations and without impairing basic environmental quality and ecosystem processes and services. This has led to the new paradigm of "industrial ecology," or IE—a field just about 20 years old and identified as one of the eight Grand Challenges of Environmental Science we are facing today.

IE is defined as the system-oriented study of the physical, chemical, and biological interactions/interrelationships both within industrial systems and between industrial and

Industrial ecology considers the environmental effects of a product over its entire life cycle. The discipline attempts to address such problems as the disposal of electronics, such as these collected at an "e-waste" dump site.

Source: iStockphoto.com

natural ecological systems. The idea of IE is based on a straightforward analogy with the natural ecological system, in which one seeks to optimize the total materials cycle from virgin material to finished material, to component, to product, to obsolete product, and to ultimate disposal. Factors to be optimized include resources, energy, and capital. It focuses on reducing the environmental impacts of goods and services and on innovations that can significantly improve environmental performance. Thus, it could help environmental policymakers address some of the core challenges of environmental policy, ranging from climate change to waste management to land use policy. In the following sections, we discuss the tools and strategies for IE, related policy developments across the globe, limitations in prevailing policies, and future perspectives.

Tools and Strategies for IE

The following are the most important tools of IE that can complement and enhance an integrated framework for sustainable industrialization:

Materials flow analysis or mass balance uses numerical data for direct inputs of materials in combination with chemical or engineering details of the processes being studied for analyzing resource flows. It is based on the fundamental physical principle that matter can neither be created nor destroyed. Therefore, the mass of inputs to a process, industry, or region equals the mass of outputs as products, emissions, and wastes, plus any change in stocks.

Life cycle assessment is used to determine the total environmental impact of a product throughout its life cycle (i.e., from cradle to grave). There are three basic stages in life cycle assessment: inventory analysis, impact analysis, and improvement analysis.

Strategic environmental assessment is conducted to ensure that the environmental consequences of proposed policies, plans, and programs are within prescribed limits and allows the integration of sustainability objectives at the earliest stage of the decision-making process.

In environmental risk assessment, the inherent hazards involved in processes or situations and the risks posed by these hazards are estimated either quantitatively or qualitatively. Environmental risk assessment includes human health risk assessments, ecological or eco-toxicological risk assessments, and specific industrial applications of risk assessment. Environmental risk assessment includes a number of steps:

- Problem formulation
- Hazard identification

- Release assessment
- Exposure assessment
- Consequence assessment
- Risk estimation/evaluation
- Risk characterization/management

Ecosystem vulnerability analysis reveals the degree of capability and sensitivity of an ecosystem to cope with the consequences of natural changes.

Eco-performance profile is the identification of energy and material-related environmental impacts generated by the company and along products' life cycles, cradle to grave. Full environmental cost accounting includes physical flow accounting and monetary flow accounting. The former deals with material and energy flows, and the latter deals with evaluation of environmental impacts, cost of damages to the environment, cost for environmental protection activities, and so on. Environmental auditing is a management tool that is a systematic, documented, periodical, objective evaluation of how well the environmental protection organization, management procedures, and equipment function in protecting the environment by facilitating the management control of environmental procedures and evaluating conformance with the company's environmental policy, legislation, permits, and so on.

Eco-mapping is a visual, simple, and practical tool to analyze and manage the environmental performance of the industry; an easy, creative, and systematic method of obtaining environmental data based on the physical reality of the industry; and a dynamic inventory of the changes in the environmental behavior of the industry in a visual format.

Total quality environmental management is a management philosophy and a set of accompanying quality improvement techniques that have been widely used in various sectors all over the world. Total quality means an ultimate goal of zero pollution, whether for the individual firm or for a cluster of companies in an industrial ecosystem, undertaken continuously for improvement across all operations by seeking to discover the reasons for poor quality performance and implementing methods to reduce and/or eliminate the causes of poor quality.

Environmental health and safety is used for reviewing legislation, recommending policies, and monitoring compliance with environmental and health and safety statutes and regulations. Environmental management system is part of a management system of an organization in which specific competencies, behaviors, procedures, and demands for the implementation of an operational environmental policy of the organization are defined. It improves the overall performance and efficiency of a company. It also facilitates cost savings through the reduction of waste and more efficient use of natural resources and avoids fines and penalties from not meeting environmental legislation by identifying environmental risks and addressing weaknesses. It improves the public image of the enterprise.

Integrated environmental strategies provide policymakers and other stakeholders with quantified data on the health, environmental, and economic impacts of selected integrated measures. Integrated environmental strategies lay the groundwork for implementation of policy, technology, and infrastructure measures with significant local and global benefits.

In addition, there are other proven strategies, such as:

- Clean technologies/products (clean development mechanism, zero waste, etc.)
- Dematerialization
- Design for environment
- Design for sustainability

- By-product exchanges
- Environmentally sound technologies
- Corporate environmental/social responsibility
- Industrial symbiosis/eco-industrial park
- Eco-industrialization

These developments signal an international shift in emphasis from managing individual manufacturing wastes and emissions to managing the overall environmental impacts of industrial sectors and of products over their life cycles. Moreover, as the importance of both information technology and the technological infrastructure are increasingly recognized, application of IE is making progress from its initial emphasis on specific products and materials to a broader emphasis on infrastructure, technological systems, and resource efficiency of the whole system.

IE: Policy Implications

Although some firms still see regulatory compliance as burdensome and attempt to reduce their costs, most large corporations and many smaller ones are taking new initiatives in managing their environmental impacts in ways that reduce their costs, increase their efficiency, lower their liabilities, and enhance their competitiveness while reducing pollution, conserving resources, and eliminating waste.

Historically, environmental regulation of industries all over the world has particularly emphasized point-source controls. In the past few years, a number of innovative policies of IE have been initiated in developed as well as developing countries. However, in the developed world, these new approaches have been adopted more quickly and more fully. In Europe, environmental policies increasingly address the overall environmental impacts of a product over its entire life cycle and seek to stimulate demand for greener products and to promote greener design and production through Integrated Product Policy. In Japan, the emerging emphasis is on the environmental design of products, extensive recycling of products, and environmental attributes such as energy efficiency and use of nontoxic materials.

However, policies of IE at the national, state, and local levels do not yet reflect the environmental problems in corporate environmental management in a broader perspective in such a way that

- policies fail to give a clear recognition of the factors that shape decisions;
- the gap is growing larger between the objectives, methods, and accomplishments of public environmental protection policies and the potential for proactive environmental management practices to achieve improvements in environmental performance;
- the complex, expensive, and inflexible command-and-control regulatory system still dominates environmental policy, which neither encourages nor rewards corporate environmental management systems beyond compliance requirements. Often the regulation for one medium increases pollution in other mediums, or restrictions on any specific category of pollutants lead to increases in other forms of pollution, and sometimes the current regulatory structure prevents the linking of industries or industrial processes for more efficient use of waste materials or transport of recycled material across certain geographic boundaries; and
- lack of understanding of motivation factor, lack of flexibility in providing regulatory relief and incentives in emergency situations, lack of stakeholder participation, mutual distrust between regulators and the business community, and lack of vision at the implementation level play a pivotal role in failure of a policy.

Future Perspectives

Bridging the gap between public policy and corporate environment management will require fundamental changes in central and state government approaches to regulation and its implementation. The challenge for policymakers is to employ a modest set of policy initiatives with a mixture of regulations and economic instruments that encourage designers to take account of rapid technological change and convert opportunities to realities, while simultaneously safeguarding environmental quality. As the shift is made from controlling pollution to preventing it and, beyond that, to achieving sustainable development, comprehensive approaches that focus on economic systems and the flow of materials and energy are needed to address the complex mix of issues raised about energy use, material choices, product and process design, interfirm relations, material and waste management, market response, and information needs. A more integrated set of policies and regulations for auditing and monitoring must use a proper framework to encourage clean manufacturing and resource efficiency; eliminate subsidies for virgin materials; put disincentives on the disposal of materials as waste; help internalize externalities; promote the interconnection of processes, plants, industries, and other enterprises to approach the objectives of IE principles; and accelerate the adoption of IE principles in their production cycles for improving environmental quality.

Policy for encouraging renewable energy should be adopted. Extending distributed renewable energy infrastructure into new regions could be cost competitive by combining smaller, highly efficient fossil fuel plants; cogeneration and energy cascading; wind; photovoltaics; passive solar; geothermal; and biomass sources.

Any policy aimed at designing for environment should attempt to allow for reuse at all possible levels but also should encourage the scenarios of reuse over reprocessing or recycling, as reuse will generally require less energy input than others, and as such will have lower overall negative environmental effects. Policy needs to encourage full development of resource recovery systems and phase out continued dependence on landfills and incinerators as the primary means of handling discards.

Policies should help to overcome institutional barriers in government. All the relevant policies should be designed to be flexible enough in a broader complex system including natural, social, and economic contexts to avoid environmental problems caused by rapid economic growth and social distress at an early stage. Also, policies should discourage the development of large resource-depleting/polluting industries in the future. Policy that can help solve potential conflicts among stakeholders should be stipulated. Partnerships, especially among policymakers, government regulators, industrialists, research analysts, environmental groups, and local communities, are the essential ingredients necessary for making IE a realistic goal for public policy.

See Also: Agenda 21; Appropriate Technology; Corporate Responsibility; Cost-Benefit Analysis; Environmental Management; Globalization; Innovation, Environmental.

Further Readings

Armstrong, Tom. “Design for Sustainability.” http://indeco.com/Files.nsf/Lookup/dfs/$file/Dfs.pdf (Accessed January 2009).

Crul, M. R. M and J. C. Diehl. “Design for Sustainability: A Practical Approach for Developing Economics.” TUDelft, United Nations Environment Programme, 2006.

Goh, Andrew L. S. "Evolution of Industrial Policy-Making in Support of Innovation: The Case of Singapore."*International Journal of Innovation and Learning*, 3/1 (2006).
Graedel, Thomas E., et al. *Industrial Ecology, International Series in Industrial and Systems*, 2nd ed. New York: Prentice Hall, 2002.
Lowe, Ernest A. "Eco-Industrial Policy." *Eco-Industrial Park Handbook* (2001).
Rondinelli, Dennis A. and Michael A. Berry. "Corporate Environmental Management and Public Policy: Bridging the Gap." *American Behavioral Scientist*, 44/2 (2000).
Thomas, Valerie, et al. "Industrial Ecology: Policy Potential and Research Needs." *Environmental Engineering Science*, 20/1 (2003).
Wolf, Steven. "A Policy Brief on Industrial Ecology." *Environmental Governance* (2004).

Gopalsamy Poyyamoli
Rasmi Patnaik
Pondicherry University

Industrial Revolution

From its earliest days late in the 18th century, the Industrial Revolution became an ongoing process that spawned continuous changes in virtually every area of society, first in England, and soon after in the United States and elsewhere. The Industrial Revolution—perhaps more aptly named industrialization—emerged side by side with evolving capitalism and continuing impulses for democracy. It shaped our economy, politics, and culture, while gradually urbanizing our geography and harnessing the land's resources to fuel factories and their outputs. It continues to be a force in the global political economy.

In general, the Industrial Revolution is credited with producing economic surpluses that led to higher living standards, improved health and nutrition, population growth, and longer life spans for large segments of the population. At the same time the Industrial Revolution was creating great wealth, however, it also was creating great poverty around the world.

From the outset, the Industrial Revolution put economic pressure on smaller, middle-class manufacturers who relied heavily on labor. Newer factories required significant amounts of capital for equipment and ultimately concentrated capital and ownership of resources and production methods into fewer hands. As a result, a class of owners emerged with the power to secure both significant influence on the government and legislation designed to help facilitate increased profitability, including lax environmental laws that allowed factory owners to transfer the costs of their pollution to the general public (externalities).

The Industrial Revolution triggered increased consumption of coal to fire machinery, unlocking massive amounts of carbon that had been stored in the Earth and releasing pollutants into the air, water, and soil. The interplay between supply and demand is evident in the implementation of increasingly efficient production techniques and a growing population that intensified resource use and created environmental problems. Warnings about resource depletion and pollution, especially after the Civil War, led to calls for conservation and some government regulation. The Conservation Movement emerged after 1900 and engaged the government in efforts to manage industrial resource use and control pollution.

In addition, the government was increasingly drawn into industrial development to ensure efficient commerce with the construction of transportation and communications systems, as well as a banking system to ensure the orderly transfer of funds. Government has played an active role in facilitating market development since the nation's beginning, through internal improvements that reshaped both settlement patterns and the landscape to facilitate market development.

Although actively assisting market development, over much of U.S. history, government generally took a passive role in resource conservation and protecting workers and communities from the environmental hazards that result from industrial processes. Minimalist government, or laissez-faire government, took a hands-off approach to any policy prescription that might infringe on an industry's profitability.

Even today, the Industrial Revolution is driven largely by the need for factory owners to be profitable. Competitive markets force continued technological change to increase production through product and process innovations that allow short-run profit increases. Over time, a system of government subsidies and incentives has evolved to assist businesses in these functions. These policies often have environmental implications because they can shape the nature and level of pollution from factories.

Industrial firms also try to boost profitability by seeking to limit wages, migrating to areas of lower labor and environmental costs, abandoning places where resources are exhausted or have become relatively too costly to exploit, and moving closer to sources of new materials or resources or lower transportation costs. They also try to ensure a supply of lower-cost energy by switching fuels and using alternative energy sources.

The emergence of a powerful industrial class, especially after the Civil War, drew political battle lines over access to and use of natural resources that persist today. Many industrial firms have grown to see conservation as an essential part of improving profitability and a significant part of positive relationships with the general public. However, there remains a seemingly unending demand for energy and other natural resources to increase output to meet the demands of a growing population. Government plays a sometimes controversial role in managing access to these resources.

The Industrial Revolution also facilitated mechanization (industrialization) of agriculture, increasing productivity, reducing the need for farm labor, and triggering a mass migration to urban areas, especially after the turn of the 20th century. In urban areas, expanding factories needed more labor and paid higher wages than farming, causing many rural areas to lose substantial parts of their population. Meanwhile, the growing urban population spread across the landscape into suburban areas, increasing demand for more goods and causing environmental degradation through the loss of prime farmland and air and water pollution. Population shifts permanently altered the political balance of the country, weakening the power of farmers and rural residents.

Increased production and new technologies, especially during the 1920s, brought on a proliferation of consumer goods, leading to the creation of a consumer economy that is a model for growth around the world. Food surpluses, new consumer goods created in massive factories, and increasing exploitation of energy supplies and the world's natural resources created comforts for a wider range of people, from the working class to the wealthy, in many countries.

Government has a mixed record in protecting industrial workers, but the post–World War II period saw about 30 years of relative peace, safety, and prosperity for many workers and business owners, with help from government tax and fiscal policies. In addition, through industrial and economic policies, government has tried to address uneven industrial development with efforts to attract plants to poorer areas through incentives, tax credits, low-cost loans, and workforce training, among other tools. Often, plants have been located in places where government environmental and workplace regulations are considered to be less imposing.

Building new factories or shifting them from one place to another was, before the 1980s, mostly a matter of domestic competition between states and regions in the United States. In the 1970s, competition for industrial jobs became more global, as firms began to seek out countries with lower wages, fewer environmental regulations, and other factors that would enhance profitability.

The effects of the Industrial Revolution continue as developing nations are drawn more closely into the global economy. The transition of these developing countries is similar to what occurred two centuries ago in Europe and North America as they moved from animal and human power to energy-dependent factories that increase productivity and increase the pace of resource use. Meanwhile, the rising demand for energy and consumer goods in these countries continues to place more demands on the Earth's natural resource base, with continuing effects of the global political economy.

See Also: Capitalism; Consumer Politics; Corporate Responsibility; Federalism; Globalization; Sustainable Development.

Further Readings

Beard, Charles A. and William Beard. *The American Leviathan: The Republic in the Machine Age*. New York: Macmillan, 1930.

Ely, Richard T. *Studies in the Evolution of Industrial Society*. New York: Macmillan, 1913.

Galbraith, John Kenneth. *The New Industrial State*. New York: New American Library, 1985.

Harvey, David. *The Urbanization of Capital: Studies in the History and Theory of Capitalist Urbanization*. Baltimore, MD: Johns Hopkins University Press, 1985.

Timothy Collins
Western Illinois University

Innovation, Environmental

Environmental or eco-innovation is about renewing products, processes, and institutions from an environmental point of view. Environmental innovation aims to reduce the (over) use of natural resources and the negative effects on ecosystems, such as emissions to air, land, and water. Innovations can be of an incremental or radical character and can take place at the level of a product or service or at the level of the sociotechnical system in which the innovation is adopted. The path of incremental innovation is commonly considered to be too piecemeal and too slow to address urgent problems such as climate change. System innovations are to be preferred, but they are hindered by institutional and societal forces. The urgency to innovate raises the question of the governability of innovation. The different views on the extent to which environmental innovation can be stimulated, and which instruments government can use to push or pull innovation, are merging into a coherent management paradigm.

Innovation focuses on the process of the development and diffusion of a novelty through society. The novelty may concern a product or service, the process by which is it produced or delivered, and the technological, social, and institutional environment in which it is developed and used. An invention becomes an innovation when it is spread through society and is used by members of that society. This process of diffusion of innovation and the adoption of it by users is considered to evolve along an S-shaped curve. The curve starts with high investments in research and development, followed by gentle slope upward, in which innovations are selected by a group of early adopters. When the majority follows, the curve steeply rises and marginal costs (change in total costs that results from producing one additional unit) in this stage are low. The curve flattens again when the innovation has become common.

In an early stage of development, green products have to doubly compete with nongreen or grey substitutes. These products already benefit from scale advantages, and they externalize environmental costs, whereas green products aim to include these costs in the price. The buyer pays for these costs, and society freely benefits from the environmental advantages. This asymmetrical allocation of costs and benefits imposes an extra barrier to the large-scale adoption of green products.

Once green products are produced on a large scale, they do not have to be more expensive than their grey substitutes, especially not when the lifetime and maintenance costs of the green product are taken into account. Responsible use of materials results in durable products that require less maintenance and/or repair and that often outlive their grey competitors. However, customers tend to base investment decisions on up-front capital costing and not on the return on investment over a longer time, which explains the continued competition between grey and green products.

Marketing the Innovation

The attitude of the end-user toward an innovation is considered to be a crucial factor in the process of diffusion and adoption. Marketing is used to position a product in the market, to identify and create an image of the innovation that attracts a potential group of early adopters. Solar panels, for example, have been marketed as high-tech gadgets, addressing a specific group of customers.

Product innovation inevitably entails process innovation, such as the use of clean technologies, renewable energy sources, nonhazardous substances, and biodegradable materials. Environmental management systems help firms to systematically reduce the environmental impact of products during production and consumption. Good housekeeping or eco-efficiency, that is, making efficient use of resources, is a well-known example of a cost-efficient form of environmental management. Product and process innovation also involve a reconsideration of the resources used: labor, monetary, and physical resources. Including external costs, for example, by improving labor conditions and increasing eco-efficiency can lead to different investment decisions. Approaches such as design-for-environment or cradle-to-cradle are examples in which product and process innovations are intertwined from the start. The design includes the end of life of a product, after which it can be disassembled and the components reused. Experiments with industrial and demountable buildings have shown that a long lifetime of products, different economic lifetimes of components, the adversity of architects against standardization, and changing legal product requirements over time make it difficult to reuse components. Changes of institutions and attitudes are needed to support the reuse of materials and components.

The environmental gains of innovation at the level of the product or service, including the way it is produced or delivered, are modest compared with the gains that can be made upstream in the manufacturing chain. For example, carbon dioxide compensation schemes for airline customers do not result in aircraft innovation. Studies of environmental innovation therefore stress the need for innovation of the system that delivers products or services.

Innovation is often classified by its effect or the "newness" of it: Innovations can be incremental or radical. Incremental or evolutionary innovations can usually be diffused and adopted without disruptive changes to the technological system in which they are produced and the society in which they are used. Radical innovations are artifacts that were unknown before and that are produced and used in new ways. Such breakthrough innovations render existing products obsolete. Incremental innovations show a high degree of path dependency and lock-in: They continue on an existing development path and block alternative paths. Radical innovations are discontinuous and involve the taking of new roads. The uncertainties and risks involved in radical innovation are far greater than those of incremental innovation, of which costs and risks are known. Investors, when making decisions, tend to rule out risks and uncertainties and thus prefer incremental innovations.

Environmental Innovation and Technology Development

Environmental innovation is closely related to technology development. New technologies can help to improve the environmental performance of products or services, processes, institutions, and systems. Within technology studies there are two competing views on what drives technological change. Technological determinism views technology development as an autonomous process that determines the nature of society. Social constructivism acknowledges the influence of groups of people on technology development. There is an interaction between technology and society. People can attribute different meanings to artifacts—human-made things—and they may value technology differently. This view also emphasizes that technology takes time to adapt to changing demands in society, or that society takes time to adapt to opportunities provided by new technologies. Developments in technology and society thus coevolve.

These two views can be recognized by the ways in which environmental innovation of products and systems is governed. Push strategies assume that a technological determinism in innovation creates demand and focuses on investing in research and development and in bringing a product to market. Pull strategies use market demands to direct research and development. In practice, environmental innovation is often a combination of push and pull. For example, a market demand for small, energy-efficient cars pulled the automobile industry to invest in these cars. Government can support this development with emission regulations and with tax advantages for hybrid-car owners.

In general, various public governance approaches to push-and-pull environmental innovation are distinguished. Command-and-control approaches make use of laws and regulations with a focus on compliance. This can push industry to adopt environment-friendly technologies or to develop such technologies. Failure to comply is penalized. Government can also create a market demand for environmental innovation. Economic instruments, such as subsidies, levies, and emissions taxes, make green innovations more attractive. The creation of markets for tradable permits aims to internalize specific environmental costs; for example, those of carbon emissions. The creation of such markets is an example of a more collaborative approach to innovation. These approaches argue that innovation is a result of a process of interaction and negotiation among stakeholders, public and private, in society. Government can influence this process, but it cannot determine it. These collaborative approaches can count on more support from industry because they respect existing power relationships, but they are criticized for supporting evolutionary developments. The interests of new firms, future technologies, and future generations are not represented as strongly in the process as vested interests.

Governing environmental innovation therefore calls for a smart combination and alternation of collaboration and command-and-control, stimulating and facilitating innovation at the level of the system and of the individual processes, upward and downward on the production chain, addressing various actors and involving technical, institutional, and societal changes.

See Also: Appropriate Technology; Ecological Modernization; Environmental Management; Regulatory Approaches; Technology.

Further Readings

Edquist, Charles, ed. *Systems of Innovation: Technologies, Institutions and Organizations.* London: Pinter, 1997.

Huber, Joseph. *New Technologies and Environmental Innovation.* Cheltenham, UK: Edward Elgar, 2004.

Mulder, Karel. *Sustainable Development for Engineers: A Handbook and Resources Guide.* Sheffield, UK: Greenleaf, 2006.

Nelson, Richard R. and Sidney G. Winter. "In Search of Useful Theory of Innovation." *Research Policy*, 6 (1977).

Porter, Michael E. and Claas van der Linde. "Toward a New Conception of the Environment-Competitiveness Relationship." *Journal of Economic Perspectives*, 9 (1995).

Rogers, Everett M. *Diffusion of Innovations*, 4th ed. New York: Free Press, 1995.

Weber, Matthias and Jens Hemmelskamp. *Towards Environmental Innovation Systems.* Berlin: Springer, 2005.

Ellen M. van Bueren
Delft University of Technology

Institutions

Institutions are persistent sets of rules and understandings that prescribe certain practices. Regimes are probably the most relevant type of institutions concerning environmental issues, defining roles, rules, and rights. The most cited definition of regimes is the one given by Stephen Krasner, in which regimes are sets of implicit or explicit principles (beliefs), norms (standards of behavior), rules (prescriptions for action), and decision-making procedures (existing practices for making and implementing collective choice). Regimes are a result of a convergence of actors' expectations in a given area. They may or may not include organizations. Unlike organizations, regimes do not exist as legal entities. Regimes are also distinct from international treaties. In some cases, international treaties have institutionalized principles and norms that were already in effect. In other cases, international treaties created the framework for certain practices to be adopted. The relevance of regimes/institutions for environmental studies is twofold: It has been the most preferred means of environmental governance, and their study allows for assessing their performance. To that effect, regimes have been classified according to their characteristics, their formation path, and the motivations underlying their creation.

As a consequence, and to better assess regimes' performance, they can be classified on the basis of the extent of their rules, on the degree of their institutionalization, on the nature of their goal, and on their overall coherence. The extent of a regime is related to the restrictiveness of its rules. Some regimes tend to pursue a "looser" approach to the issue in question, whereas others are "tighter," defining in detail the rules members have to comply with. Regimes also vary according to their degree of institutionalization. Some regimes are laid out in formal agreements, conventions, or treaties, and even predict the creation of organizations. Others are just customary practices based on generally accepted principles. A regime's ultimate goal can also vary, ranging from a simple assertion of a common problem (e.g., Convention on Climate Change) to a specific plan of action (e.g., maritime pollution regime), or to a straightforward prohibition of some activity (e.g., Convention on International Trade in Endangered Species of Wild Fauna and Flora). As for regimes' coherence, prevention and precaution are the basic principles cross-cutting most environmental regimes. Often, however, regimes' rules and procedures denote a constant tension between a state's sovereignty and the international commitments agreed on. As a consequence, more coherent regimes are more easily and widely accepted and complied with than less coherent ones. One can trace these different characteristics, and especially the (in) coherence of the regime, to the path and motivations that led to their formation.

Most existing environmental regimes are a result of a negotiated process, resulting from the integrative nature of environmental issues and the impossibility of unilaterally solving most environmental problems. This negotiation process can follow different paths. There are regimes that are a result of "patchwork" (piecemeal track), others that are a consequence of an evolution of practices and behaviors (evolutionary track), and others still that were created on a more "constitutional" basis (contractarian track). Most environmental regimes emerged along a piecemeal (e.g., Law of the Seas) or evolutionary path (e.g., special economic zones principle), although more recent efforts have attempted to follow a contractarian track (e.g., Antarctica regime). Depending on the track followed, regimes will be more or less coherent, directly influencing their performance and effectiveness.

Motivations and Effectiveness

The motivations underlying the emergence of different regimes are also an important factor when studying regimes' effectiveness and impact. Most motivations can be classified as power, interest, or knowledge. A regime that is created based on power usually implies a certain degree of "imposed acceptance," as it is created based on a unilateral decision or on the exertion of power by one (or few) actors. Since the basis of power concerning environmental issues is often unclear, some argue that actors engage in regime creation to coordinate their behavior so as to prevent suboptimal outcomes from occurring. Clearly, most environmental regimes can be traced to interest issues: pollution prevention, sustainable management and exploitation of marine resources, or endangered species protection. None of these questions could be dealt with unilaterally, and the transaction costs associated with dealing with these issues are usually so high that it only makes sense if a considerable number of players, or a certain group of crucial players, participate.

Regime formation can also be understood in terms of knowledge. Because interests are often unknown or not fully specified, consensus on causal understandings and policy alternatives become crucial in the regime process of formation and evolution. In the environmental arena, knowledge-based explanations have become increasingly useful, as environmental regimes formation has been driven essentially by scientific understandings that are then incorporated into environmental policy. Epistemic communities—networks of knowledge-based communities with an authoritative claim to policy-relevant knowledge within their domain of expertise—have been crucial to the role scientific knowledge has had in environmental regime negotiations. Because of the scientific complexity of most environmental problems, regime negotiations and changes are directly dependent on scientific and technological progress. New information can determine the obsolete character of a regime or the need to create a new one, in addition to changing or interlinking existing ones. Environmental regimes' evolution has been vitally dependent on scientific information and estimation to change its rules and procedures, becoming looser or tighter depending on the input received and its credibility.

It is undeniable that environmental issues are tightly interconnected, but the myriad of existing environmental regimes tends to dissociate these issues and provide issue-specific or issue-areas' principles and norms. Issues are separate topics that appear in the agenda of negotiations. Even though environmental concerns are simultaneously present in the international agenda, and undeniably ecologically linked in the minds of the actors, these topics are not necessarily connected. Of course, actors can pool, and have pooled, different concerns together justified by their constitutive interrelationship. The ocean regime(s) is a clear example of this dynamic: Actors recognized that oceans' specific issues and their related parties' activities constituted a cluster of interdependent concerns. The process of formation and evolution of environmental regimes is directly related to the definition of issue/areas. Issue-specific negotiations tend to deal with topics on which there is an accepted body of knowledge, and the narrower the scope of the topics being negotiated, the higher the level of consensus on efficient solutions. Because environmental problems are, by definition, complex, there seems to be a tendency to restrict the scope of each environmental item on the international agenda to make the issues more tractable.

Linking Issues

Nevertheless, and in spite of the issue-specificity of most environmental regimes, precisely because of their complexity, one can identify a simultaneous reverse tendency toward issue-linkage. The degree of uncertainty in the field no longer assures experts that the solutions adopted will lead to the outcomes expected. As a result, some have tried to overcome this problem by linking issues, in search of knowledge in adjacent fields that might contribute to broader and more effective and efficient solutions. Still, one cannot argue that issues will be linked as a result of environmental complexity. The motives underlying the creation of regimes are mixed, and consensus on knowledge is hardly ever enough. Convergence of interests and distribution of power also contribute to the formation of regimes.

Two last aspects should be noted concerning environmental regimes. First, despite the existence of several regimes, linked or not, environmental regimes can be assumed to have convergent interests. Even when they differ in their structure, degree of institutionalization, goals, means, or effectiveness, they still reflect similar underlying principles, such as international cooperation, subsidiarity, science and precaution, and openness. Nevertheless, the coexistence of different regimes affecting the same issue/area or different parts of one issue/area raises obvious concerns about the effectiveness and the hierarchy of authority each portrays in case of conflict.

Second, a close relationship between environmental and economic regimes has gained increasing relevance, because economic and environmental regimes' goals overlap without, however, being congruent. For instance, when it comes to international trade, several environmental regimes either have provisions that affect trade directly (e.g., endangered species regime) or include implementation instruments that are trade measures (e.g., Basel Convention), or because they encompass such a comprehensive framework of action, trade impacts are included (e.g., climate change regime). The issue at stake here is that environmental regimes usually have clear economic impacts, but their structure is not defined by those impacts; conversely, economic regimes also usually have clear environmental effects, but their structure is not defined by those effects either.

See Also: Basel Convention; Convention on Biodiversity; Corporate Responsibility; Organizations.

Further Readings

Haas, Ernst B. "Why Collaborate? Issue-Linkage and International Regimes" *World Politics*, 33/3 (April 1980).

Haas, Peter M. "Epistemic Communities and the Dynamics of International Environmental Co-Operation." In *Regime Theory and International Relations*, edited by Volker Rittberger. Oxford: Clarendon Press, 1993.

Hasenclever, Andreas, et al. *Theories of International Regimes*. Cambridge, MA: Cambridge University Press, 1997.

Krasner, Stephen. "Structural Causes and Regime Consequences: Regimes as Intervening Variables." In *International Regimes*, edited by Stephen Krasner. Ithaca, NY: Cornell University Press, 1983.

Young, Oran R. *International Cooperation: Building Regimes for Natural Resources and the Environment*. Ithaca, NY: Cornell University Press, 1989.

Paula Duarte Lopes
University of Coimbra

Intergovernmental Panel on Climate Change

The Intergovernmental Panel on Climate Change (IPCC) was established by the World Meteorological Organization and by the United Nations Environment Programme in 1988. It is based in Geneva, Switzerland. As an agency of the United Nations (UN), the IPCC seeks to use the information it provides as a tool for promoting its human development goals. It is considered by many to be the most important source of information on climate change. It seeks to be a model for scientific interaction and government policymaking so that it can be policy relevant, but not policy prescriptive.

The reason for creating the IPCC was to provide decision makers and other parties interested in climate change with an open, transparent source of objective information about climate change. To have an agency that was as above the partisan use of scientific investigations as possible, the IPCC was organized as a scientific organization with open government participation. As a scientific body, it is officially a multigovernmental body in which many nations participate at various levels. When the IPCC meets to discuss findings, to hear reports, and to consider adopting reports from different parties, various governments participate in the decision making and in subsequent reviews.

The IPCC's constituency is composed of governments of the World Meteorological Organization and UN Environmental Programme member states. It is funded through the IPCC Trust Fund, which receives contributions from many different governments. The money enables IPCC to support the work of scientists and to provide research in poor countries of the world.

Those who contribute to IPCC reports include hundreds of scientists at universities, research institutes, or other centers of study around the world. They contribute by being authors of scientific papers, making reports of research findings, acting as reviewers, and in other ways.

Research, monitoring climate change, or other scientific experimentation is not a part of the mission of the IPCC. The role of the IPCC is to assess the latest technical, scientific, and impact literature in the world that is focused on climate change. Impact literature includes literature that discusses the economic effects of climate change, sociological studies, and public opinion studies as well as other information that is relevant. For example, findings could include reports that global warming is beneficial to some people who prefer warmer weather to the cold or rainy weather they usually experience in their location. Or in the case of rainfall shifts, it causes some regions that have historically been arid or desert places that experience droughts to increase in rainfall, whereas other areas may have reductions of rainfall. These changes affect the attitudes people have and therefore their support for or opposition to actions that could reduce transportation and industrial greenhouse gas emissions. Impact literature also could discuss the lack of willingness or open opposition many people may have to paying higher taxes to fund greenhouse gas reductions.

The IPCC examines the whole body of literature that is currently being produced that deals with climate change. In its examination, the IPCC not only evaluates the scientific and socioeconomic impact of the literature but also seeks to understand risks that anthropogenic global warming may entail. For example, climate change that produces global warming could affect agricultural production through increased droughts. In contrast, climate change that brings another little ice age would drastically affect the most industrially productive regions of the world and lead to problems with global industry.

The reports IPCC issues make every attempt to present information, findings, conclusions, or other reporting in as objective and neutral form as possible. It also seeks to have the highest scientific and technical standards used so that it will provide open, objective, and transparent information. In this way it can keep the volume of political controversy at a low level. Without this approach, political controversy would interfere with the common goal of all people on Earth, which is to live in a healthy environment.

The work of the IPCC is organized in panels that meet in plenary sessions. The panels are called Working Groups, of which there are three. Working Group 1 deals with the scientific basis of climate change. Working Group 2 reviews impact, vulnerability, and adaptation arising from climate change. And Working Group 3 deals with migration that arises from climate change. In addition, there is a Task Force on National Greenhouse Gas

Inventories. All together, the membership and staff of these panels constitute a sizable bureaucracy.

The head of the IPCC Bureau is the chairman, who along with the cochairs of the three IPCC Working Groups and the Task Force Bureau on National Greenhouse Gas Inventories, prepares the report. The membership of the IPCC Bureau is currently at 30 members. It has 2 cochairs and 12 members.

The IPCC prepares an Assessment Report ever five to six years. Members of the panels are usually elected for the period necessary for the preparation of the report. The reports are prepared by teams of authors who have been selected for their expertise. They have also agreed to follow strict procedures. The first-stage review of the two-stage review process is performed only by experts. The second-stage review is performed by experts and by governments. Accepting the reports is the work of the plenary sessions, when the summaries of the reports are approved by policymakers on a line-by-line basis.

The phenomenon of climate change is a natural phenomenon. The climate of the Earth is its longer weather patterns. Historically, the climate of the Earth has had periods much warmer than now, one of which was the age of the dinosaurs, and it has had at least four ice ages. These climate changes may have been caused by enormous volcanic explosions or by asteroids hitting the Earth or by some other cause such as variation in average temperatures resulting from the variation in sunlight levels caused by the wobbling of the Earth in its orbiting over many thousands of years. The slight variations that the Earth receives in sunshine in its cycle of orbits over thousands of years are called Milankovitch cycles. If the Milankovitch cycle theory is correct, then it is a factor in climate change. Regardless of the cause, the fact is that climate change is indeed a natural phenomenon unless human behavior plays a role in climate change.

Human climatic impacts are known from historical examples. In some localities the local climate has been changed by human action; for example, some tropical islands, when first populated by humans, had rainforests. However, after the humans destroyed the forests, the islands became semiarid. So the major question has become whether the current apparent changes in the Earth's climate are anthropogenic (caused by humans) or whether the climate change that to many since the early 1970s has appeared to be global warming is a natural variation. Even if the current phenomenon of climate change is natural instead of anthropogenic, it still affects human life on Earth. Therefore, the objective consideration of the current investigation and debate of all aspects of this issue are of global importance.

The definition of climate change used by the IPCC refers to a statistically significant variation in average weather conditions. Its definition recognizes that climate change is the result of natural causes, and possibly human activity. In contrast, the definition used by the UN Framework Convention on Climate Change considers climate changes to be caused by human activities, although it says that climate variability is the result of natural causes. The difference opens the way for political controversy in the case of the UN Framework Convention on Climate Change's definition.

To deal with climate change in an objective way to mitigate its negative effects, the World Meteorological Organization and the UN Environmental Programme established the IPCC in 1988. As a scientific body, the IPCC seeks to be an information center for reports on climate change. This includes objective information about the causes of climate change. In addition, it seeks to identify, to quantify, and to understand potential environmental and socioeconomic consequences of climate change before they occur. The strategy of research is therefore proactive as well as informational.

When the IPCC develops its reports, it does so based on scientific evidence and in the light of the often opposing viewpoints of scientists over the method and results of scientific investigations. To be comprehensive, the IPCC seeks contributions from experts from the scientific community around the world. It also seeks contributions from any and all relevant disciplines. In addition, it uses as often as is reasonable all industrial literature and traditional practices in a two-stage review process by experts and governments.

The reports that the IPCC issues are usually in a policy-ready format. However, the IPCC prepares them and issues them in a policy-neutral manner. Ultimately, it is governments that respond to the issues of climate, which are for all societies a public affair that affects every person in every country. The policy portion of an IPCC report is the "Summary for Policymakers." By accepting its report, governments are acknowledging that the scientific content of the report is valid for making policy decisions.

IPCC reports are issued at regular intervals. When issued, they become a standard work of reference for experts, students, and policymakers. The first IPCC Assessment Report was issued in 1990 in Sundsvall, Sweden. Working Group 1 reviewed many topics including greenhouse gases, aerosols, observed climate changes, and many other scientific findings. It concluded from the reports that anthropogenic greenhouse gases were causing climate change. The rate of change was extrapolated to calculate a global temperature increase of several degrees centigrade in only a few decades.

The report of Working Group 2 reviewed in summary form the scientific understanding of climate change caused by global warming. Its major concern was the environmental impacts on the polar ice caps, on oceans levels, and other important consequences for the total environment.

Working Group 3 reported strategies for dealing with climate change. The strategies were to meet the needs of agriculture and industry as well as the health of people and nature. Under consideration were food and water supplies and other resources. It also considered the kinds of effects the different scenarios of climate change would have on nature and humans. In addition, it considered the effect of emissions of greenhouse gases, especially in the industrialized world and in the United States. A final point was a review of countries that were participating and strategies for enlisting a much wider range of all the countries of the world.

The 1990, the IPCC Assessment Report played an instrumental role in the formation and adoption of the UN Framework Convention on Climate Change. The Convention was opened for signature in 1992 at the Rio de Janeiro Summit meeting. The UN Framework Convention on Climate Change went into force in 1994 and afterward provided the framework for addressing climate change.

The IPCC Second Assessment Report was issued in 1995. The information it contained was used in 1997 to develop the Kyoto Protocol. The Kyoto Protocol was rejected by the United States and later by many other countries in what has become a matter of conflict over industrial emissions that are viewed as contributors to global warming. However, the countries that would have felt the greatest negative effect if the Kyoto Protocol were adopted were the most industrialized. The consequences for their economies would have been very significant, whereas the rules of the Kyoto Protocol would not have had any effect on the growing industrial economies of China, India, and other Third World countries.

The administration of President George W. Bush refused to sign the Kyoto Protocol agreement. Subsequently, it was rejected by some European countries and was strongly opposed by many leaders in Congress as well.

The IPCC's Third Assessment Report was issued in 2001. The Third Assessment Report presented a collective picture of climate change that was, in effect, global warming. However, the Third Assessment Report came under criticism from several quarters. One of these challenged the validity of the report by saying that it had been weakened to pander to opposition from some governments.

The Fourth Assessment Report was issued in 2007. It reported that global warming was going to hit the peoples and ecologies of the Third World hard because these were the countries least able to adjust to the environmental changes. An example offered a review of the effect of Hurricane Katrina (2005) on the poor in New Orleans, Louisiana.

The implications and the political resistance of the decision makers were found to be unacceptable to many scientists. Several publicly stated that they would never participate in the creation of an Assessment Report again.

Others found the whole process flawed because the Assessment Reports are supposed to be scientific consensus documents. However, the need to persuade governments to accept the summary reports creates a political process that softens scientific finding as part of political compromises. Instead of being a political reconciliation process, many scientists saw it as a political trading process. Critics seemed to ignore the fact that the scientists have the right to exercise a veto.

Those who support the two-step process in which the second step is a political step believe that it is important because it means political compromises can lead to political accomplishments. These types of supporters believe that politics are necessary to gain universal or near-universal governmental support. So when governments approve the reports line by line, they are acquiring ownership of the reports so that the reports cannot be dismissed as "mere" science. Political debates are inherent in the process because there are many interested parties representing many facets of the issue of climate change.

Among the problems the Assessment Reports face is media alarmism and opposition that is willing to deny or dismiss scientific facts. Others have pointed to serious omissions of scientific reports that are counterevidence to the claims of climate change. This has led critics to allege that the whole game is about politics and not science.

Another political problem is IPCC's increasingly obvious partisanship. This behavior has given grounds for those who were not supporters to dismiss the IPCC or to reject its reported conclusions as partisan and not objective.

Political opposition to IPCC claims climate variability is real, and a serious threat has occurred because there are those who see the whole issue as a conspiracy to crush the economy of the United States. Opponents in this category will have caused great damage if the worst-case scenarios occur because of climate change. These opponents will have hampered the full implementation of mitigating strategies.

In the politics surrounding the IPCC there are political opportunists who are really ideologues. These apparent supporters of the scientific conclusion in the Assessment Reports are using them to push an agenda to gain political control. Hysteria is a favorite political tool for these ideologues. Using exaggerated claims about global warming disasters would allow politicians to hastily enact policies that would give control to those whose decisions would be made politically and not scientifically. It is thus seen as an advocacy organization.

A recent political problem for IPCC emerged when it was awarded the Nobel Peace Prize in 2007, which was received by its chairman, R. K. Pachauri. However, his cooperation with a number of left-wing or progressive think tanks that are really policy advocates has led critics to assert that the IPCC is a partisan supporter and not an objective scientific organization.

Responses to critics who support the IPCC point to the fact that the IPCC is an intergovernmental agency and not a global forum of all scientists. However, claims to objectivity are thereby weakened in the interest of politics.

See Also: Global Climate Change; Kyoto Protocol; UN Framework Convention on Climate Change.

Further Readings

Bolin, Bert. *History of the Science and Politics of Climate Change: The Role of the Intergovernmental Panel on Climate Change*. Cambridge. MA: Cambridge University Press, 2007.

Gray, Vincent. *Greenhouse Delusion: A Critique of "Climate Change 2001."* Essex: Multi-Science, 2004.

Jochem, Eberhard, et al. *Society, Behavior and Climate Change Mitigation*. New York: Springer, 2001.

McCarthy, James J., et al., eds. *Climate Change 2001: Contribution of Working Group II to the Third Assessment Report of the Intergovernmental Panel on Climate Change*. Cambridge, MA: Cambridge University Press, 2001.

Metz, Bert, et al., eds. *Carbon Dioxide Capture and Storage: Intergovernmental Panel on Climate Change*. Cambridge, MA: Cambridge University Press, 2005.

Metz, Bert, et al., eds. *Climate Change 2001: Contribution of Working Group III to the Third Assessment Report of the Intergovernmental Panel on Climate Change*. Cambridge, MA: Cambridge University Press, 2001.

Metz, Bert, et al. *Methodological and Technological Issues in Technology Transfer: A Special Report of the Intergovernmental Panel on Climate Change*. Cambridge, MA: Cambridge University Press, 2000.

Michaels, Patrick J. *Shattered Consensus: The True State of Global Warming*. Lanham, MD: Rowman & Littlefield, 2006.

Parry, Martin and Timothy Carter, eds. *Climate Impact and Adaptation Assessment: A Guide to the IPCC Approach*. Herndon, VA: Earthscan/James & James, 1998.

Reay, Dave. *Climate Change Begins at Home: Life on the Two-Way Street of Global Warming*. New York: Palgrave Macmillan, 2005.

Watson, Robert T., ed. *Climate Change 2001: Third Assessment Report of the Intergovernmental Panel on Climate Change*. Cambridge, MA: Cambridge University Press, 2002.

Watson, Robert T., et al., eds. *Land Use, Land-Use Change, and Forestry: A Special Report of the Intergovernmental Panel on Climate Change*. Cambridge, MA: Cambridge University Press, 2000.

Zedillo, Ernesto, ed. *Global Warming: Looking Beyond Kyoto*. Washington, D.C.: Brookings Institution Press, 2007.

Andrew Jackson Waskey
Dalton State College

International Whaling Commission

Since its inception in late 1946, the International Whaling Commission (IWC) has evolved from what was essentially a whalers club to an international organization attempting to

Antiwhaling activists photographed these Japanese vessels taking a minke whale in the Pacific on February 8, 2009.

Source: Wikipedia

protect some of the most highly migratory species on the planet. Over the years, however, efforts to truly protect these aquatic mammals within the cetacean family have been severely hampered. IWC membership is voluntary, and key rules are riddled with loopholes that allow nation-states to continue whaling if they file protests within a 90-day window or simply label their hunts as "scientific research." Furthermore, the IWC has no authority to enforce any of its decisions through penalties—and even these ineffectual regulations do not apply to nonmembers.

Established by the International Convention for the Regulation of Whaling during its December 1946 Washington, D.C., meeting, the IWC's initial charge was a simple one: "provide for the proper conservation of whale stock and thus make possible the orderly development of the whaling industry." From its first meeting in London in 1948, the IWC came to be known as a big-game shooting club, one in which exclusive members met annually to set dates for opening season, restrict catches from nursing mothers and undersized whales, and establish quotas. All sporting rules were officially implemented, IWC members often failed to adhere to these guidelines. Quotas and minimum sizes were supposed to be determined scientifically, to ensure conservation was given equal consideration to short-term economic gain. For example, the largest whale declines occurred in the Antarctic, where over 66,000 whales were killed during the peak 1961–62 season. The population of blue whales in the Southern Hemisphere now stands at approximately 2,300, down from pre-whaling estimates of over 250,000. In its early years, however, the scientific committee created to provide this data was largely ignored.

That began to change as public opinion galvanized first in the United States and then in a number of other countries from Australia and New Zealand to Brazil and Chile. Perhaps the 1972 Stockholm Conference (United Nations Conference on the Environment) serves as the first global example of this public mindset shift as 52 nation-states unanimously adopted the U.S. proposal for an immediate moratorium on commercial whaling. Given the continued whaling-state advantage within the IWC, though, whaling did not stop.

Nevertheless the tide had indeed begun to turn in favor of whale-conservation states, even within this body. The United States developed domestic legislation to further facilitate this process. First the 1971 Pelly Amendment to the 1967 Fisherman's Protective Act authorized the president to ban importation of fish products from any nation-state impeding an international fishery program. Then the 1979 Packwood-Magnuson Amendment to the 1976 Fishery and Conservation Management Act dictated the secretary of state cut at least half the U.S.-water fishing allocation rights of any nation-state found to negatively affect the IWC.

These efforts were buttressed by nongovernmental organizations such as Greenpeace, particularly as their zodiacs sped out to shield whales from the whalers and popular imagination shifted from a 19th-century Herman Melville mentality of the ferocious Moby Dick wreaking havoc to one where whales themselves were the sympathetic characters in the unfolding drama. Still other nongovernmental organizations such as the World Wildlife Fund upped the ante further by recruiting new members to the IWC in exchange for paying their membership dues, even at times providing million-dollar funds for panda reserves, as was the case with China.

Ten years later, in 1982, despite continued opposition, enough support existed within the IWC to create the 1986 moratorium on commercial whaling. Despite this progress, those such as Japan and Norway, which lodged objections within the 90-day window, were legally allowed to continue whaling. Flouter states such as Iceland also skirted the rules by pulling out of the IWC altogether and forming a 1992 potential replacement to it along with Norway, Greenland, and Faroe Islands in the form of the North Atlantic Marine Mammal Commission. When Iceland did rejoin the IWC in 2002, it immediately began scientific whaling and in 2006 resumed commercial whaling. An assorted mixture of non-state actors were also actively engaged in whaling, essentially modern-day pirates seeking to supply the lucrative whale-meat food market in Japan. Finally, the moratorium applied only to commercial whaling, meaning whaling under scientific research claims like those of Japan and aboriginal subsistence hunting like the Inuit in Alaska or the Chukotka in the Siberian region of Russian Federation were still allowed.

Similar problems exist when it comes to the two sanctuaries established by the IWC. One of these is in the Indian Ocean. It was established in 1979 thanks to the small island state Seychelles and their hope to protect whale breeding grounds off their coast. The other, created in 1994, is the Southern Ocean Whale Sanctuary encompassing 50 million square kilometers surrounding the continent of Antarctica. Each has its status reviewed and thus is susceptible to change by the IWC every 10 years, but each has continued to receive sanctuary designation, despite disappointing failures to add a South Atlantic or South Pacific sanctuary to these partly protected areas. Such qualifiers are important, as the designation of sanctuary again only applies to commercial hunting, allowing states like Japan the freedom to continue their “scientific research”—and fill tables from the tony restaurants of Tokyo to the school lunchrooms of Osaka.

Looking to the future, the IWC continues to be hamstrung by an acrimonious and decades-long dispute between whale conservation nation-states such as the United States and whaling countries like Japan, Norway, and Iceland. New membership in the IWC remains ground zero for this battle, as the body continues to grow annually and its 84 members in early 2009 represented a roughly 50–50 split among whaling and nonwhaling states, effectively ensuring political deadlock given the three-fourths majority needed to make substantial changes in the IWC.

Conservationists fear this singular focus comes to the detriment of other pressing issues in terms of improving whale survival rates. They argue that populations decimated in the 20th century can only recover if the full range of threats to whales are addressed, including ship strikes, gear entanglement and by-catch, the effects of overfishing on whale food resources, toxic waste disposal, and climate change. Yet paralysis within the IWC over the commercial whaling moratorium prevents any significant headway in these arenas as well.

See Also: Biodiversity; Environmental Nongovernmental Organizations; Marine Mammal Protection Act; Precautionary Principle; Tragedy of the Commons; Transnational Advocacy Organizations.

Further Readings

Black, Richard. "Time for Peace in the Whaling World?" BBC News, June 19, 2008. http://news.bbc.co.uk/go/pr/fr/-/1/hi/sci/tech/7463633.stm (Accessed February 2009).
Day, David. *The Whale War*. San Francisco: Sierra Club Books, 1987.
Dean, Cornelia. "The Fall and Rise of the Right Whale." *New York Times* (March 17, 2009).
International Whaling Commission. http://www.iwcoffice.org/index.htm (Accessed February 2009).
Kraus, Scott D. and Rosalind M. Rolland, eds. *The Urban Whale: North Atlantic Right Whales at the Crossroads*. Cambridge, MA: Harvard University Press, 2007.
Melville, Herman. *Moby Dick*. New York: Harper and Brothers, 1851.
Porter, Gareth, et al. *Global Environmental Politics*. Boulder, CO: Westview, 2000.
World Wildlife Fund. "Whales, Whaling, and the International Whaling Commission." June 2008. http://assets.panda.org/downloads/wwf_position_iwc60_final.pdf (Accessed February 2009).

Michael M. Gunter, Jr.
Rollins College

Intrinsic Value

Many people claim that nature has intrinsic value. Sometimes called inherent value or worth, intrinsic value falls within the philosophical domain of metaethics—the meaning and status of moral language. Although there can be confusion about what the term *intrinsic value* means, working through this confusion is central to projects in environmental ethics to address environmental problems.

Many people claim that values are wishy-washy, subjectively messy, and best left out of serious scientific and policy discussions about environmental issues. Values cannot be objectively measured and quantified, and value preferences seem to be no different than preferences for different flavors of ice cream. However, to claim that discussions about values are not valuable is itself a value claim. In a factual sense, science can only describe and explain what the environmental issues are and predict what might happen if such and such happens first. Science cannot tell us what we should or should not do. We must make ethical judgments that stem from our stated values, enter into dialogue with each other, and arrive at some form of intersubjective agreement about what to do about environmental problems. Values thus play a central role in the resolution of all environmental problems.

No one denies that nature has instrumental value for people. Nature consists of natural resources that can be consumed or used. But does nature also have intrinsic value? To address this question, philosopher Richard Routley devised a thought experiment that became known as the "last man argument." In this thought experiment, Routley asks us to imagine an Earth where everyone has died except for one man. Before this man dies, he goes about eliminating animals (painlessly) and plants—every living thing he can. Has this last man done anything morally objectionable? If you answer no, then you supposedly subscribe to an anthropocentric (human-centered) environmental ethic in which the only kinds of things that have intrinsic, noninstrumental value are humans. If you answer yes, then you supposedly subscribe to a nonanthropocentric environmental ethic in which nonhuman nature (or parts of nonhuman nature) has intrinsic, noninstrumental value.

Beyond anthropocentrism, where one locates intrinsic value determines the kind of environmental ethic to which one subscribes. If nonhuman animals have intrinsic value, this is what environmental philosophers call a zoocentric (zoology-centered) environmental ethic; this is sometimes called sentientism (focused on the capacity to experience pain and/or pleasure) or psychocentrism (centered on having a psychological makeup). If nonhuman animals and plants have intrinsic value, this is a biocentric (life-centered) environmental ethic. If holistic biological and ecological entities such as species and ecosystems have intrinsic value, this is an ecocentric (ecology-centered) environmental ethic.

Philosopher Dale Jamieson calls intrinsic value the "gold standard" of morality because what has intrinsic value has ultimate moral value. But what precisely is intrinsic value? It is commonly used in at least four different senses. First, intrinsic value can simply mean noninstrumental value. Something has instrumental value if it can be used for something else. That something else has intrinsic value. Thus, money has instrumental value for people who have intrinsic value.

Second, intrinsic value can be a criterion for moral standing. For something to count morally or be what philosophers call morally considerable, it must have intrinsic value. Intrinsic value in this case is the ticket that gets something admitted into the moral community. The German philosopher Immanuel Kant (1724–1804) claims that human beings have intrinsic value and thus are ends in themselves.

Third, something that has intrinsic value can be understood to have inherent value because its value depends entirely on what inheres in the thing itself. The British philosopher G. E. Moore (1873–1958) defines intrinsic value in this sense, claiming that it is a nonrelational property—that is, its value depends on the existence of nothing else. In this sense, something that has intrinsic value stands in contrast to what has extrinsic value, understanding extrinsic value to be externally dependent on something else.

Fourth, something that has intrinsic value can have what philosophers call objective value in the sense that its value is independent of the valuations of valuers. This is sometimes called mind-independent value because the value really exists in the object independent of whether humans perceive it or not. To put a twist on an old question, does a tree in a forest that has never been perceived by people have intrinsic value? If you answer yes, this tree has objective intrinsic value.

The first and second senses of intrinsic value seem to encapsulate a similar idea that something that has intrinsic value is an end in itself because it has its own moral standing (second sense) and because it is where instrumental value terminates (first sense). The third and fourth senses seem to encapsulate another similar idea that something that has intrinsic value is self-sufficient because its value does not depend on anything else (third sense) and because its value does not depend on human valuers (fourth sense).

The fourth sense of intrinsic value is an instance of what philosophers call a realist account because the value really exists—in this case, independent of valuers. The idea that there are values independent of valuers strikes some people as an odd idea. Perhaps there still is intrinsic value, but its existence depends on the existence of valuers. This is what is known as a subjectivist account in that the value is projected as an attitude from a human subject; in this case, something can be valued intrinsically. Environmental philosophers question whether a realist account of intrinsic value or a subjectivist account of intrinsically valuing is more satisfactory and makes for a better metaethical grounding of an environmental ethic.

Some environmental pragmatists claim that we should give up on the notion of intrinsic value altogether because it is too theoretically problematic and not pragmatically useful for developing an environmental ethic. They suggest that all value is extrinsic—dependent on something else—and all we have are interrelated webs of value. Nonpragmatists worry that if everything has only this kind of value, then it seems that everything—including people—only has value for the sake of something else; it is not clear what this something else is.

This entire discussion of intrinsic value might seem to be theoretically nebulous and less helpful when thinking about the environment and environmental problems. All discussions about what we should do in response to environmental problems, however, involve value judgments. Getting clear about what the values are remains an important step toward solving the problems.

See Also: Deep Ecology; Ecocentrism; Green Discourse; Pragmatism.

Further Readings

Butler, William F. and Tim G. Acott. "An Inquiry Concerning the Acceptance of Intrinsic Value Theories of Nature." *Environmental Values*, 16/2:149–68 (2007).

Jamieson, Dale. "Values in Nature." In *Morality's Progress: Essays on Humans, Other Animals, and the Rest of Nature*. New York: Oxford University Press, 2002.

O'Neill, John. "Meta-Ethics." In *A Companion to Environmental Philosophy*, edited by Dale Jamieson. Malden, MA: Blackwell, 2001.

Routley, Richard. "Is There a Need for a New, an Environmental Ethic?" In *Environmental Philosophy*, edited by Don S. Mannison et al. Canberra: Australian National University, 1980.

Mark Woods
University of San Diego

Iron Triangle

The iron triangle is a metaphor for the strong and closed relationships between group representatives of specific interests, government bureaucrats, and politicians. Decisions made within this triangle dominate policymaking in a particular area. Iron triangles or subgovernments could especially be recognized in policymaking in the United States in the 1950s and 1960s. In the 1970s, juridical laws opened up the iron triangles by providing citizens and stakeholders access to the policy process. Policymaking was a process in which no longer three, but multiple actors participated, including environmental interest groups. In political science and public policy, iron triangles are considered to be a form of closed decision making of the past, but the term is sometimes still used to explain the hegemony of unsustainable practices, in spite of the availability of appropriate, sustainable alternatives and juridical means to enforce interest representation. Although environmental interest groups made abundant use of the opportunity to participate in policymaking, they have not always succeeded in influencing policies.

Iron triangles flourished in the United States in the two decades following World War II. Policies concerning a particular policy area, such as water, agriculture, public works, and defense, were made in closed communities consisting of three groups: (1) government bureaucrats in executive positions; (2) a legislative committee; and (3) a corporate interest group. The relationships between these groups were reciprocal and mutually enforcing. Members of Congress who were part of congressional committees or subcommittees had close relationships with the upper-level bureaucrats of the agencies that executed policies. The committees and subcommittees formulated these policies and granted the agencies budget for implementation. The executive agencies had considerable discretionary powers; programs that they had to implement were formulated in broad terms by Congress, and these agencies could influence the policy during implementation. Interest groups affected by the policy lobbied the committee and the agency to secure their interests. In return for a policy that favored their interests, they supported the policy as implemented by the agency and provided electoral support to the congressional committee members. The relationships between the upper-level bureaucrats and the interest groups were long-lasting and usually outlived the four or eight years in office of members of Congress. The closed, reciprocal, and continuing relationships provided stability to the policy process.

The Iron Triangle as Metaphor

The iron triangle is a metaphor that emphasizes the strength and the impermeability of the tripartite relationships between upper-level executives, elected politicians, and organized interests. Even the presidency had difficulty accessing the decisions made within the triangle. Although these closed relationships between a small number of actors were a phenomenon in the 1950s and 1960s, they became known as iron triangles in the 1970s. The metaphor was especially suitable to describe the policy process in the federal system of the United States. In Europe, policymaking in particular areas was characterized by closed and reciprocal relationships between a small group of actors as well, but there were often more than three groups of actors involved, and the relationships between these groups were more complicated than the direct returns by which relationships in U.S. iron triangles were characterized. In addition, in the federal system of the United States, government was almost entirely made up of subgovernments, whereas in Europe central government had a stronger role and usually had some influence on subgovernment politics. European policymaking could therefore better be described and explained by models that acknowledged the multiplicity of groups of actors, their interests, and the complexity of their interdependent relationships, such as the (neo-) corporatist model or the issue network model, than by the simple model of the iron triangle.

In the 1960s, policy processes were criticized by political scientists and by the public interest organizations, such as the civil rights movement and environmental interest groups, for excluding interests and for blocking policy change and innovation. In addition, descriptions of the policy process showed how organizational routines and power struggles among bureaucrats influenced policymaking to the detriment of the goals and values that the policies were supposed to support. In the late 1960s and early 1970s, new laws were passed that aimed to secure the representation of interests in the policy process.

The 1969 New Environmental Policy Act was the first of environmental laws in the United States and one of the first in the world. Many other environmental laws were modeled after this act. In the United States, it was followed by the 1970 Clean Air Act, the 1972 Water Pollution Control Act, and the 1973 Endangered Species Act. These acts ensured environmental interest representation in two ways: environmental impacts of proposed activities and of appropriate alternative activities had to be reported on and assessed before making decisions, and citizens had the right to participate in this process. This cleared the road for environmental interest groups to participate in the policy process, and when access was still denied, they could sue to enforce access.

No Guarantees

Despite the juridical provisions to ensure interest representation, environmental protection is not always guaranteed. The safeguards provided by environmental acts are of a procedural nature. When a decision has been made by an agency and environmental interest groups challenge the decision in court, courts can only ascertain whether the environmental interests have been allowed access and whether they have been included in the Environmental Impact Assessment and in the Environmental Impact Statement. Courts have no means to check the extent to which environmental interests have been taken seriously in the process, and they have no instruments to verify the substantive outcome of the Environmental Impact Assessment and the Environmental Impact Statement.

In highly specialized areas, executive agencies and environmental interest groups have neither the knowledge about the subject concerned nor the resources to develop or hire such knowledge. Often, they have to rely on the information provided by a corporate actor who wants to initiate a project. There are no neutral actors who can give information on the environmental effects of the proposed project and about appropriate alternatives and their effects. In cases like these, despite the formal access, it turns out to be difficult to influence the policy process, and iron triangles are still suspected to dominate the course and outcome of policymaking. For example, Hayden (2002) showed how the process of licensing hazardous waste facilities depended on the information provided by the industry. The highly specialized industry that produced hazardous, toxic, and radioactive waste was organized into big power blocks, each consisting of many corporations. This constituency gave representatives of these blocks a powerful voice in the policy process. The knowledge advantage of industry reinforced this power—they had to deliver the input for the Environmental Impact Assessment and Environmental Impact Statement. In court, judges could only determine whether procedures were followed; they lacked the means to assess the substance of the decisions made.

Although there might be policy processes in the United States and in Europe that continue to be dominated by tight-knit relationships between a small number of actors, theories of public policy showed how, in general, policy processes were more complex than pictured in the iron triangle model. Policy processes are often much more dynamic and complex than can be captured by the iron triangle model, or have become so over the years; for example, more issues are addressed by more varied groups of actors, and these processes take place in multiple arenas at a time. Models that are better capable of capturing the complexity and dynamics of policymaking, such as network models, have replaced the simple model of the iron triangle.

What these models of the policy process have in common, and what was perhaps most emphasized by the iron triangle model, is that policymaking is influenced by organized interests that have not been elected and that are beyond democratic control. In addition, organized corporate interests tend to have a more powerful constituency than organized environmental interests, and this gives them an advantage in the policy process.

However, sometimes corporate interest groups blame environmental interests as being subject to newly established iron triangles, consisting of reciprocal relationships among researchers, government, and media. These triangles communicate a sense of urgency that releases substantial flows of funding for research and for policy programs concerning environmental issues.

See Also: Environmental Justice; Environmental Nongovernmental Organizations; Environmental Protection Agency; Policy Process; Power.

Further Readings

Freeman, J. *The Political Process: Executive, Bureaucrat, and Legislative Committee Relations*. New York: Random House, 1955.

Gormley, William T., Jr. "Regulatory Issue Networks in a Federal System." *Polity*, 18 (1986).

Hayden, F. Gregory. "Policymaking Network of the Iron-Triangle Subgovernment for Licensing Hazardous Waste Facilities." *Journal of Economic Issues*, XXXVI (June 2002).

Heclo, Hugh. "Issue Networks and the Executive Establishment." In *The New American Political System*, edited by Anthony King. Washington, D.C.: American Enterprise Institute, 1978.

Jordan, A. Grant. "Iron Triangles, Woolly Corporatism and Elastic Nets: Images of the Policy Process." *Journal of Public Policy*, 1 (1981).

Lowi, Theodore J. *The End of Liberalism: Ideology, Policy and the Crisis of Public Authority*. New York: Norton, 1969.

Lunch, William M. "Science, Civil Rights & Environmental Policy: A Political Mystery in Three Acts." Paper presented at Western Political Science Association Annual Meeting, Portland, Oregon, 2004.

Meiners, Roger E. and Bruce Yandle, eds. *Taking the Environment Seriously*. Lanham, MD: Rowman & Littlefield, 1993.

Ellen M. van Bueren
Delft University of Technology

Kuznets Curve

An environmental Kuznets curve (EKC) is a statistical artifact that, for a single country, sample of countries, or other political regions, shows graphically the estimated relationship between a specific measure of environmental quality—say, the concentration of sulfur dioxide for a specific set of air quality stations—and some measure of per capita income. The device got its name from a famous economist, Simon Kuznets, who discovered a relationship between the distribution of income and income growth in developing countries. Kuznets found that income distribution in developing economies begins highly skewed but becomes more equal with higher per capita income. When mapped with a measure of income distribution measured on the vertical axis and per capita income on the horizontal axis, the Kuznets curve looked like an inverted U.

When EKCs are estimated and drawn with a measure of environmental pollution on the vertical axis and per capita income on the horizontal axis, an EKC enables one to infer what happens to environmental quality when incomes rise or fall. The relationship may indicate that growth in income is associated with environmental decay. Alternately, the EKC may show that rising incomes are associated with improved environmental quality. In fact, EKCs can and do show a variety of relationships for different countries, different time periods, and different pollutants. Furthermore, it is possible to find several different pollution/income relationships for differing income levels when focusing on one country for different time periods or for a sample of countries for one time period.

Zones along an EKC that show a changing environment/income relationship for the same country sample are found in a growing body of evidence that supports the notion that some, but not all, EKCs are shaped like an inverted U for a number of important environmental pollutants and activities. The accompanying theory suggests that human communities in early stages of development are more concerned about keeping body and soul together than about clear sunsets and clean water. As incomes increase, and more basic human needs are met, the rate of environmental deterioration declines. Finally, deterioration ends with additional income growth, which is to say the inverted U reaches its peak. From that point forward, rising incomes are associated with improved environmental quality. One may think of the first EKC zone, where incomes are low and rising, as representing a race to the bottom. The zone after the peak may be thought of as representing a race to the top.

It is important to keep in mind that not all EKCs conform to the inverted U shape. Indeed, some are linear, suggesting that environmental decay continues systematically with income growth across all levels of income. Other estimated EKCs appear to strap together two functions and look more like an inverted U with a tail that shows a second phase of growth in deterioration after environmental quality has recovered to a high level.

EKCs emerged on the scene in the 1990s, when Eugene Grossman and Alan Kreuger, two Princeton economists, reported a strong statistical relationship between some commonly used measures of environmental quality and per capita income for a cross-section of countries. Their innovation brought important focus to debates about the North American Free Trade Agreement (NAFTA). At the time, it was clear that no one really knew the relationship between income growth, which was a fundamental basis for NAFTA support, and environmental quality. Many thought industrialization would lead to decay. Others thought higher incomes would lead to environmental improvement. Grossman and Krueger saw the environmental arguments opposing NAFTA as a refutable hypothesis that could be tested with data. To do this, they assembled panels of World Bank data on air quality and per capita gross domestic product (GDP) data for a sample of 42 countries for sulfur dioxide emissions and 19 for smoke and estimated statistical models to find the relationship between per capita GDP and environmental quality. Their results supported the notion that lower-income countries did accept higher levels of air pollution and indicated that higher-income countries were associated with lower, not higher, levels of air pollution. Grossman and Krueger identified the turning point—the range of per capita GDP where the race to the bottom ended and the race to top begins. The range of per capita GDP at the turning point was roughly $7,800 to $9,800 in 2008 dollars for sulfur dioxide and smoke. No turning point was found for suspended particulates, which they also estimated. Interestingly enough, Mexico's per capita GDP at the time fell within the EKC turning point range. One could infer that if NAFTA generated higher GDP for Mexico, then the people of Mexico would demand improved air quality, at least for sulfur dioxide and smoke.

Literally hundreds of EKC studies have been published, and scores have been reviewed by Bruce Yandle, Madhusudan Bhattarai, and Maya Vijayaraghavan. These include studies that have focused on water quality, deforestation, carbon emissions, and a host of other measures of environmental quality. Early EKC research was hampered by a lack of time series data for individual countries. The passage of time has improved the situation: It is now possible to examine statistically individual countries as well as groups of countries for extended time periods. In an important 2007 study, Kuheli Dutt found a turning point for per capita carbon emissions for a large sample of countries. Evidence that human communities across the world are reducing carbon emissions as incomes rise offers a bit of optimism for those who are concerned about carbon emissions and climate change. The evidence also indicates the importance of income, which provides the means to pay for environmental control.

Critics argue EKCs fail to take into consideration several key components that can alter the environment/income relationship. EKCs do not account for trade relationships, whereby developed countries are improving environmental quality by exporting much of their manufacturing base rather than over environment concerns. The spatial distribution and variability of environmental pollutants and income within a country are also not shown; hence, the EKC often represents only mean conditions. EKCs also assume that the environmental degradation occurring during the low-income or developing phase is irreversible and relationship is unidirectional. Many times the environmental variable being analyzed is seen as isolated, independent and not part of the larger environmental system that may be impacted in other ways.

As statistical artifacts, EKCs do not tell us what human communities actually do to change their treatment of the environment when incomes allow them to do so. Yet we may think of each point on a single country's EKC as being associated with a set of institutions that have been developed to manage and protect the environment. Each point represents a set of property rights institutions, which may vary dramatically across countries and human communities. Some countries, for example, may prefer public property rights that are managed by government. Others may prefer the use of private incentives that cause people to adjust their use of the environment. Still others may prefer the use of regulatory property rights in which a government agency issues permit rights to approved environmental users. EKCs may help us to understand the relationships between human communities and environmental quality.

See Also: Deforestation; Ecological Economics; Environmental Management; Institutions; North American Free Trade Agreement; Regulatory Approaches.

Further Readings

Dutt, Kuheli. "Governance, Institutions and the Environment-Income Relationship: A Cross-Country Study." *Environment, Development and Sustainability* (2008). http://www.springerlink.com/content/661x3u658wk81507/ (Accessed March 2009).

Grossman, Gene M. and Alan B. Krueger. "Economic Growth and the Environment." *Economic Journal*, 110:353–77 (1995).

Stern, David I. "The Environmental Kuznets Curve: A Review." *Economics of Nature and the Nature of Economics*, edited by Cutler J. Cleveland, David I. Stern, and Robert Costanza. Northhampton, MA: Edward Elgar Publishing, Inc., 2001

Yandle, Bruce, et al. "Environmental Kuznets Curves: A Review of Findings, Methods, and Policy Implications." RS-01-1a. Bozeman, MT: Property and Environment Research Center, 2004.

Bruce Yandle
Clemson University

Kyoto Protocol

The Kyoto Protocol is a set of rules establishing emissions reduction as a part of the United Nations Framework Convention on Climate Change (FCCC). The UNFCCC is an international treaty designed to organize, standardize, and fund global efforts to minimize the harm of climate change, especially through the reduction of greenhouse gas emissions. The treaty was developed at the 1992 Earth Summit and now has nearly 200 countries as signatories. However, the treaty itself, though legally binding and in force, did not stipulate specific requirements for its signatories to hasten negotiations over the framework of the treaty and allow room for the summit's other business. Instead, such specifics are left to the protocols that are negotiated and adopted at subsequent COPs (Conferences of Parties), which meet annually. These protocols are also legally binding and are in force for a specific length of time—however, to be bound by such a protocol, a country must sign it as well as the UNFCCC. The Kyoto Protocol has eclipsed the UNFCCC in popularity in large part because the United States is a signatory to the UNFCCC but has refused to ratify the Kyoto Protocol, thus making it a member of the Convention that is not legally bound by its current requirements. The fact that the United States has not withdrawn from the UNFCCC entirely, however, indicates that the administrations that rejected Kyoto—Clinton and Bush—may have expected that the country would be willing to ratify the protocol at some point in the future, or that a protocol succeeding Kyoto would be more amenable. In any event, the United States' noncompliance has drawn significant international criticism, but at the same time its withdrawing from the negotiations (attending COPs in an observing capacity) has made those negotiations proceed much more cleanly and efficiently, as U.S./European Union conflicts were the most prominent obstacles to consensus.

The UNFCCC is concerned principally with greenhouse gas levels in the atmosphere and with the effects of greenhouse gases on the climate and the world's human populations. At the time of its inception, 1990 levels of world greenhouse gas emissions were chosen as a benchmark relative to which goals would be set. For instance, the Kyoto Protocol calls for the European Union to reduce its emissions to 8 percent less than 1990 levels. Gauging these emission levels is done through greenhouse gas inventories. Emission inventories can be constructed through a number of methodologies and are regularly assembled by government, scientific, and industry agencies to assess various factors and inform decisions. Greenhouse gas inventories, specifically, are the sum of the emissions under consideration and the sinks that offset them. The greenhouse gas inventory used under the Kyoto Protocol tracks carbon dioxide, methane, nitrous oxide, sulfur hexafluoride, hydrofluorocarbons, and perfluorocarbons. Allowing sinks to be included in the construction of greenhouse gas inventories was actually a victory for U.S. negotiators, despite the United States' nonratification.

There are three categories of membership in the UNFCCC: the developed country groups of Annex I and Annex II and developing countries. Annex I countries are obligated to reduce greenhouse gas emissions below specific levels set by the protocols. Annex II countries are a subset of the Annex I and are expected to help pay the costs of developing countries. Central to the UNFCCC is this idea of "common but differentiated responsibilities." Developing countries are expected to work toward emissions reductions as is feasible without interfering with their development and may voluntarily join Annex I, but no requirements are placed on them—a fact that some critics oppose, as long-term environmental reform requires all countries to reduce their greenhouse gas emissions. At the moment, however, the mechanism of reducing emissions does have an undeniable economic effect, perhaps short term. One reason for excusing developing countries from responsibility is that most of the greenhouse gas emissions past and present that are responsible for causing the problem the UNFCCC and other entities seek to solve originated in the developed nations of Annex I.

The Kyoto Protocol entered into force (becoming binding for its signatory parties) on February 16, 2005, after years of negotiations. The first two annual COPs were primarily planning and administration-oriented, assessing existing scientific data and setting mid-term goals. COP-3 convened in Kyoto in December 1997. The protocol itself was agreed to over the 10-day period of the COP; more than seven years passed before it went into force because of prolonged debate over its implementation, which dominated the intervening COPs.

The Kyoto Protocol set legally binding emissions levels limits for the first emissions budget period of UNFCCC, 2008–12. Those limits, which represented the target average that must be achieved in that period for compliance, varied from country to country within Annex I as part of the "common but differentiated responsibilities," with most countries required to target a reduction to levels about 7 percent less than 1990 levels. The signing period extended from March 16, 1998, to March 15, 1999, after which time article 25 of the protocol stipulated that it would enter into force and become binding on its signatories "on the ninetieth day after the date on which not less than 55 parties to the Convention, incorporating parties included in Annex I which accounted in total for at least 55 percent of the total carbon dioxide emissions for 1990 of the parties included in Annex I, have deposited their instruments of ratification, acceptance, approval, or accession." A protocol that was not going to be followed by most of the signing parties responsible for most of the emissions being quelled, after all, would serve little purpose. The 55th party to ratify Kyoto was Iceland, on May 23, 2002, but the 55 percent requirement was not met until November 18, 2004, when Russia ratified it.

COP-4, meeting in 1998 in Buenos Aires, established a two-year plan for Kyoto's implementation, but that goal was not met because COP-6 (in 2000 at The Hague) degenerated into arguments between the United States and the European Union and was suspended for 8 months until July 2001. The crux of the argument was over the U.S. proposal to allow carbon projects (such as reforestation) to generate credit that would allow countries to satisfy emissions reductions requirements by "buying them down" rather than actually reducing emissions. Many felt this was an unsustainable solution and contrary to the spirit and goals of the UNFCCC. COP-6 was suspended at a critical juncture, as it took place shortly after the U.S. elections, while the results of the presidential election were still being determined. Though President Bill Clinton had not sent the Kyoto Protocol to the Senate for ratification, he had not explicitly refused to do so but had only put it off indefinitely. His vice president, Al Gore, was the Democratic nominee for the presidency and well known for his environmental concerns. Had the election turned out in his favor, the Senate might have been given the chance to discuss Kyoto (which is not to say they would have ratified it). Instead, Republican governor George W. Bush was elected Clinton's successor, and one of the actions of the first 100 days of his presidency—typically viewed as setting the tone for an administration—was to explicitly reject the ratification of Kyoto.

COP-6 resumed in Bonn, Germany, in July 2001, a couple of months after the United States rejected Kyoto's ratification but opted not to withdraw from the UNFCCC. Though other countries had debated the protocol, none had rejected its ratification. The United States opted to send its representatives to the second half of COP-6 in an observatory capacity, remaining uninvolved in the negotiations—which proceeded much more smoothly, though ironically the U.S. proposal for carbon projects was approved, resulting in the creation of the Joint Implementation and Clean Development Mechanism (CDM) projects. Further, as long as carbon projects were carried out primarily domestically, there was no limit on how many carbon credits they could generate. The terms *carbon footprint* and *carbon neutral*, part of the language surrounding the idea of this balance sheet of emissions and offsets, entered the general public's lexicon largely as a result of Kyoto Protocol discussions.

Clean Development Mechanism

The CDM is a project established under the Kyoto Protocol that allows signatory countries to invest in approved carbon projects that reduce emissions in developing countries—which are not required by the protocol to reduce their emissions, although they are encouraged to do so—in return for increased emissions allowances in their own country. Such projects must be approved by the CDM Executive Board and must be created as a result of this program—an Annex I country cannot simply "buy out" an existing project in a developing country to generate credits but must invest in or initiate a project that would not otherwise exist. This is an important requirement, because it is the only way to make the underlying, U.S.-supported argument work: that it is the net effect on global emissions that matters and that reducing atmospheric greenhouse gases through carbon projects is as beneficial as reducing emissions levels.

To create a CDM project, the country first negotiates with the developing nation that will host the project (many developing nations actually opposed the CDM during COP-6 on the grounds that the industrialized world should clean up its own mess first) and must then demonstrate to the CDM Executive Board the project's "additionality"—that is, that it satisfies the aforementioned condition that it is a new project that would not exist if not for the incentive provided by the CDM. A third party is consulted to evaluate the proposed project's ability to produce measurable long-term results, and the board then determines whether to approve the project.

Approved projects generate certified emission reductions (CERs), better known as carbon credits, measured in units of metric tons of "carbon dioxide or its equivalent." The executive board issues these credits based on the measured results of CDM projects.

There is ongoing debate about what exactly constitutes additionality, and future accords, or the protocol that succeeds Kyoto, may define it more specifically. For instance, one view of additionality is that a project only satisfies the requirement if it would not have occurred without the CDM; another supports "financial additionality," which includes under its umbrella projects that may have been proposed or explored but were unable to secure funding or find a means of financial viability and that became financially viable as a result of the CDM and the applicant industrialized country.

As of summer 2009, the CDM Executive Board had approved 1,699 carbon projects around the world, including TransMilenio, the public bus system in Bogota, Colombia—a streamlined public transit system that has reduced the overall emissions levels of the city by encouraging use of buses instead of commuting and by replacing the earlier and less efficient transit system.

Joint Implementation

The Kyoto Protocol also permits the generation of carbon credits and CERs through the creation of joint implementation projects—carbon projects based in Annex I countries. Most of these projects are in Annex I countries with economies in transition; Russia and Ukraine, which are highly industrialized but undergoing significant economic transition comparable to some developing nations, host most such projects.

Emissions Trading

CDM and joint implementation are two of the three methods sometimes called "flexible mechanisms" for dealing with the emissions reduction requirements of Kyoto. Although the most straightforward way of reaching those requirements is, clearly, by reducing emissions, this is sometimes impractical for a country—or less practical than other options that, in the final analysis, seem to be just as beneficial, at least in the 2008–12 period. Common to all flexible mechanisms is that they seek the same goal as reducing emissions within a country: reducing the overall emissions levels globally.

The third such flexible mechanism is emissions trading, which essentially treats all countries' permitted emissions, expressed in CERs, as tradable commodities. If country A is sure it will reduce its emissions to 1 CER below the level required of it, and country B wants to reduce its emissions to 1 CER above the level required of it, country B can purchase country A's "extra" CER and avoid facing repercussions for exceeding its emissions limit. The total number of CERs in the pool—the total amount of global emissions—remains the same as it would if both countries had hit their target exactly. Of course, the amount is higher than it would be if country B had not been able to purchase that spare CER and had put in extra effort to hit its target. To work efficiently and beneficially, the emissions trading system depends on a number of things: First, it works ideally when the possibility of selling extra CERs motivates a country to reduce its emissions more than it would otherwise, and second, trading presumably only occurs when the cost of a CER is less than the cost of either reducing emissions by one CER (for the potential customer nation, that is) and less than the cost of noncompliance. A situation of comparative advantage can develop in which if it is cheaper for emissions to be reduced in country A than in country B, it is in both countries' economic interest for country A to reduce emissions well below its target level and sell those extra CERs to country B.

The same system, outside the purview of the UNFCCC, is in place in many political jurisdictions, where businesses are the trading entities rather than countries and the law or industry regulations are the relevant authorities that set emissions-level targets.

One of the effects of emissions trading is to make clean and renewable energy more competitive in the marketplace—a fact that eased the UNFCCC's approval of the emission trading system.

As of this writing, the United States is the only country that has signed the Kyoto Protocol but does not intend to ratify it.

See Also: Global Climate Change; UN Framework Convention on Climate Change.

Further Readings

De Leo, Giulio A., et al. "Carbon Emissions: The Economic Benefits of the Kyoto Protocol." *Nature*, 413 (2001).

Hilsenrath, Jon E. "Environmental Economists Debate Merit of U.S.'s Kyoto Withdrawal." *Wall Street Journal* (August 7, 2001).

"Kyoto Protocol to the United Nations Framework Convention on Climate Change." http://unfccc.int/resource/docs/convkp/kpeng.html (Accessed July 2009).

Oberthur, Sebastian and Hermann E. Ott. *The Kyoto Protocol: International Climate Policy for the 21st Century*. New York: Springer, 1999.

Victor, David G. *The Collapse of the Kyoto Protocol and the Struggle to Slow Global Warming*. Princeton, NJ: Princeton University Press, 2004.

Bill Kte'pi
Independent Scholar

Land Ethic

The land ethic invites humans to include land—soil, water, plants, and animals—in their moral calculations as they consider different courses of actions. It was developed and argued for by Aldo Leopold in *The Sand County Almanac*, first published in 1949.

Leopold rejects the traditional view of human dominance over nature. In its place he proposes what he titles a "land triangle." As Leopold explains, this triangle is made up of many different layers. Each of the layers shares one single characteristic. Members of any given layer are alike in the type of food or energy that they consume. Each layer of soil, and in the layers above it, there are plants, insects, birds, reptiles, and mammals in ascending order. Humans, since they are not true carnivores, are not at the apex of the triangle. Leopold suggests that if one of the layers of the triangle were to collapse, all of the layers above it would also collapse. Thinking of nature in the form of the land triangle suggests that humans are dependent on the land and not dominant over it. Leopold calls this a community of independent parts.

Recognizing the human relationship with the land is not Leopold's end goal, however. The land ethic requires individuals to directly consider this land community as an independent consideration in their ethical dilemmas and debates. Traditionally, humans have only considered other humans in their moral calculations. A utilitarian approach asks humans to choose an option that provides the greatest amount of good for the greatest number of people. Other approaches to solving ethical dilemmas ask people to consider whether their actions would hurt anyone. Over time, humans have extended this type of ethical consideration to animals. Leopold's land ethic asks humans to consider the land community. In other words, a person should ask him- or herself if an action would produce the greatest good for the land community or whether the action would directly hurt the land community. This approach focuses on the ecosystem.

There is a passage in the land ethic in which he invites readers to consider whether or not an action will affect the "integrity, stability, and beauty of the biotic community." The action is right, he says, if it upholds these values.

See Also: Biodiversity; Domination of Nature; Ecocentrism; Intrinsic Value.

Further Readings

Leopold, A. *Round River; From the Journals of Aldo Leopold.* New York: Oxford University Press, 1953.

Leopold, A. *A Sand County Almanac, and Sketches Here and There.* New York: Oxford University Press, 1949.

Leopold, A., et al. *Aldo Leopold's Wilderness: Selected Early Writings by the Author of* A Sand County Almanac. Harrisburg, PA: Stackpole Books, 1990.

Leopold, A., et al. *For the Health of the Land: Previously Unpublished Essays and Other Writings.* Washington, D.C.: Island, 1999.

Jo A. Arney
University of Wisconsin–La Crosse

Limits to Growth

Limits to Growth is a phrase coined by Donella Meadows and collegues in their 1972 book, *The Limits to Growth: A Report for the Club of Rome's Project on the Predicament of Mankind.* The research, commissioned by the Club of Rome, models the world system to examine the consequences of economic growth and development when that growth does not account for the scarcity of the Earth's resources and its capacity to absorb wastes. The model, World3, was designed to examine the impact of accelerating industrialization, rapid population growth, food production, depletion of nonrenewable resources, and environmental degradation. The report concludes that humans will inevitably overshoot the Earth's ability to support the total human population, or carrying capacity, if growth continues unsustainably. The result would be economic collapse and a decline in human population. However, humans have the ability to avert potential disaster by achieving equilibrium between human demands for scarce resources and population growth at the global scale. The concept of *Limits to Growth* is often described as Neo-Malthusian, since the model assumes exponential growth in human populations.

Population growth, rising consumer demands, industrialization, and environmental degradation strain the world's resources since the world's population is also growing in prosperity. The authors reached the conclusion that the limits to growth would be reached within the 21st century, citing the most likely outcome as sudden and unrecoverable decline in human population and industry. In the years after its publication, critics were quick to attack. American economist Robert Solow, one of the book's greatest critics, cited weak data to support the authors' claims and conclusions. Bjørn Lomborg, in *The Skeptical Environmentalist*, repeatedly denounced *Limits to Growth* for missing the mark on the stated prediction of running out of oil by 1992. The authors in their own criticism stated that their confidence in numerical parameters was mixed; however, their confidence in the basic qualitative assumptions, predictions, and conclusions was strong.

The neoclassical economics argument would suggest that economic forces spur technological solutions prior to any impending doom predicted by *Limits to Growth*. Scarcity of natural resources drives prices higher. As prices increase, society begins to conserve its resources and to seek out innovations to enhance the efficiency of the resource use or to substitute away from that resource in favor of another. In essence, economists focused on relative scarcity.

For example, as oil reserves are depleted, the market will observe increases in price for oil. When oil prices rise, people begin to use less oil. In Europe, taxes on oil serve to impose a much higher price per gallon of oil than in the United States. The difference is noticeable in that the average European consumes much less oil than the average American. When prices are higher, the demand for alternative resources increases. Wind, solar, geothermal, and other energy resources are becoming important alternatives to nonrenewable resources.

Today, *Limits to Growth* is receiving renewed attention as global climate change and the failure of economic systems dominate public discourse. American economist Joseph Stiglitz, who once criticized *Limits to Growth*, has recently agreed that oil is underpriced relative to the cost of carbon emissions from the burning of fossil fuels. The observation is consistent with the criticism that economic models do not account for the finite nature or absolute scarcity of the Earth's resources or its finite ability to absorb wastes without consequence. Moreover, in 2008, resource use and costs of the resource use have continued to rise with increasing demand from a growing and industrializing population.

Absent of market signals, humans will have difficulty in identifying when resources become scarce in absolute terms. For instance, environmental goods and services such as clear air, clean water, and waste absorption are provided by the Earth's biogeochemical cycles, and they are not traded in the global marketplace in the traditional sense. When these systems are compromised or degraded, market mechanisms do not signal to consumers to conserve water, reduce pollution into the atmosphere, or reduce waste production. Likewise, signals may not be present to spur on innovation and substitution away from these vital resources in favor of another is not possible.

Global climate change perhaps represents the best example of how the economic system does not provide an appropriate signal to consumers to signal the potential collapse of a biophysical system. Nonrenewable resources, oil in particular, are facing increasing demands as the economies of industrializing countries advance. In particular, oil demands from China and India have increased exponentially in recent years. The increase in carbon dioxide emissions represents a by-product or externality of the process that affects sea level, temperature, and precipitation patterns that have consequences for human settlement and activity. Because the negative consequences are not bought and sold in a market or captured in the form of a pricing mechanism, the marketplace cannot adjust to a perceived scarcity.

The purpose of the *Limits to Growth* report was not to make accurate predictions, but rather to examine the relationship between exponential population growth and the Earth's finite resources. The report suggested that disaster was inevitable, but as a solution, several recommendations were made. Zero population growth and zero expansion in the use of materials would potentially avert impending doom.

Humans are constantly seeking ways to fight against the limits of a finite system rather than living within them. Technological optimism is described as dangerous. One of the report's authors, Dennis Meadows, was awarded the Japan Prize from the Science and Technology Foundation of Japan for his work on *Limits to Growth*. *The Limits to Growth* influenced important policy decisions since its publication. It was cited in "The Global 2000 Report to the President by the Council on Environmental Quality and the United States Department of State." The report began discussions about future trends in climate change, energy scarcity, loss in biodiversity, and population growth.

See Also: Club of Rome; Ecological Economics; Steady State Economy; Tragedy of the Commons.

Further Readings

Dragos Aligica, Paul. "Julian Simon and the 'Limits to Growth' Neo-Malthusianism." *The Electronic Journal of Sustainable Development*, 1 (2009).

Hall, Charles A. S. and John W. Day. "Revisiting the Limits to Growth After Peak Oil." *American Scientist* (March/April 2009).

Lomborg, Bjørn. *The Skeptical Environmentalist: Measuring the Real State of the World.* Cambridge, MA: Cambridge University Press, 2001.

Meadows, Donella H., Dennis L. Meadows, and Jørgen Randers. *The Limits to Growth: The 30-Year Update.* White River Junction, VT: Chelsea Green Publishing Company, 2004.

Meadows, Donella H., Dennis L. Meadows, Jørgen Randers, and William W. Behrens III. *The Limits to Growth: A Report for the Club of Rome's Project on the Predicament of Mankind.* New York: Universe Books, 1972.

Jessica Kelly
Millersville University

M

MALTHUSIANISM

In his *An Essay on the Principle of Population* (1798) and subsequent revisions, British economist Thomas Robert Malthus set forth the political and economic concept that unchecked population growth outpaces the means of subsistence, or that population tends to increase exponentially or geometrically, whereas food supplies increase arithmetically or linearly. The result of unchecked population growth inevitably leads to economic catastrophe. Critics of Malthus's theory claim that progress in science and technology will inexorably solve the problems of scarcity of resources, particularly the food supply, as humans develop methods to increase productivity or substitute away from scarce resources in favor of less scarce or renewable resources. Malthus's controversial ideas continue to have salience today as Malthusians, or neo-Malthusians, apply Malthus's mathematical model to the increasing demands placed on the Earth's finite resources by the exploding human population.

Primarily based on his observations of demographic trends during the Industrial Revolution, Malthus argued that excessive population growth would inevitably be controlled by positive checks on populations when food shortages were imminent. Positive checks include war, famine, disease, and poverty, and these checks primarily afflicted the poor or lower class. He argued that it was the tension between people and the scarcity of resources that led to misery and conflict in society. To avoid such misery in the lower class, Malthus argued that preventative checks acted in a way to stave off such misery. These preventative checks included postponement of marriage, birth control, and abortion. He preached self-control and social responsibility in averting the disaster of unchecked population growth. Each of the checks would reduce the size of the human population or slow its growth.

At the time of publication, Malthus was heavily criticized for blaming the victims of the lower class for their misery and for combining discussions of scientific reasoning with lectures on moral behavior. Notably, Karl Marx asserted that it was the injustices of the capitalistic system that created the societal ills of poverty, famine, and disease among society's most vulnerable, not population growth. Moreover, Marx claimed that the technological innovations of society would avert the doom of food shortages predicted by Malthus. Malthus's doomsday prediction was proven wrong in the short run when the technological advances in the green revolution allowed for massive increases in agricultural productivity. However, his theory garnered a substantial following that continues today and is frequently revisited for its rational argumentation.

Malthusianism implies that there is a carrying capacity for the human population, or a limit to the size of the human population that the Earth can support. Below that capacity, the population will increase. Above that capacity, the population will decline. No one knows exactly what the maximum size is for the Earth to sustainably support the human population, and the figure depends a great deal on the standard of living and consumption patterns of that population. Some have argued that the carrying capacity has already been exceeded by the consumption-based lifestyles and habits of the people of the world's industrialized nations. The growing trend of consumerism in rapidly industrializing countries like India and China accelerates the demand on the Earth's finite resources, and the carrying capacity is surpassed further. Others suggest that population growth could continue well into the future with some modifications to human behavior. Malthus argued, incidentally, that society would be reduced to subsistence conditions when populations exceeded agricultural production.

Notable proponents of Malthusianism or Malthusian-like theories include the Club of Rome, Robert Kaplan, Lester Brown, Anne and Paul Ehrlich, and Garrett Hardin. In addition, Malthus has been credited with influencing Charles Darwin's natural selection theory. Today, Malthus's theory continues to resonate in ecological economics, environmental studies, and political economy.

One famous example of Malthus' continued importance in environmental discussions includes the debate between ecologist Paul Ehrlich and economist Julian Simon that has come to be known as the Ehrlich–Simon wager. In what would be a philosophical face-off between the Malthusian and the cornucopian, the pair publicly debated the question of the future welfare of the world by placing a 10-year bet on the change in prices of copper, chromium, nickel, tin, and tungsten on the pages of notable scholarly journals and popular media in 1980 and after. Ehrlich, resting on the concept of resource scarcity, argued that the prices would inevitably rise. However, he noted that the five metals were poor indicators of the state of the Earth's ability to support a growing population. Rather, he suggested that it would be more appropriate to examine the changes in soils, forests, and water quality. Simon argued that prices for the five metals would fall as people substituted for other materials. He advocated that the Earth would not run out of resources. Simon won the wager because three of the five metals declined in price over the course of the 10-year period; however, the flaw was in the selection of the indicators and the use of price as the measure of scarcity. Had the bet included measures for environmental degradation and/or absolute values of resource availability, Ehrlich would have won. As such, Ehrlich offered a second wager on the changes in the ecological conditions. Simon declined to participate in the second wager. What the wager illustrates for Malthusians and others is that the economy is ineffective in signaling the declining of resource pools.

Once-denounced doomsayers, Malthusians such as American environmentalist Lester Brown are increasingly gaining respect in popular society with claims of food crises that threaten global stability. Images of disappearing ice shelves, starving polar bears, and frequent natural disasters are held up as the latest evidence of global climate change. As more questions are raised about the future of the natural world, Malthusian arguments continue to have importance in discussions of the relationship between humans and their effect on the environment.

See Also: Club of Rome; Industrial Revolution; Limits to Growth; Sustainable Development; Tragedy of the Commons.

Further Readings

Brown, Lester. "Could Food Shortages Bring Down Civilization?" *Scientific American Magazine* (May 2009).
Ehrlich, Paul R. *The Population Bomb*. New York: Ballantine Books, 1978.
Ehrlich, Paul R. and Anne Ehrlich. *Betrayal of Science and Reason: How Anti-Environmental Rhetoric Threatens Our Future*. Washington, D.C.: Island Press, 1996.
James, Patricia. *Population Malthus: His Life and Times*. New York: Routledge, 1979.
Malthus, Thomas R. *An Essay on the Principle of Population, or, A View of Its Past and Present Effects on Human Happiness: With an Inquiry Into Our Prospects Respecting the Future Removal or Mitigation of the Evils Which It Occasions*, edited by Patricia James. Cambridge, MA: Cambridge University Press, 1992.

Jessica Kelly
Millersville University

Marine Mammal Protection Act

The Marine Mammal Protection Act (MMPA), 16 U.S.C. 1361 et seq., enacted in 1972, protects all marine mammals from being taken within waters controlled by the United States and by U.S. citizens on the high seas. It also forbids importation of marine mammals and marine mammal products into the United States. Marine mammals protected by the act include cetaceans (whales, dolphins, and porpoises) and pinnipeds (seals, sea lions, and walruses). Other protected marine mammals include polar bears, sea otters, and manatees.

The National Oceanic and Atmospheric Administration (NOAA) Fisheries within the Department of Commerce is responsible for the protection, conservation, and recovery of approximately 160 marine mammal stocks of whales, dolphins, porpoises, seals, and sea lions. Its duties include: (1) engaging in scientific research to assess marine mammal stocks and determine whether those stocks are depleted; (2) creating and implementing conservation plans for species designated as depleted; and (3) administering permit programs and authorizations that allow limited takings of marine mammals. The U.S. Fish & Wildlife Service protects walruses, manatees, otters, and polar bears. The Animal and Plant Health Inspection Service of the Department of Agriculture regulates treatment of marine mammals kept in captivity. Agency efforts to protect marine mammals are coordinated by an independent federal oversight agency, the Marine Mammal Commission.

The Marine Mammal Protection Act safeguards animals like the Florida manatee, shown in this 2008 U.S. Fish & Wildlife Service photograph.

Source: U.S. Fish & Wildlife Service

Ecosystem Management

Congress passed the MMPA because of scientific and public concern that the marine mammal populations were declining as a result of human activities and were reaching the point at which some species were in danger of extinction. The MMPA adopted a national policy that marine mammal species or stocks must not be permitted to be depleted, that is, allowed to fall below their optimum sustainable population level, and that conservation measures should be taken to replenish marine mammal species or stocks. Most important, the MMPA established a national policy to manage marine mammal species and population stocks to maintain the health and stability of the marine ecosystem. This was the first time that a government made conservation of healthy and stable ecosystems as important as the conservation of individual species.

Other MMPA Innovations

The 1972 MMPA had other innovative aspects never included before in resource legislation. The act created a single comprehensive federal program to replace the patchwork of state-run programs, despite the fact that fish and wildlife were traditionally matters left to the states. In a move that presaged the protection of the Endangered Species Act (ESA) for distinct population segments, the MMPA extended protection beyond species and subspecies to population stocks. The act shifted the burden from resource managers to resource users to show that proposed taking of living marine resources would not adversely affect the resource or the ecosystem. By establishing the concept of "optimum sustainable populations," the act aimed to maintain populations that would ensure healthy ecosystems. Management of marine species had previously been managed to produce "maximum sustainable yield" to ensure that the species replenished itself for an adequate harvest in subsequent years. In focusing on factors other than sustainable harvest, the MMPA foreshadowed adoption of "optimal yield" as a national standard for marine fisheries under the Magnuson Fisheries Conservation and Management Act in 1976. Finally, the MMPA established a moratorium on the taking of marine mammals in U.S. waters and directed federal agencies to seek changes in the Whaling Convention and the North Pacific Seal Convention consistent with the MMPA.

Banning Take of Marine Mammals

The MMPA was one of the first efforts to implement the recommendation of the United Nations Conference on the Human Environment at Stockholm to place a 10-year moratorium on commercial whaling to allow whale stocks to recover. Although Congress anticipated that the ban in the MMPA on all taking of marine mammals would be short term while stocks recovered and sufficient scientific information was developed to allow whaling and commercial takes of other marine mammals, the moratorium imposed by the International Whaling Commission in 1982 has never been lifted.

In protecting marine mammals from being taken in U.S. waters or by U.S. citizens on the high seas, the MMPA defines "take" to mean "to hunt, harass, capture, or kill" any marine mammal or to attempt to do so. The inclusion of harassment within the MMPA take prohibition was groundbreaking and was subsequently extended to all endangered and threatened species in the ESA.

General Authorizations and Exceptions

Alaska natives who live on the Alaskan coast can take marine mammals for subsistence use or to create and sell "authentic articles of handicrafts and clothing" without a permit or other authorization, so long as the taking is not "accomplished in a wasteful manner." However, if a species or stock becomes depleted, the secretaries of Commerce and the Interior may regulate the taking of a depleted species or stock, regardless of the purpose for which it is taken.

Scientific research that has an incidental effect on marine mammals limited to level B harassment (potential to disturb) has a "General Authorization" and may be undertaken without securing an incidental take permit provided the researcher files a Notice of Intent.

Incidental Take Permits

Permits to take marine mammals may be sought for takings incidental to commercial fishing, educational or commercial photography, and other nonfishing activities, for scientific research, for enhancing species survival and recovery, and for public display at licensed institutions such as aquaria and science centers.

The MMPA allows incidental take permits to be issued for activities other than commercial fishing, even if the take involves small numbers of depleted as well as nondepleted marine mammals, provided (1) there will be a negligible effect on the species; (2) there will not be an adverse effect on the availability of the species for taking for subsistence uses by Alaska Natives; and (3) NOAA Fisheries prescribes regulations detailing methods of taking, monitoring, and reporting requirements. The last requirement may be waived if the proposed activity results in only harassment and no serious injury or mortality is anticipated.

The MMPA provisions allowing marine mammals such as whales to be taken for scientific research and for Alaska subsistence are consistent with the International Convention for the Regulation of Whaling ("the Whaling Convention"). The remaining provisions for incidental takes are also consistent with the Whaling Convention because they do not contemplate intentional killing of whales but only harassment incidental to other legitimate activities such as educational and commercial photography.

The 1994 MMPA amendments established a new approach to incidental takes by commercial fisheries. This system requires that NOAA Fisheries prepare stock assessments for all marine mammal stocks in U.S. waters. Based on those stock assessments, NOAA Fisheries must develop and implement take reduction plans that are at or below their optimal sustainable population levels as a result of interactions with commercial fisheries.

Depleted Species and Conservation Planning

Marine mammals listed under the ESA automatically qualify as depleted under the MMPA. The majority of the 28 marine mammal species or stocks designated as depleted by NOAA Fisheries under the MMPA are also listed as endangered or threatened species; however, six species or stocks have been designated as depleted prior to being listed as endangered or threatened. Thus, the MMPA can set population recovery efforts in place for marine mammals before they reach the point of being endangered or threatened. Three species are currently candidates for designation as depleted. Another 36 species managed by NOAA Fisheries are not depleted, not candidates, or have been delisted. Seven of the nine marine mammal species managed by the U.S. Fish and Wildlife Service are listed as endangered or threatened. The two exceptions are the northern sea otter & the Pacific walrus.

Two species of marine mammals with conservation plans prepared by NOAA Fisheries under the MMPA have subsequently been listed as endangered or threatened species—the Cook Inlet beluga whale of Cook Inlet and the Southern Resident killer whales—and recovery planning is being done under the ESA. NOAA Fisheries has completed a conservation plan for a third species, the Northern fur seal (Pribilof Island/Eastern Pacific), which is not listed as endangered or threatened. However, NOAA has not prepared conservation plans as yet for five other depleted species, even though four species have been listed as depleted for 15–30 years. Data available for three of these species indicate that their stocks are remaining flat, and insufficient data are available for one species. Although conservation plans have not been prepared, NOAA Fisheries has created two programs to address excess dolphin mortality: the Dolphin-Safe program, which focuses on reducing yellowfin tuna fishing–related dolphin mortality by developing alternative fishing methods that do not involve dolphins, and the Dolphin Energetics Program, which focuses on determining whether energetics limitations associated with fishing may be contributing to the observed lack of recovery of fishery-associated dolphin stocks.

The remaining species that does not yet have a conservation plan, the AT1 transient killer whales, was not listed as depleted until 2004. These killer whales use Prince William Sound, and their population has been cut in half since the 1989 *Exxon Valdez* oil spill, with a number of confirmed deaths occurring during the 1990s and the remainder of the missing whales presumed dead.

See Also: Biodiversity; Endangered Species Act; Fish & Wildlife Service; International Whaling Commission; Stockholm Convention.

Further Readings

Curnutt, Jordan. *Animals and the Law: A Source Book*. Santa Barbara, CA: ABC-CLIO, Inc., 2001.

Marine Mammal Commission. http://mmc.gov/ (Accessed February 2009).

Marine Mammal Protection Act. http://www.nmfs.noaa.gov/pr/pdfs/laws/mmpa.pdf (Accessed February 2009).

NOAA Fisheries. Marine Mammal Regulations, 50 CFR Part 216. http://www.nmfs.noaa.gov/pr/pdfs/fr/50cfr216.pdf (Accessed February 2009).

NOAA Fisheries, Office of Protected Resources. http://www.nmfs.noaa.gov/pr/ (Accessed February 2009).

NOAA Marine Fisheries. "The Tuna-Dolphin Issue." http://swfsc.noaa.gov/textblock.aspx?Division=PRD&ParentMenuld=248&id=1408 (Accessed February 2009).

U.S. Fish & Wildlife Service. Marine Mammal Regulations, 50 CFR Part 18. http://www.fws.gov/habitatconservation/50CFR18.pdf (Accessed February 2009).

U.S. Fish & Wildlife Service, Division of Habitat and Resource Conservation. "Marine Mammals." http://www.fws.gov/habitatconservation/marine_mammals.html (Accessed February 2009.)

Susan L. Smith
Willamette University College of Law

MEGACITIES

Megacities are generally recognized as metropolitan areas whose populations exceed 10 million people. Their sheer size and density make them major engines of economic growth and opportunity, but at the same time their complexity, burden on natural ecosystems, and dramatic socioeconomic disparities make them increasingly vulnerable to natural and anthropogenic hazards. As of 2004, it was estimated by the United Nations that there were 20 megacities in the world, most concentrated in Asia and the global South. Although megacities are significant generators of economic activity, most are not considered globally significant "global cities" capable of influencing markets and political events on that scale.

It is not so much the size of these cities that is the cause of their social and environmental impacts but rather their speed of urbanization that has outpaced the ability of their governments to respond. London, for example, took over a century to grow from 1 million to 7 million; the same rate of growth has swelled some African cities over a single generation. As a result, residents in megacities may suffer from insufficient built and social infrastructure, as well as official corruption and high rates of crime.

It must be recognized that, despite the hardships and hazards associated with many megacities, for their many residents and prospective migrants these are outweighed by the opportunities they afford. However, the rate at which the world is gaining megacities, as well as the rate at which they are growing, makes them a significant global concern.

When considered against the overall history of urbanization, the rise of megacities has not just been extraordinarily rapid but also geographically concentrated. In 1950, only New York, London, Paris, Tokyo, and Shanghai housed 5 million people or more. By 2001, there were 18 cities with populations of 10 million or more, 12 of which were in Asia. By 2015, Asia will have 14 of the Earth's estimated 22 megacities.

Although it is true that populations have been urbanizing rapidly all over the world, the pace of growth of megacities reflects some specific global processes. Although some of this urban population growth is a result of natural population increases and high fertility rates, what is more significant is the tremendous scale of the rural-to-urban migration. These migrations are the result of deteriorating economic and environmental conditions in rural areas. Deforestation, drought, war, and conflict are major factors. Export-oriented restructuring of the global economy has made traditional farming and small-scale production no longer viable, encouraging millions of people to migrate to their nations' larger cities in search of employment, education, and a better future for their children.

Megacities are growing especially quickly in Asia and south Asia, where a number of metropolises have exceeded 10 million people. Many of these large cities, like this one in India, suffer from poor infrastructure and extreme economic inequity.

Source: iStockphoto.com

Megacities offer their residents and migrants the opportunity to participate in the economy largely through the informal sector—consisting of everything from food vending to pedicabs to website design—which is a major factor in some nations' gross domestic product and intersects significantly with the formal economy.

There are some conceptual issues that make the term *megacity* problematic. One is that the threshold of population size has fluctuated over the years, from 5 million (Asian Development Bank) to 8 million (United Nations, 1980) to 10 million (United Nations, 2001). The other issue is that a definition may need to consider the functional boundaries of an urban area, rather than simply a formal jurisdiction. Based solely on administrative boundaries, a city may have fewer than 10 million people, but when seen as an agglomeration of multiple but essentially seamless urban areas, the region can be seen as a megacity. Otherwise, such distinctions may not be all that helpful: A city slightly smaller than some international standards for a megacity may still face all the problems of a larger center, and so for all practical purposes it may be considered by some to be a megacity. Conversely, some smaller centers may have worse environmental, health, and social outcomes than some megacities. There is also the fact that national economic and political capacity has a significant role to play in determining whether or not a megacity will be functional or dysfunctional. This is why some argue for more layered criteria for the designation of a megacity—one based on features, functions, and characteristics of the city itself: Its demographic variables, resources, economic and industrial structure, administrative and governance considerations, and role in the global urban system.

The pace of population growth in megacities creates huge demand for housing, which can only be met by the expansion of informal squatter settlements, or slums. These developments are generally considered noncompliant both in terms of land tenure and construction, as they are often built with substandard, short-lived materials that become highly vulnerable to extreme weather, floods, or earthquakes. Population densities are generally quite high. Slum areas also suffer from a lack of access to electricity, cooking fuel, surfaced roads, lighting, and proper waste disposal. Given the mass of economic activity in megacities, these infrastructural deficits lead to extreme congestion of people and goods.

Of particular concern is the widespread lack of access to potable water and sanitation, which contributes to the spread of disease and high mortality rates. In addition to the squalor of their slums, megacities may also suffer from other forms of environmental degradation, such as deforested hillsides, aquifer depletion, and heavy air pollution. Furthermore, because of the scale and intensity of the pollution created in megacities, their ecological impacts can extend considerably beyond their boundaries and indeed can be global in reach.

Along with the insecurity of tenure are found other forms of social exclusion and dislocation. Megacities are characterized by dramatic divisions between extreme wealth and poverty. Wealth becomes concentrated in guarded, privatized enclaves, and the high degree of social fragmentation contributes to crime and the potential for social unrest.

However, such conditions may be found in many large cities throughout the world and are not exclusive to megacities. What makes these challenges so acute for megacities is the sheer scale and complexity of urban governance: governments have difficulty maintaining their control over urban economic and legal functions or obtaining the consent of the governed.

In the face of these and other issues, a number of initiatives have been put in place to address the challenges of megacities. Some megacities have implemented the phased provision of infrastructure to informal settlements, rather than condemning them for their noncompliance. The governance of megacities can also be improved by decentralizing and strengthening local governments and by using microloans to finance self-help projects facilitated by nongovernmental organizations. Congestion can be relieved by investments in public transit systems. Most significant have been efforts to recognize land tenure by granting "squatters" title to their properties so that they may participate in real estate markets.

To counter the negative trends associated with megacities and to "unleash the potential" of cities, the United Nations Population Fund recommends that instead of discouraging migration to cities, urban policies need to be directed at securing the rights of poor people to the city, where they can access education and employment, and that planning needs to be not just people-centered and proactive but actually "explicit" in its recognition of the land needs of poor people, supplying them with access to water, sewers, power, transport, and other necessities.

See Also: North–South Issues; People, Parks, Poverty; Urban Planning.

Further Readings

Daniels, P. W. "Urban Challenges: The Formal and Informal Economies in Mega-Cities." *Cities*, 21/6:501–11, 2004.

Fuchs, Roland J., ed. *Mega-City Growth and the Future*. Tokyo: United Nations University Press, 1994.

Laquian, Aprodicio A. *Beyond Metropolis: The Planning and Governance of Asia's Mega-Urban Regions*. Baltimore, MD: Johns Hopkins University Press, 2005.

United Nations Population Fund. *State of World Population 2007: Unleashing the Potential of Urban Growth*. New York: United Nations, 2007.

Michael Quinn Dudley
University of Winnipeg

Millennium Development Goals

The United Nations Millennium Declaration was adopted in September 2000 by world leaders at the United Nations headquarters in New York. One hundred eighty-nine states committed themselves to a series of targets known as the Millennium Development Goals (MDGs), the majority of which are to be achieved by 2015. The overall aims of the MDGs are to support the reduction of poverty and to improve quality of life. There are eight key, interconnected goals, with a number of associated targets:

1. Eradicate extreme poverty and hunger
2. Achieve universal primary education
3. Promote gender equality and empower women
4. Reduce child mortality
5. Improve maternal health
6. Combat HIV/AIDS, malaria, and other diseases
7. Ensure environmental sustainability
8. A global partnership for development

The first goal, the eradication of extreme poverty and hunger, has three key targets associated with it. The first is to halve the number of those in the developing world who live on less than $1 a day by 2015. The second target is to achieve full and productive employment and decent work for all, including women and young people. The third target is to halve the proportion of people who suffer from hunger (measured between 1990 and 2015).

The second goal, achieving universal primary education, has a target of ensuring that both boys and girls complete a full course of primary education by 2015. Evidence suggests that the cost of education, location, and availability of educational facilities, particularly for those living in rural areas or urban slums, can prevent attendance or completion of primary education. In some regions, a combination of these factors, alongside cultural beliefs, act as a further barrier to the participation of girls in primary education.

The third goal, the promotion of gender equality and empowerment of women, is closely related to the second goal but places an emphasis on gender equality within primary, secondary, and tertiary education. There are three main measures within this goal: ratios of girls to boys in primary, secondary, and tertiary education; the proportion of women in waged employment in the nonagricultural sector; and the proportion of seats held by women in national parliament. There are a number of reasons for existing gender inequalities: traditional gender roles may require women and girls to fetch water in places where there is a lack of local access to water; also, as discussed earlier, cultural values may prevent girls being sent to school, especially where there is a cost associated with education. Practical factors also play a part, including distance and a lack of private sanitation facilities within schools. In the developing world, women take up around 66 percent of work as unpaid/self-employed. In terms of parliament, women make up less than 10 percent of members of parliament in one-third of all countries. By promoting gender equality from an early age, it is hoped that greater opportunities for women will be created.

The fourth goal, the reduction of child mortality, is one of a number of health-related goals. Progress is measured in terms of the under-five mortality rate, the infant mortality rate, and the proportion of 1-year-old children immunized against measles. Mortality rates vary by region; however, the United Nations suggests that a child born in a developing country is over 13 times more likely to die within the first 5 years of life than a child born in a developed country. A number of factors are thought to explain this: deaths from preventable diseases such as measles, malaria, and pneumonia; deaths from malnutrition; access to basic healthcare; and access to water and sanitation. It is also thought that poor maternal education (in combination with the factors listed above) can increase the risk of child mortality.

The fifth goal, the improvement of maternal health, is the second explicit health-related goal and also has a clear link to goal three. There are two facets of this goal: first, the improvement of neonatal and postnatal facilities, and second, access to family planning services. Progress is measured against two key targets: first, to reduce the maternal mortality ratio by three-quarters between 1990 and 2015, and second, to achieve universal access to reproductive healthcare by 2015. These goals are measured by the maternal mortality ratio, proportion of births attended by skilled health personnel, contraceptive prevalence rate, adolescent birth rate, antenatal care coverage, and unmet need for family planning. The overall aim of this goal is therefore to improve the health of expectant mothers, but also to enable and to encourage family planning.

The sixth goal, to combat HIV/AIDS, malaria, and other diseases, is the third health-specific goal. There are three key targets associated with this goal: (1) to have halted and begun to reverse the spread of HIV/AIDS by 2015; (2) to achieve universal access to treatment for HIV/AIDS for those who need it by 2010; and (3) to have halted incidences of malaria and other major diseases and to have begun to reverse such incidences by 2015. There are a number of indicators associated with these targets: HIV prevalence among the population aged 15–24 years; condom use at last high-risk sex; the proportion of population aged 15–24 years with comprehensive correct knowledge of HIV/AIDS; the ratio of school attendance of orphans to school attendance of nonorphans aged 10–14 years old; the proportion of the population with advanced HIV infection with access to antiretroviral drugs; the incidence and death rates associated with malaria; the proportion of children under age 5 years sleeping under insecticide-treated bed nets; the proportion of children under age 5 years with fever who are treated with appropriate antimalarial drugs; the incidence, prevalence, and death rates associated with tuberculosis; and the proportion of tuberculosis cases detected and cured under directly observed treatment short course.

The seventh goal, to ensure environmental sustainability, has four main objectives (some of which are more clear-cut than others). The first is to integrate the principles of sustainable development into national policies and programs and to reverse the loss of environmental resources. The second is to reduce the loss of biodiversity, with a significant reduction in the rate of loss by 2010 (and seven indicators to measure this in different environmental areas such as forestry, marine resources, and species threatened with extinction). The third is to halve by 2015 the proportion of people without sustainable access to safe drinking water and basic sanitation (which clearly links to the health measures listed above). The fourth is to improve the lives of at least 100 million slum dwellers (again, a measure closely related to the health goals and also to the education goal).

The eighth goal, to develop a global partnership for development, has four main elements: (1) to further develop an open, rule-based, predictable, nondiscriminatory trading and financial system; (2) to address the special needs of the least-developed countries, for example, addressing tariffs and quotas that support least-developed countries' ability to export, cancellation of official debts, and so on; (3) to address the specific needs of both landlocked and small island developing countries; and (4) to ensure that debt problems of developing countries are dealt with in a sustainable, comprehensive manner.

Limitations of the Goals

Progress since 2000 has been varied, and the MDGs are not without their critics. In broad terms there are four main targets of criticism: the focus and methods of measurement, the generic approach to the goals, the ability and commitment of countries to make policies that will further the goals, and that the MDGs can be seen to promote the existing state of economic inequalities.

First, as with any suite of indicators, there are criticisms over the focus of the goals—what they measure, what they do not measure, and how they measure it. For example, one criticism is that there is a lack of focus on human rights—such as freedom of speech or association—and no mention of indigenous or minority groups. Equally, there are criticisms over the definition of poverty used both within and to direct the MDGs, with a suggestion that it is too narrow in focus. Also, Amir Attaran suggests that although some indicators such as those considering poverty are straightforward and relatively simple to measure (even if they are questionable in nature), others, such as the health goals, are much more problematic. He suggests that in the poorest countries, births and deaths are not registered systematically or completely. As a result, the health indicators are likely to be incomplete, and at worst, misleading. He suggests that this problem exists throughout the health suite of indicators, with the lack of data in some regions and questionable data collection practices in others raising questions over the robustness of the evidence presented.

Second, the global-based targets, for example, to halve those living in poverty, are criticized for being too generic, given differences in progress and development status of different regions. For example, in 2008 the United Nations reported that the goal of reducing absolute poverty was within reach for the world as a whole; however, although rapid development in Asia enabled this goal to be met, it masked limited reductions in absolute poverty in other areas such as sub-Saharan Africa. Equally, there have been improvements in the reduction of hunger (although this has been hampered by rising food prices), but once again, these successes are very much region-specific: China halved the proportion of underweight children between 1990 and 2006, whereas there was only limited progress in sub-Saharan Africa. Critics such as William Easterly argue that this approach fails to recognize progress within sub-Saharan states, and indeed turns successes into negative news stories. In summary, he argues that for some states the MDGs are underambitious, whereas for others they are too ambitious and do not recognize local or regional factors.

Third, where a country lacks the institutional capacity to drive policy change, or indeed has cultural barriers that prevent it, for example, around gender-based targets and goals, progress toward the MDGs may be limited. A factor that may exacerbate this is that the goals are not binding, and states are not punished for lack of progress.

Fourth, there is some criticism that the MDGs promote the existing state of economic inequalities. Jose Maria Sison argues that they do not address the structural inequalities that have led to the current situation and are "imperialist" and prescriptive in nature.

Conclusion

In conclusion, the MDGs provide a global view of poverty and quality of life in the developing world. Progress has been made toward a number of the targets; however, this varies greatly by region, and it may be that rapid development by some countries masks poverty and quality-of-life issues in the least-developed countries. Nevertheless, the MDGs do mark a significant level of progress within the international community, bringing together in a single package many of the most significant international commitments made throughout the 1990s.

See Also: Agenda 21; Equity; Gender; Globalization; North–South Issues.

Further Readings

Attaran, A. "An Immeasurable Crisis: A Criticism of the Millennium Development Goals and Why They Cannot Be Measured." *Public Library of Science: Medicine,* 2:10, 2005.

Easterly, W. "How the Millennium Development Goals Are Unfair to Africa." Global Economy and Development Working Paper 14. Washington, D.C.: Brookings Institution, 2007.

European Commission. *European Commission Report on Millennium Development Goals 2000–2004.* Brussels: European Commission, 2004.

United Nations. *High-Level Event on the Millennium Development Goals, United Nations Headquarters, New York, September 25, 2008 FACT SHEETS Goals 1–8*. New York: United Nations Department of Public Information, 2008.
United Nations. *The Millennium Development Goals Report: 2008*. New York: United Nations, 2008.
United Nations. *United Nations Millennium Declaration*. New York: United Nations, 2000.

Carolyn Snell
University of York

Montreal Protocol

The Montreal Protocol on Substances that Deplete the Ozone Layer is an international treaty that is designed to protect the ozone layer. The ozone layer is, in a word, the Earth's "sunscreen" that absorbs ultraviolet radiation, thus allowing life to exist on land. The Montreal Protocol was designed to phase out ozone-depleting substances (ODSs)—substances that had been created for various purposes, most famously for use in aerosols, refrigeration, and other forms of air-conditioning. The treaty was signed in 1987 and entered into force in 1989. Since then, the treaty has been amended four times: the London Amendment (adopted in 1990, entered into force in 1992), the Copenhagen Amendment (adopted in 1992, entered into force in 1994), the Montreal Amendment (adopted in 1997, entered into force in 1999), and the Beijing Amendment (adopted in 1999, entered into force in 2002). These amendments allowed for adjustments in protocol rules to accelerate the phase-out of certain ODSs. The Montreal Protocol now touts 191 signatories—a large increase from the original 46 participating countries.

Early History of the Ozone Layer Issue

Stratospheric ozone layer concentrations are rather small, making up only 8–10 parts per million of air at around 15–35 kilometers in altitude. A healthy ozone layer would average only 3 millimeters in thickness if compressed and measured at the Earth's surface. In the atmosphere, it is much less concentrated. However, this small concentration of ozone buffers ultraviolet radiation and aids in the regulation of the Earth's temperature and air circulation. In short, without the ozone layer, life on Earth could not exist. In 1974, chemists F. Sherwood Rowland and Mario Molina published an article predicting that chlorofluorocarbons (CFCs) would deplete the ozone layer. Molina and Rowland discovered that CFCs would persist long enough in the atmosphere to reach the stratosphere, where they could then be broken down by ultraviolet radiation that would release chlorine from the CFC molecule. These chlorine atoms could then potentially break down large quantities of ozone. Molina and Rowland would be awarded the Nobel Prize in Chemistry for their groundbreaking work on the ozone layer issue. However, early scientific research on the effect of CFCs on the ozone layer was hotly contested by the chemical industry. Yet, after Molina and Rowland testified before the U.S. House of Representatives in 1974, considerable funding was provided to tackle the problem. In 1976, the U.S. National Academy of Sciences confirmed the link between CFCs and ozone layer loss, leading to a flood of research on modeling the exact effect of CFCs on the ozone layer.

Based on this and other scientific evidence, in 1985—the same year that the first ozone hole was discovered in Antarctica—the Vienna Convention for the Protection of the Ozone Layer was established to initiate international cooperation in research on the potential of ozone layer loss resulting from human activity. Although not a legally binding agreement, the Vienna Convention set the stage for creating a framework for an international response to ozone layer protection. As late as 1987, the chemical industry still denied the validity of the scientific findings on CFCs, evidenced by the Alliance for Responsible CFC Policy's testimony before the U.S. Congress. Nevertheless, in 1987, the Montreal Protocol was ratified by 29 countries and the European Community, representing 83 percent of world CFC consumption.

Organization and Innovations of the Montreal Protocol

The protocol is considered innovative for having included a lag time for phase-outs of ODSs in developing countries—10–15 years, depending on the chemical. Less-developed countries (LDCs) are noted in Article 5, and developed countries often described as "non–Article 5" countries. Another innovation is the protocol's Multilateral Fund, a financial mechanism created with the recognition that developed countries should be responsible for helping LDCs eliminate ODSs. The Multilateral Fund has an independent secretariat and an executive committee with equal representation from industrialized countries and LDCs. United Nations agencies were put in charge of implementing projects for LDCs funded by the Multilateral Fund, such as the World Bank, the United Nations Development Programme, and the United Nations Industrial Development Organization. To date, the Multilateral Fund has funded over 5,000 projects in over 140 countries.

All parties to the agreement are required to report annually to the ozone secretariat on their production, import, and export of each chemical they have agreed to phase out. An implementation committee assesses annually the parties' progress and makes suggestions to the parties on how to handle any noncompliance issues. Another innovation is that the protocol includes an amendment provision, which allows new chemicals and institutions for monitoring them to be added without the need for any tedious national ratification process. This is also considered very adaptive: Article 6 requires that parties reassess the control measures based on available scientific, environmental, technical, and economic information at least every four years. Scientific experts report any updated information to the ozone secretariat one year before each assessment so that parties can amend the protocol to fit new findings. Essentially, with Article 6 the signatories created the Technology and Economic Assessment Panel (TEAP) and the five original Technical Options Committees (TOCs). Article 6 is seen as essential for allowing the TEAP, the protocol's scientific body that provides assessments of ODSs and their alternatives, and its TOCs to make adjustments to their suggestions to parties based on current science and technology.

Periodically, the TEAP and its TOCs provide updates on their research, including updates on the estimated effect of the protocol on ozone layer recovery, updates on the estimated ozone-depleting potential of certain ODSs, advice for acceptance or rejection of nominations for exemptions to phase-outs from nominating parties, and suggestions on many other matters. It is widely believed that the protocol's four amendments were made possible because of the guidance of the TEAP and TOCs.

There is an official Meeting of the Parties (MOP) held annually, which is preceded by annual open-ended working groups (OEWGs) designed to prepare parties for the MOPs.

The ozone secretariat facilitates the plenary meetings, helping the president-elect and vice presidents keep the meeting flowing, providing translation of documents in United Nations languages, drafting the agenda, providing guidance on procedures, and other housekeeping activities. "Extraordinary meetings" are allowed under Rule 4.3 but were not needed until 2004. Countries not signatories to the protocol, nongovernmental organizations, and other nongovernmental groups may attend MOPs and OEWGs (as long as no party objects to their attendance), but they have no voting power. Each party holds one vote, with a two-thirds majority prevailing for adoption of adjustments and amendments to the protocol. For other matters (e.g., decisions), decision is by consensus. All participants have the opportunity to speak in plenary in the order that they request the floor. However, nonparties and observers are typically allowed to speak only once the floor is clear, and then only if time permits. Committees and ad hoc working groups designed to discuss particular issues can be established during the meetings if the parties decide they are needed. For example, in 2004, a Methyl Bromide Ad Hoc Working Group was established during the Geneva OEWG to discuss the upcoming U.S. proposal for "multi-year critical use exemptions" for the methyl bromide phase-out.

Chemicals Covered Under Montreal Protocol Provisions and Issues of Compliance

The initial provisions included five types of CFCs and two halons—which contain bromine—that would be controlled, with production reduced on an incremental basis. CFC production and consumption would be frozen at 1986 levels, with 20 percent reductions occurring in 1993 and 30 percent reductions in 1998. Halons would undergo a production freeze in 1992. As stated, what became one of the most notable and successful provisions of the protocol was the phase-out schedule for developing countries, which was to be delayed by 10 years as long as their CFC consumption remained very low—a principle commonly known in international environmental law as "common but differentiated responsibility."

The protocol was designed to encourage ratification by all countries by placing trade restrictions on nonparty countries: Signatories were not allowed to import ODSs or products containing such substances after a short period of time. To avoid trade discrimination suits from the General Agreement on Tariffs and Trade, any country that abided by the protocol provisions but had not signed the treaty would be considered a signatory in terms of trade. Such restrictions on trade are thought to be the real teeth of the Montreal Protocol, discouraging free-rider behavior. Another innovation of the Montreal Protocol was that it would become active with only 11 countries (holding two-thirds of CFC global production) as signatories. This was likewise designed to encourage expedited membership by all countries.

The proposed alternatives were mainly hydrochlorofluorocarbons, or HCFCs—themselves depleted ozone, but at lower levels (at 2–10 percent that of CFCs). Therefore, the early success of the protocol was not a determinate one but one that bought time for a future transition to ozone-free production and complete ozone layer recovery. The 1990 London Amendment to the protocol, after all, resulted in a nonbinding agreement that would not require phase-out of HCFCs until the year 2040. The 1992 Copenhagen Amendment would see that time span reduced to 2030. By 1992, it became clear that growth levels of CFCs were beginning to slow down, but not to decrease. In 2007, the Parties agreed to freeze HCFC consumption and production in 2013 and to begin their phase-out in 2015. The expedited phase-out of HCFCs was largely a result of their higher global warming potential compared with CFCs.

Impacts and Criticisms of the Montreal Protocol

Since the ratification of the Montreal Protocol, levels of CFCs have either leveled off or declined. Indeed, 95 percent of CFCs have been taken out of production processes. Halon concentrations are still increasing, but at a slower rate than previously. Also, levels of compliance among parties have been extremely high. Scientific research predicts that, without the Montreal Protocol, by 2050 even the middle latitudes of the Northern Hemisphere would have lost half of their ozone layer, and the Southern Hemisphere would have lost 70 percent. Because of the high level of compliance and cooperation among parties, it is no exaggeration to state that the Montreal Protocol is widely considered the most successful global environmental treaty in history.

Some criticisms among scholars point to the fact that the ozone issue is not completely resolved, as is often portrayed in the media and some scholarship. Critics point to the 2003 ozone hole that was reported over the Antarctic, which had reached the second-largest size in history. Environmental nongovernmental organizations such as Greenpeace reminded parties at the 2005 2nd Extraordinary Meeting of the Parties of the ozone hole that surprisingly appeared over central Europe that same year. In short, parties are accused of not moving fast enough, of letting economic concerns supersede environmental concerns, and of not paying attention to the signs of uncertainty given to us by the natural world.

Major criticisms focus on several lingering issues:

- A sizable quantity of CFCs used via the protocol's "essential use exemptions" remains in use, and CFCs will be produced in some LDCs until 2010.
- International nongovernmental organizations such as the Environmental Investigation Agency have exposed illegal CFC and HCFC trading in LDCs, making achievements in reduction less significant.
- Because of their ozone-depleting potential and global warming potential, HCFCs should have been phased out much sooner than even the 2015 deadline. In addition to being greenhouse gases, HCFCs are also ozone depleting.
- Ozone-depleting methyl bromide "critical use exemptions" still remain sizable, totaling over 5,000 metric tons in the United States alone in 2007 (down from almost 10,000 in 2005), and close to 700 metric tons in the European Union in 2007.

Because of the complex links between global climate change and ozone layer depletion, scholars and scientists have begun looking for ways to link these concerns in international environmental decision making. As a result of incoherence between ozone and climate change decision making, sometimes successes in the Montreal Protocol, such as the accelerated HCFC phase-out, can have an inadvertent negative effect on global climate change efforts in the Kyoto Protocol. The connections between these issues are extremely complex; however, the Montreal Protocol shows that governments, business, and the scientific community can come together, with pressure exerted from civil society groups, to resolve these important global environmental issues—which are perhaps the most important we face today as a species.

See Also: Environmental Nongovernmental Organizations; Global Climate Change; Kyoto Protocol.

Further Readings

Andersen, Stephen O. and K. Madhava Sarma. *Protecting the Ozone Layer: The United Nations History*. London: Earthscan, 2003.

Parson, Edward. *Protecting the Ozone Layer: Science and Strategy*. Oxford: Oxford University Press, 2003.

United Nations Environment Programme (UNEP). *Handbook of the Montreal Protocol on Substances That Deplete the Ozone Layer*, 7th ed. Nairobi: Secretariat of the Vienna Convention for the Protection of the Ozone Layer & the Montreal Protocol on Substances that Deplete the Ozone Layer, UNEP, 2007. http://ozone.unep.org/Publications/MP_Handbook/index.shtml (Accessed February 2009).

van der Leun, Jan C. "The Ozone Layer." *Photodermatology, Photoimmunology & Photomedicine*, 20 (2004).

Brian J. Gareau
Boston College

NIMBY

NIMBY, an acronym for "not in my backyard," is used to describe the collective or individual opposition to the siting of locally undesirable land uses (LULUs), such as incinerators, landfills, nuclear power plants, lead and zinc smelter plants, and other hazardous waste facilities. Because LULUs potentially expose local residents to some external cost, NIMBYists challenge the siting of such facilities in communities on the grounds that the facilities may present unusually high risks to public health, personal property, and the natural environment. NIMBY represents the democratization of land use planning decisions, as individuals mobilize in response to public and private decision making surrounding land use. However, NIMBY is also criticized as an irrational, emotional, self-interested, or unethical response from those who are unwilling to share in the costs associated with living in an industrialized society. The spread of NIMBY across the United States, in particular, has resulted in what many in business and in government would describe as a political gridlock in which industry is impeded by widespread fear embedded in communities.

A girl stands before an incinerator in a rural community. The 1976 Resource Conservation and Recovery Act compelled industries to dispose of waste safely, but community NIMBY campaigns have forced many incinerator operators to find alternative locations.

Source: iStockphoto.com

Although NIMBY is often used in the pejorative, citizens engaged in land use planning and decision making raise critical issues and concerns about the effect of private and public development that potentially alters the community in important ways. To some degree, citizen participation in planning may help to improve the environmental, cultural, and social justice impacts on the community. Just as critical as the development in industry itself, citizens exercise their rights to participate in democratic processes and their civic responsibilities to respond in constructive ways to that development.

NIMBYism gained popularity within the U.S. environmental movement after a wave of environmental awareness swept the United States in the 1960s and 1970s about the dangers of hazardous materials. Several high-profile cases, including the near nuclear disaster at Three Mile Island near Harrisburg, Pennsylvania, and the public health emergency and urban planning disaster of Love Canal in Niagara Falls, New York, led to increasing public awareness about the dangers to public health from hazardous materials. Public reaction to the partial core meltdown at Three Mile Island in 1979 and subsequent NIMBY campaigns are held as the most important reasons why nuclear plant construction declined precipitously in the United States through the 1980s and 1990s.

After the passage of the Resource Conservation and Recovery Act of 1976, which regulates the storage, transportation, and disposal of hazardous wastes, industries were made responsible for safely managing industrial by-products and forbidden from the casual disposal practices that were previously the norm. Yet with the new environmental legislation, industries have faced resistance to their attempts to safely process or dispose of the hazardous materials in accordance with Resource Conservation and Recovery Act. Operators of incinerators and landfills are frequently forced to find alternative locations to site their facilities when communities launch successful NIMBY campaigns. As NIMBYism becomes more prevalent, the failure to site hazardous waste facilities creates a backlog of waste processing and disposal and increases the long-distance transportation of hazardous wastes to facilities in other communities. The consequences of preventing the development and the distant siting of the facilities may actually exacerbate threats to the environment.

Narrowly defined, NIMBY is limited to the opposition to the siting of LULUs locally, and as such, NIMBYists are not inherently against the production of hazardous wastes or the facilities to dispose of them. Rather, pure NIMBYists would prefer that the facilities are simply located anywhere but near their homes. Scholars of environmental justice have been critical of NIMBY campaigns in that LULUs are more likely to be sited in communities where grassroots mobilization and community opposition are minimal. It has been argued that NIMBY may contribute to continued environmental racism, as LULUs are more likely to be sited in poor or minority neighborhoods where NIMBY campaigns are less likely instead of in white, middle-class, and/or elite neighborhoods, where NIMBYists are more prevalent.

NIABY, the acronym for "not in anyone's back yard," became wider in practice as the siting of hazardous waste facilities in poorer communities increased in the face of NIMBY campaigns elsewhere. Members of local organizations that had success with their own NIMBY campaigns began to adopt the attitude that hazardous waste facilities should not be sited anywhere, in principle. These organizations mobilized to assist others in their efforts against hazardous waste facilities. NIABY, although helping to eliminate environmental injustices in the siting of these facilities in poorer neighborhoods, does not address ways to reduce hazardous wastes in the industrial cycle. The demands for products of the industrial cycle continue in the United States, and therefore, hazardous waste by-products are also produced. Yet, the widespread success of NIMBY and NIABY has made it nearly impossible to site new facilities in the United States. As a result, efforts to export these wastes from the United States are increasing, raising ethical and renewed environmental justice concerns about the exportation of hazardous wastes abroad to developing countries. New measures to pressure industry to engage in source reduction, or reduction of the hazardous waste by-product production, are being advocated by NIMBYists.

The Mothers of East Los Angeles and the Citizen's Clearinghouse for Hazardous Waste are two notable NIMBY organizations in the United States that have successfully thwarted efforts to halt proposals to site hazardous waste facilities in several communities. Mothers of East Los Angeles has worked to oust plans for a Clem-Clear plant, which would have treated cyanide and other hazardous chemicals near a high school, and another hazardous waste incinerator in East Los Angeles on the grounds that both were too close to their homes and schools. Mothers of East Los Angeles has subsequently expanded its efforts to demand environmentally sound land use planning in other communities. Nationwide, the Citizen's Clearinghouse for Hazardous Waste has organized grassroots movements in several communities, where the group has a near-perfect record of stopping hazardous waste siting in each community with which they have worked.

NAMBI, the acronym for "not against my business or industry," represents the response of businesses and industries to communities that oppose the development of hazardous waste facilities. The term is reactionary in that it is meant to be equally inflammatory, or at least as rhetorical, as NIMBY. Increasingly, businesses are seeking partnerships with communities to reduce NIMBY syndrome. NIMBY campaigns are being launched in response to land uses other than hazardous waste facilities, including prisons, low-income housing projects, drug rehabilitation facilities, domestic violence shelters, and transportation infrastructure projects, among others. Notable to the environmental movement are the NIMBY campaigns against "green energy" projects such as wind farms. Cape Wind is the first offshore wind farm project in the United States. Located off the shore of Massachusetts in Nantucket Sound, the wind farm represents the apparent contradiction of competing demands in an industrializing society. Surveys of residents reveal overwhelming support of alternative energy, such as wind energy; however, responses also indicate that people are not supportive of wind farms that potentially ruin the aesthetic value of scenery, reduce property values, harm wildlife, and produce noise pollution.

Although NIMBY has been an important contribution toward the democratization of land use planning in the United States and elsewhere, critics argue that NIMBY is undemocratic, in that negotiations are nearly impossible under the threats of litigation. Federal and state governments have developed resources and tools for businesses and communities to work together to avoid the stalemates and barriers that so often paralyze the process of what would otherwise be democratic land use decision making. Fair Share programs have been implemented in regions to equally distribute the negative, but necessary, hazardous waste facilities and development projects among communities to ensure that no one community is more heavily burdened than another.

See Also: Anti-Toxics Movement; Environmental Justice; Environmental Movement; Participatory Democracy.

Further Readings

Frederiksson, Per G. "The Siting of Hazardous Waste Facilities in Federal Systems: The Political Economy of NIMBY." *Environmental and Resource Economics,* 15/1 (2000).

Heiman, Michael. "From 'Not in My Backyard!' to 'Not in Anybody's Backyard!' Grassroots Challenges to Hazardous Waste Facility Siting." *Journal of the American Planning Association,* 56/3 (1990).

Kraft, Michael E. and Bruce B. Clary. "Citizen Participation and the NIMBY Syndrome: Public Response to Radioactive Waste Disposal." *Western Political Quarterly,* 44 (1991).

Lake, Robert W. "Volunteers, NIMBYs, and Environmental Justice: Dilemmas of Democratic Practice." *Antipode,* 28 (1996).
Livezey, Emilie Tavel. "Hazardous Waste." *Christian Science Monitor* (November 6, 1980).
Popper, Frank. "The Environmentalist and the LULU." *Environment,* 27/2 (1985).
Weisberg, Barbara. "One City's Approach to NIMBY: How New York City Developed a Fair Share Siting Process." *Journal of the American Planning Association,* 59/1 (1993).

Jessica Kelly
Millersville University

Nonviolence

The global environmental movement is composed of diverse groups working toward a sustainable and healthy world. They run the gamut from legislative lobbying groups to local grassroots organizations, and each draws from an array of strategic tools, including nonviolence, to achieve their goals.

Nonviolence is a complex philosophy that has produced a diverse array of proactive practices, referred to as nonviolent direct actions, such as sit-ins, walk-outs, tax withholding, hunger strikes, slowdowns, stalling, noncooperation, delaying, withholding support, signs, speeches, marches, prayer circles, strikes, street theater, demonstrations, and civil disobedience (breaking the law). According to nonviolence scholar Gene Sharp, there are 198 methods of nonviolent protest and persuasion. Nonviolence is usually associated with local small groups reacting against decisions or actions by powerful groups and is often used by those less powerful because nonviolent action provides alternatives to what appear to be untenable situations, empowers individuals and their groups, and allows them to have a voice and make decisions in political and social venues that they normally would not have access to. The most well-known example is the U.S. African-American community in the 1950s South who chose nonviolent actions such as sit-ins, marches, and boycotts rather than armed revolution, resulting in the Civil Rights Act of 1964. White authorities possessed sufficient power to overwhelm a revolution but were thrown into confusion and lost part of their power through the use of nonviolence.

Nonviolent environmental protests include such actions as giving speeches and marching in the streets, as demonstrated by this antinuclear protestor in Lille, France, in April 2008.

Source: Wikipedia/Benh Lieu Song

Civil rights activists relied heavily on their Christian faith to resist the force that was used against them. This Christian-based nonviolence was a bridge between early Christian pacifism, Mahatma Gandhi's 20th-century interpretation of nonviolence, and contemporary nonviolent philosophy. According to David McReynolds of the War Resistors League, nonviolent philosophy states unequivocally that no one person or group can be absolutely certain they know the whole truth. Nonviolence is an effort to do battle with injustice without risking hurting the opponent, each of whom is a unique individual possessing some kernel of truth. For this reason, according to nonviolence theory, interactions between people in opposing camps must include respect, compassion, love, and understanding. Respect and empathy have the capacity to create dialogue and coalitions of unlikely groups, which is an essential part of nonviolence. Nonviolent action, despite its public appearance of contentiousness, must include an open, honest invitation to hear each others' views and look for common ground, common goals, and connections that could help resolve the conflict.

Conflict over societal change is inevitable. Conflict and change are painful, and often that pain falls to the very people who are experiencing some injustice already. Conflict can be resolved in many ways; however, violence provides the opportunity for only one reaction—more violence. Nonviolence provides multiple choices for different actions, as well as the opportunity to find an acceptable solution, or at least the ability for participants on each side to walk away with dignity—if they choose.

How Nonviolence Works

Nonviolence insists that to be successful, it must be a grassroots activity and empower the entire community involved. Violence relies on a hierarchical authoritarian structure, which usually disempowers women, the elderly, and those who are not strong. Nonviolence activity, in contrast, is supposed to fold everyone in, including former opponents. Nonviolence is meant to change how people think about themselves by empowering those who felt they had none and allowing the opposition opportunities to know and respect the members of the disempowered group.

Nonviolent activists try to create conditions under which the opponent is free to try different behaviors and to find alternative solutions. In 1973, women in an Indian village in the Himalayas stopped a contractor from clear-cutting the forest around their village. They talked the workers into leaving by threatening to hug the trees and die with them. The women refused to leave in spite of physical threats, and eventually the woodcutters abandoned that part of the forest. Meeting the same form of tree-hugging resistance in other parts of the forest, the contractor was unable to log one tree. The women's actions—refusing to commit violence and placing themselves as barriers to the trees—changed the terms of the conflict and allowed the laborers to save face by declaring their commitment to not harming women. If the women had committed violence against the laborers, the authorities would have felt justified in choosing violent retaliation, even against women. The tree-hugging method of preventing clear-cutting became known as the Chipko movement, which spread throughout India and had its largest political victory when Indira Gandhi, then Indian Prime Minister, signed a 15-year ban on green felling.

The Chipko women were aware of and used nonviolent guidelines set out by Mahatma Gandhi in his struggles to gain Indian independence from Britain. Contemporary nonviolent activists use similar guidelines, which vary slightly in wording but not intent from group to group. Guidelines include an attitude of openness and respect toward everyone; using no violence, verbal or physical, toward any person; not destroying or damaging any property;

not carrying anything that could be construed as a weapon; not using alcohol or drugs during an action; and accepting the consequences of breaking the law—civil disobedience.

The various groups involved in the environmental movement use a diverse array of tools, including media coverage, lobbying efforts, research, and writing as well as nonviolent action. Some groups like Greenpeace, the Rainforest Action Network, the Energy Action Coalition, the Ruckus Society, and the Climate Action Network are more dedicated to using direct action, including civil disobedience, in their repertoire. Several groups like Earth First!, Earth Liberation Front, and Sea Shepherd have redefined nonviolence action to include destruction of property, calling it monkeywrenching and eco-tage. Unlike traditional nonviolence, the perpetrators refuse to take responsibility for their actions, choosing instead anonymous sabotage. There is a dispute between various factions concerning the undermining of the original concept of nonviolent action, the association of new more violent groups with groups dedicated to traditional nonviolence, and the efficacy of limited violence over nonviolence.

What Nonviolence Has Achieved

Regardless of whether a group adds the extra dimension of eco-tage, environmental groups around the globe repeatedly use nonviolent direct action in all its creative forms as a tool to create a more environmentally healthy world. There have been defeats, partial successes, and also significant successes.

Australia's Green Ban Movement (1971–75) succeeded in saving hundreds of acres of open space and historic landmarks through a coalition of suburban women and the NSW Builders Laborers Federation, a trade Union. The builders union threatened to stop work on the developers' other sites if they continued planning the destruction of various open spaces.

In 1975, a year-long occupation of a proposed nuclear power plant site in Wyhl, Germany, prevented the developers from even starting building. A coalition between the farmers, the townspeople, and local university staff, faculty, and students created a continuous occupation of the site until the government canceled construction.

From 1976 until 1989, multiple direct actions, including civil disobedience, by a collection of affinity groups called the Clamshell Alliance took place at the Seabrook, New Hampshire, nuclear power plant. Initially meant to prevent the plant from going online (which it did not do), eventually the sheer volume of protest and public sentiment made the government cancel plans for a second reactor. The organizing work, affinity group structure, strategies, and tactics of the Seabrook organizers assisted the Abalone Alliance to blockade the Diablo Canyon nuclear power plant. Although the plant powered up, the furor, cost, and public support for the activists created the opportunity for the Nuclear Regulatory Commission to refuse all nuclear power applications for the rest of the century.

Greenpeace, a grassroots environmental group, has performed multiple direct actions including chasing Japanese whaling ships to prevent illegal whaling, "corking" coal-fired plants' smokestacks to bring attention to carbon fuel pollution, placing wind turbines across Chicago's Michigan Bridge to educate people about alternative energy solutions, and hanging a "Stop Global Warming" sign from a crane near the state department, where climate talks were taking place.

Another grassroots group, the Rainforest Action Network, supported a local logging blockade by the First Nation people of Grassy Narrows in Northwestern Ontario, Canada, while assisting them in direct lobbying of logging industries. The last company logging in Grassy Narrows declared its intention to stop logging that area in 2008.

Between 1978 and 1983, direct action, combined with massive publicity, created enormous public support to stop the Franklin Dam in Tasmania, Australia, from being built. One novel technique to gain government attention was a campaign to encourage voters to write "No Dams" on their ballots. From 5 percent in the first election, the campaign built momentum until 42 percent were writing it on their ballots. Despite massive public protest, dam construction was scheduled. In 1982, a blockade, including civil disobedience, drew 2,500 people, and around 50 people a day arrived over the summer. By 1983, plans for the dam were shelved.

These examples are only a few of the many small and large direct actions that have changed the tone and direction of the global environmental movement. The use of nonviolence as a philosophy and practice for creating social change is part of a continual debate among activists and theorists. The search for methods to equalize power relationships, provide alternatives, find creative methods of taking a principled stand, truly engage the opposition in an equal conflict, and present the opposition with other, more constructive choices on how to respond has returned activists repeatedly to nonviolent direct action and probably will continue to do so.

See Also: Environmental Movement; Nuclear Politics; Participatory Democracy.

Further Readings

Doherty, Brian. *Ideas and Actions in the Green Movement.* London: Routledge, 2002.

McReynolds, David. "Philosophy of Nonviolence." http://www.warresistersleague.org/store (Accessed June 2009).

Powers, Roger S. and William B. Vogele, eds. *Protest, Power, and Change: An Encyclopedia of Nonviolent Action From ACT-UP to Women's Suffrage.* New York: Garland, 1997.

Sharp, Gene. *Waging Nonviolent Struggle: 20th Century Practice and 21st Century Potential.* Boston: Extending Horizon Books, 2005.

Susan Birchler
Northern Virginia Community College

North American Free Trade Agreement

The North American Free Trade Agreement, or NAFTA, formed the largest regional free trade economy agreement between Canada, the United States, and Mexico. NAFTA represented the continuation of a trend established by the 1989 U.S.–Canada Free Trade Agreement. The interest in forming the free trade zone was advanced by a perceived need to balance the harmonization that led to the formation of the European Union and by reforms in Mexico that made it more compatible with the economies of the United States and Canada. In proposing the expansion to be included in the regional economy, Mexico was able to rely on the precedent set by the U.S.–Canada Free Trade Agreement. In exchange for improved access to the largest market available through the formation of

NAFTA, Mexico accepted the responsibility to develop environmental, labor, and other business regimes in keeping with the United States and Canada.

The development of the NAFTA zone was in some respects a logical development of historical patterns of trade between the United States and its immediate neighbors to the north and the south. Canada and Mexico have traditionally been the first and second suppliers of food to the United States, respectively. Although the trade between Canada and Mexico has been relatively smaller than each country's trade with the United States, the agreement advanced the harmonization of the markets of the three major trading partners.

The conclusion of the agreement and its implementation in 1994 led to broad consequences for the participating political systems and markets. Amplified by the forces of a globalizing economy, the unified North American Free Trade market for goods, services, and resources irreversibly modified international and local boundaries and redistributed power. Perhaps more so for Mexico than for the United States and Canada, governments at all levels in the region were required to reconfigure to equalize market access to eligible participants.

Through NAFTA, by international agreement, the reconfiguration is visible in the amendments by which domestic legal regimes included the terms of NAFTA. For instance, the patent laws in Mexico were amended to provide consistency with U.S. and Canadian patent protection. Subject to professional licensing provisions, designated professionals were granted mobility rights to provide services in all jurisdictions where the agreement applies. There are also particular provisions for foreign direct investment, trade-related intellectual property, labor, and the environment.

In the context of economic growth and a consumer economy, a redistribution of locations of production and consumers may have been inevitable. With the region extending to the continent, the scale of the redistribution to locations with lower cost of production, including employment costs and larger sales demand, became correspondingly greater. Some forms of protection became less effective, and local actions now have continental implications and enforceability.

Some initial presumptions on which the parties negotiated the agreement continue to be concerns. Reasonably accessible wealth increase in all NAFTA jurisdictions continues to be a prerequisite to increase in market size and profitability. In a market economy, access to ownership and investment in all forms of property is critical. The ability to trade in property—whether real (land), intellectual (including trademarks, patents, and copyrights), or personal (being all things other than land or intellectual property)—is essential to the agreement. The agreement is founded on the expectation that increased trade is healthy and indefinite growth is possible.

The development of the agreement may be seen as an example of the ambivalent relationship between accommodation and opposition. Here, the terms of the agreement represent an attempt to identify and accommodate incidents of conflict. The very inclusion of the agreements on the environment and labor were the result of such processes.

Since the implementation of NAFTA, the evolution of the globalized economy, the growing acceptance of climate change as a fact, and the recent economic collapse, the significance of the agreement has also been transformed. Considered from the dialectic of reform and revolution, the current realities continue to provide the traditional choices if only in an unprecedented manner. Protectionist regulations between the parties may now be in violation of the order established since NAFTA came into force. Other international agreements such as the General Agreement on Tariffs and Trade (GATT) may also apply. In this sense, the concept of sovereignty was affected by the trade interests.

The linkage between trade and the environment in the NAFTA side agreement on the environment represents a historical development. Previously, the United States and Canada had declined to link international trade and environment in agreements, including the 1988 Canada–U.S. Free Trade Agreement. At least some of the resistance was an attempt to prevent a "race to the bottom," in which production is relocated to spaces that have the lowest environmental protection regime and reduced labor costs arising out of conditions unacceptable in other locations.

Events such as the 1991 U.S.–Mexico tuna fishing case at the GATT panel illustrate the issue of trade and environment. The case was taken by Mexico to the panel when the United States banned the import of tuna caught by Mexican fishing vessels using means prohibited to U.S. vessels. The regulations applicable to the U.S. vessels were established out of concern for the excessive dolphin deaths that were caused by the prohibited means of fishing. From an exclusively trade-based perspective, the GATT panel ruled that the U.S. prohibition was a trade barrier contrary to the GATT.

As a result of NAFTA and subsequent events, the inclusion of environmental concerns in trade has become more easily accessible, even if unintended or unforeseen at the time of its negotiation. The establishment of the link between trade and the environment may have already inhibited the commodification of goods and services where environmental concerns may outweigh profit. Such developments may be noted in the triple bottom line analyses used by some enterprises in which environmental and community impacts are part of financial reports. In this regard, the commodification of any goods or services that could be included within the terms of NAFTA is approached with a caution that was previously unnecessary.

Further to the concept of the triple bottom line reporting, the formation and implementation of NAFTA also illustrates the heightened role that transnational business interests achieved in the formulation of national and international policy and legal regimes. Traditionally, international agreements were concluded by negotiations between sovereign states. By the time of the NAFTA negotiations, state and nonstate actors were involved and the agenda was not set exclusively by national interests. The nonstate actors are sometimes identified by the term *nongovernmental organizations*, which may include or be in addition to transnational corporations (TNCs). In the NAFTA negotiations, the role of the representatives of the sovereign states included facilitating the agenda and no longer acting as the sole authors.

The evolution of international trade as being the jurisdiction of the state to one of shared interests with nongovernmental organizations including TNCs may be illustrated by considering the production chain and industrial districts. Before NAFTA, in the apparel and garment industry, U.S.-based TNCs were able to structure their business to take advantage of free trade zones in Mexico and the Caribbean and the preferential access granted in law to businesses importing garments assembled in these countries. The preferential treatment pre-NAFTA was limited to articles assembled in the designated non-U.S. locations. The restriction to "assembling" the articles protected U.S.-based preassembly jobs and disadvantaged subcontracting preassembly stages of production. NAFTA eliminated this impediment and permitted all stages of production to be located anywhere in the jurisdiction of the signatories. As a result, since NAFTA came into force, garment businesses could subcontract previously U.S.-based processes such as the cutting of the cloth as well as the assembly of the items to anywhere in Mexico and not just the designated free trade zones.

For all the strength of the trade regulations and consistency among NAFTA jurisdictions, local enforcement of the rules has been more resistant to consistency. Local enforcement authorities, perhaps not surprisingly, may be more susceptible to interests that dominate their perspective. Resistance to enforcement has not been unique to any particular authority, as is evident in the United States' failing to enforce compliance when trade tribunal decisions decided against U.S.-based trade interests in the softwood lumber dispute with Canada.

The effects of NAFTA and the integration of trade regimes in the region have produced uneven results. A localized analysis reveals that benefits and costs are distributed with corresponding differences that cannot be generalized either across a single industry or the entire region. From the environmental interest, a similar differentiation and detail applies.

To some extent NAFTA institutionalized free trade and structured the subsequent policy and political debate. Since then and at the time of writing, there appears to be a cataclysmic loss of credit liquidity between TNCs and financial institutions on a global scale. To an extent, the resulting dependence of the TNCs on the state directs attention to the limits of credit. The same analogy can be applied to the growing awareness of carbon footprints and an acceptance of the view that the environment can no longer be treated as an unbounded source of credit against which humanity can continue to draw without consequences. This confluence presents anew the choice and dialectic of reform or revolution.

See Also: Agriculture; Deforestation; Domination of Nature; Endangered Species Act; Endocrine Disrupters; Environmental Protection Agency; Gender; Green-Washing; Groundwater; Indigenous Peoples; Land Ethic; Suburban Sprawl; World Trade Organization.

Further Readings

Cohen, M. J. and J. Murphy, eds. *Exploring Sustainable Consumption: Environmental Policy and the Social Sciences*. Amsterdam: Elsevier Science/Pergamon, 2001.

Duquette, Michel. "The NAFTA Side-Agreement on the Environment: Domestic Politics in the Making of a Regional Regime." *Canadian Review of American Studies*, 27/1 (1997).

Fox, Annette Baker. "Environment and Trade: The NAFTA Case." *Political Science Quarterly*, 110/1 (1995).

Gallagher, Kevin P. *Free Trade and the Environment. Mexico, NAFTA and Beyond*. Palo Alto, CA: Stanford University Press, 2004.

Lester de Souza
Independent Scholar

North–South Issues

The idea of the North and South—sometimes called Global North and Global South to prevent confusion with countries' domestic northern/southern divisions—as socioeconomic categories rather than just geographic ones was first formulated by West German Chancellor Willy Brandt in 1980. Brandt described a line encircling the globe at roughly 30 degrees North latitude, dipping down to place Australia and New Zealand above the line. Above the line was the North: rich, industrialized, developed nations, the First and

Second World. Below the line was the South: poorer, developing or undeveloped nations, often former vassals of the North during the Colonial era, the Third World. The terminology is sometimes found wanting, and many "Southern" nations like Brazil are difficult to categorize as "developing," given their high incomes, levels of industrialization, and participation in the global economy.

Though coined during the Cold War, the North/South terminology essentially reflects a post–Cold War division. During the Cold War, the First/Second/Third World terminology was more common, with the capitalist Western bloc constituting the First World, the communist Eastern bloc the Second World, and unaligned developing nations (potential and actual sites of proxy wars between the first two alliances) the Third World. One reason "developing nations" and "Global South" have replaced "Third World" is because the Third World label made that segment of the world sound like an afterthought, a spare part; at the same time, it lost some of its specificity when it was unofficially adopted to describe impoverished conditions within the First World. That association with poverty and starvation was never intended by French economist Alfred Sauvy, who first coined the term *Third World* (inventing First and Second World by reverse engineering)—rather, he intended a parallel with the French term *tiers etat*, the Third Estate, the workers, peasants, and middle class. "Like the Third Estate," Sauvy wrote, "the Third World has nothing and wants to be something," a connotation of movement and ambition that is perhaps retained by "developing nations" more than "Global South."

The different needs of the North and the South have been an obstacle in international trade and may prove a problem in the worldwide negotiations to repair and revise the global financial system in the aftermath of the 21st-century financial crisis. The Doha negotiation round of the World Trade Organization, for instance, is as of 2009, several years behind schedule and likely several more years away from completion, in part because of the matter of agricultural subsidies. Although developing nations have been urged away from protectionist policies by developed nations since the Bretton Woods era, the use of agricultural subsidies in the European Union and the United States has acted as a trade barrier against agricultural goods from developing countries. By inflating the profits and artificially dampening the prices of domestic agricultural goods, agricultural subsidies make it difficult for foreign producers to compete—which, in turn, encourages those producers to use cost-lowering methods, many of which (such as the use of pesticides, genetically modified organisms, and chemical fertilizers to increase yields and decrease crop loss) are in opposition to the goals of sustainability and environmental safety that are espoused in the North. The South depends on the North as customers, in effect, but is arguably sometimes placed in the position of deciding between the product the customer asks for and the product the customer is actually willing to buy.

BRICs

In contemporary usage, much of the geographic north is sometimes elided from the socioeconomic North, notably Russia and the People's Republic of China. This elision is one reason for the increasing popularity of the term *BRICs* to refer to Brazil, Russia, India, and China—large, fast-growing, resource-rich, industrialized nations that are not easily grouped with either the North or the South, and that share common concerns in international trade and politics as a result of that common difference. The term *BRICs* was introduced by investment bank Goldman Sachs in 2001 in a report discussing forecasted economic trends for the first half of the 21st century.

Collectively, the BRICs nations account for a quarter of the world's land and two-fifths of its population. The Goldman Sachs prediction was that because of their rate of growth and the economic potential offered by their resources, thus far imperfectly harnessed because of various domestic and historical issues, the BRICs nations by 2050 could be among the most dominant economies in the world, and collectively would constitute the most economically powerful group. What is interesting about the history of these countries and the terminology in the decade since the Goldman Sachs report is that although their pre-21st-century shared history is slim—the countries are spread out around the globe with few cultural or ethnic ties—they have in fact become increasingly allied as a direct result of that report.

In 2009, the first BRIC summit was held on June 16, with the heads of state of the four nations convening in Yekaterinburg, Russia, making official the shorthand term introduced by Goldman Sachs 8 years earlier. Foreign ministers of the countries had met the previous year, and annual summits are expected to continue. Among the issues discussed were international trade and the ongoing difficulty with the Doha Round of negotiation in the World Trade Organization, as well as issues pertaining to the United Nations and the ongoing world food price crisis.

The summit was an explicit challenge to the North–South model and the dominance of the Western nations of the North. Russian president Dmitri Medvedev called for a new international monetary system that would rely on regional currencies rather than on the U.S. dollar to slow or prevent the spread of economic shocks, which use of the dollar has failed to do; this point was left out of the official declaration produced by the summit, and it is speculated that China may not be prepared to abandon the dollar's popular use as an international currency.

See Also: Global Climate Change; Globalization.

Further Readings

The Brandt Commission. "The Brandt Report." http://www.stwr.org/special-features/the-brandt-report.html (Accessed July 2009).

The Brandt Commission. *Common Crisis North–South: Cooperation for World Recovery.* Cambridge, MA: MIT Press, 1983.

Therien, J. P. "Beyond the North–South Divide: The Two Tales of World Poverty." *Third World Quarterly,* 20/4 (1999).

Bill Kte'pi
Independent Scholar

Nuclear Politics

The term *nuclear politics* encompasses two broad areas: nuclear weapons and civilian nuclear technologies. Those areas are linked in a number of ways, and disputes regarding their linkages are themselves part of nuclear politics. Political issues surround the development, manufacturing, testing, deployment, proliferation, and policies for use of nuclear weapons. Civilian nuclear energy is political in terms of regulation, public health and safety implications, environmental impact, waste disposal, financing and risk insurance, competition with other energy sources, and connections to weapons development and proliferation. Other industrial and medical applications of nuclear energy present less prominent controversies.

The controversial Diablo Canyon nuclear power plant drew years of protests and civil disobedience, including a blockade, before it eventually opened in 1982.

Source: Wikipedia

Scientists first alerted government officials to the military potential of atomic energy in the late 1930s. Internal policy decisions then led to weapons programs in a number of nations that would become principal actors in World War II. Nuclear energy received little public attention until the end of the war, when the bombings of Hiroshima and Nagasaki dramatically opened a public conversation. Throughout the Cold War period, the nuclear arms race between the United States and the Soviet Union was a primary driver of international politics and of domestic politics within the industrialized nations. Political controversies encompassed the development, testing, and deployment of nuclear weapons; proposals and negotiations for test-ban, arms control, and nonproliferation treaties; civil defense programs; antiballistic missile defense systems; and the Reagan administration's Strategic Defense Initiative (known colloquially as the "Star Wars" program). The end of the Cold War shifted attention to concerns including further nuclear weapons proliferation, tensions between other nuclear-armed nations such as India and Pakistan, and the growing potential for nuclear terrorism. International and domestic tensions now surround possibilities for new nuclear weapons development, existing and proposed missile defense systems, and the future of weapons deployments by the various nuclear-armed nations.

There are important links between nuclear weapons and environmental politics. Concerns about the health effects of fallout from nuclear tests emerged in the mid-1950s, leading to widespread public protests that helped motivate the Limited Test Ban Treaty of 1963. In his 1982 manifesto *The Fate of the Earth*, Jonathan Schell argued that a full-scale nuclear war would produce profound environmental devastation that might render the planet unfit for human life. "Nuclear winter" scenarios derived from computer models soon reinforced Schell's troubling forecast. After the end of the Cold War, public awareness increased regarding the environmental damages and health risks resulting from the industrial processes of nuclear weapons production. Along with those material hazards, the culture of secrecy and expertise surrounding nuclear weapons has had broader consequences for democracy, inhibiting public access to information and participation in policy debates. Those influences have contributed to a technocratic and expertise-driven model of decision making that has affected environmental politics beyond the nuclear domain.

Civilian nuclear energy programs emerged as a by-product of military applications, and public perceptions of civilian nuclear energy remain haunted by that connection. Some nuclear technologies are dual-use technologies that can support weapons production, as well as civilian power programs. The United States has avoided developing some technologies that would facilitate expanded, long-term commitments to nuclear power, such as breeder reactors, because of their potential for increasing the proliferation of nuclear weapons. Mixed-oxide reactor fuels remain controversial in the United States because of their mixed implications for proliferation: Although they provide an approach to disposing of excess plutonium by using it as reactor fuel, they also open possibilities for increased circulation of plutonium in the global nuclear economy. Some critics argue that virtually all civilian nuclear power technologies can contribute to proliferation.

Promoting International Development

The U.S. Atoms for Peace program, initiated by President Dwight Eisenhower in 1953 during the early stages of the Cold War, sought to offset concerns about nuclear weapons by promoting the international development of civilian nuclear power. Programs in a number of nations led to the first commercial reactors, with the first fully civilian power reactor beginning operation at Shippingport, Pennsylvania, in December 1957. Two months earlier, a fire at the British Windscale military production reactor had released significant radioactivity into the environment. Although the Windscale fire provoked concerns and a government inquiry, it did not affect public attitudes as strongly as would later accidents. The U.S. Atomic Energy Act of 1946 assigned questions of civilian nuclear safety to the newly created Atomic Energy Commission. From its inception the Atomic Energy Commission managed two missions that were often in tension: the simultaneous promotion and regulation of commercial nuclear power. The Energy Reorganization Act of 1974 abolished the Atomic Energy Commission and transferred its regulatory responsibilities to the new Nuclear Regulatory Commission. A series of reactor safety studies providing the quantitative framework for safety regulation has been criticized by a range of environmental and public interest groups. Most recently, the Nuclear Regulatory Commission has undertaken a State-of-the-Art Reactor Consequence Analyses program to modernize its approach to reactor safety. Some critics of the Nuclear Regulatory Commission maintain that its regulatory approach favors the interests of the nuclear industry, whereas industry advocates view the commission as overly influenced by environmental and public interest groups. Radiation safety standards regulating public and worker exposures have also been ongoing sites of conflict.

In the culture of technological enthusiasm that characterized the 1950s and early 1960s, advocates promoted civilian nuclear energy as a solution to the demands of an increasingly energy-intensive society. Most early policy debates were conducted within a closed community of technical, governmental, and industrial actors with little public participation. Government subsidies were required to support a costly and risky technology that was not economically competitive with other energy sources. The U.S. Price-Anderson Act, first passed in 1957 and most recently renewed in 2005, addressed the reluctance of industry to bear the financial risk of nuclear accidents. The act shifts responsibility to the federal government for financial damages above specific limits while requiring the nuclear industry to maintain a fund insuring lesser risks. Critics assert that the nuclear industry would not be sustainable without such a public assumption of risk.

Growing Public Opposition

Public opposition to nuclear power grew in Europe and in the United States throughout the 1960s and 1970s, with concerns expressed regarding safety, environmental impact, and economic competitiveness. Some critics argued, and continue to argue, that nuclear power is inherently undemocratic because of its technical complexity, centralization, and reliance on government support. Protests at nuclear sites in France and Germany during the 1970s drew tens of thousands of participants, some resulting in violent clashes between police and protestors. Significant protests also took place at the U.S. Seabrook and Diablo Canyon nuclear sites beginning in 1976.

The highly publicized accidents at Three Mile Island (1979) and Chernobyl (1986) exacerbated these challenges to commercial nuclear power. Antinuclear campaigns and public protests became more frequent, more widely supported, and more strategic, organized by issue-specific groups and by international environmental groups such as Greenpeace and Friends of the Earth. Green parties in Europe adopted antinuclear positions, and opponents often linked civilian nuclear energy to debates over nuclear weapons policies. These political developments and prevailing economic analyses led a number of European nations to reverse their plans to expand nuclear power, some adopting timelines for phasing out their programs. France was a notable exception and continues to be cited by nuclear advocates as an example of a successful national commitment to nuclear energy.

In the first decade of the 21st century, nuclear power received renewed consideration as a result of rising concerns about energy costs, energy security, and global climate change. Industry advocates promoted plans for a "nuclear renaissance" or "new nuclear build," linked to new reactor designs addressing long-held safety concerns. The U.S. Energy Policy Act of 2005 provided new subsidies and risk insurance for nuclear plant construction and simplified the approval processes for new plant licenses and extensions of existing licenses. The European Union and a number of its member nations began reexamining the trend away from nuclear power.

Current debates over civilian nuclear energy involve a number of key issues. Emerging green energy technologies, such as solar and wind power, are increasingly competitive in the energy market, and decision makers must weigh investments in nuclear energy against investments in those alternative sources. Critics maintain that many analyses do not adequately address costs and risks across the full nuclear fuel cycle, from uranium mining through disposal of nuclear wastes. Although the U.S. Nuclear Waste Policy Act of 1982 requires the federal government to establish a permanent repository for spent fuel and other nuclear wastes, technical and political controversies continue to impede the proposed Yucca Mountain site in Nevada. Nuclear fuel reprocessing is controversial in part because of its potential for facilitating nuclear weapons proliferation. Proliferation concerns also surround the increasing globalization of the nuclear energy economy, illustrated by the U.S. Global Nuclear Energy Partnership, a separate U.S.–India agreement for nuclear energy cooperation, and ongoing nuclear cooperation between Russia and Iran.

See Also: Department of Energy; Participatory Democracy; Risk Assessment; Risk Society.

Further Readings

Balogh, B. *Chain Reaction: Expert Debate and Public Participation in American Commercial Nuclear Power, 1945–1975*. Cambridge, MA: Cambridge University Press, 1991.

Dalton, R. J., et al. *Critical Masses: Citizens, Nuclear Weapons Production, and Environmental Destruction in the United States and Russia.* Cambridge, MA: MIT Press, 1999.

Kinsella, W. J. "One Hundred Years of Nuclear Discourse: Four Master Themes and Their Implications for Environmental Communication." *Environmental Communication Yearbook*, 2:49–72 (2005).

Schell, J. *The Fate of the Earth.* New York: Knopf, 1982.

Smith, J., ed. *The Antinuclear Movement.* San Diego, CA: Greenhaven Press, 2003.

Taylor, B. C., et al., eds. *Nuclear Legacies: Communication, Controversy, and the U.S. Nuclear Weapons Complex.* Lanham, MD: Lexington, 2007.

William J. Kinsella
North Carolina State University

ORGANIZATIONS

Organizations are social arrangements created to pursue collective goals by a group of members. Organizations generally possess legal personality and are therefore able to enter into contracts, to own property, or to sue and be sued. They are physically located and encompass bureaucratic and budgetary structures, managed by different governing bodies. Their activities are developed within a framework predefined by its members, who are regularly assessed and planed. Because organizations embody persistent sets of rules and understandings that prescribe certain practices, they are also institutions, although not all institutions are organizations.

Organizations can be classified according to different criteria, one of them being the area within which their activities are developed: economic, religious, educational, cultural, environmental, and so on. Probably the most pertinent classification, however, is based on the nature of its members. Organizations can be (inter-)governmental (GOs) or nongovernmental (NGOs). In the first case, members are representatives of one or several states; in the second case, members do not represent states but, rather, individuals or groups of individuals outside government. This distinction is quite important because of the implications concerning the organizations' activities and their financial support. Both GOs and NGOs, national or international, adopt and implement their members' principles, objectives, and framework of action. However, in the first case, the members represent their governments' policies and programs, whereas in the second case, the members do not represent state entities but, rather, more atomized and diffuse interests.

Another structural distinction relates to the resources available to each type of organization. GOs usually rely on state budgets and public funds, including intergovernmental funding, such as provided by the International Monetary Fund or the European Union, to pursue their mission, whereas NGOs try to rely as much as possible on private funds, such as donations or campaign fund-raising, which are irregular and hard to predict. As a result, GOs are theoretically more stable than NGOs. However, in the last decades, for instance, the United Nations, an intergovernmental organization, has struggled with the infringement of member-states' financial obligations, producing instability. NGOs, in contrast, have been criticized for their stability strategy: Most of them apply regularly to public entities for funding. In some cases, their autonomy has been questioned because of their overdependence on governmental resources.

Organizations, in general, perform certain technical functions. They act as clearinghouses for information, becoming a hub to disseminate and receive information for and by its members and others. This role is supported by its permanent secretariat and provides the members value added concerning updates and guiding principles on the issue at stake. Organizations are also coaxers, as they can persuade members and nonmembers to adopt certain rules and conducts of behavior through membership and by extending their application to members' relations with nonmembers. Another function organizations usually perform is being a catalyst for further work. Their legal personality makes them actors in the international and national systems.

Motivations

However, the motivations underlying their creation also speak to the more general functions performed by organizations. Independent of the type of organization, the motivations for their creation are basically the same as the ones underlying the formation of any other institution: power, interest, and knowledge. Organizations can be instruments of their members, states, or individuals, reflecting their interests and/or the struggle of power between them, in which decisions and activities may become a continuation of a certain member or group of members' agendas, interests, and power. As a consequence, in this case, organizations are seen as just another means of certain members or groups of members exerting power, tapping into the organizations' functions to their own benefit.

Interests can also motivate the creation of organizations to address collective action problems. Certain issues are impossible or extremely difficult to be dealt with unilaterally. As a consequence, some organizations are created with the intent to address a common and/or pervasive problem jointly and make members co-responsible. The creation of an organization for this effect is considered the most efficient way of facing the situation. Finally, organizations can also be created in an attempt to build collective consensus and shared understandings based on knowledge and perceptions concerning causes and solutions. It should be noted that none of these motives are excludable. Often, even if the process was triggered by power, interest, or knowledge considerations, organizations change and evolve throughout time, and at different moments and depending on the issue, these motivations can influence with different intensities the life of the organization and its outcomes.

The understanding of organizations cannot be dissociated from certain pervasive debates. The first one questions whether an organization can be considered an actor or can only be considered a framework for action (agency-structure debate). In other words, whether an organization can be an independent actor with its own agenda, power, and interests, or whether it is always a mere reflection of its members' agendas, power, and interests. A second debate has to do with the possibility of change in organizations. Because they have administrative and budgetary structures, some argue that organizations are path-dependent; that is, they are stuck with the motivations underlying their formation and their structures. The argument is that it becomes extremely costly to change an organization because it is set in its ways and, as a consequence, only an external shock can change an organization. Others argue, on the contrary, that although recognizing a tendency toward path-dependency, organizations can change through internal incremental processes and can reinvent themselves by redirecting their goals and introducing new instruments.

Last, but not least, several organizations have been accused of suffering from a democratic deficit. Their members are not elected and are not democratically accountable to anyone; as a consequence, one does not have democratic channels to complain, contest, or participate in these organizations' decisions and activities. Governmental organizations, as long as the governments they represent are democratic, are usually not afflicted by this problem. Intergovernmental organizations, however, are usually under fierce criticism concerning their undemocratic performance. The members are represented by individuals who are usually not elected and, as a consequence, are not democratically accountable. The counterargument is that they represent a government and consequently reflect that government's level of democracy. This issue is directly related to the debate on whether organizations are actors or not. Because if organizations are considered actors, then the democratic deficit becomes much more pertinent, as independent of representing democratic governments, the organization itself should be democratically accountable. Finally, NGOs are also very vulnerable when it comes to assessing their democratic nature. On the one hand, they represent civil society and, as such, constitute a democratic means of individuals participating in politics. However, on the other hand, they are also not elected and are not democratically accountable. In other words, they do not represent the people—even when they claim to represent, for instance, environmental concerns—who elected them, and what if someone with environmental concerns would rather be represented by a different NGO—how can that be addressed?

Environmental Regimes

These issues become much more complex when one analyzes the environmental arena. Although NGOs have been the preferential means of civil society participation in environmental issues, states have preferred to create international regimes rather than organizations. These environmental regimes have, however, come to perform several of the functions attributed to organizations. Several environmental regimes have secretariats, and with the help of digital technology, they are able to perform the same informational-hub functions usually attributed to organizations (e.g., Convention on Biological Diversity). There are also some examples of environmental regimes being able to coax ratifying states to adopt certain measures, such as the Montreal Protocol on Substances that Deplete the Ozone Layer. The ratification of the protocol and its subsequent amendments has not only influenced its members' codes of conduct but also affected its members' relations with third parties (Article 4). Finally, different environmental regimes, although supported by different organizational structures, have been a catalyst for developing further work (e.g., UN Framework Convention on Climate Change). Consequently, these regimes end up gaining enough momentum, energy, and power to perform certain functions usually attributed to organizations, facing the same debates on agency structure, path-dependency, and democratic deficit.

It should be noted that the strategy to resort to the creation of regimes rather than organizations and to tap into already-existing organizations' structures to promote environmental interests and concerns has actually demonstrated itself to be a quite effective means of providing visibility and salience to environmental issues. At this time, most states have environmental organizations and are party to dozens of environmental treaties and regimes. Simultaneously, environmental NGOs have proliferated at the national, transnational, and international levels, putting pressure not only on governments and intergovernmental organizations but also on multinational corporations, globalizing local concerns, transnationalizing common problems, and localizing global challenges. Although there is no one international environmental organization, there is a binding web of environmental commitments, often associated with economic, cultural, and human rights commitments, that generate a different sort of environmental governance that is not structurally hierarchical but that has been slowly becoming embedded in the political arena.

See Also: Environmental Nongovernmental Organizations; Institutions; Montreal Protocol; Transnational Advocacy Organizations; UN Framework Convention on Climate Change.

Further Readings

UN Framework Convention on Climate Change. http://unfccc.int/2860.php (Accessed August 2009).

Young, Oran R. *International Cooperation: Building Regimes for Natural Resources and the Environment*. Ithaca, NY: Cornell University Press, 1989.

Paula Duarte Lopes
University of Coimbra

Participatory Democracy

Liberal democracy is defined by free and fair elections, universal suffrage, rule of law, and protection of individual rights. Public participation is often limited to voting or nonbinding public hearings, rather than agenda setting and decision making—particularly in environmental policy, where scientific and technical expertise is often necessary to comprehend the issues as framed by professionals and policymakers. Perhaps as a result, the social, health, economic, and other costs associated with environmental degradation have been disproportionately concentrated among those who are least represented among decision makers: the poor and people of color.

Participatory democracy, in contrast, is characterized by engagement in civil discourse to develop consensus on policies that meet the needs, values, and priorities of diverse citizens. Through persistent engagement, participants gain a greater appreciation for the potential implications of policy choices. Thus, individual freedoms may be better balanced with the associated ecological, health, social, and economic costs.

The common understanding of "democracy" as a political system in which citizens have the right to select representatives through regular, fair, popular elections and wherein decisions are made by "majority rule" reflects liberal democracy. Liberal democracy is defined by the protection of individual rights through representation and juridic order. Government is "utilitarian"; its primary role is to minimize barriers to individual freedom and to ensure the greatest aggregate good. Liberal democracy is, therefore, consistent with market capitalism, and global democratization efforts predicated on this model equate conversion to a market economic system with democratic transformation. Economic reform precedes democratization on the presumption that increased economic security and resulting access to education will foster a better informed citizenry that is prepared to engage effectively in the democratic process.

Observers note, however, that in nations where economic reform precedes democratization, political engagement becomes the specialized purview of the wealthy and formally educated. These concerns are addressed through voter registration and "get out the vote" campaigns and public hearings on legislative initiatives, the results of which are seldom binding on decision makers, who ultimately determine which concerns and values to reflect in policy. Moreover, those who share the values and backgrounds of political leaders are most likely to participate and least likely to challenge their rationale.

In the realm of environmental policy, in particular, issues are often framed in highly technical terms. Thus, extensive education or professional expertise may be necessary to participate effectively in the debate as defined by professional resource managers and policymakers, despite the existence of myriad human values and concerns surrounding nature and natural resources, creating critical questions regarding the inherent rights of citizenship and the role of technocrats versus citizens in defining acceptable solutions. Public participation has often been defined in terms of the professionals' interests, such as to improve efficiency and cost-effectiveness or to reduce public resistance. These "instrumental participation" strategies can create the appearance of democratic process without either genuine inclusion of local values, needs, and interests, or building community capacity toward self-determination and independence.

Participatory democracy seeks to balance individual freedoms with the societal responsibility of meeting the fundamental needs of all citizens and to better reflect the full diversity of public values and concerns in identifying appropriate outcomes and acceptable risks. Rather than relying on instrumental participation to appease an otherwise relatively disengaged public, participatory democracy calls for "transformative participation." Transformative participation represents a fundamental redistribution of power wherein citizens engage in debate leading to consensus around policies that can meet their collective needs and interests. Governmental and public agencies are not absolved of accountability to those constituencies as in "decentralized" models of democracy (e.g., right libertarianism), however. They are responsible for carrying out policy in accordance with the public will and for ensuring that citizens have the skills, information, and resources needed to engage actively in local self-governance and to make informed decisions.

There is a complementary shift in the responsibilities of citizens. For participatory democrats, civic engagement is a fundamental responsibility of citizenship. Liberal democracy undermines the very concept of citizenship by overemphasizing individual rights with no attendant responsibilities, save to comply with the rule of law. Lack of public involvement in voting and electoral politics, they argue, results from the lack of real influence represented by those activities. In contrast, empowered to make binding decisions, citizens will engage in active learning, creation of a shared community vision, and understanding of acceptable risks and benefits.

Dominant models of participatory democracy share a common understanding of "community" as a collective of citizens voluntarily contributing to a common vision for the collective good, and "participation" as engagement in dialogue through which consensus on such a vision emerges. These definitions simplify debates around community and participation in favor of the expectation that members share a common notion of their community's interests and can be expected to work collectively to achieve them. They presume a homogeneity of values and power within communities and, in the effort to balance power between community and policymakers, fail to adequately address social and economic stratification within communities. In response to this critique, participatory democrats argue that engagement in civic debate will increase familiarity with the diversity of citizens, values, and implications of policy choices, fostering increased accountability and mutual responsibility in decision making. Those who fail to develop increased responsiveness to these concerns are apt to be in the minority and, thus, to have less influence in collectively formulated decisions than they may have currently.

See Also: Anarchism; Capitalism; Ecological Imperialism; Environmental Justice; Globalization; Green Neoliberalism; Social Ecology; Sustainable Development.

Further Readings

Barber, Benjamin R. *Strong Democracy: Participatory Politics for a New Age*. Berkeley: University of California Press, 1984.

Fischer, Frank. *Citizens, Experts and the Environment: The Politics of Local Knowledge*. Durham, NC: Duke University Press. 2000.

Nelson, Nancy and Susan Wright, eds. *Power and Participatory Development: Theory and Practice*. London: Intermediate Technology Publications, 1995.

Roussopoulos, Dimitrios C. and George Benello, eds. *Participatory Democracy: Prospects for Democratizing Democracy*. Montreal: Black Rose, 2005.

Kerry E. Vachta
Wayne State University

PCBs

Polychlorinated biphenyls (PCBs) are a group of synthetic chemicals that can have severe health effects on humans. There are many different types of human-made PCB molecules, but they have similar chemical structures, with the formula of $C_{12} H_{10-x} Cl_x$. PCBs were once popular in industry because of their chemical stability, flame resistance, and insulating properties, and were used in a variety of manufactured products. Although there has been much public and legal attention given to their toxic nature, which has resulted in a ban on their production in many countries since the 1970s, PCBs endure in the environment. PCBs are classified as persistent, bioaccumulative, and toxic chemicals. They are considered persistent because they remain in the environment for a long time and break down very slowly. They are soluble in fats, which also makes them bioaccumulative, meaning that these chemicals concentrate, or build up, as they move up the food web.

There are no known natural sources of PCBs, and their appearances range from oily liquids to waxy solids. Before being banned in many countries, these mixtures were commonly used in the electrical industry as cooling fluid or lubricants for transformers and capacitors because they are stable compounds and do not burn easily. In addition, PCBs were used in paints and caulk materials because they increased the durability of paints and provided resistance to water, heat, and chemicals. They have also been used in a wide variety of other materials, including lightbulbs, flame retardants, and carbonless copy paper.

PCBs enter the environment in various ways and have been found in air, water, and soil. When they were used in manufacturing, they entered air, water, and soil during their use and manufacturing processes, as well as their disposal. Also, accidents, fires, and spills have led to PCB contamination. In addition to PCBs that are already present in the global environment, they are still released from improper dumping of industrial wastes and leaking of older electrical transformers that contain PCBs.

Some factories have deposited PCB contaminants into bodies of water over many years. Given that PCB can contaminate the soil, surfacewater like rivers and lakes, groundwater, and even air, the remediation of this pollution is complex. In Pittsfield, Massachusetts, for example, PCB contamination is widespread, and the causes are multiple. For one, when mixtures spilled in the factories of General Electric, workers would commonly put soil on the spill to absorb it. This almost never-ending stream of topsoil was given to the community as a benefit, and it was used in gardens, for building houses and schools, and in parks and playgrounds. In addition, underground lakes were formed by over a million pounds of contaminated oil that was dumped down drains at factories, which resulted in the contamination of a large river and the groundwater. In addition to such localized examples of extreme PCB contamination, there have been multiple international large-scale incidents involving PCBs.

PCBs do not degrade readily, so they persist in the environment. In addition, PCBs bind strongly to soil and can be carried by water or air. In bodies of water, they attach to the sediment, which is the loose material that settles at the bottom, and they can stay there for a long time. They also spread through aquatic habitats as they become absorbed by bottom-dwelling organisms and small fish, and then begin to pass along the food web and accumulate in larger fish and marine mammals. This process is called bioaccumulation, which means that toxins are stored in the tissue of animals. PCBs can also travel through air, and thus the environmental transport of PCBs is nearly global as PCBs move through the atmosphere.

In addition to exposure through contaminated air, soil, and water, food is a major source of human exposure to PCBs, especially in fish and fatty foods like animal meats. PCBs have been found to be damaging to the immune, reproductive, endocrine, and nervous systems in humans and in wildlife. PCBs are classified as a "probably human carcinogen," which means that they are considered to likely be cancer causing. In addition, they are highly lipophilic, or fat soluble, and are accumulated in the body and not excreted. Because they are fat soluble, they are found in breast milk and can cross the placenta. When PCBs enter a body, they mimic estrogen and therefore are particularly dangerous to women of childbearing age. Many states that have PCB-contaminated waters have health advisories that outline the maximum number of meals of local fish that people should eat, and these numbers are even more limited for children and women of childbearing age.

There are numerous bodies of water in the United States that are known to be contaminated by PCBs, and many remediation efforts have been undertaken to remove contaminated soil and sediment and keep it contained. In these lakes, rivers, and bays, PCBs are generally not equally distributed but are found in concentrated "hot spots" with large deposits. For example, in New York, the Hudson River is one of the more contaminated sites in the United States and has been designated by the Environmental Protection Agency as a Superfund site. That designation means that it is a toxic site in need of cleanup, and the responsible party is required to pay for the remediation of the environmental conditions. The management of such remediation efforts is complicated and charged with political, social, economic, and emotional responses. Dredging or excavation is one common way of successfully removing PCB contamination, as the sediment, rocks, and debris at the bottom of a body of water are removed and cleaned, and there are many technologies for doing this. Additional possibilities for remediation include capping, natural attenuation, and the use of special enzymes to help break down parts of the PCB molecules. When an area is remediated through dredging and shoreline excavation, important considerations include the reestablishment of wetlands and revegetation of disturbed areas.

There is often much controversy surrounding the ways to remediate PCB contamination in communities that are affected. One example of this is in New York State, where there has been a long struggle between GE and the Environmental Protection Agency about whether or not contamination of the Hudson River should be cleaned up and, if so, how. For decades GE has contended that dredging the river of the PCBs will make pollution worse; however, environmentalists and community activists emphasize that PCB-contaminated sediment needs to be dealt with to remediate the health and environmental effects. Risks to be considered in these types of situations include societal, cultural, and economic effects of having contaminated environments, as well as the risks to human heath and local as well as global ecologies. However, given that Superfund legislation stipulates that the polluter must pay for the cleanup, companies like GE undertake sophisticated advertising approaches to attempt to sway public support in favor of not remediating the contamination. Cleanup of bodies of water like the Hudson River is expensive, with the Hudson Valley cleanup estimated to cost between $500 million and $1 billion for the treatment and disposal of contaminated sediment and treatment of the water.

See Also: Clean Air Act; Clean Water Act; Environmental Protection Agency.

Further Readings

Agency for Toxic Substances and Disease Registry, Division of Toxicology. "ToxFAQsTM." http://www.atsdr.cdc.gov/toxfaq.html (Accessed May 2009).

Encyclopedia of Earth. http://www.eoearth.org/article/PCBs (Accessed May 2009).

Greene, M. J. "The Hudson River PCB Cleanup: A Light at the End of the Tunnel." *Clearwaters,* 1/4 (Summer, 2008). http://www.clearwater.org/pcbs/ (Accessed May 2009).

National Academies Press. "A Risk Management Strategy for PCB-Contaminated Sediments" (2001). http://www.nap.edu/catalog.php?record_id=10041 (Accessed May 2009).

United States Environmental Protection Agency. http://www.epa.gov/waste/hazard/tsd/pcbs/index.htm (Accessed May 2009).

Christina Siry
Manhattanville College

People, Parks, Poverty

The creation of national parks began in the 19th century, a time when frontier expansion was coming to an end in North America. This was the expansion of modern European-oriented American society across the landscape of American indigenous peoples. A similar expansion previously occurred in the modernization of Europe. This is a process also known as the enclosure of the commons—the expansion of the Western world into the realm of rural societies and non-Western societies. Both Yellowstone and Yosemite National Parks were areas previously occupied by indigenous peoples. In fact, in both areas, indigenous peoples were evicted to give way to modern and scientific park management. In 1851, the U.S. armed forces entered the Yosemite Valley to remove the Ahwahneechee people from the valley. Once the valley was cleared of indigenous people, this gave way to the new settlers arriving in the area. In the midst of these profound changes, in 1864, President Abraham Lincoln signed a bill granting Yosemite Valley and the Mariposa Grove to the state of California. Later, in 1890, John Muir helped establish Yosemite National Park.

Especially since the 1980s, indigenous peoples have been vocal in their insistence on participating in plans for the conservation of regions where they live. The photo shows an indigenous village in the Amazon rainforest.

Source: iStockphoto.com

From a North American point of view, national parks are set up for the purpose of protecting an untouched nature: the wilderness, understood as an area without human presence. However, the facts of Yellowstone and Yosemite National Parks show that such pristine nature had long ceased to exist and that the areas that were being turned into national parks had been occupied by indigenous peoples for a few millennia.

Another perspective is that of the European conservationist, who did not expect an untouched nature, but included the possibility of human presence in a landscape to be protected. These two points of view have been the extremes of the park management enthusiasts ever since.

In 1940, an international convention on Nature Protection and Wildlife Preservation in the Western hemisphere was the precedent to further developing the concept of reserves and protected areas that would occur in the following decades. During the 20th century, the proposal for national parks spread through the world and gradually enriched its content into a diverse network of production forests and national parks. The park management developments in North America and Europe provide the theoretical backbone, but the movement became international with the advent of park enthusiasts the world over. However, the debate on how to secure nature conservation while ensuring adequate resource use by rural populations had not been solved.

By 1962, the United Nations prepared its first list of protected areas, including over 1,000 areas; by 2003, this list had reached 102,102 sites, covering 18.8 million square kilometers. The development of these protected areas into national protected systems has helped conserve some rare ecosystems, while setting up agreements with local populations that initially felt constrained or were even evicted from some of these areas.

In the 1980s, the indigenous peoples' actions regarding human rights, particularly in tropical regions, set a new agenda for the following decade. A milestone was the 1989 statement by Amazonian indigenous peoples that they should be the first people to be considered for any conservation in the Amazonian forests. This call to action occurred in meetings in Iquitos, Peru, and in Washington, D.C., by the Amazonian Indigenous Peoples Confederation, and helped establish a new relationship between the creation of protected areas and indigenous peoples. Similarly, indigenous peoples and environmental activists from southeast Asia and India voiced their disagreement with the fate of the populations of the world's rainforests and demanded their participation in protected areas development and policy.

The 1992 4th World Congress on National Parks and Protected Areas, held in Caracas, Venezuela, February 10–21, 1992, aimed to define the role of protected areas in a healthy relationship between people and the rest of nature. The 1,840 participants from 133 countries were posed with questions such as: How can protected areas contribute in sustainable ways to economic welfare without detracting from the natural values for which they were established? How can local people be provided with more of the benefits of conservation, therefore becoming supporters of protected areas?

A major change occurred in the following decades in support of more people-friendly protected areas—something more in the line of the Man and the Biosphere Program, launched in the 1970s to improve the relationship of people with their environment globally. This program's promotion of a World Network of Biosphere Reserves as a vehicle for knowledge-sharing, research and monitoring, education and training, and participatory decision making provided a greater space for indigenous peoples and other rural populations in the management of protected areas. As such, the protection of many new areas involved the local population in the design, set-up, and management.

In many cases these changes meant the design of new protected areas to include successful examples of resource harvesting and conservation. These new protected areas' framework emphasized early indigenous peoples' involvement, their participation, and even comanagement experiences. For indigenous peoples this is a great achievement in the democratization of protected areas management.

The question then arises of how to adequately understand the relationship between these protected areas, the economic spaces that they represent, and the modern free market economy to assess the utility of the definition of poverty. If poverty is measured economically, that is, living on less than $1 or $2 a day, then poverty exists in protected areas. However, these natural landscapes provide rural populations with food, shelter, entertainment, and health benefits. How poor is a population that lives in an area where money is almost not needed and where survival depends on nature and harvesting it adequately? An economics poverty framework approach therefore seems too narrow to adequately represent the richness of life outside urban areas.

However, the priorities of globalization form a trend that reinforces the existing dynamic in the connection of the Amazon to national societies and, ultimately, to the global marketplace. Amazonian development priorities raised by governments in the 21st century (from Chavez to Garcia, from Lula to Morales) share the view of developing infrastructure and communication networks to reinforce the migration trend, looking for new frontier expansion toward protected areas, and facilitating access to natural resources in the region. This would involve resource-hungry masses moving toward protected areas under the banner of development and modernization.

See Also: Biodiversity; Biosphere; Indigenous Peoples; Participatory Democracy.

Further Readings

Adams, William M. and Jon Hutton. "People, Parks, and Poverty: Political Ecology and Biodiversity Conservation." *Conservation and Society*, 5/2 (2007).

Chape, S., et al. *2003 United Nations List of Protected Areas*. Gland, Switzerland: IUCN; Cambridge: UNEP-WCMC, 2003.

Ghimire, Krishna B. and Michael Pimbert, eds. *Social Change and Conservation: Environmental Politics and Impacts of National Parks and Protected Areas*. London: Earthscan, 1997.

Carlos Antonio Martin Soria
Instituto del Bien Comun

Petro-Capitalism

Petro-capitalism is a capitalism that hinges on the production, exchange, and consumption of petroleum. Petro-capitalism is central to the development of today's society. Geopolitical analysts describe the dynamics of petro-capitalism in terms of competition over access and rights to petroleum-producing areas around the world, a competition linked to issues of resource scarcity and energy security. Political economists, in contrast, argue that the interactions between local histories, larger processes of material transformation, and unequal power relations have shaped petroleum's role as a "strategic commodity" in the current capitalist world order. Yet more traditional political economists argue that attention must be paid to how and under what conditions competing petroleum producers (capitalists) and petroleum-producing countries control overproduction and oversupply of petroleum.

Petroleum is the basic unit (commodity) of petro-capitalism. The logic of capitalism is to produce surplus value—to expand or create new capital from already-existing capital. This occurs through the process of capital circulation: new capital (e.g., profit) is formed when raw materials and labor are combined, exchanged in the market, and reinvested in more production. In the case of petro-capitalism, petroleum is natural capital, the raw material that is transformed through an arrangement of labor, scientific knowledge, and technologies to become an exchangeable commodity that will generate the production of further commodities.

The exchange value of petroleum is determined by the "law of value," the terms of trade that dictate the quantity of any other commodity for which petroleum will be exchanged. The law of value, in turn, is shaped by perceptions of value, social dynamics (e.g., competition, collaboration), and the cost of commodity production. Petroleum is a commodity of extremely high use value because of its wide range of uses and very high energy fuel properties. It is a raw material fundamental to the motion of global capital today—a necessary object without which the production of a multitude of other commodities cannot proceed. Yet it is petroleum's exchange value (not its use value) that makes it matter to capitalism—a value that is recognized through money, the universal means of exchange in today's capitalist society.

The monetary value of petroleum is shaped by its demand, its scarcity, and the conditions of production (e.g., access to the resource). The high demand for petroleum indicates that it is a desirable commodity. Petroleum provides 90 percent of the energy consumed in motorized transportation—one of the fundamental characteristics of capitalist society—and its derivatives are found in most objects consumed in daily life, from fertilizers to plastic to lipstick. Petroleum's abundance, in contrast, is difficult to ascertain. Some analysts argue that petroleum reserves around the world are limited and in decline, whereas others argue that there is no petroleum scarcity problem. This uncertainty contributes to instability in petroleum prices and, at times, has increased the exchange price of petroleum. Petro-capitalists—firms that profit from the extraction and trade of petroleum—play a fundamental role in setting the price of petroleum. For the most part, they are able to control rates of extraction and the quantities and qualities of petroleum available in the market, thus shaping public perceptions about its abundance or scarcity.

Petro-Capitalism and Competition

Competition is a basic characteristic of the capitalist mode of production and dominates the concerns and activities of capitalists. Petro-capitalists are concerned with the uncertainty that the commodities produced will be bought and the risk that the cost of petroleum production will not be covered—both of which would lead to a loss of capital rather than the generation of surplus. The petro-capitalist thus has to be a competitive producer, which can occur through investment of more capital to promote growth (e.g., through technological innovations or locating new oilfields), cutting production costs (e.g., increasing labor exploitation, relocating production to other areas with lower production costs), avoiding overproduction of petroleum, and reducing competition. Some capitalists will be successful, concentrating capital in the hands of a few. Others will go bankrupt or be absorbed by the successful ones, centralizing capital. Competition for petroleum among capitalists is fundamentally competition to control the flow of capital to prevent overproduction and the expansion of competition. Ultimately, control over the flow of capital is control over surplus value.

Competition also emerges from the conflicting interests of extractive petroleum companies and petroleum-producing states. By virtue of their sovereignty over petroleum-rich lands, some countries lease petroleum rights to petroleum companies. This takes place through a concessionary politic in which states grant rights of use to parts of their territory to allow capitalists to extract petroleum (i.e., territorial concessions). In return for the right to extract petroleum, landlord states receive rent: payments from petroleum-extracting firms in the form of royalties and taxes. Rent is shaped by the terms of surplus-value distribution agreed on by the extractive firm and the landlord state, both of which will attempt to extract the most value from the arrangement.

Modern petroleum production is marked by these dynamics of competition. In the 1850s, petroleum production took off in the United States as a powerful group of local companies acquired property rights to U.S. oilfields, developed technological innovations, and effectively reduced competition. Following the Civil War, these companies became giant corporations that dominated national and transnational petroleum production and captured the rapidly growing national petroleum market. The largest of these, Standard Oil (now ExxonMobil), was eventually divided into 38 separate companies to break down petroleum monopolies. These companies eventually folded (as a result of costs of production, inability to compete, or lack of access to production areas) and merged into what is referred to as the "Seven Sisters," an oligopolistic industry that dominated the market and created spheres of influence that allowed them to produce petroleum at low costs.

By the 1960s, petroleum reserves in the United States were being depleted or were inadequate to meet growing petroleum demand. U.S. firms sought petroleum in other regions of the world, particularly in the Middle East. Confrontation between petroleum-producing states and petroleum companies developed soon after, as petroleum states saw an uneven distribution of profits that favored petroleum companies. A cartel of major petroleum-exporting countries formed in 1960, the Organization of Petroleum-Exporting Countries, to obtain more favorable terms of trade. In 1973, as a war between Israel and its Arab neighbors developed, the major Arab petroleum-producing countries cut back on their petroleum production and stopped selling to countries deemed pro-Israel, including the United States. Through this and other initiatives, the Organization of Petroleum-Exporting Countries negotiated a larger share of petroleum revenues. Some countries moved to acquire partial or complete ownership of their petroleum industries in an effort to retain some control, which led to an era of partnerships between the Organization of Petroleum-Exporting Countries and major petroleum companies that lasted until the end of the 1970s. More recently, landlord–capitalist petro-partnerships have inspired numerous criticisms, as they are seen as corrupt, promoting uneven development and perpetuating a geopolitical world order that engenders violence, dispossession, and militarized conflict.

Effects of Petro-Capitalism

State-company partnerships offer insights into the uneven effects of petro-capitalism. The economies of the major petroleum-producing countries today are largely dependent on petroleum and have little diversification. These countries are referred to as "petro-states"—countries in which local economies and national policies are dominated by the influx of "petrodollars" (money earned by a country through the sale of petroleum). Because petroleum extraction is capital (rather than labor) intensive, and because export to consumer countries requires little linkage into the rest of the local economy, production tends to create enclave industries with strong ties to foreign companies. Nationalizations of the petroleum industry in the 1970s were an attempt to challenge these extractive enclaves. Yet, as petroleum rents are derived from outside the country and channeled through an externally oriented and financed enclave sector, profit distribution has been difficult to reshape. Studies on petroleum income typically refer to these sorts of socioeconomic effects as "the resource curse": despite the production of significant surplus value, petro-capitalism—through its concessionary political economy and unequal income distribution—has generated significant poverty among residents of petroleum-extracting regions, as well as domestic instability, and has locked national economies into hydrocarbon dependency.

Another form of uneven development associated with petro-capitalism stems from its concessionary political economy. Petro-capitalism operates through an "oil complex" (a configuration of transnational firm, state, and communities where petroleum is extracted) that is territorially constituted through petroleum concessions. Critics of petro-capitalism define territorial concessions as a form of wealth dispossession (or primitive accumulation) that operates through the transfer of property rights to petroleum companies, even if the lands in question are already being used by local residents.

The logic of petroleum production, along with the circulation of petro-capital and primitive accumulation have challenged customary forms of community authority, systems of ethnic identity, and the functioning of local state institutions in petroleum-producing areas. In many cases, the presence and activities of petroleum companies, along with states that protect their right to invest, have engendered land disputes, popular mobilization, and agitation among locally affected communities seeking to gain access to company rents and compensation revenues to reduce inequality.

See Also: Capitalism; Commodification; Resource Curse.

Further Readings

Boal, Iain A. [Retort]. *Afflicted Powers: Capital and Spectacle in a New Age of War*. London: Verso, 2005.

Coronil, Fernando. *The Magical State: Nature, Money, and Modernity in Venezuela*. Chicago: University of Chicago Press, 1997.

Labban, Mazen. *Space, Oil and Capital*. London: Routledge, 2008.

Shelley, Toby. *Oil: Politics, Poverty and the Planet*. Dhaka: University Press, 2005.

Yergin, Daniel. *The Prize: The Epic Quest for Oil, Money, and Power*. New York: Simon & Schuster, 1992.

Gabriela Valdivia
University of North Carolina at Chapel Hill

Pinchot, Gifford

Gifford Pinchot (1856–1946) shaped the 20th-century conservation movement by creating the profession of forestry in the United States, advancing the fledgling discipline of "scientific forestry," developing the U.S. Forest Service into a potent federal agency, defining conservation itself in utilitarian terms, and by dint of his tireless advocacy, making conservation central to the agenda of progressive politics from the presidency of Theodore Roosevelt through the New Deal of Franklin Roosevelt.

Historians have identified Pinchot with "utilitarian conservation," an approach emphasizing government regulation, rational planning, and scientific methods applied by "experts" to ensure the efficient and sustainable use of natural resources. Pinchot's views are typically contrasted with the "preservationist conservation" of John Muir (1838–1914), which deified nature and sought to withdraw wilderness areas of unique beauty and recreational worth from commercial uses, specifically logging, grazing, and hydro development. Both men lived in an America haunted by resource exhaustion, as land, water, and forests disappeared into the vortex of industrialization. It was an era in which George Perkins Marsh, Charles Sargent, and John Wesley Powell predicted that without proper management, America's abundantly wooded and watered lands would become desolate barrens, the remnants monopolized by the privileged few, deepening poverty and intensifying class conflict.

Pinchot was born into the ranks of these privileged few, the son of James and Mary Pinchot, wealthy New Yorkers, their fortune founded by grandfather Cyrille Pinchot, a lumberman who stripped many of Pennsylvania's forests. Turning away from the cut-and-run lumbering of Cyrille's generation, James and Mary read, absorbed, and pressed into their son's hands George Perkins Marsh's *Man and Nature* (1864). Marsh argued that deforestation had devastated great civilizations, threatening even that of the United States. Early in his life, Pinchot fell in love with wild nature, abhorring the waste, disorder, and inefficiency of contemporary lumbering practices. Accordingly, Pinchot gladly assented to his parents' ambitions that he become a forester.

Graduating from Yale in 1889, Pinchot found schools of professional forestry solely in Europe, where professionally trained foresters had transformed ecologically complex and diverse old growth forests into orderly and profitable forest monocultures. In a whirlwind tour of European forestry, Pinchot enrolled briefly in France's L'École Nationale Forestière and worked with the Forstmeister of Zurich's municipal forests but learned the conceptual framework of scientific, "fiscal" forestry from Sir Dietrich Brandis, British India's former chief forester. Brandis' influence is reflected in Pinchot's belief that forestry is tree farming, with the goal of a sustained, profitable yield of the forest crop. Pinchot's utilitarianism was a calculated choice: he hoped to win public support for conservation by making it meet market tests, by making it pay, but with the common good in mind. Accordingly, he defined conservation as the state-regulated, sustainable use of natural resources for "the greatest good for the greatest number, for the longest run." For Pinchot, conservation was the paramount democratic value because wise resource management would ensure equality of opportunity, defined not as equal access to political power but as access to material comfort and prosperity by the many, not just the wealthy few.

Returning from Europe as a very novice forester, Pinchot parlayed his public lectures, writings, and his brief, unprofitable management of George W. Vanderbilt's North Carolina forests into an appointment in 1896 to the National Forest Commission, charged with making recommendations for managing the forest reserves created under the 1891 Forest Reserves Act. The 1891 Forest Reserves Act and the 1897 Forest Management Act revolutionized U.S. land policy by authorizing the president to withdraw public lands from settlement and sale, reserving them as forest reserves, thereby reversing a century-old policy of disposing of lands to raise revenue and settle the frontier. In 1897, Pinchot was appointed chief of the Department of Agriculture's Bureau of Forestry, a tiny, underfunded agency with few employees, and because the forest reserves were under the purview of the Department of Interior, no forests to manage. The assassination of President William McKinley in 1901 reversed Pinchot's fortunes by propelling into the presidency Theodore Roosevelt, an ardent conservationist. In 1905, Roosevelt transferred the forest reserves from the Department of Interior to Pinchot's fledgling U.S. Forest Service in the Department of Agriculture.

Pinchot and Roosevelt enjoyed a symbiotic relationship: Roosevelt was Pinchot's patron in building a national forestry system, and Pinchot shaped Roosevelt's conservation policies and served as a lightning rod, allowing Roosevelt to take credit for popular conservation policies, deflecting criticism away from Roosevelt for unpopular ones. In 1907, western congressmen, alarmed by Forest Service control of western lands, attached to a popular agriculture bill a rider requiring congressional approval before creating additional national forests. Roosevelt, before signing the bill, designated 16 million acres of new national forests, unleashing western rage on Pinchot, but not so much on Roosevelt. When Roosevelt dodged, deciding to flood Hetch Hetchy Valley in Yosemite National Park to ensure a water supply for San Francisco, Pinchot supported it against John Muir's determined opposition. Pinchot was less successful with Roosevelt's successor, William Howard Taft, who fired Pinchot in 1910 over Pinchot's brawl with Interior Secretary Richard Ballinger over Alaskan coal leases.

Pinchot's tenure in the Forest Service was brief, but stellar, and his influence on the agency was long lasting because he was able to build a national conservation organization whose members were bound to Pinchot by friendship, personal loyalty, and Pinchot money. Favoring impartial civil service examinations over traditional patronage practices, Pinchot recruited into the Forest Service a cadre of civic-minded foresters, inspiring in them an esprit de corps, commitment to public service, and deference to Pinchot's direction. Many of these young foresters were Pinchot's associates from Yale; all were professionally trained, sometimes at Pinchot's personal expense. Pinchot's success in managing the national forests was also a result of his system of leasing and user fees, whereby the forests could be exploited by private interests for a market price, thereby creating a bureaucratic resource management regime—one in which "public goods"—trees, water, grass—could be sold to the highest bidder and extracted under Forest Service regulation and supervision. Ironically, the national forests rarely have been as profitable as Pinchot predicted. Nevertheless, this system allowed Pinchot and his foresters to cooperate with special interests—cattlemen, lumbermen, and railroads—thereby winning over key congressmen, and, when necessary, decentralizing decision making to the district level, flexing Forest Service policies to accommodate local needs and sensitivities, avoiding conflicts that the Forest Service could not win, but always retaining Washington's ultimate authority on policy questions.

After leaving the Forest Service, Pinchot tempered his utilitarian view of forests as "crops" and forestry as "tree farming," developing, during his successful governorship of Pennsylvania, a more inclusive conservation ethos, one recognizing the ecological and recreational values of forests and allied with the cause of social and economic justice.

See Also: Conservation Movement; Forest Service; Utilitarianism.

Further Readings

Balogh, Brian. "Scientific Forestry and the Roots of the Modern American State: Gifford Pinchot's Path to Progressive Reform." *Environmental History*, 7/2 (2002).
Hays, Samuel P. *Conservation and the Gospel of Efficiency: The Progressive Conservation Movement, 1890–1920.* Cambridge, MA: Harvard University Press, 1959.
Miller, Char. *Gifford Pinchot and the Making of Modern Environmentalism.* Washington, D.C.: Island, 2001.
Pinchot, Gifford. *Breaking New Ground.* New York: Harcourt, Brace, and Company, 1947.

James Bruggeman
Independent Scholar

Policy Process

Environmental policy is the outcome of the activities of various political actors known as the policy process. These actors include citizens, interest groups, and elected officials, among others, that differ in their interests, beliefs, and access to resources. Policy process activities occur over both short and long periods of time, creating both incremental and nonincremental outcomes. In addition, they may span geographical and political territories. These factors make for a complex policy process that has thus been viewed and described through models that are based on such factors as sequence, beliefs, time, crisis, and geography, among others.

One way of conceptualizing the policy process is through the life cycle of public policy. The stages in the policy life cycle include issue identification, agenda setting, policy formulation, legitimation, implementation, and evaluation. In the initial stage of issue identification, the general public, news media, scientists, and interest groups express environmental concerns that require government action. During the agenda-setting stage, environmental problems are defined and policy solutions are proposed for addressing the problem as defined.

The policy alternatives that are successful in being placed on the agenda are dependent on the manner in which the problem is defined and the solutions that citizens and groups organize themselves in support of. During policy formulation, policy alternatives are introduced to the legislative process by way of legislative committees, bureaucratic agencies, and nongovernmental professionals that specialize in developing policy proposals for recommended actions. Successful policy proposals are then enacted, or legitimized, usually through legislative procedure. Implementation then occurs when funds are established through budgetary allocation and bureaucratic agencies manage those funds in the execution of the aims of the policy. In the final stage of evaluation, the implementation of an environmental policy is systematically assessed to determine a policy's effectiveness in addressing an environmental issue and the societal outcomes that can be directly and indirectly linked to a policy.

The advocacy coalition approach examines policy formulation and implementation by focusing on the policy subsystem, or the governmental and nongovernmental actors most involved in a policy issue area. The constraints and resources that are external to the subsystem bind subsystem actors. These include static elements such as rules and sociocultural values, and dynamic elements such as changes in public opinion and socioeconomic conditions. The network of actors in a policy subsystem can be grouped into advocacy coalitions, or groups of actors that share common belief systems. These belief systems are characterized by fundamental normative claims of political philosophy from which policy positions, strategies, and decisions emerge. Advocacy coalitions thus compete to control the subsystem through the formulation and implementation of policies that are aligned with their belief systems.

At the implementation stage, decisions are made regarding institutional rules, resource allocations, and appointment. Policy outcomes stimulate learning over time as coalitions revise their policy positions and change their strategies. The fundamental normative claims remain intact, however, and thus the static nature of belief systems ensures that the learning process creates only small changes in public policy over periods of a decade or longer. Larger changes are usually the result of factors external to the subsystem that alter coalition dynamics by changing the nature of constraints and resources.

Environmental problems, the policy solutions that address the problems, and the politics that surround the issue area may be thought of as three separate and distinct streams that characterize the policy process. These streams of problems, policies, and politics may remain separate and distinct components of an issue over time until a window of opportunity—an event that brings significant change to the conditions surrounding the issue—brings the three streams together to create a policy change.

Similarly, over long periods of time policies may be viewed as incremental improvements on previously accepted policies. This may be viewed as a function of competitive stability among competing groups. However, this equilibrium may dissolve as a result of instability in group power dynamics, as well as changes in public opinion, creating nonincremental policy changes. Thus, the policy process may be viewed through a lens of static equilibrium followed by punctuated change.

Policy innovations are policies that are new to the governments that adopt them. These innovations are not necessarily policy inventions, as other governments may have previously adopted them. Very often the adoption of a policy innovation is the result of a decision that is based on the actions of other governments. This movement of policy innovations from one governmental body to another is known as policy innovation diffusion. The process describes the spread of a particular policy innovation across a geographic area. Various factors drive the process of innovation diffusion.

Some are internal to the adopting government, such as the social, political, and economic conditions within its boundaries that favor a policy innovation. Others factors are external to the government, stemming from sources beyond government boundaries. For example, leadership on the part of other governments in a policy issue area is an external influence. Similarly, states that are close in geographic proximity may create competitive advantages or opportunities for learning through public policies, stimulating horizontal diffusion across borders. Subnational units such as state and local governments often participate in vertical diffusion by adopting the policies of the national government. Networks of communication among elected officials may also serve to influence policy diffusion through interactions by which policy learning occurs.

These approaches are only a few examples of the many ways in which the policy process may be conceptualized. Additional approaches focus on such factors as how political actors are guided by political institutions, or the rules that frame political interactions and create both incentives and constraints for political actors. These examples, however, reflect the variety of approaches and complexity of these processes.

See Also: Cost-Benefit Analysis; Institutions; Political Ideology; Regulatory Approaches.

Further Readings

Anderson, James E. *Public Policymaking: An Introduction.* Belmont, CA: Wadsworth Publishing, 2005.

Birkland, Thomas. *An Introduction to the Policy Process: Theories, Concepts, and Models of Public Policy Making.* New York: M. E. Sharpe, 2005.

Dye , Thomas. *Understanding Public Policy.* New Jersey: Prentice Hall, 2002.

Kingdon, John W. *Agendas, Alternatives, and Public Policies.* Upper Saddle River, NJ: Longman, 2002.

Sabatier, Paul. *Theories of the Policy Process: Theoretical Lenses on Public Policy.* Boulder, CO: Westview, 1999.

Stone, Deborah A. *Policy Paradox: The Art of Political Decision Making*, revised ed. New York: Norton, 2001.

Thomas D. Eatmon, Jr.
Allegheny College

Political Ecology

Political ecology is an empirically based approach, perspective, or research agenda engaged by a broad spectrum of academics to study environmental change and its relationship with humans. Political ecology is also a critique against dominant explanations of human–environment issues that seeks to demonstrate how environmental conditions, relations, and conflicts are contingent outcomes of power, as well as instrumental to political and ecological change. Political ecologists argue that the nature, causes, and effects of environmental change (e.g., climate change, food scarcity, land degradation, deforestation, pollution, water shortages, biopiracy, and biodiversity loss) are simultaneously political and ecological, social and biophysical and, thus, need to be examined in relation to the political-economic and ecological processes that frame them.

The way in which the political economy–ecology relationship is studied varies according to disciplinary background and research interests. Journal articles, edited volumes, and books containing the term *political ecology* in their titles have proliferated in recent years across academic disciplines, suggesting significant differences in the way the term is used. The analogy "a tree with deep roots and many branches" is an adequate description for political ecology, as this approach builds on a diversity of disciplinary contributions and methodologies.

Although interest in the relationship between people and their environments has a long and rich history in anthropology and geography, the term *political ecology* officially came into use in the mid-20th century. Among early political ecologists, many drew on the intellectual contributions of cultural ecology to examine human–environment relations. Drawing on the ecosystem concept—all parts of an ecosystem (organic, inorganic, biome, habitat, human, and nonhuman) are interdependent and interact through control mechanisms, energy pathways, and feedback loops to maintain and regulate the system—cultural ecologists have sought to better understand human adaptation to environmental change. This school of thought provided a sophisticated body of theory and research to demonstrate how subsistence people in isolated regions maintained adaptive structures and cultural mechanisms (embodied in various ritual, symbolic, and religious practices) with respect to their environment.

Although this functionalist approach has enjoyed significant following, it also has been criticized for treating human populations as isolated from global economic and political processes and for drawing on a Malthusian conceptualization of environmental problems; a human population that exceeds the carrying capacity of the environmental system that supports it faces starvation, mortality, and degraded resources as a consequence.

The concept of a self-regulating human ecosystem as a unit of analysis also has been challenged, as it is difficult to reconcile with the reality of labor migrations, commodity production, market participation, and the introduction of new technologies that can fundamentally alter human–environment relations.

Critiques against cultural ecology propelled researchers to locate other ways of explaining environmental change and its relationship with human behavior. This shift took place during the 1970s and 1980s, when interest developed across various disciplines in exploring human and cultural responses to hazards and disasters. The new tendencies criticized the study of isolated/subsistence communities in equilibrium with their physical environment and sought to examine the effect of markets, social inequalities, and political conflicts and to analyze forms of social and cultural disintegration associated with the incorporation of local communities into a modern world system. Some scholars drew on organic analogies of adaptation and response. Although systems thinking was central to this scholarship, scholars were also sensitive to the role of cultural perceptions and questions of organizational capacity and access and how availability of information mediates adaptation to environmental change.

Other scholars sought to reinvigorate the study of human–environment relations through the analytical contributions of political economy, bringing forward questions on the role of social differentiation, exploitation, and agrarian transformation and the effects of international markets on the rural poor in the developing world. Proponents of a more political economy–infused approach drew on Marxism in the social sciences to advance concepts of control and access to resources, marginalization, relations of production, surplus appropriation, and power. These scholars criticized the concept of adaptation and functionalist systems and advocated the need to engage with social theory to locate and explain local responses to environmental change historically, socially, and politically. Their critiques focused on the treatment of "human" and "environment" as discrete objects subject to observation and the simplification of the social character of human–environment relations.

Notwithstanding the development of these critiques, the contributions of systems thinking and Malthusian approaches (e.g., notions of carrying capacity, overpopulation, and environmental degradation) continue to significantly influence the characterization of human–environment relations and their management today. This is evident in projects of environmental conservation in the Third World, for example, through the creation of conservation territories. This approach begins with the assumption that environmental degradation results from the mismanagement of resources by local populations and concludes that the solution is to introduce rational planning by "specialists" (e.g., agronomists, scientists) or exclude humans from access to resources to protect the latter.

Just as the vision of a "static" and "rational" human actor in cultural ecology was criticized, notions of stable and bounded environments have also been revised. Some political ecologists, drawing on biogeographical and "new ecology" approaches, argue that environmental change is remarkably more complex than previously thought and that there is a need for reformulating the current state of knowledge about nature and its relation to human societies. These researchers draw on theoretical developments within ecology to recognize the role of nonequilibrium conditions and nonlinear change processes for better human–environment management, as well as the need to pay attention to how the "nature" of environmental change—the frequency, rate, scale, and degree of change—matters to the analysis of human–environment relations.

The premise is that not all environmental change is the same and that this recognition must be the foundation for environmental management debates (e.g., environmental conservation and sustainable development). Some scholars have focused on critically examining theoretical perspectives and empirical findings from ecological science; specifically, the political forces behind different accounts of "ecology" as a representation of biophysical reality.

Poststructuralist approaches also have introduced an emphasis on the complexity of human society to argue that cultural politics, positionality, and symbolic meaning play a substantial role in human–environment relations. A number of political ecologists are interested in exploring how "the environment" and "environmental problems" are discursively constructed, with an emphasis on a critical perspective toward modernist notions of objectivity and rationality and the relationship between power and scientific knowledge, and the recognition of the existence of multiple, culturally constructed ideas of the environment and environmental problems. Others have focused on identifying and studying relations of power among actors, with specific attention to the role of gender, class, and/or race in shaping inclusion and/or exclusion in environmental management. Given these developments in political ecology, it has become widely recognized that material analyses in political ecology cannot be conducted in the absence of or separately from assessments on the role of power relations.

Political ecologists employ a multitude of methodologies and have often shown that the combination of diverse methodologies can offer important insights into the conceptualization of human–environment relations. From "chains of explanation" (tracing social relations of production from the local to the national and global scales to explain environmental change) to "progressive contextualization" (starting with the actions/interactions of individuals and showing their place within complexes of causes and effects) to network approaches that trace the connections and the quality of relationships that give shape to existing socionatural situations, political ecologists draw on a variety of approaches to understand what is happening here ("the local") in relation to everywhere else.

Despite the diversity of methodologies employed, multiscalar analysis is considered a hallmark of political ecology. In general, political ecologists approach human–environment questions through:

- Political–economic analyses that examine the roles of and interactions between the state and the market on environmental outcomes. These approaches are grounded on the notion that the relationship of nature and society is dialectical; nature transformation cannot be understood without consideration of the political and economic structures and institutions within which the transformations are embedded.
- Historical analyses that focus on interpreting society–nature relations in the past, how and why those relations have changed (or not) over time and space, and the significance of those interpretations today. Historical analyses share an affinity with environmental history and historical geography but explicitly attempt to view nature in light of social issues and the political forces constraining both. Historical ecology, land-use/land-cover (landscape) change, colonial legacies, and geohistorical revisionism fall within the broad spectrum of historical analyses.

- Ethnographic analyses that highlight the diversity of perspectives on human–environment relations and resource conflicts among various actors operating at multiple scales. Ethnography pays particular attention to the symbolic meaning ascribed to resources and how these are part and parcel of struggles over control and access to material resources. Ethnography interrogates the role of ideology and the importance of socially constructed identities (class, race, gender, and ethnicity) in shaping the perceptions of actors that share a common social position, for example, farmers, women, scientists, policymakers, or indigenous peoples. Ethnography provides a critical medium for exploring the dynamics of cultural politics that animate environmental conflicts and how symbolic struggles over the meaning, definition, and categorization of rights, responsibilities, and benefits are at the core of material struggles over the environment.
- Discourse analyses that explore and aim to make visible the ways in which "the environment" and "environmental problems" are discursively constructed. These analyses emphasize a critical perspective toward modernist notions of objectivity and rationality. They interrogate the relationship between power and scientific knowledge and recognize the existence of multiple, culturally constructed ideas of the environment and environmental problems.
- Ecological field studies that draw on vegetation transects, plot sampling, soil tests, analysis of remotely sensed imagery, or use of geographic information systems to identify the direction and cause of ecological change. Ecological methodologies have a strong temporal component, are sensitive to the role of biophysical processes in shaping human–environment relations, and incorporate ecological findings as a source of empirical evidence.

Given the diversity of theoretical approaches and methodologies employed by political ecologists, it is not surprising to find numerous internal debates about proper methodologies, research agendas, and analytical foci as political ecology matures. For example, although some scholars have argued for more focused attention to politics, others argue that biophysical factors—despite existing independently of society—appear to have been wrongly relegated to the position of discursive construction, and thus to have a secondary role to politics. Yet others argue that political ecology should move beyond the Third World and should also become a lens through which to analyze human–environment relations in the First World. For others, political ecologists are dangerously overemphasizing constructionism and tending toward relativism. These internal debates highlight the need to address the ecological and the political dimensions of environmental issues in a more balanced and integrated manner.

Political ecologists have taken to heart the normative commitment to offer better explanations for environmental change and its relationship with humans. Broad themes engaged by political ecologists include:

- A critique of knowledge about environment (or how we know environment). One approach followed by political ecologists has been to question "environmental orthodoxies." Analyses focus on (1) how politics and power render some definitions of environment and environmental change more significant than others; (2) the role of situated knowledges in shaping resource governance politics; (3) nonequilibrium ecology critiques of environmental conservation practices, specifically, questioning notions of equilibrium, stability, and balance in "nature" and how understanding the environment under these terms can lead to further social inequality; and (4) the temporal and spatial dimensions of the scale of observation (how we measure and regulate environment) and its relationship to the scale of phenomena.
- A renewed focus on nature's agency (or how the environment affects humans).
- Many political ecologists draw on a materialist philosophy and acknowledgment that material reality (e.g., the physical environment) exists independent of human action, but that

our knowledge about it is always situated and partial. Although some analyses focus on nature's agency as socially mediated, others examine how nature's materiality—for example, biophysical characteristics, biology and behavior, and ecological associations—mediate human relations, such as possibilities of governance, management, and contestation.

- The evaluation of environmental management practices (or how humans act on the environment). A large number of political ecologists are interested in examining the conflicts that arise in relation to environmental conservation and management practices. Their work has centered on: (1) examining how social imaginaries, representations, and perceptions shape the regulation of socioecological dynamics; (2) questioning the theoretical underpinnings and effects of "technological fixes" (e.g., how environmental change may be perceived as a problem in need of a solution), a critique leveled particularly at conservation science; (3) deconstructing assumptions about behaviors and motivations toward environment by local populations (e.g., a critique of blame for a "wrong" focused on "the local" has often moved to a blame on capitalism as a destructive process); and (4) exploring questions of access to natural resources and governance of property (e.g., who is entitled to what claims and rights).
- Ethics and social justice. Political ecologists tend to follow humanist principles, exploring social and environmental changes with a normative understanding that there are better, less coercive, less exploitative, and more sustainable ways of doing things. Although any notion of "social justice" is historically contingent and culturally specific, among political ecologists it includes a respect for cultural differences, customary rights and ways of knowing the world, as well as an equitable mode of resource distribution, economic opportunity, and political representation. Critiques of development, capitalism, and conservation science, for example, aim to bring human–environment analyses closer to an explicit engagement with ethics. These works, rooted in political ecology's tradition of a dialectical treatment of human–environment relations, emphasize (1) questions of environmental justice (e.g., the uneven distribution of environmental "goods" and "bads"); (2) the importance of looking at the legitimacy of moral and ethical concerns for determining the outcomes of environmental conflicts; and (3) remaking our moral relationship with nonhumans.

See Also: Domination of Nature; Ecology; Environmental Justice; Malthusianism; North–South Issues; People, Parks, Poverty; Power.

Further Readings

Biersack, Aletta and James B. Greenberg. *Reimagining Political Ecology (New Ecologies for the Twenty-First Century)*. Durham, NC: Duke University Press, 2006.

Blaikie, Piers M. and H. C. Brookfield. *Land Degradation and Society.* London: Methuen, 1987.

Carney, Judith Ann. *Black Rice: The African Origins of Rice Cultivation in the Americas.* Cambridge, MA: Harvard University Press, 2001.

Forsyth, Tim. *Critical Political Ecology: The Politics of Environmental Science.* London: Routledge, 2003.

Neumann, Roderick P. *Making Political Ecology.* London: Hodder Arnold, 2005.

Nietschmann, Bernard. *Between Land and Water: The Subsistence Ecology of the Miskito Indians, Eastern Nicaragua.* New York: Seminar, 1973.

Paulson, Susan and Lisa L. Gezon. *Political Ecology Across Spaces, Scales, and Social Groups.* New Brunswick, NJ: Rutgers University Press, 2005.

Peet, Richard and Michael Watts. *Liberation Ecologies: Environment, Development, Social Movements.* London: Routledge, 2004.

Robbins, Paul. *Political Ecology: A Critical Introduction.* Malden, MA: Blackwell, 2004.
Watts, Michael. *Silent Violence: Food, Famine, & Peasantry in Northern Nigeria.* Berkeley: University of California Press, 1983.
Zimmerer, Karl S. and Thomas J. Bassett. *Political Ecology: An Integrative Approach to Geography and Environment-Development Studies.* New York: Guilford Press, 2003.

Gabriela Valdivia
University of North Carolina at Chapel Hill

Political Ideology

Debates in ecology revolve around substantive matters, such as the causes or reality of climate change and how, or if, to manage urban growth and other forms of development. But it is important to recognize that these debates are not purely substantive but, rather, nested in mainstream and nascent political ideologies, which act as a lens through which to interpret facts, frame arguments, and position policy interventions.

Political ideologies are collections of beliefs, ideals, principles, and conceptions concerning relations between the individual and society. They constitute the basis of belief for political parties, as well as the intellectual foundation for multiple discourses on society, government, and the limits of state power, in addition to broader values such as freedom, responsibility, ethics, justice, and liberty.

Given their universal reach and potency, political ideologies are highly influential in all contemporary policy debates, including those revolving around ecological issues. Everything from recycling campaigns to "ecotaxes" to funding for public transit can face barriers born of politically motivated ideological beliefs, and addressing such barriers requires an awareness of, a sensitivity to, and dialogue with stakeholders holding diverse belief systems.

Many mainstream political ideologies relate to theories of distributive justice, those beliefs oriented to determining the just distribution of a range of goods in a society. Broadly considered, they seek to understand the extent to which the state bears responsibility for this distribution and what ethics and principles should guide it.

These principles help define, for example, the appropriateness of imposing regulations, taxation, and incentives designed to change social outcomes, many of which can be seen to infringe on what are considered to be personal choices. In the case of policy debates over green issues, this can include the desirability of encouraging different residential and employment locations and modes of transport.

Theories of distributive justice are framed in terms of determining who or what receives benefits or, in contrast, bears the burdens imposed by public policy decisions. In other words, they ask, what is the scope of the benefits or costs in terms of constituencies affected? What pattern or set of criteria guides this distribution (e.g., efficiency, equality, priority, or sufficiency)? What is the "currency" involved (e.g., remuneration, benefits, land, etc.)?

One of the best known of these theories is utilitarianism, which considers those outcomes to be most desirable when they maximize good consequences for the greatest number of people. Theoretically, the scope of utilitarianism encompasses everyone; the pattern it adopts is based on considerations of equality and egalitarianism; and the "currency" it is most concerned with is pleasure, happiness, and the satisfaction of preferences.

There is much to admire in utilitarianism: it is fundamentally humanist and progressive, in that it does not require appeals to either religious morality or the weight of tradition. In circumstances where an oppressed majority suffers under the domination of an elite minority, it can also guide calls for liberation from oppression.

However, critics point out that utilitarian principles are less appropriate when the majority of people in society are already benefiting from present arrangements, when the greatest level of pleasure and satisfaction of preferences are being met but at the expense of a minority of people who can take little or no advantage.

In the case of debates in green politics, we can see the grave limitations to utilitarian thinking as ecosystems become ever more degraded as preferences and pleasures are maximized.

For example, a utilitarian view on suburban sprawl and cars would hold that as most people seem to like suburban living and widespread automobility, then these should be supported by land use regulations and public policy. Denying people the ability to make these choices—through such policy directions as Smart Growth—would result in a large amount of unhappiness and should therefore be discouraged.

In contrast, deontological approaches—derived from Immanuel Kant's ethics—argue for the rightness of actions being determined solely from the principles of the actor. Morally correct actions are not those that are correct because of the particulars of their consequences but, rather, they derive from what should be universal ideals that must hold true for everyone.

In ecological ethics, deontological—or categorical—imperatives are those that we must pursue without consideration for how they might benefit some individuals. Nature has intrinsic value on its own, without reference to its utility for meeting human needs. It therefore becomes a categorical imperative to protect it.

In the case of suburban expansion, then, it does not matter to a deontologist that many people like suburban living; it must be slowed so that farmland and natural areas may be preserved.

However, the extent to which debates in green politics lean toward conceptions of utility or deontology depends not so much on the cases themselves but on dominant political ideologies such as liberalism, conservatism, Marxism, feminism, progressivism, libertarianism, neoliberalism, and communitarianism.

The term *liberalism* is the source of some confusion not only because of how the term has evolved over the centuries but also for its oftentimes pejorative connotations, particularly in the American context.

Classical liberalism, as articulated by such Enlightenment writers as John Locke and Adam Smith, advocated the expansion of individual rights, limitations on the power of government, and the flourishing of free markets based on robust property rights. The government in a liberal economic regime has no mandate to intervene and redistribute wealth despite the existence of inequality and poverty. In other words, classical liberalism is quite consistent with the beliefs commonly associated in the 21st century with conservatives.

"New" liberalism, in contrast, arising from U.S. President Franklin D. Roosevelt's New Deal, favors a more interventionist government to redress inequalities, to fund institutions seen as essential for the public good, and to regulate the free market. A public "safety net" consisting of welfare, pensions, and healthcare are all generally considered core elements of new liberalism. This conception is how liberalism is now most commonly understood, and it is most often contrasted with so-called conservative values as they relate to the economy, society, government, and faith.

It is important to recognize that "conservatism" has no substantive ideological program of its own: It simply refers to any movement to "conserve" or protect a tradition or principles thought to be of value and at risk of unwarranted change or erosion. Forces of conservatism may be found in any economic regime, any form of government, and any faith. This is why we can speak of social, fiscal, and religious conservatives.

This diversity is reflected in green debates. Although conservatives are often seen leading the fight against proenvironment regulations, there are some conservatives who take quite the opposite view. Recently coined "crunchy conservatives," they are just as skeptical of big business as they are of big government and want to discourage the aesthetic void of suburbia and preserve the beauty of natural ecosystems. Economic arguments for urban growth controls and public transportation—that they contribute to economic growth and result in efficiency gains—may be quite attractive to some conservatives.

More commonly, however, conservative politicians, writers, and activists have resisted government regulations on industry and are especially protective of property rights that may be seen as infringed on by environmental protection measures. Smart Growth policies and publicly funded public transportation systems are often favorite targets of some conservatives, as they are seen in the first case as restricting individual liberty to live where and how one chooses, and in the second as a poor use of tax dollars.

This tension between public and private goods lies at the heart of neoliberalism, which promotes to the extent possible open and free markets, private investment, and free trade. Neoliberalism emphasizes a dramatically reduced role for government and—significantly for environmental concerns—turning over to private interests many of the functions formerly undertaken—or owned—by governments. The result in the global South has included notorious examples of commodifying resources normally considered public goods such as water and natural services such as regional biodiversity.

Progressives, in contrast, are more inclined to be associated with calls for strengthening government and, in particular, environmental policies. Oriented as it is toward privileging no tradition over another, and viewing them as having no inherent value of their own, progressivism seeks any change to processes and systems that can more directly lead to improvements in social justice. Progressives are associated with the political left and causes such as those of organized labor and the environment—although it should be noted that these causes have hardly been uniformly congenial to one another. However, progressives have managed to create stronger labor–environmentalist coalitions in recent years, building a consensus that moving to a more ecologically sustainable economy will build a more equitable and socially just one as well.

This search for social justice lies at the heart of Marxist and feminist critiques. In contrast to the later conceptions of the "tragedy of the commons," in which self-interested individuals overexploit natural resources, Marx conceived of natural resources ideally as property to be held communally, and not to be owned or exploited by private interests. In classical Marxism, man is an active agent in transforming nature. However, as capitalism overexploits its own base of production, Marx's collaborator Friedrich Engels posited a "revenge of nature," in which the "use" value of natural resources becomes degraded. Marxist and neo-Marxist critics remind us that our ecological crisis cannot be seen without reference to capitalism, class disparities, and power relations.

Similarly, ecofeminist thought focuses on processes of domination: that the ideologies of power surrounding capitalism, imperialism, and masculinity have dominated and exploited both nature and women. The assumed right of "man" in Western culture to dominate nature is seen by ecofeminist writers as closely related to his assumed right to dominate women, as well as the right of "advanced" economies to dominate the global South. Like a doctor seeking to find the cause of an illness, ecofeminist writers believe that before we can remedy our ecological crisis we must first understand the underlying social processes that led us to the crisis in the first place.

For many environmental activists motivated by such critiques, governments are seen as key to achieving various progressive goals, through promulgating stronger environmental and labor regulations and reining in exploitative and environmentally destructive capitalist practices.

Such interference by the state is, however, anathema to libertarians, who feel that the scope of government should be limited only to protecting and enforcing through sufficient police and military functions the individual's ability to pursue life, liberty, and property. The only defensible role of government is to defend these three rights, and it may not tax the people to fund and build public goods such as roads, parks, and public transportation. Although government may legitimately create laws to protect people from other people, it may not institute laws to require someone to help someone else, or to protect the individual from their own bad choices or habits. Like conservatives, libertarians are adamant about the sanctity of property rights and favor the unfettered operation of the free market, not regulation, to determine land use decisions.

A specific libertarian criticism concerns so-called takings, under which the state may confiscate through eminent domain or otherwise infringe on one's private property to serve some public interest, and for which owners must be compensated. The conception of takings was considerably broadened by the 2004 passage of Oregon's Ballot Measure 37, which recognized as a taking any reduction in property value that could be tied to an environmental regulation.

However, pollution is regarded by libertarians as a form of theft and must be compensated for by those responsible, not by the taxpayer. Governments are also seen as greater polluters than corporations but are unjustly protected from redress by "sovereign immunity," which should be revoked. Unnecessary government subsidies are also identified by libertarians as a significant source of environmental degradation by encouraging wasteful and inefficient farming and resource extraction practices. Finally, libertarians share with antisprawl activists a rejection of many land use zoning ordinances that both dictate what may be done with one's property and at the same time promote the inefficient use of land.

An even more minimalist role for government is sought by anarchists. Contrary to the popular conception of anarchy as lawlessness and destruction, early proponents such as geographer Peter Kropotkin saw anarchism as a means for achieving self-organization and communalism in small-scale communities in which decision making is achieved through direct democracy. The writings of both Kropotkin and Murray Bookchin have been highly influential on the development of anarchistic and ecological thought. Bookchin articulated the notion of social ecology—the idea that environmental degradation cannot be viewed in isolation but, rather, is the consequence of inequitable and dysfunctional social relations.

The reduction of society to an appropriate scale is the goal of communitarianism, which sees as inadequate in liberal theories of justice the valuation of community. Shared practices and understandings are emphasized: Contrary to classical liberalism or libertarian conceptions, the individual is not seen as entirely free but is instead embedded in a community, the priorities of which supersede those of the individual. Infused with deep utopian aspirations, communitarian responses to the ecological crisis include eco-villages and other forms of intentional, self-reliant communities.

As may be seen, achieving environmental sustainability should not be viewed as a problem that can be resolved simply by instituting the right mix of technologies and financial policies. Instead, environmental values are viewed dramatically differently by competing—but not mutually exclusive—political ideologies. Major arguments for or against certain forms of development are nested in major streams of political thought, which frame both problem identification and potential solutions.

Engaging in environmental debates then requires negotiating competing visions of what constitutes a good society and working toward solutions that are seen to achieve a balance among the needs of individuals, groups, whole societies, and nature itself.

See Also: Ecofascism; Ecofeminism; Suburban Sprawl; Utilitarianism.

Further Readings

Abel, Steven W. "Presentation on Oregon Ballot Measure 37." http://www.stoel.com/showarticle.aspx?Show=1826 (Accessed July 2009).

Bookchin, Murray. *The Ecology of Freedom: The Emergence and Dissolution of Hierarchy.* Oakland, CA: AK Press, 2005.

Burkett, Paul. "Marx's Vision of Sustainable Human Development." *Monthly Review,* 57/5:34–62 (2005).

Dreher, Rod. *Crunchy Cons: The New Conservative Counterculture and Its Return to Roots.* New York: Three Rivers, 2006.

Eckersley, Robyn. *Environmentalism and Political Theory: Toward an Ecocentric Approach.* Albany: State University of New York, 1992.

Hackett, Steven C. *Environmental and Natural Resources Economics: Theory, Policy, and the Sustainable Society.* New York: M. E. Sharpe, 1997.

Hubbard, F. Patrick. "'Takings Reform' and the Process of State Legislative Change in the Context of a 'National Movement.'" *South Carolina Law Review,* 93:109–21 (1998).

Kalof, Linda and Terre Satterfield. *Earthscan Reader in Environmental Values.* London: Earthscan, 2005.

Kropotkin, Peter and George Woodcock. *Evolution and Environment (Collected Works of Peter Kropotkin).* Montreal, Quebec, Canada: Black Rose Books, 1996.

Kymlicka, Will. *Contemporary Political Philosophy: An Introduction,* 2nd ed. New York: Oxford University Press, 2002.

Michael Quinn Dudley
University of Winnipeg

Politics of Scale

The term *scale* refers to a spatial level of analysis, representation, reference, or articulation. Scale operates through the social, economic, and political processes of human geography as well as through the biological and ecological processes of physical geography. Though the conventional scalar classifications are often viewed as static, discrete, and hierarchical, many scholars within and outside the academy have noted the dynamic, interrelated, and constructed nature of standard scales of reference. Rather than being ontological givens, scales have come to be recognized as inherently political in nature. The term *politics of scale* gained purchase within the context of economic globalization and the resistances and strategies calling for socially just and ecologically sustainable alternatives. Recently, as neoliberal economic structures, states, supranational agencies, and numerous grassroots environmental movements around the world negotiate access to and protection of ecological resources, the politics of scale emerge as central to the most pressing environmental debates.

Despite its contested meanings and slippery modes of application, the political potency of scale is attested to by its ubiquity in popular spatial analysis. Within the present social, political, and economic landscapes, invocations of global and local, household and national pervade geographic articulations. For instance, critics of the dominant form of globalization have interrogated the global scale and its lack of accountability. Meanwhile, nongovernmental organizations strive to influence international policies so as to aid village and neighborhood communities. Agriculture, another key example, has undergone enormous transformations around the world over the past few generations, and scalar language has been used to promote, explain, justify, critique, and decry these shifts, as well as to imagine and advocate its alternatives. Though its operations remain abstract and subtle, scalar language has always and continues to configure and reconfigure power relations, and for this reason it demands continued analytical attention—particularly from those addressing environmental crises, policies, and opportunities.

Until recently, the term *scale* paralleled physical geography's notion of ratio of distance, attempting to map cartographic proportion onto social phenomena. This scalar configuration partook of government levels, stacking such familiar categories as "nation-state" atop "region" atop "local." Economic geography reworked the scalar framework, trumping "nation-state" and even United Nations "supranational" with the uber-large-scale "global economy." Power frames this scalar analysis, serving as the link between various levels and comprising a top-down chain of command.

Geographic scale, historically, has operated alongside its cartographic counterpart to measure the level or spatial magnitude of analysis, thus operating with regard to such trope-scales as region and nation-state. Accordingly, its existence was taken for granted as pregiven. Since the 1980s, however, geographical theorists have argued that scale is the ongoing product of construction, negotiation, and contestation. It did not precede its own conception, representation, and articulation, and as such, the question is not how scales affect social processes but how actors and activities make and employ scalar claims.

As a consequence, many argued that the process of scale-producing and the relationships between and among scales become their defining characteristics. Neil Smith introduced the metaphors of "scale-jumping" and "scale-bending," whereas Eric Swyngedouw proposed the conflation of "globalization" to describe the key interstitial realm between fossilized scalar categories. Andrew Herod describes the global as the "multi-local," rethinking each scale as an entry point into a scalar process. These and other theorists emphasized the constructed, as well as fluid and multidimensional attributes of scalar configuration, and worked to mitigate lingering ladder connotations and democratize scalar differentiations.

Meanwhile, the scalar articulation of "local" has recently become a strategic stage from which to launch a critique of or cultivate an alternative to neoliberal globalization, a means of advancing, for instance, alternative agricultural policies and manifestos of extrication. The local scale has been embraced by many around the world as a means of defying the disempowerment of the rural, the Third World, and the realm of social reproduction in general. Grassroots strategies toward environmental justice involve applying the political geography of scale to reenfranchise the previously devalued scales of the neighborhood, the household, and the body.

Accordingly, within the growing discourse on "globalization" and its mounting counterpart of "globalization from below," emerges the nebulous, yet vigorous rallying cry of local. Though the adjective remains politically complex and intellectually contested, nevertheless, local foods, local economy, local politics, and local knowledge all speak to a revisioning of dominant conceptions of economy, agriculture, development, and progress. "Local" epistemologies call for a reconstitution of geographic primacy and periphery—fundamental shifts in paradigmatic reorientation of reference and representation.

Nevertheless, the nested vision—underlying even emancipatory scalar language—implies a linearity of subjection that necessarily favors spatial scope, according power to size and thereby undervaluing the agency of such scales as the household. Recently, S. Marston et al. go even further to expunge the whole concept of scale, as the hierarchical languages of moving "upward" and "downward" between scales remains axiomatic, looking down on and thus disempowering lived social life; hence, "the need to expose and denaturalize scale's discursive power." This would overcome the binary of the transcendent, causal global and the effected, mundane local, which obscures the fact that even the chief executive officers live and work in various local spheres, as life itself exists on the local level—or more accurately, on many interrelated local levels.

Scalar imaginaries continue to exert, hide, exacerbate, and reenvision power through space. Scale figures as a seemingly innocuous, and thus effective, geographical tool of prioritization and enframing, endowing advantage, primacy, and legitimacy to some perspectives and not others. Actors experience, advocate, articulate, construct, and contest various positions through scalar language—which is then intrinsically laden with authority and agency. The politics of scale come to the fore as these scales of organization and governance become more fluid and consciously chosen, highlighting movements within and between scales, wherein, for instance, local communities contesting dam construction use the Internet to forge political links of solidarity with other locales around the world who are also fighting dam displacements.

Politics of scale imply a collective perspective: Scales do not exist in isolation, but rather are mutually constituting—the city defining itself as a scale of reference in relation to the neighborhoods within it, or to the nation-state beyond it. Scale refers to a common experience, though it retains its barriers of exclusivity as well as inclusion. Regional scale, for this reason, is emerging as a political tool of self-description and alignment—with Appalachia emerging as a scale of resistance against mountaintop removal, the Amazonian against deforestation, the Sahel against desertification, and centers of world origin and diversity against genetically engineered seed drift and contamination. Concurrently, universal scales of reference have emerged through environmental politics of scale—such as the global predicament of climate change and peak oil.

Questions of scale cannot be extricated from the politics that frame and reproduce them. The processes of scalar (re)configurations and (re)productions serve as the key sites of struggle and strategy that define environmental politics; this is where the ecological becomes political. Scales of reference themselves may be ultimately ephemeral, contested epistemological constructions, but the invocation and deployment of scale remains powerful, with political and ecological consequences.

See Also: Globalization.

Further Readings

Herod, A. and M. Wright, eds. *Geographies of Power: Placing Scale*. Malden, MA: Blackwell, 2002.

Marston, S., et al. "Human Geography Without Scale." *Transactions of the Institute of British Geographers*, New Series, 30/4:416–32 (2005).

Sheppard, E. and R. McMaster, eds. *Scale and Geographic Inquiry: Nature, Society and Method*. Malden, MA: Blackwell, 2004.

Garrett Graddy
University of Kentucky

Postmaterialism

The concept of postmaterialism was introduced in 1971 by Ronald Inglehart, who suggested that Western societies were experiencing a "Silent Revolution" in their values. The concept has been extremely influential in the study of public values and political behavior, including the study of environmental attitudes. The thesis suggests that there are two kinds of values characterizing advanced industrial societies: materialist values, which emphasize physical or economic security, and postmaterialist values, which emphasize self-affirmation, quality of life, democratic participation, social expression, and equality. Proponents of the thesis suggest that, as a result of economic development, advanced industrial societies are becoming more postmaterialist with each successive generation. The wide-ranging debate and criticism generated by the postmaterialism thesis has focused on both theoretical issues concerning the nature of values and methodological issues concerning measurement.

The Postmaterialism Thesis

The thesis is based on two hypotheses. First, the "scarcity hypothesis" suggests that a person's priorities reflect their socioeconomic environment. In general, greatest subjective value is placed on things in short supply. So in times of economic deprivation people focus on immediate needs such as food, money, and shelter. In times of prosperity, other, nonmaterial needs take precedence. This is similar to Abraham Maslow's concept of a hierarchy of needs. Second, the "socialization hypothesis" suggests that this relationship is not an immediate adjustment; rather, a person's basic values reflect the conditions prevailing during their formative, preadult years.

Increased security and improvements in the standard of living since the two world wars have led to decreased anxiety over basic needs. The postmaterialism thesis suggests that this new environment changes how people evaluate their own well-being. As perceived security increases, the materialist emphasis on economic and physical security decreases, and people increasingly emphasize postmaterialist priorities, such as freedom, self-expression, and quality of life. Proponents of the thesis suggest that advanced industrial societies have been undergoing a gradual transition from a materialistic to a postmaterialistic value orientation since World War II.

Measuring Postmaterialism

Since the thesis was first proposed, Inglehart and others have been developing ways to measure support for postmaterialist values among individuals and populations. The thesis was first tested in surveys conducted in 1970 in Great Britain, France, West Germany, Italy, the Netherlands, and Belgium. Investigators presented participants with four social goals, of which two reflected materialist priorities (limiting inflation and maintaining order) and two reflected postmaterialist priorities (increasing the public role in decision making and protecting freedom of speech). Participants were asked to select the two goals they considered most important. Participants selecting the two materialist goals were classified as materialist, those selecting the two postmaterialist goals were classified as postmaterialist, and those choosing any other combination were classified as mixed. For some later studies, this scale was expanded, and participants were asked to select 6 of 12 possible goals. Indexes of this kind have been included in several large-scale surveys, including the World Values Survey and Eurobarometer. Respondents' answers are sometimes aggregated by country to make cross-national comparisons.

Using these indexes, Inglehart has accumulated a body of findings, based on surveys of more than 60 countries, that suggests that Western societies have become increasingly postmaterialist since World War II. Similar trends have also been identified in some middle- and low-income countries. The postmaterialism thesis suggests that this is a result of new generations replacing old ones. Some studies have found that young people tend to be more postmaterialist than older people, as the thesis would suggest. However, other studies have challenged this relationship. The 2000 World Values Survey shows the highest proportion of postmaterialists in Australia (35 percent), followed by Austria (30 percent), Canada (29 percent), Italy (28 percent), Argentina (25 percent), and the United States (25 percent). The lowest proportions of postmaterialists are found in Pakistan, the Russian Federation, Tanzania, India, and Hungary, which all have 2 percent or less. (These data are drawn from a 2004 sourcebook by Ronald Inglehart and others.)

There have been extensive debates within the academic literature concerning the validity of the postmaterialism concept and the means by which it can be measured. These arguments have often concerned technical issues, such as the use of the statistical procedure of factor analysis. The ordering of survey questions is another area of concern. One especially controversial issue is the method by which participants express their priorities when shown a list of postmaterialist and materialist goals. It has been argued that the postmaterialist–materialist polarization is an artifact created by the method of ranking responses, and that participants should rate, rather than rank, the importance of the goals. However, this remains a topic of debate. Some also question whether the index is a valid measure of enduring political and social values, as responses may depend on the economic context within which questions are asked.

Implications and Effect

The thesis has been influential in the study of public values and political behavior. It has been argued that postmaterialism is linked with the decline of class-based politics, as shifting values make traditional socioeconomic divisions less relevant. It has also been suggested that postmaterialists tend to hold core values such as social justice, freedom, and equality, whereas materialists value wealth, national security, and social order. Some claim that the countercultural movements of the 1960s represented the first manifestation of the new value system, and that postmaterialism played a crucial role in the rise of new social movements, with postmaterialists being much more likely to participate in these movements than materialists.

Inglehart has suggested that environmental concern could be linked to postmaterialist values, and this relationship has become commonly accepted. In some cases, environmental concern has even been used as an indicator of postmaterialism. A study by Quentin Kidd and Aie-Rie Lee confirmed that postmaterialists tend to be more concerned about the environment than materialists, regardless of the economic development level of their society, and this is also supported by other findings. Investigators have suggested the postmaterialist transition as a potential explanation for the popularity of green political parties in certain countries. However, the relationship of postmaterialism and environmentalism has been challenged by some researchers, including Steven R. Brechin and Willett Kempton. One investigation, by Andrew Knight in 2007, found that postmaterialist values were not significantly related to attitudes to genetic modification. The relationship remains a topic of debate.

The postmaterialism thesis has been extremely influential in sociology, politics, and related disciplines, attracting a widespread following but also some critics. Many researchers have accepted the logic of the thesis but question to what extent the values are enduring and their importance to people's broader political beliefs and behaviors. Others pose more fundamental challenges to the concept and the techniques used to measure it. Despite this, the postmaterialism thesis has become one of the most extensively analyzed trends in postwar politics, and the attention focused on it shows no sign of diminishing.

See Also: Counterculture; Environmental Movement; Green Parties; Political Ideology.

Further Readings

Davis, Darren W. and Christian Davenport. "Assessing the Validity of the Postmaterialism Index." *The American Political Science Review,* 93/3 (September 1999).

Inglehart, Ronald. *The Silent Revolution: Changing Values and Political Styles Among Western Publics*. Princeton, NJ: Princeton University Press, 1977.

Inglehart, Ronald F., et al. *Human Beliefs and Values: A Cross-Cultural Sourcebook Based on the 1999–2002 Value Surveys*. Mexico City: Siglo XXI, 2004.

Sarah Hards
University of York

Power

In the simplest terms, power refers to the capacity to get work done—an effective means to an end. In physics, power is defined by units of work over a period of time. However, in terms of political philosophy, dozens of theories contribute to understanding how different subjects hold influence on one another. In simplest terms, power can be understood through a typology based on coercion or co-optation. In terms of environmental politics, more abstract theories of power come into play, including theories on the fundamental political relationship between society and nature, contested political relationships with nature within society, and the complexities behind the dynamic understandings and interpretations of human–environment relationships. Power, it should be noted, is not a tangible "thing" that can be captured. In fact, its definition is a highly contested work in progress by activists, scholars, and even nature itself. The best approach to understanding power in environmental politics is through an investigation of major theories and a practical application of such theories to a particular situation.

Powers of Coercion and Co-Optation

In everyday politics, power is often understood in terms of the effectiveness needed to accomplish a goal. One enduring tradition of political analyses focuses on the governing state and how the state effectively governs. In raw terms, the state can govern through coercion ("hard power") or co-optation ("soft power"). The melding of these two strategies yields structural power. Structural power, which is discussed in greater detail here, refers to institutionalized relationships within a political framework.

Powers of coercion, or hard power, are often violent. In the "carrot and stick" analogy, coercion is the stick. The exercise of hard power involves laws and limitations, and it is backed by the threat of violence. Powers of co-optation, or soft power, are often directed at the "hearts and minds." It involves defining and granting rights like education, safety, and inclusion in decision making. Soft power aims to build a relationship of desire and dependency between the state and the subjects, so that the governed believe that the sacrifice of their own self-determination is actually a benefit to them. When hard and soft powers are working together in harmony, structural power is produced. Structural powers are the enduring economic, political, and social relationships that are upheld by ideology. For example, orthodox Marxists can often trace nearly every relationship among individuals, government, and other institutions to a rationality based on the best arrangement for capitalism. Marx explains structuralism best, noting, to paraphrase, that men make history—just not history of their own choosing.

Power as an Inscribed Capacity

In traditional political analysis, hard power is backed by the threat of force. The ability, or inscribed capacity, to exercise violence defines this controlling, coercive relationship. This arrangement is sometimes referred to as the "power over" relationship. In environmental politics, the manifestation of hard power is the rationality and ability for humans to command and control elements of nature, including human interaction with nature. From a simple rational perspective, humans hold this power over nature because of their ability to apply technology in the control of nature. In a figurative forest, humans have the technological ability to cut down trees, extinguish natural fires and set unnatural ones, hunt animals, harvest produce, and so on. Essentially, nature is a resource that is subject to the use of humans. Philosophical perspectives on this relationship range from Judeo-Christian narratives about human dominion over nature to various economic philosophies that view nature as a resource that must be both exploited and conserved.

A critical take on this perspective of power is the discussion of ecological managerialism, or eco-managerialism. As a practice, eco-managerialism places the natural environment into a framework of governance, subjected to the decisions of humans. Environmental concerns are identified, discussed, analyzed, and ultimately subjected to a management plan with measurable outcomes. Often, eco-management plans are society's response to widespread degradation or concerns about overuse of resources. The ability to control nature through the application of knowledge, is eco-managerialism's central tenet. Timothy Luke suggests that the rationale for exercising such power is explained economically. The degradation of nature is the direct result of capitalism's pursuit of wealth and the need for resources (land and raw material). One of the contradictions of capitalism is that humans use and abuse nonrenewable resources, which threatens the stability of capitalism. In addition, humans use and abuse renewable resources beyond the rate of natural renewal and replenishment. Eco-managerialism seeks to contain this contradiction through control of resources, and the performance of nature in this regard provides reason to maintain management programs in perpetuity.

Power as a Resource

As noted previously, discussions of politics often refer to power as a resource that can be drawn on to accomplish work. In contrast to the "power over" viewpoint, this "power to" definition recognizes that different societal groups may be differentially empowered in terms of decision-making capacity. Within an environmental framework, empowerment refers to the process through which marginalized groups gain power to control local environments. The call for empowerment often focuses struggles of environmental justice, debates on the inclusion of women and children in environmental decision making, and anticolonial movements that seek to enhance local control over natural resources.

The environmental justice movement is a fusion of social justice and environmentalist ethics. In the 1980s, scholars presented evidence that African Americans and other minority populations were disproportionally exposed to toxic waste. Although other studies dismissed these claims, the movement gained traction through dozens and dozens of anecdotal cases of environmental regulations not being enforced in poor, ethnic communities and disconnections between people of color and the environmentalist movement. Furthermore, legacies of racism, institutionalized in urban space, created an environment of disproportionate siting of toxic waste facilities in African-American and minority neighborhoods and tended to limit the ability of exposed residents to "vote with their feet," or move out of these neighborhoods. Although grassroots efforts to publicize these problems often led to on-the-ground successes, it was the formal recognition of the environmental justice movement by the U.S. government that signaled success. In 1994, President Bill Clinton signed Executive Order 12898, which acknowledged the movement, and called on federal agencies to consider the environmental implications of their projects on historically discriminated groups. Later, the U.S. Environmental Protection Agency adopted a stance that included "fair treatment" and "meaningful involvement" of all parts of society (regardless of race, ethnicity, or class) in decision making with negative environmental consequences.

For some, having a seat at the decision-making table is insufficient. Full and total local control of local resources is seen as the ultimate goal. This sentiment is particularly strong in antiglobalization movements and indigenous rights movements that seek local political autonomy from states that have been ignorant of or abusive to their needs. One notable movement is that Zapatista movement of Mexico, in the southern state of Chiapas. On January 1, 1994, armed rebels in Chiapas seized villages, towns, and government buildings in attempt to start a national revolution. The armed insurgency was short-lived, but it gained enough political support to survive into formal negotiations with the Mexican state. In 1996, the San Andreas Accords were signed by both the Zapatista rebels and President Ernesto Zedillo. This agreement ostensibly grants indigenous populations of Mexico (especially in Chiapas) autonomy, or self-control, over local resources. Although these accords have been ineffective, the Zapatista movement has been successful in autonomizing villages in Chiapas, regardless of the Mexican state. Throughout the years, Zapatista rebels have stressed that their movement is not about inclusion but about actual self-governance and self-control of their resources.

These two cases bookend a wide range of thought on empowerment. On the one end, empowerment is defined by political recognition of disproportionate power, seen in cases of environmental racism. On the other end, a social justice movement seeks total and complete autonomous control of local resources and development. In between these poles are limitless numbers of strategies for governmental authority and local populations negotiating schemes of participatory planning. Summarily, environmental empowerment does not necessarily represent an end but is a continually evolving process that reflects unique contingencies and variable capacity.

Power as Discourse

Although Michel Foucault is often referred to a poststructuralist, his work more often challenged the legitimacy of structural power, rather than the concept of structures themselves. As already discussed, power is typically seen as a top-down relationship between that which holds the capacity to govern and that which is subjected to such governance or is a political resource that can be captured, managed, controlled, and so on. Even in terms of soft power, the state (structure) is in the control of individuals. Foucault defines power as a system of control (structural, in a sense) but argues that it is manifested through regimes of knowledge that discipline individuals. This concept of "power-knowledge" is predicated on *discourse*, a term Foucault uses synonymously with thoughts, traditions, teachings, and behaviors of society. The proliferation of a normative discourse, he argues, is internalized by individuals who then discipline themselves in a manner of how things should be. Discourse is understood as always being a matter of practice and always represented by the actions of society's individuals. Early in his own work, Foucault put forth this theory in an attempt to understand the society's control of sexuality and mental health through definitions of normality and deviance. By the end of his career, he began to adapt his power-knowledge work into grander theories of "governmentality." Essentially, Foucault argues that governance is most effective when individuals (subconsciously) govern themselves through normative behavior.

Foucault's work on governmentality focused on political relationships within society; however, scholars have adapted his fundamental theory to environmental management. In *Environmentality*, Arun Agrawal (2005) argues that central state control of forests provokes local unrest from forest users who do not like being told what to do. Through archival research and interviews, he contends that local user behavior can be understood as the product of state-produced knowledge (economic analyses of forest products), rationalized by local community government. In *Lawn People*, Paul Robbins (2007) is a bit more blunt in describing why highly educated American suburbanites dump chemicals into their lawns. Despite their awareness of the health hazards posed by lawn chemicals, these upper-class homeowners felt compelled to maintain bright, healthy turfgrass lawns because it was a matter of community identity. In both of these works, the environmental subjectivity was not determined simply from a power hierarchy but from the personal internalization of society's norms.

Final Complications

A complication to many understandings of power is "nature." Political scholars may be well versed on contemplating relationships within society, or between institutions, but know very little about the behavior of Earth's flora and fauna. Conversely, biologists, geologists, and other environmental scientists often prefer to work in an apolitical environment so as to not present any bias in their work. From a practical standpoint, environmental planners need to give equal attention to both the human and the environmental sides of research. Understanding the contingencies inherent probably provides the best chance for projects to achieve material success for both society and nature. Critically, scholars should think about incorporating nature into explanations of human behavior—not simply from a determinist perspective, but as a relationship of dependency in which both humans and nature are intertwined and dependent on each other.

Even the most seasoned scholars struggle with understanding the concepts of power. Aside from the contingences of various societies, there is the matter of new insight stemming from constant inquiry, research, and critique. James Scott's *Seeing Like a State* (1998) analyzes several environmental development projects and their failures to achieve progressive outcomes. In doing so, he turns to a theoretical toolbox that includes analyses of power as discourse, problems of ignorant and ineffective governance, and local rebelliousness—all within a familiar framework of critiquing structural state policies. Thus, no single theoretical perspective can possibly explain, in totality, the political relationships between states, societies, and nature. In fact, the best explanations typically draw on multiple theories.

See Also: Domination of Nature; Ecological Imperialism; Environmental Justice; Participatory Democracy; Political Ecology.

Further Readings

Agrawal, Arun. *Environmentality*. Durham, NC: Duke University Press, 2005.

Bullard, Robert. *Dumping in Dixie*, 3rd ed. Boulder, CO: Westview, 2000.

Dahl, Robert. *Who Governs? Democracy and Power in an American City*, 2nd ed. New Haven, CT: Yale University Press, 2005.

Foucault, Michel. "Governmentality." In *The Foucault Effect*, edited by G. Burchell et al. Chicago: University of Chicago Press, 1991.

Luke, Timothy W. "The Practices of Adaptive and Collaborative Environmental Management: A Critique," *Capitalism Nature Socialism*, 13/4 (2002).

Lukes, Steven. *Power: A Radical View*, 2nd ed. New York: Palgrave McMillan, 2005.

Robbins, Paul. *Lawn People*. Philadelphia: Temple University Press, 2007.

Robbins, Paul. *Political Ecology: A Critical Introduction*. Oxford: Blackwell, 2004.

Scott, James A. *Seeing Like a State*. New Haven, CT: Yale University Press, 1998.

Derek Eysenbach
University of Arizona

PRAGMATISM

Environmental pragmatism represents one school of environmental philosophy, a larger discipline concerned with the relationships among humans, other animals, and their environments. Environmental pragmatists also are active in environmental ethics, a subfield within environmental philosophy that considers moral principles and rights relative to nonhumans and to the environment in general. During the early 20th century, the views of influential conservationists and regional planners—Liberty Hyde Bailey, Aldo Leopold, Louis Mumford, and Benton Mackaye—were influenced by the classical American pragmatism of William James, Charles Pierce, George Herbert Mead, and John Dewey. By the mid-20th century, the appeal of pragmatism waned in philosophical circles, only to be revived by "neopragmatists"—philosophers such as Richard Rorty, Jurgen Habermas, Richard Bernstein, and Hilary Putnam. Recently, a number of contemporary environmental philosophers and legal scholars—environmental pragmatists, or "ecopragmatists"—have appropriated elements of both classical and revived American pragmatism. Environmental pragmatists seek practical strategies for conducting open-ended inquiry into particular environmental problems, recognizing the plurality of public interests inherent in environmental disputes, and therefore, attempting to resolve, if not transcend, the inflamed rhetoric and "either/or" choices that often deadlock environmental debates.

The classical American pragmatism of the late 19th and early 20th centuries rejected the fatalistic evolutionary determinism of Herbert Spencer and William Graham Sumner, particularly their claim that intractable economic, social, and racial laws determine human relations and potential. Pragmatists redefined evolution's social implications by asserting that humans, though inextricably embedded in their environments like other organisms, can actively learn from their experience, systematically inquiring into their circumstances and using the resulting knowledge to better adapt to their environments and ameliorate social problems. Furthermore, classical pragmatist epistemology and ethics are radically empirical, positing that knowledge and value result from these human interactions with and within their environment, and therefore, the truth and efficacy of ideas and values depend on their practical results—their usefulness—including the emotional satisfactions that they give individuals. Because people learn from experience and derive values by testing their actions and ideas against their resultant consequences, classical pragmatists contend that truth and values are plural, indeterminate, variable, and fallible because what is consequentially good or true for an individual at one time may not be good or true for other individuals at other times. Pragmatists, therefore, are critical of all metaphysical and ethical absolutes and immutable, universally applicable principles.

In late-19th-century America, a society that mythologized the self-made man, William James' pragmatism explained how, against daunting odds, individuals might creatively bootstrap themselves up from humbler origins. In early-20th-century America, a more corporate society seeking alternatives to possessive individualism, John Dewey's communitarian pragmatism showed how societies might reform themselves by using experimental methods to solve social, and even environmental, problems. It was little wonder, therefore, that in the first half of the last century, conservationists and regional planners were attracted to pragmatism's emphasis on the environment, experimental science, collaborative action, and social and environmental reform. Accordingly, Liberty Hyde Bailey's rural environmental education, Aldo Leopold's land ethic, Louis Mumford's participatory regional planning, and Benton Mackaye's integration of wilderness preservation with rural community development initiated what Ben Minteer calls a "third way tradition" in American conservation thought—a pragmatic and civic environmentalism that stands midway between the ecocentric, "nature-first" orientation of John Muir's preservationist philosophy and the "human-first" emphasis of Gifford Pinchot's utilitarian conservationism. A small but influential group, these civic environmentalists were just as concerned about the corrosive effects of industrialization and urbanization on the health of American democracy as they were about their effects on the health of the land. Always mindful of citizens' needs, desires, and civic capacities, they advanced their urban, regional, and wilderness planning as tools for creating a robust civic life and healthy landscapes, balancing the needs of human, animal, and plant communities and expanding citizens' capacity to collectively discuss, debate, and decide issues affecting these communities.

Contemporary environmental pragmatism, the heir to this "third way civic environmentalism," consists of allied and intersecting ideas rather than a single, unified viewpoint. Nevertheless, several common threads run through the work of pragmatic philosophers and ethicists as diverse as Mark Sagoff, Richard Rorty, Ben Minteer, Bryan Norton, Kelly Parker, and Andrew Light, as well as "legal pragmatists" such as Keith Hirokawa and Daniel Farber, all of whom adopt pragmatist approaches to environmental ethics, law, and policy. Environmental pragmatists seek to devise practical strategies for use by policymakers, land managers, public officials, and citizens to resolve environmental debates and solve environmental problems. Because human efforts to conserve the Earth and its biodiversity involve widely differing places, shifting biophysical realities, and culturally diverse peoples, they argue that no single ethical program, no single set of policy principles, fits all cases. Wildfire policy for the California suburbs, for example, must be different than that for the Canadian wilderness. Furthermore, in making and implementing environmental policy, we might never know with certainty what the best choices are, but, as Bryan Norton argues, we can take actions that reduce future uncertainty while allowing for future corrections based on empirical examination of the consequences of our actions.

Environmental pragmatists also embrace the plurality of views and values arising in most environmental debates, interpreting disagreements as opportunities and expressing skepticism that universally held values or cut-and-dried rules can produce acceptable answers to messy environmental policy dilemmas. In accepting plural values, environmental pragmatists differ from deep ecologists, biocentrists, or ecocentrists, such as Holmes Ralston and J. Baird Callicott, who seek indisputable foundations for policy in nature's intrinsic values—infallible groundings, such as the integrity of ecosystems or the rights of nonhuman animals, that can be advanced to support claims in environmental policy deliberations. Holding fast to such fixed and immutable values, environmental pragmatists argue, discourages discussion, debate, accommodation, and ultimately, resolution of environmental controversies. Instead, moral claims and public policies on behalf of nature must be developed through thoughtful and creative engagement with human communities, enquiring into their everyday norms and traditions, rather than deriving policies by reasoning back to transcendent rights and first principles. Formulation of environmental policy, according to environmental pragmatists, must begin not in an appeal to intrinsic rights of nature but, rather, with intrinsic community values: vernacular traditions of resource stewardship, shared experiences, and sense of place, as well as concern for future generations. In Bryan Norton's adaptive management approach, these vernacular values, rooted in experience, become the grist for robust democratic deliberation and debate, facilitated by natural resource experts and managers—deliberations that seek to integrate public values, experimental science, and potentially conflicting perspectives, rather than simply appealing to the intrinsic rights of nature in the vain hope of "changing consciousness" and, thereby, environmental policy.

See Also: Conservation Movement; Deep Ecology; Ecocentrism; Urban Planning; Wilderness.

Further Readings

Farber, Daniel A. *Ecopragmatism: Making Sensible Environmental Decisions in an Uncertain World*. Chicago: University of Chicago Press, 1999.
Light, Andrew and Eric Katz, eds. *Environmental Pragmatism*. New York: Routledge, 1996.
Livingston, James. *Pragmatism, Feminism, and Democracy: Rethinking the Politics of American History*. New York: Routledge, 2001.
Minteer, Ben A. *The Landscape of Reform; Civic Pragmatism and Environmental Thought in America*. Cambridge, MA: MIT Press, 2006.
Norton, Bryan G. *Sustainability: A Philosophy of Adaptive Ecosystem Management*. Chicago: University of Chicago Press, 2005.

James Bruggeman
Independent Scholar

Precautionary Principle

The precautionary principle, which has become a general principle of international law, says that if the environmental consequences of human action may be serious and irreversible, efforts should be made to avoid or lessen them. It is based on the classical virtue of "prudence" and embraces the folk wisdom of "better safe than sorry" and "look before you leap."

Although uncertainties about the consequences of human behavior have always existed, they have become more significant in recent times because of the growing scope, complexity, and hazard of human activities. This means it is becoming more vital to be able to prevent the harm these activities might do, even without being sure what that is.

Environmental regulations generally aim to prevent known risks rather than anticipate and prevent uncertain potential harm. It is when the risk is uncertain because either the probability of damage is uncertain and/or the extent of damage is uncertain that the precautionary principle applies. When the risk is unknown, then risk assessment is not an appropriate tool to use because the probability of harm cannot be quantified. The precautionary principle fills the gap.

United Nations Educational, Scientific and Cultural Organization's (UNESCO) World Commission on the Ethics of Scientific Knowledge and Technology defines the precautionary principle as follows: "When human activities may lead to morally unacceptable harm that is scientifically plausible but uncertain, actions shall be taken to avoid or diminish that harm."

"Morally unacceptable harm" refers to harm to humans or the environment that is

- threatening to human life or health, or
- serious and effectively irreversible, or
- inequitable to present or future generations, or
- imposed without adequate consideration of the human rights of those affected.

The judgment of "plausible" should be grounded in scientific analysis. Analysis should be ongoing so that chosen actions are subject to review. "Uncertain" may apply to, but need not be limited to, causality or the bounds of the possible harm. "Actions" are interventions that are undertaken before harm occurs that seek to avoid or diminish the harm. Actions should be chosen that are proportional to the seriousness of the potential harm, with consideration of their positive and negative consequences, and with an assessment of the moral implications of both action and inaction. The choice of action should be the result of a participatory process.

In the past, many products and processes have been marketed without prior approval or any requirement that the manufacturer show evidence that they will not harm human health or the environment. Similarly, many activities and developments have been undertaken without the need for developers to show they will not have an adverse environmental impact. Traditionally, it has been up to consumers, environmentalists, or government authorities to make a convincing scientific case that such activities or products were harmful before they could be regulated. The thinking was that regulations constrained economic activity and should only be justified if there was scientific proof that such activity would cause harm. This is a "wait and see" approach, in which the burden of proof is on those asserting that damage is being or will be done.

These days, certain activities require developers to prepare environmental impact statements or assessments, and some products, such as pharmaceutical drugs, pesticides, and food additives, must gain approval before they can be marketed. In these cases, it is initially assumed that the activities in questions or the products may be hazardous or environmentally damaging—the burden of proof has been shifted to the manufacturer or developer, who needs to produce scientific evidence that the activities or products are safe to get approval. Although we say the burden of "proof" has been shifted, proof is not actually required, just a convincing case—supported by scientific evidence—that the product or activity is safe.

This shifting of the burden of proof from one party to another, for example, from the regulatory authority to the polluter, is only one element of the precautionary principle. However, the fact that those proposing an activity have to show it is safe before it is approved—rather than the government needing to show it is unsafe before it can be restricted—is an important aspect of the precautionary principle.

In practice, the burden of proof has generally only been shifted for new products and activities, where there is a long history of harm arising from similar products and activities. Existing products are generally "presumed safe" until the harm they do becomes irrefutable. This bias is partly based on the assumption that it is cheaper and more politically acceptable to prevent new products being manufactured than it is to ban existing products, and it is easier to prevent new developments than to dismantle existing ones. Similarly, synthetic substances may require licenses, but natural substances are assumed safe, even if they are added in unnatural quantities to the environment. Yet the precautionary principle also applies to existing products and natural substances added by humans to the environment.

Those proposing new environmental regulations still often have the burden of making a watertight scientific case that the proposed regulations are necessary to protect human health or the environment. This gives opponents the opportunity to undermine the justification for such regulations by emphasizing the uncertainties in their scientific evidence.

What the precautionary principle does is to ease the standard of proof so that scientific evidence of possible harm is sufficient to prompt regulatory action without proof being required. It is no longer sufficient to raise doubts about whether the harm will happen to prevent an activity or product from being regulated. In this way, the balance between environment and economic development is shifted a little more toward environmental protection.

The precautionary principle has two parts: the political decision whether to act, which requires

- identification of potential adverse effects that threaten the desired level of protection now or in the future, when
- these adverse effects are caused or exacerbated by human activity, and
- scientific evaluation of such effects shows they are plausible and likely, and
- the exact risk cannot be determined because of scientific uncertainty, and
- postponing action will make effective action more difficult later on, and the measures to be taken if action is decided on.

The precautionary principle achieved widespread recognition after it was incorporated into the Rio Declaration on Environment and Development decided at the 1992 United Nations Conference on Environment and Development in Rio. In 1993, the Treaty of Maastricht required European Community countries and the European Commission to base environmental policy on the precautionary principle. In 1999, the Council of the European Commission urged the commission to ensure that future legislation and policies were guided by the precautionary principle.

The precautionary principle has been incorporated into many international laws and almost all recent international treaties that aim to protect the environment. However, international courts are still reluctant to accept it as a legal principle. It is, however, widely accepted as a principle with similar standing to that of sustainable development.

The precautionary principle has been incorporated into national laws in several countries, including Germany, Belgium, and Sweden, and it has influenced several court judgments. In France, it has even been included in the nation's constitution, as part of an environmental charter. This gives the principle priority over other legislation.

The legal system in English-speaking countries is less conducive to the incorporation of broad principles and tends to be based on specific rules and regulations. In the United Kingdom, for example, the precautionary principle is not included in statutory law. Nor has the precautionary principle made much headway in United Kingdom courts. However, it has been included in discussion papers and government policy statements.

The precautionary principle is controversial in the United States, where corporate interests have succeeded in spreading confusion about what the principle means and implies. Opponents argue that the precautionary principle is unscientific, that it can be triggered by irrational concerns, that it aims at an unrealistic goal of zero risk, and that it will result in the banning of useful chemicals and preventing technological innovation. Excessive caution, it is argued, leads to paralysis and stagnation.

In actual fact, the precautionary principle cannot be applied without scientific evidence of harm. Nor does the precautionary principle aim to reduce risk to zero but, rather, to avoid or mitigate likely harm. The measures to be adopted to achieve this are not dictated by the precautionary principle, and there is no requirement on the part of the precautionary principle to ban anything, although decision makers may decide that a ban may be appropriate in certain circumstances. The precautionary principle does not preclude technological innovation—it calls for innovation to be redirected toward the development of safer technologies.

See Also: Equity; Policy Process; Regulatory Approaches; Risk Assessment; Risk Society; Uncertainty.

Further Readings

Andorno, Roberto. "The Precautionary Principle: A New Legal Standard for the Technological Age." *Journal of International Biotechnology Law,* 1 (2004).

Beder, Sharon. *Environmental Principles and Policies*. Sydney: UNSW Press and London: Earthscan, 2006.

de Sadeleer, Nicolas. *Environmental Principles: From Political Slogans to Legal Rules*. Oxford: Oxford University Press, 2002.

World Commission on the Ethics of Scientific Knowledge and Technology. *The Precautionary Principle*. Paris: United Nations Educational, Scientific and Cultural Organization (UNESCO), March 2005.

Sharon Beder
University of Wollongong

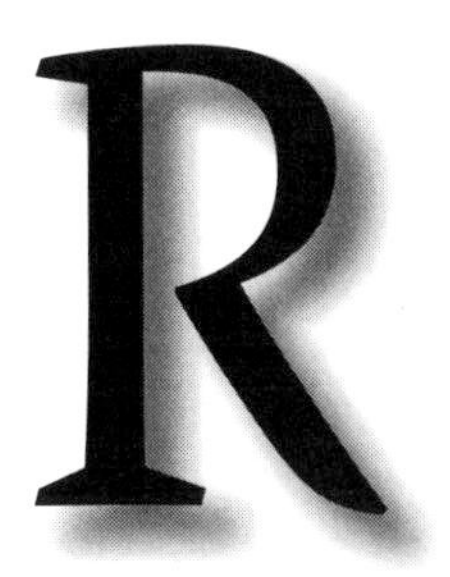

Regulatory Approaches

Markets coordinate the exchange of goods and services between producers and consumers. Under an ideal market arrangement, resources are efficiently allocated in such a way that no person can be made better off without making another person worse off. Among other conditions, this requires that all costs and benefits are accounted for by the participants of the transaction so that the true value of goods is reflected. When these conditions are not met, costs and benefits are passed on to third parties, and inefficiencies result. Environmental pollution can be thought of as the product of inefficient markets, where the burden of environmental protection is passed on to third parties and the true value of goods and services is not reflected in their market exchange. These shortcomings in market allocation, also referred to as market failure, serve as economic rationale for government intervention into markets.

The regulatory approach that governments choose to address market failures depends on the type of market failure that is to be addressed. Market failures associated with environmental pollution are often problems of public goods, externalities, and information asymmetry. Public goods are goods that are characterized by nonrivalry, nonexcludability, or both. Nonrivalry occurs when more than one consumer can enjoy the benefits of a good without reducing the benefits for other consumers. Nonexcludability occurs when a single person cannot control the use of the good. Public goods often have varying degrees of rivalry and excludability. For example, many consumers enjoy the benefits of clean air without infringing on the enjoyment of others, and no single person can control its consumption. Therefore clean air is both nonrivalrous and nonexcludable. An ocean fishery, however, is a public good that is nonexcludable in its accessibility but rivalrous in the scarcity of its fish population. Private markets fail to provide for public goods, and thus the inefficiencies of market failure occur under these circumstances.

Negative externalities are costs that are passed off to a third party that has not consented to a voluntary exchange. As the third party must assume these costs, the true value of the good is no longer reflected in the exchange, resulting in a market failure. For example, pollution from automobile emissions that present costs to society is an example of negative externalities that are passed from the consumer to society. An oil company may pollute the air of the surrounding community, passing on the costs of production from the producer to society. Positive externalities may also result from market transactions. The upkeep of an

In the United States, nearly one-third of all fossil fuel goes toward residential uses, so voluntary reductions by consumers would have a significant effect on overall consumption.

Source: iStockphoto.com

outdoor area may provide aesthetic value to the surrounding community, although it has not contributed to the costs of maintenance. In this way benefits are received by those who are third parties to market transactions.

Information asymmetries are a third type of market failure that occurs when producers and consumers have unequal information about the quality of a good. This presents a problem for inefficiency, as the true value of the good may not be revealed in the transaction. Producers may choose to pass on costs to uninformed consumers, who unknowingly pay a price that is higher than an alternative price under conditions of complete information. For example, in recent years many producers have marketed products as having environmentally friendly characteristics. Lacking the ability to assess the credibility of these claims, consumers often buy products with the belief that they are contributing to the environment, when in many circumstances no such contribution is made at all. Without the intervention of a third-party supervisor, the producer might continue to sell a product at a price that does not reveal the consumers' true willingness to pay for the good, further contributing to environmentally damaging production practices. Several regulatory approaches are available for government intervention to address these market failures. These include incentives for voluntary actions, conventional command-and-control regulations, and market mechanisms such as carbon taxes and permit systems. Each approach varies in its ability to mitigate environmental impacts while meeting competing societal goals of economic efficiency, distributive equity, and political feasibility.

Voluntary Approaches

Voluntary approaches to regulation provide producers and consumers with incentives to reduce environmental pollution. Unlike other forms of regulation, here producers and consumers are able to make their own decisions about the costs that they will incur in providing pollution prevention. Producers may find it advantageous to increase the energy efficiency of their products or processes, often decreasing costs and increasing profits by increasing their visibility as environmentally conscious entities and cutting the cost of production. Likewise, consumers may feel a moral obligation to buy products that reduce carbon emissions or may pursue monetary gain through cost-of-living reductions resulting from energy efficiency and tax incentives provided by voluntary regulation.

The Energy Star program, created by the United States government in 1992, is one example of a voluntary policy approach that allows producers who meet certain energy-efficiency standards to apply the Energy Star seal to their products. This logo provides energy-efficiency information to consumers about the product's energy demands. Eco-labeling programs similarly overcome information asymmetry by providing third-party oversight from nongovernmental organizations. Other policies allow consumers to take advantage of offered tax breaks that reduce income taxes for those households that invest in home improvements that are energy efficient.

The environmental impacts of these programs have the potential to be very high. Residential consumption alone accounts for almost one-third of fossil fuel consumption in the United States. A decision by these consumers to adopt energy-efficient appliances would make a significant contribution to reductions in energy consumption. However, by placing energy-efficiency decision making in the hands of producers and consumers, there is a lot of uncertainty involved in the amount of greenhouse gas reduction that will actually occur. Voluntary policies are economically efficient because they provide information to a consumer that decreases transaction costs. The distributional equity and political feasibility concerns are also low because these policies give consumers and producers decision-making authority and spread the burden of environmental sustainability across the entire spectrum of voters. Therefore these policies are attractive, but they should be considered alongside other policies, given the high level of uncertainty involved in meeting the desired environmental targets.

Prescriptive Regulation

The second type of policy that can be used to mitigate greenhouse gas emissions is conventional regulations, also known as "command-and-control" regulations. Governments hold a unique power among societal entities in that they have a monopoly on the use of force to coerce consumers and producers to behave in a specified manner. For example, governments can use this power to place a tax on the emissions from a producer's manufacturing process. This tax provides an incentive for the producer to adopt technologies that reduce the amount of emissions released from the process. This approach is effective as long as the total cost of adopting new technologies is less than the cost of emission taxes. Otherwise, the best alternative for producers would be to pay the taxes on their emissions. Standards such as minimum renewable energy requirements from energy and transportation producers are other examples of regulatory instruments that help to stabilize pollution. These regulations typically target producers and set a cap on emissions. Because of limited technologies in monitoring emissions such as carbon dioxide, these regulations are usually applied to stationary sources of emissions rather than mobile sources such as vehicles. Some policies do regulate mobile emissions though, such as the Corporate Average Fuel Efficiency standards in California, which require cars and light trucks to meet a minimum standard of fuel efficiency.

The environmental impacts of such policies are many times more predictable than those of voluntary regulations. The government has the flexibility to decide on the harshness of regulation to the extent that the policies are politically feasible. Once the limits are set, the effectiveness of the regulation on reducing environmental impacts is a function of the ability of the government to monitor and enforce the regulation. However, conventional regulation is the least economically efficient of all regulatory approaches. These regulations force producers and consumers to make decisions that they normally would not make. For example, when Corporate Average Fuel Efficiency standards result in costly design and manufacturing changes, producers are forced to incur these costs to comply with regulations. To maintain their profits, producers must then pass these costs on to consumers, who pay higher prices when purchasing automobiles. In response to the increased prices, consumers may then choose to buy older-model cars that are less expensive. This in turn reduces the demand for more-expensive, newer-model cars that have been manufactured according to regulation. The result is decreased incentives for both consumers and producers. To the extent that these policies target a particular sector, the burden of emissions reduction is unevenly distributed to that sector, thus raising distributional equity concerns. The political feasibility of such regulations similarly depends on the targeted sector, the degree to which the regulation poses significant costs to industries in the sector, and the extent to which the sector is sufficiently organized to fight against the regulation.

Market Mechanisms

Market mechanisms are the third type of regulatory approach in dealing with market failures associated with environmental pollution. Two mechanisms, taxes and property rights, may be used in solving the problem of public goods. Taxes increase the incentive to reduce pollution by increasing the cost of pollution to reflect its true value. This approach has been recommended by many economists to address the problems of global warming. Although the technology to monitor emissions of carbon dioxide is not well developed, the emissions that result from the combustion of fossil fuels can be directly observed by the carbon content of the fuel. Thus a tax can be applied based on the level of fossil fuel consumption. An "upstream" tax is applied to fossil fuels at their point of entry into the market. This approach significantly reduces the number of actors involved in the process. A "downstream tax" that targets electric utilities, petroleum refineries, and households significantly increases the amount of red tape involved in the process. Because the downstream actors will pay higher costs regardless of where the tax is levied, the upstream tax is favored by most economists.

Property rights transform the public good of air quality into a private good. By auctioning or allocating a certain number of permits, or the right to pollute the atmosphere, the ability to emit carbon dioxide becomes rivalrous and excludable (more than one consumer may neither purchase the good nor consume the good simultaneously), and hence able to be bought and sold on the market, increasing the incentive to provide for the good. When the number of permits that may be distributed are finite (thus making a permit different from a tax), then the system is commonly referred to as a "cap-and-trade" system. Over time the cap is reduced, hence reducing the total number of permits in the market and the total level of emissions. Although carbon taxes do not exist in the United States, a number of regional initiatives are working to develop carbon cap-and-trade markets that would share the properties of both carbon taxes and permit systems. Both systems offer increased economic efficiency over command-and-control mechanisms, as those who can reduce their emissions and optimize costs will do so and those who cannot will pay for their emissions. This arrangement offers the largest benefits to society at the least cost.

Taxes have the potential to force significant environmental impacts by reducing levels of emissions. Under some scenarios, taxes could lead to an even larger reduction in emissions than quantified targets. However, there is greater uncertainty in the absolute environmental impact than in a permit system, as this depends on the ability to know the true cost of emissions reduction, which is still uncertain from a lack of historical experience and the inability to forecast future costs and emissions. From an economic standpoint, taxes are much more efficient than permits. Studies have shown that taxes may be implemented at one-fifth of the cost of a permit system. Although taxes would decrease profits for fossil fuel suppliers, these costs would be passed on to electric utilities and consumers. In addition, the revenue collected from taxing could be recycled to give fossil fuel suppliers cuts in corporate taxes that distort markets by decreasing the incentive to invest. This is known as revenue neutrality, or revenue recycling. This would significantly decrease the costs of the tax system. In addition, as the costs of emissions reductions and the future level of emissions are uncertain, carbon taxes impose costs at the margin over time. A distributional equity concern would present itself, however, in that electric utilities, petroleum refineries, and consumers would incur higher costs passed down from fossil fuel producers unless tax relief was distributed to these groups as well. Carbon taxes, however, are not politically feasible. Businesses oppose the transfer of money to the government, and environmental groups prefer the assurances of quantified limits. The failure of the Clinton tax initiative in the mid-1990s is a good example of the low political feasibility of taxes.

Permit systems increase the level of certainty in emissions reduction by capping the level of emissions and offering a finite number of permits that decrease over time. As mentioned previously, however, permit systems are much less efficient economically than tax systems. This is for two reasons: the uncertainty in the future cost of mitigation and the uncertainty in future emissions levels. Permits hold the level of emissions reduction constant, so that the cost of meeting that level may highly fluctuate over time. Taxes instead hold the cost constant over time and allow the total amount of emissions reduction to fluctuate. Because we know there is so much uncertainty in the cost of mitigation, as well as future levels of emissions, taxes provide a more cost-effective approach, allowing the cost to increase only marginally over time. The distributive equity involved in a permit system may change according to whether the permits are allocated or auctioned. Permits are more politically feasible than taxes because they have the potential to achieve the same ends as taxes but hide the costs.

See Also: Ecological Economics; Ecotax; Equity; Green-Washing; Policy Process.

Further Readings

Goodstein, Eban. *Economics and the Environment.* New York: John Wiley & Sons, 2007.

Keohane, Nathaniel and Sheila Olmstead. *Markets and the Environment.* Washington, D.C.: Island Press, 2007.

Weimer, David and Aidan Vining. *Policy Analysis: Concepts and Practice,* 4th ed. Upper Saddle River, NJ: Prentice Hall, 2004.

Thomas D. Eatmon, Jr.
Allegheny College

Resource Curse

The resource curse posits a negative relationship between resource abundance and economic performance. Also known as the "paradox of plenty," it suggests that nations with bountiful quantities of extractable natural resources such as oil, timber, and precious

A woman working on a farm in Nigeria, a country whose oil wealth is thought to be an example of a resource curse. Its oil industry spurred a self-determination movement by the indigenous Ogoni people.

Source: World Bank

metals are associated with disappointing rates of economic development and are more likely to become mired in political violence.

Intellectual interest in this phenomenon began in the 1920s and 1930s, sparked by the decline in Latin American economies hit hard by a global slump in commodity prices. Postwar scholarship suggested that this negative relationship between resource wealth and economic performance was independent of price fluctuations. Economists tested this hypothesis using repeated regression analysis, which compared resource intensity and economic growth data. Nations with abundant natural resources (such as Venezuela, Nigeria, and Guyana) saw their economies expand slower than those nations (such as Taiwan, Singapore, and Korea) that started out with little resource wealth. These findings confirmed that high resource intensity correlates with ragged economic growth.

The economic mechanisms underpinning this phenomenon have been subject to intense debate. Scholars argue that export booms caused by the discovery of natural resources distort long-term economic growth by making other financial sectors (such as manufacturing and tourism) less attractive, leading investment dollars away from these slow-growth enterprises. Resource wealth is also linked with "Dutch disease," whereby a booming resource sector inflates the exchange rate, reducing the competitiveness of other industries. Labor demands for resource extraction are generally for unskilled labor, pushing wages up and decreasing the incentive for education. Large natural resource endowments tend to result in little diversification, lower overall education rates, and few spin-offs into the broader economy.

Natural resources can further undermine political governance. States undergoing resource booms are flooded with more revenue than they can manage effectively. Many spend and borrow heavily on the strength of their resource, leaving them vulnerable to commodity price fluctuations. In this way, the volatility of commodity markets can overwhelm normal budgeting procedures and weaken state institutions.

Resource wealth also tends to make governments less accountable and less democratic. This occurs though a number of mechanisms: First, the rentier state effect, in which resource rents decrease the comparative importance of tax revenues, allowing states to govern based on networks of patronage rather than meeting the needs of citizens. Second, the repression effect, in which governments use resource revenues to repress or co-opt any opposition threats to their political power. Third, the mechanization retarding effect, in which heavy reliance on resource extraction fails to bring about the socioprofessional and cultural transformations that tend to promote a strong, independent civil society. The result is shortsighted and risk-averse states that prioritize preserving the status quo over promoting development or democratization. Governments that receive most of their wealth from natural resources thus tend to become more autocratic and less accountable to their citizens. Natural resources also play a role in weakening the power of national governments by increasing the likelihood of political rupture. Resource wealth weakens the state's territorial control by increasing regionalization: Groups living among these resources generally want a disproportionate amount of the gains and feel resentful at "their" wealth being extracted by the national authority. At the same time, resource wealth strengthens the viability as well as the will for secession by providing those in resource-rich areas with the capital to launch armed struggle. Oft-cited examples of resource wealth acting as a catalyst for political fragmentation include the secession movements of the Ogoni people in Nigeria, the Katanga region in the Democratic Republic of the Congo, and the Cabinda Province in Angola.

One of the most common examples of the resource curse is the 40-year conflict that engulfed the southern African nation of Angola. Widespread violence first broke out in the Portuguese colony in 1961 as various factional groups began waging independent campaigns to topple colonial rule. The Portuguese departed in 1974, ceding control to the three main independence groups in a power-sharing agreement. This deal broke down almost immediately, and the independence war morphed into a civil war. After years of fighting fueled by neighboring countries and Cold War tensions, two groups were left fighting: the MPLA (Popular Movement for the Liberation of Angola) and UNITA (National Union for the Total Independence of Angola).

The prolonged conflict in Angola could not have lasted as long as it did were it not for the nation's resource wealth in the form of oil and diamonds. The MPLA—which emerged victorious in a series of contested elections—captured 80 percent of the nation's offshore oil revenues and used these revenues to import more than US$5 billion in arms through the 1990s. UNITA's struggle was similarly underpinned by resource wealth: Alluvial diamonds clustered in UNITA territory offered a convenient storage of wealth to advance their military operations against the MPLA. Experts estimate that UNITA harvested between US$3 billion and $4 billion in diamonds between 1992 and 2000. The conflict in Angola thus pitted oil and diamond revenues against one another. Dividends from these resources financed both sides of this decades-long conflict.

Recent critiques suggest that the prevalence of this correlation between resource abundance and economic underdevelopment might be exaggerated. Some scholars have inverted the causality of the resource curse, arguing that poor governance leads to an overreliance on resource revenues, rather than the other way around. When an unstable or repressive regime scares off foreign investment and stalls domestic industries such as tourism and manufacturing, natural resources remain. Resource extraction becomes the default sector, which continues to function after other industries have evaporated. Within this analysis, resource dependence becomes a symptom, rather than a cause, of economic decline.

Other researchers emphasize methodological shortcomings that might result in an overexaggeration of the resource curse phenomenon. Two major criticisms have emerged: First, the proxy for estimating resource endowments—primary exports divided by gross domestic product—skews results in favor of the resource curse. When a different variable is used—the relative abundance of natural resources in the ground, for instance—resource wealth correlates with slightly higher economic growth and fewer armed conflicts. A second line of critique focuses on the deterministic nature of the resource curse. The empirical examples used to underpin the formulation of this phenomenon were studied for periods of no more than 40 years—too short to reveal other factors such as history or culture that might have shaped characteristics of political governance. A longer-term approach reverses claims that a reliance on resource rents undermines democracy: Countries that were democratic remained democratic after finding natural resources, and countries that were authoritarian mostly remained authoritarian.

The most prominent exception to the resource curse hypothesis is the southern African nation of Botswana. Botswana was the 25th poorest country in the world when it declared independence from Britain in 1966. Diamonds were discovered soon after, and the country laid claim to the fastest-growing economy in the world between 1966 and 1989. Botswana is now classified as an upper-middle-income country boasting some of the highest socioeconomic indicators on the continent.

Botswana avoided the resource curse by limiting external debt and investing wealth accrued through diamonds into other sectors of the economy. The government practiced sound management: It refused to increase spending when commodity prices were high, invested heavily in the service and manufacturing sectors, and managed the exchange rate to avoid excessive oscillation. Botswana relied on a Sustainable Budget Index to measure the ratio between consumption expenditures and nonresource revenues, ensuring that the Sustainable Budget Index remained less than 1 so that natural resource capital was not consumed. The government proved adept at avoiding external debt and using public expenditures to counter the boom–bust cycle of natural resource markets. It avoided the curse by using resource rents as a source of investment, rather than a source of public expenditure. The case of Botswana suggests that the relationship between resource abundance and poor economic performance is not deterministic: Prudent financial management can help mitigate its detrimental effects.

See Also: Environmental Management; People, Parks, Poverty; Political Ecology.

Further Readings

Brunnschweiler, C. N. and E. H. Bulte. "Linking Natural Resources to Slow Growth and More Conflict." *Science,* 320/5876 (2008).

Humphreys, Macartan, et al. *Escaping the Resource Curse.* New York: Columbia University Press, 2007.

Le Billion, Phillipe. "Angola's Political Economy of War: The Role of Oil and Diamonds, 1975–2000." *African Affairs,* 100 (2001).

Ross, Michael L. "The Political Economy of the Resource Curse." *World Politics,* 51/2 (1999).

Sachs, Jeffrey D. and Andrew M. Warner "The Curse of Natural Resources." *European Economic Review,* 45 (2001).

Matthew Schnurr
Dalhousie University

Revolving Door

The "revolving door" refers to the interchange of personnel, usually between industry and government, but also between lobby groups or think tanks and government, as well as between the media and government. The problem is that government officials can be unduly influenced, by either their previous employers or potential future employers, and this undermines the effectiveness of governments in regulating to protect the environment.

There is also a revolving door between environmental groups and the industries they criticize that can give rise to similar conflicts of interest, whereby environmental groups can be unduly influenced by business interests. What is more, the revolving door can help business interests gain unearned environmental credibility.

Industry and Government

Regulatory agencies find it attractive to hire people from industry because of their expertise in the industry and their ability to liaise with the industry. However, business executives and lobbyists who take government positions often maintain their industry sympathies in their new jobs. In some cases they even seek government jobs so as to further industry interests. This can contribute to regulatory capture, whereby an industry has undue influence over a regulatory agency or government ministry. In the case of environmental problems, this undermines both the development and also the enforcement of strong environmental protection laws.

As an example, there is a long history of a revolving door between Monsanto and the U.S. agencies that regulate its activities. William Ruckelshaus became a director of Monsanto after being head of the Environmental Protection Agency (EPA). Linda Fisher worked for the EPA, then headed Monsanto's Washington lobbying office before becoming deputy administrator of the EPA. She then became a vice president of government affairs at Monsanto, followed by vice president at DuPont. She is now assistant administrator for pesticides and toxic substances of the EPA.

Despite efforts to regulate against the revolving door in the United States, it became far more common during the Bush administration, when President George W. Bush appointed unprecedented numbers of corporate executives and business lobbyists to government posts, including regulatory agencies with power over decisions affecting the environment. For example, Philip Cooney became chief of staff of the White House Council on Environmental Quality for George W. Bush after serving as a lawyer and lobbyist for the American Petroleum Institute. In his government position he amended government scientific reports to downplay the role of greenhouse gas emissions in causing global warming. After he resigned his government position, he was hired by ExxonMobil.

In addition, senior government bureaucrats and politicians often look forward to a postgovernment career in business. It is in the interests of businesses, lobbyists, and think tanks to hire former government officials and politicians because they know how the political system works, have valuable policy experience and inside knowledge, and still have influence with their former colleagues.

Public Citizen's Congress Watch found that in the United States, between 1998 and 2005, 42 percent of the members of the House of Representatives who left government and 53 percent of senators who left government became lobbyists. Regulations stop them from directly lobbying their former colleagues for only 1 year. After that, they are able to turn their political careers into extremely lucrative lobbying careers.

Politicians who do favors for particular corporations or business interests can also look forward to highly paid corporate executive positions, board memberships, or lucrative consultancies after they retire from political life. Even if such job offers are not explicitly promised, it is in the interests of those looking forward to a postgovernment career in business not to displease the business community while in office, and this means they tend to take a business viewpoint while still in government.

The premier of New South Wales in Australia, Bob Carr, is one of many Australian examples of the revolving door. In 2005, two months after leaving office, he became a part-time consultant to Macquarie Bank for a fee reported to be A$500,000 per year. The investment bank had been involved in many of the state's infrastructure projects during Carr's premiership, including controversial private toll roads. In his new position, Carr lobbied the state government in favor of privatization proposals on behalf of the bank. Macquarie Bank has appointed several former Australian politicians to well-paid positions in the bank.

Environmental Groups and Industry

The existence of a revolving door can also give rise to a lack of independence on the part of career environmentalists. There are numerous examples of activists who now work for the industries they once opposed. They are attractive to corporations and industry associations that wish to present an environmentally benign image to the public; because they have a reputation for caring for the environment derived from their previous positions in environmental groups, they are often assumed to have environmental interests at heart, despite their change of employment.

In particular, high-profile environmentalists have valuable green credentials that industry interests can use in their green-washing efforts. For example, Trisha Caswell was executive director of the Australian Conservation Foundation before she became chief executive of the Victorian Association of Forest Industries—the main lobby group for logging interests.

Former environmental activists are especially attractive to lobbying and public relations (PR) firms because of their inside knowledge of how environmentalists think and operate, as well as their expertise in environmental campaigning, which requires an understanding of how the public relates to environmental issues. When PR giant Burson-Marsteller hired the former chairman of Friends of the Earth in the United Kingdom, Des Wilson, as director of public affairs and crisis management, he was "reckoned to be one of the highest paid people in PR." He subsequently became a director of corporate and public affairs for BAA.

Burson-Marsteller also hired Peter Melchett in 2002, following his leadership of Greenpeace UK. Melchett, whose father was chair of British Steel, had previously had a career in the United Kingdom's Labour government. Following his time at Burson-Marsteller, Melchett became an environmental consultant whose clients include Burson-Marsteller and Walmart.

Environmental activists can also parlay their green credentials and experience in the environmental movement into a lucrative consulting career, hiring themselves out to clients wishing to green-wash their image, oppose environmental groups, or undermine environmental measures being proposed by governments.

For example, Patrick Moore was a cofounder of Greenpeace and director of Greenpeace International from 1971 to 1986 and head of Greenpeace Canada from 1977 to 1986. Moore now has his own consultancy, Greenspirit, and is on the board of and spokesman for the industry-funded Forest Alliance of British Columbia (set up by Burson-Marsteller), from which he champions clear-cutting as an environmentally appropriate method of forestry. His clients have included mining giant BHP, Canada, the National Association of Forest Industries, the Canadian Mining Association, the Canadian Pulp and Paper Association, Westcoast Energy and BC Gas, and the BC Hazardous Waste Management Corporation. Moore has also been employed to promote nuclear energy by the Clean and Safe Energy Coalition, which is a front group for the Nuclear Energy Institute, set up with the help of PR-giant Hill and Knowlton and millions of dollars from the nuclear industry.

Paul Gilding was head of Greenpeace Australia and then Greenpeace International before he started his own consultancy, Ecos, in 1995. Past and present clients of Ecos include companies and industries that have been targeted for their environmental misdeeds by environmental groups including Greenpeace.

Environmentalists who are on the lookout for a postenvironmentalist career tend to take more moderate stances on environmental issues and avoid being overly confrontational with corporations that damage the environment. Over time, as the larger, better-funded environmental groups have become more institutionalized, they have tended to hire career-minded professionals rather than enthusiastic activists. This has encouraged the revolving door between environmental groups and industry.

The institutionalization of large environmental groups has also seen an increase in the number of industry executives joining environmental groups at the top level. For example, in 2004 WWF-Australia appointed Greg Bourne as their chief executive officer. Bourne's previous career had been with the oil company BP, including a position as regional president of BP Australasia.

Some environmental groups that depend on corporate funds are particularly amenable to the involvement of current corporate executives in their operations. On occasion, corporations lend executives to the Nature Conservancy, for example, as in the case of Georgia Power, which loaned Gordon Van Mol from its External Affairs Department to be a member of the Nature Conservancy's development team for a year.

See Also: Environmental Movement; Green-Washing; Policy Process.

Further Readings

Beder, Sharon. *Global Spin: The Corporate Assault on Environmentalism*, 2nd ed. Devon, UK: Green Books, 2002.

Public Citizen's Congress Watch. "Congressional Revolving Doors: The Journey From Congress to K Street." *Public Citizen*, July 2005.

"Revolving Door." Center for Responsive Politics. http://www.opensecrets.org/revolving/index.php (Accessed November 2008).

Revolving Door Working Group. "A Matter of Trust: How the Revolving Door Undermines Public Confidence in Government—And What to Do About It." Revolving Door Working Group, October 2005.

Sharon Beder
University of Wollongong

Risk Assessment

Risk assessment is the first step in environmental risk analysis, a tool designed to assist decision makers in making environmental policy decisions. Risk assessment is meant to provide a science-based methodology with clearly defined steps to give decision makers the necessary background on which to base their decisions. Risk assessment is intended to allow decision makers to separate the fact-based elements of environmental issues from "political," value-based decisions. The extent to which it is able to do this is debated, but risk assessment has been widely adopted by national and international institutions as a model for helping decision makers formulate policy on complex scientific issues.

The classic model for risk assessment was first set out in the United States in a 1983 report, *Risk Assessment in the Federal Government: Managing the Process*, developed by the National Research Council Committee on Institutional Means for Assessment of Risks to Public Health. This report, widely known as the "Red Book," defined risk assessment as "the determination of the probability that an adverse effect will result from a defined exposure." The definition is commonly expressed as an equation, $R = H \times E$, or risk equals (probability of) hazard times exposure.

This method for determining what degree of hazard may ensue from human activities derives from techniques that were developed by toxicologists for assessing hazards from chemicals and food additives. The National Research Council published the Red Book in response to congressional requests for assistance in applying techniques for decision making about environmental risk in a comprehensive and uniform manner across a range of policy areas and federal agencies.

This type of risk assessment, also called "probabilistic risk assessment," is meant to define a standardized, step-wise process for solving this equation. The phases of ecological risk assessment include (1) hazard identification, or problem formulation, which involves choosing goals and desired end points, models to be used, and analysis method; (2) measurement and characterization of exposure and effects through dose response assessment and exposure assessment; and (3) risk characterization, which involves describing, integrating, and summarizing the previous steps. Questions for the risk assessor include: Is the risk chronic? How severe is the risk? Who or what is affected, and over what time period? Risk characterization ideally includes a statement of level of confidence, remaining uncertainties, and interpretation of adverse effects.

Techniques used in risk assessment include laboratory studies and field surveys. Laboratory studies, such as animal assays to assess dose-response relationships, are usually more controlled but less applicable to responses in the environment. Field surveys are more difficult to control but allow measurement and analysis of biological responses in the environment. Other techniques include models of varying levels of complexity, categorical rankings, and characterization of exposure and effects data. Some of the issues of concern are what data to use, collect, and fit into models; what statistical techniques to use in the analysis; whether to use indexes (aggregated data); and how long to collect and analyze data.

The goal of risk assessment is to create a tool for regulators that can be broadly used and is not subject to politically driven manipulation. It is meant to create a transparent process for decision making. This was of particular concern at the time of the Red Book's publication, because political interference into environmental decision making had become a major issue during the Reagan administration as a result of widespread concerns about political issues overriding scientifically justified decisions. William Ruckleshaus, head of the Environmental Protection Agency during the Nixon administration, returned to the chief Environmental Protection Agency position and laid out a plan for "science-based" risk assessment.

Ruckelshaus noted the importance of maintaining risk assessment as a process driven by science, with risk management to follow as the "political" portion of the risk analysis process. Although both he and the National Research Council recognized that the fact/value split foreseen by the separation between the risk assessment and risk management process was more a conceptual than a completely achievable distinction, efforts to keep the two as discrete components in the risk analysis process were still an important goal.

Ruckelshaus also highlighted the need for effective risk communication between risk assessors and the public throughout the process.

More recently, questions as to the adequacy of probabilistic risk assessment as recommended in the Red Book model have arisen, particularly for cases where uncertainty makes hazard identification and exposure assessment difficult. When there are high levels of uncertainty involved in the science used in risk assessment, values are more likely to infiltrate decisions. Furthermore, a reliance on quantitative data, especially where such data involve a number of judgments (as is the case with high levels of uncertainty), can obscure socioeconomic, cultural, and ethical concerns. In response, critics suggest that instead of claiming an impossible level of accuracy, both scientists and decision makers should be clear and explicit about uncertainties, allowing the public to make decisions based on what is unknown as well as on what is known.

Other recommendations for improving risk assessment include making the entire process more deliberative, particularly in addressing uncertainty. Some suggest that the entire process of conventional risk analysis needs to be rethought, with preference given instead to approaches such as precautionary approaches, alternatives assessment, decision-tree analysis, and life cycle analysis that focus on avoiding hazards earlier in the process, before exposure can occur.

Risk assessment can, in fact, form part of these other approaches. Risk communication and risk management can also address some of the concerns raised during (or about) the risk assessment part of risk analysis. The process of risk assessment and risk analysis continues to evolve, with increasing attention to iterative and adaptive approaches.

See Also: Cost-Benefit Analysis; Environmental Management; Institutions; Technology; Uncertainty.

Further Readings

Andrews, Richard N. L. "Risk-Based Decision-Making." In *Environmental Policy: New Directions for the Twenty-First Century*, edited by Norman J. Vig and Michael E. Kraft. Washington, D.C.: CQ Press, 2003.

EPA Risk Assessment Portal. http://www.epa.gov/ncea/risk/ecological-risk.htm (Accessed January 2009).

Glickman, Theodore S. and Michael Gough, eds. *Readings in Risk,* 4th printing, 1995 ed. Washington, D.C.: Resources for the Future, 1990.

Mayo, Deborah G. and Rachelle D. Hollander, eds. *Acceptable Evidence: Science and Values in Risk Management.* New York: Oxford University Press, 1991.

Morgan, M. Granger, et al. *Uncertainty: A Guide to Dealing With Uncertainty in Quantitative Risk and Policy Analysis.* Cambridge, MA: Cambridge University Press, 1990.

National Research Council. *Risk Assessment in the Federal Government: Managing the Process.* Washington, D.C.: National Academy, 1983.

Ruckelshaus, William D. "Risk, Science, and Democracy." *Issues in Science and Technology,* 1/3:19–38 (1985).

Ruckelshaus, William D. "Science, Risk, and Public Policy." *Science,* 221:1026–28 (1983).

Russell, Milton and Michael Gruber. "Risk Assessment in Environmental Policy-Making." *Science,* 236/4799:286–90 (1987).

Slovic, Paul, ed. *The Perception of Risk.* London: Earthscan, 2002.

Anna Milena Zivian
University of California, Santa Cruz

Risk Society

Coined by scholar Ulrich Beck, the risk society thesis is that risk is a key organizing principle in contemporary society. Beck's *Risk Society; Towards a New Modernity* (1992), and subsequent *World Risk Society* (1999), has become a significant and highly relevant synopsis of the interaction of humanity and its environment and the consequence this has for institutional change and political dynamics. The underlying message of *World Risk Society* is that an older industrial society, whose basic principle was the distribution of goods, is being replaced by an emergent risk society that is structured around the distribution of hazards.

This analysis distinguishes between three epochs of modernity, which include premodernity, industrial or first modernity, and finally, second or late modernity, which is defined as a reflexive modernity. It is the transitional stage between the latter two epochs of modernity that are the focus of Beck's work.

Epochs of Modernity

In general, the concept of modernity has been employed as a shorthand diagnosis for a broad set of social relations and processes that typify Western societies. Beck's understanding of modernization is representative of established observations of this period. Historically, modernization refers to the organization of social life that emerged from Western Europe during the 17th century. Technological advancements in many areas, including transportation, communications, and weaponry enabled early modern nations to establish colonizing empires. From the epicenter of Western Europe, modern practices radiated outward, conflicting with and ultimately subsuming, or juxtaposing, traditional forms of social organization.

Among others, Beck tells us that the processes of modernization are built on the Enlightenment notions of progress and scientific certainty, which holds a hegemonic position in creating the knowledge on which progress is founded. During this time, humanity is raised above other species, separating them from the natural global processes, to which previous generations of humanity had been bound. Centralized human settlement and intense urbanization produced new forms of social interaction and innovation, giving legitimacy to the purity of the scientific process.

Even in the latter stages of the 19th century, following intense industrialization when issues such as environmental despoliation, disease, and hunger were being witnessed on a massive scale, science was, paradoxically, still empowered to solve these problems. Built on this faith in science, unitary political structures in the form of nation-states developed policies that jostled for increasing access to, and control of, the world's resources. Moreover, there was the increasing success of a capitalist market system that was a driving force of political philosophy in the modern age. Here, the acquisition of wealth was abstracted from its environmental base through the development of a monetary system.

For Beck, however, the 21st century's recognition of the different forms of risks caused by unfettered scientific development and its inappropriate application by political structures was creating space for another form of modernity. According to Beck, modernity had turned inward and was questioning its most central tenets, creating a stage of reflexive modernity. Reflexive modernity is a complex collection of processes operating at many different levels within society. Reflexive modernity is a recursive turning of modernity on itself, in essence becoming its own theme.

Risk in a Risk Society

Beck refers to three primary forms of risk in the risk society. These are nuclear power, environmental despoliation, and genetic technologies. These areas have been variously expanded on over the course of Beck's work to include topical and emerging risks such as global warming. In a risk society it is important to understand that risks are not the same as destruction; instead, they are the probability of destruction. The probability of risk exists on a curve between security and destruction, and so once destruction has occurred, risks will cease to exist.

The probability of an occurrence for a given hazard, and the degree of susceptibility of the exposed to the source of that hazard, produces the variable vulnerability. So risk is a systematic way of dealing with hazards and insecurities induced and introduced by modernization itself. Combined with these two elements a third can be added: this is the capacity of humanity to deal with these hazards. Risk as presented in *World Risk Society* can be understood as follows:

$$\text{Risk} = \text{Hazard} \times \text{Vulnerability/Capacity}.$$

Beck is adamant that it is the perception of threatening risks that determines thought and action in a risk society. For Beck these risk perceptions create a "manufactured risk." That is, risks that are created both in a literal sense through inappropriate human action on the Earth and that are dependent on decisions made by people, firms, state agencies, and politicians. Although there is often a blurring of the boundaries between manufactured and natural risk, risks in a risk society are overwhelmingly focused on the socially contextual and mediated dimensions of risk. Beck uses the term *relations of definition* to highlight this point. These relations of definition are created by the dominant institutions, predominantly represented by government, firms, science, law, and the media.

Beck argues that new arenas and new institutions must be built where various experts and lay people may meet to negotiate questions of what now constitutes risk and how it should be dealt with. He claims that the nature of politics itself is changing; if the centralized nation-state politics of first modernity no longer applies, the focus needs to be on other political formations.

Reflexive Modernity

Reflexive modernity throws the central tenets of the industrial process into disarray. Science and technology no longer form the "holy grail" of knowledge formation, the relationship between established science and unconventional knowledge has become blurred, and the infiltration of the political into the scientific process disturbs the boundaries of expert and lay knowledge. In a reflexive modernity, global institutions and individuals are competing for political space. These observations outline the basic tenets of a reflexive modernity. With this said, what constitutes a reflexive modernity is complex and dynamic. There is particular confusion over the nature of reflexivity. The essence of the misunderstanding over reflexive modernity occurs when considering whether reflexivity represents, first, a purposeful knowledge-based action, which may be termed *reflection*, or second, should be considered as the unintended consequence of modernity that is *reflexivity*, a situation that is further complicated by a lack of distinction between the two. This issue is further confused, as Beck's position on this is dynamic. In earlier writings Beck alludes to a definition of reflexivity that is created by the unforeseen externalizations of the modern world that are reshaping the central components of modernity and notions of progress. However, as Beck's work develops, a softening of this position becomes evident, with the reflexive and reflective domains beginning to converge. In *World Risk Society,* Beck refers to society's response to global risks, which indicates a raising of awareness and a heightened reflection on risks in society.

Through processes of globalization in a risk society, the idea of the emergence of a reflexive modernity is applicable to both the global and the local levels of analysis through the rise of "cosmopolitan politics." Fundamentally, the idea of cosmopolitanism within a risk society is that there is a new relationship between the global and the local levels that does not easily fit into national politics.

Global and Local Connection

Beck understands the emergence of a reflexive modernity at the global and the local level in the following way. First, Beck argues for macroanalysis of a reflexive modernity through looking at the changing formations of global governance. These are most visible in the fluctuating architecture of the nation-state and the integration of broader social actors into conventional governance structures. For Beck, new forms of global risk are fundamentally altering the nation-state structure in both the developed and the developing world. Beck suggests that the nation-state has lost a degree of power, as risks have become transnational and have resulted in multinational coalitions and governance regimes that attempt to address these risks. Global warming is an example of a global risk that cannot be combated by a single nation-state.

Second, at the microlevel, reflexive modernity exists through altered social relations and the reconstitution of traditional networks of social cohesion. This is occurring both at the local governance level and on an individualized basis. At the local level within a risk society, the notion of subpolitics is used to define changing political structures and social interactions. According to Beck, the complex nature of risk has created a detachment from the conventional political base of the nation-state, bypassing the political parties and bureaucratic organs. Such activities can take on a number of guises; for example, the guerrilla policies of early Greenpeace activities or similar tactics employed by groups such as Fathers for Justice. Another dimension of subpolitics can include political mobilization of the public through the formation of events such as citizen juries, deliberate polls, or consensus conferencing. An important part of the reorganization of politics at the local level is the idea of individualization.

Initially, it is important to emphasize that individualization in the risk society does not represent the common-sense understanding of the term that would indicate a state of isolation or autonomy of the individual in society. Individualization in a risk society represents a reconstitution of the relationship between institutional structure and the individual. To exemplify this relationship Beck uses the term *institutionalized individualization.* Beck suggests that in a risk society the reorganization of the individual with relation to society occurs in three stages: First, the "liberating dimension," where the individual is removed from historically prescribed forms of commitment, such as marriage, family, and religious obligations. Second, there is the "detachment dimension," where the individual experiences a loss of traditional security with respect to faith and the fundamental guiding norms of society. These initial dimensions are followed by what Beck terms the *reintegration dimension*; here, there is a reembedding, a new form of social commitment and social control in the welfare state. To this three-dimensional model Beck grafts two subsequent factors. These are "objective" (life situation), and "consciousness" (identity, biography, personalization). These final two factors constitute a fourth dimension, which is used in varying combinations with the first three to produce a model that proposes six variations of individualization.

The analysis of contemporary social and political processes offered in a risk society has been criticized on a number of points. A significant number of these criticisms argue that the risk society does not take account of empirical evidence that points to alternative points of view. As such, the idea of a risk society has stimulated a significant amount of empirical work on the relationship between humanity and its environment.

See Also: Globalization; Sustainable Development; Uncertainty.

Further Readings

Beck, Ulrich. *Risk Society; Towards a New Modernity*. Newbury Park, CA: Sage, 1992.
Beck, Ulrich. *World Risk Society*. Malden, MA: Polity, 1999.
Borne, Gregory. *Sustainable Development: The Reflexive Governance of Risk*. Lampeter: Edwin Mellen, 2009.
Mythen, Gabe. *Ulrich Beck: A Critical Introduction to* The Risk Society. London: Pluto, 2004.

Gregory Borne
University of Plymouth

S

Sagebrush Rebellion

In 1979, roughly a dozen states in the American West launched the Sagebrush Rebellion (SBR), demanding the transfer of public lands managed by the U.S. Bureau of Land Management to state control. The dispute was rooted in early land ordinances wherein western territories relinquished claims to public land titles in exchange for statehood. By 1979, the federal government owned roughly 96 percent of Alaska; 86 percent of Nevada; over one-half of Utah, Idaho, and Oregon; and over one-third of California, Arizona, and Wyoming. Although a land transfer remained the central focus, reasons for the rebellion included conflicts over federal versus state power, urban versus rural concerns, and environmental conservation versus economic development. In all likelihood, a complex combination of these factors contributed to the rebellion. By the early 1980s, the Sagebrush Rebellion had dissipated, replaced, in part, by the Wise Use Movement. Even so, the central role of public lands in the changing dynamics of the West ensures that Sagebrush concerns continue to simmer just below the surface.

Initially, the federal government dispersed significant amounts of land through legislation (e.g., Homestead Act, 1862) and land grants (largely to railroad companies) to promote private development and generate revenue. In response to reports of corruption, and in line with the onset of the U.S. conservation movement, the General Revision Act of 1891 authorized the president to establish forest reserves from public land, signifying a shift in land management policies toward federal retention. Throughout much of the 1900s, disputes erupted over timber, irrigation, grazing, and mineral resource development. To address competing claims for public lands, Congress established a bipartisan Public Land Commission in 1964. The commission recommended that the federal government retain control of public lands for the common good, using multiple-use management policies, with attention to environmental quality and maintenance. The report's recommendations led to the Federal Land Policy and Management Act or Organic Act of 1976, establishing land retention as the federal government's primary approach to land management.

Despite federal retention, access to public lands remains available through a user-fee system. Nonetheless, rebellion supporters, primarily oil, coal, ranching, mining, and timber industries, argued that the bureaucracy of the federal government precluded westerners from using the land. New environmental legislation, most notably the Wilderness Act of 1964 and the National Environmental Policy Act of 1969, either forbade resource extraction altogether or required Environmental Impact Assessments before development—requirements thought to be unduly burdensome for those deriving their livelihoods from public lands. A primary target of the rebellion was the Carter administration's environmental policy, which emphasized conservation, reduced the number of grazing permits on public lands, and expanded wilderness designations.

In response to new land management policies and to mounting pressure from powerful industries, western lawmakers initiated a series of Sagebrush legislation. In 1979, Nevada governor Robert List signed Assembly Bill 413, declaring all Bureau of Land Management lands within its borders under state control. Thereafter, Utah senator Orrin Hatch introduced S1680, the Western Lands Distribution and Regional Equalization Act, which called for the return and title of unappropriated public lands west of the 100th meridian. Under the act, each state would oblige multiple-use management and be required to organize a Land Commission to oversee management. Sagebrush rebels advanced a legal argument based on "equal footing," claiming that because western states gave up so much land with statehood, they were not equal with other states, and thus, the federal government had no claim to public lands.

SBR supporters also framed the issue of public lands as one of states' rights, cultivating an image of a distant and oppressive bureaucracy managing the West with little knowledge of local issues. Rebellion supporters, pointing to federal mismanagement, argued that public lands would be more efficiently managed for economic benefit if controlled by states or private interests, as they would be more familiar with western land issues and personally invested in their management. Sagebrush rebels also cited lost revenue from property taxes and user fees as reason to transfer control of public lands to the states.

The SBR faced stiff challenges from national environmental organizations, such as the Sierra Club and the National Audubon Society. Opponents critiqued the rebellion as a greed-motivated land grab on behalf of big industry, citing state constitutional mandates to maximize profit on public lands as evidence that states would sell the land to private interests. Environmentalists argued that privatization would result in continued degradation and a loss of public access to open space, as industry would simply exploit the lands for quick profit.

Opponents to public land transfers also argued that states were too susceptible to manipulation by big business and lacked both financial and personnel resources necessary for adequate land management. Critics dismissed states' legal claims, arguing that public lands were acquired through conquest or purchase and were therefore never actually state lands. Finally, opponents pointed to the loss of yearly subsidies and payments in lieu of taxes (totaling over $100 million per year), as well as the high cost of management as financial incentive not to transfer control of public lands.

Additional opposition emerged from an increasingly urbanized western population not dependent on the land for economic opportunities but wanting to preserve open space and recreational opportunities. New urban residents played a crucial role in the conflict by raising concerns about land grabbing. Furthermore, urbanization undercut the political economic power held by extractive industries, fueling rebels' claims that policymakers undervalued the rural West. Specifically, rebels argued that the federal government's land management policies jeopardized western livelihoods by blocking access to resources necessary for economic survival.

The SBR assisted the presidential campaign of Ronald Reagan, who rode to power on the pretension of western roots. Rebels believed that Reagan's election and the subsequent nomination of James Watt to Secretary of the Interior would boost their cause. Yet, from the outset, Secretary Watt opposed land transfers, endorsing instead a "Good Neighbor

Policy." Watt loosened federal involvement in the West through decentralization, allowing greater local control over population growth and resource development. Consequently, support for public land transfers diminished. The rebellion suffered further when James Watt resigned as Secretary of the Interior in 1983, amid controversy and accusations of misconduct, taking Reagan's support for the rebellion with him.

Despite the rebellion's loss of momentum, concerns raised by the SBR persist. In 1995, several bills designed to rearrange public land management went before Congress on "county supremacy" claims. In 1996, a U.S. District Court ruling reasserted federal control over public lands, effectively quashing hopes of a transfer. Yet, in the final days of George W. Bush's administration, the Bureau of Land Management sold over one million acres of Utah's public land to oil and shale industries. With the election of Barack Obama in 2008, the momentum of the public lands debate could swing back in favor of environmentalists, with Obama pledging to maintain federal control of public land management. However, unless policymakers address the ideological clash between federal management of natural resources and rugged western individualism, public land disputes will likely continue.

See Also: Bureau of Land Management; Federalism; Forest Service; Wilderness; Wise Use Movement.

Further Readings

Cawley, McGreggor R. *Federal Land, Western Anger: The Sagebrush Rebellion and Environmental Politics*. Lawrence: University Press of Kansas, 1993.

Chomski, Joseph M. *The Sagebrush Rebellion: A Concise Analysis of the History, the Law, and Politics of Public Land in the United States*. Anchorage, AK: Legislative Directory Office, 1980.

Graf, William. *Wilderness Preservation and the Sagebrush Rebellions*. Lanham, MD: Rowman & Littlefield Publishers, 1990.

Kristen Van Hooreweghe
City University of New York

Silent Spring

In the history of the environmental movement, Rachel Carson is a name that carries a lot of influence, as well as controversy. She is credited with starting the environmental movement that is still going strong today, as it was her cautionary tale about the pesticide DDT (dichlorodiphenyltrichloroethane) titled *Silent Spring* that helped foster environmental awareness in people. Carson and her book received a lot of publicity for the dangers she predicted would occur to the environment and humans as a result of using DDT. Her book came out at a time when pesticides were recognized as reason for the progress that America saw after World War II. Bashing pesticide use was a very controversial act, and Carson saw a lot of opposition for it.

By the time Carson's *Silent Spring* was published in 1962 she was already an established author and scientist. She received her master's degree in zoology from Johns Hopkins University and went on to work in the Bureau of Fisheries (now known as the Fish and

Wildlife Service) in Washington, D.C. She started out writing science radio scripts but eventually acquired a full-time position as an aquatic biologist. Carson was the second woman to be hired by the Bureau of Fisheries for a nonsecretarial position. To pursue her second passion—writing—as well as to earn some extra money, she started writing articles for newspapers on marine zoology. In time, she left her position at the bureau to pursue a writing career. She published two books before writing *Silent Spring*, one of which made the best-seller list and won her a handful of literary awards. However, Carson's true fame came when she decided to use her writing career as a platform to express her doubts and worries about the widespread use of the pesticide DDT.

Swiss chemist Paul Hermann Müller discovered the effectiveness of DDT in 1939. After successfully controlling Colorado potato beetles on crops, Müller tried out the chemical on the lice found on war refugees and had equally positive results. The chemical was quickly patented in 1940, and field tests were started. In the field tests DDT was successful at stopping small typhus epidemics in Egypt, Mexico, and Algeria. The new pesticide was also responsible for decreasing cases of malaria in parts of the United States, India, Italy, and Venezuela. Müller could not have discovered the effectiveness of DDT at a more opportune time. In the beginning months of World War II, Allied doctors feared that more soldiers would be dying from insect-borne diseases than from bullets. The War Production Board began encouraging the manufacture of DDT, and by 1943 the pesticide was on army supply lists, and a year later it was on navy supply lists as well.

The "Wonder Insecticide"

By the end of the war, DDT was the most publicized synthetic chemical in the world. It had been dubbed the "wonder insecticide" for its great success at protecting soldiers from malaria, typhus, dysentery, and typhoid. People were excited about the chemical and were eager to use it for agricultural purposes. In 1945, the U.S. Department of Agriculture called DDT a "two-edged sword" and warned the public that if the chemical were improperly used, it could be extremely menacing. A few short months after that statement was released, however, DDT became available for civilian use. The Food and Drug Administration established that 7 parts per million of DDT on food was safe for humans, and the United Nations supported large-scale spraying of the pesticide to lower incidences of malaria and to increase food production in developing nations. By the time Müller received the Nobel Prize for Medicine in 1948, any qualms or questions about DDT had seemingly been forgotten.

Rachel Carson was not so easily swayed by the Food and Drug Administration's hasty research. She did not believe that a chemical that had such an effective impact on bugs was safe for the environment, and more important, for humans. She approached publications such as *Reader's Digest* about letting her research the negative side of DDT and its widespread use, but she was turned down because of the chemical's popularity. Carson was determined to show the dark underbelly of pesticide use and decided to forgo magazines and newspapers by writing a book on the subject instead. Carson began to conduct research with the help of some scientists she met while doing oceanographic research and with support from her friends. She had compiled so much research that it proved to be a roadblock to getting her work published. She received rejection letters that stated the book was "too great an undertaking on the basis of the material submitted." Undeterred by this setback, Carson made revisions to her work and eventually caught the interest of several journals as well as *The New Yorker*. The magazine had offered to help condense and publish some selected chapters from *Silent Spring*. These chapters were released as a three-part series on June 2, 1962. Carson's work ignited editorial commentary throughout the country. Houghton Mifflin ordered 100,000 copies of *Silent Spring* to be printed as a result of the interest that the articles had generated among the public. Her book opened people up to the idea that something they had thought of as a miracle chemical might actually be harmful to the environment, and more important, to themselves. This was the first time anyone had publicly presented the idea that there was a flip side to chemical agriculture. Few people had considered that altering an ecosystem could have an effect that reached all creatures, including humans. *Silent Spring* described how the technological progress that had contributed to so much prosperity following the war was in conflict with natural processes.

Not Everyone Loved It

Despite being a literary success, Carson's book encountered much opposition. She was criticized for using scare tactics to drive home her message. Her opponents accused her of exaggerating the truth and only telling one side of the story to make her case. One of the most controversial claims found in her book was that DDT was causing a high rate of cancer among children. Critics argued that she was incorrectly presenting statistics. She declared that the incidences of cancer among children were rising. However, as her critics were quick to point out, the incidences of other diseases were decreasing among children. Carson had failed to provide some context when comparing the statistics for cancer rates to those of other diseases. Overstatements, holes in the truth, and lack of context such as the cancer claim were seen as a way to generate good public relations for her cause—and for her book. Carson was viewed by her opponents as a spinster and an ardent feminist who was more concerned with having a best-selling novel with than public health. The book was released not long after the tranquilizer thalidomide was found to cause birth defects, which made it seem that Carson was using the failure of this chemical to validate the claims she was making about DDT in her book.

The message in *Silent Spring* was interpreted as completely antipesticide, which angered food growers and distributors, as well as many others involved in the agricultural industry. Chemical agriculture was seen as one of the greatest advances in technology of the time, so Carson's book seemed blasphemous. She was accused of being antipesticide and supporting the health of insects over the health of humans. This was a serious charge against Carson, considering that DDT had been extremely effective in reducing cases of malaria and typhus in various places around the world and was credited with the rapid progress the United States saw after World War II. Food production and efficiency had gone up significantly since the advent of DDT. Those in the industry argued that trace amounts of chemicals were the price that had to be paid in exchange for more efficient food production, and in turn, lower food prices. The Food and Drug Administration had previously established limits to the amount of pesticide that could be found in food, thus supporting this argument against Carson. To further disprove Carson's point, the agricultural industry agreed that pesticides do in fact disrupt the balance of nature but argued that this disruption of nature helped humans more than it harmed them.

Lasting Impact

Despite the numerous attempts to disprove and discredit Carson, her book *Silent Spring* has had a tremendous effect on society and on the government. After the book came out,

President John F. Kennedy ordered the President's Science Advisory Committee to examine DDT and its effects. What the committee reported was that there are many faults in the Department of Agriculture's pesticide-approval process. A pesticide that had been denied by the Department of Agriculture could still be used because of loopholes in the process. This finding backed Carson's warning about the indiscriminate use of pesticides.

This led to a push in all levels of government for the regulation of pesticide use. More than 40 bills were passed in different state legislatures on this issue following the release of *Silent Spring*, and in 1970 the Environmental Protection Agency was created to handle environmental issues that face the public, such as pesticide use. The book also garnered attention from governments around the world. It was published in 13 European countries as well as in Japan, Iceland, and Brazil, and inspired governmental action and environmentalism in these countries. Before the release of the book there was no such thing as an "environmentalist." Concern for the environment and how humans affect it did not exist until Carson voiced her concern about DDT and its effect on nature and humans. Whether all of the charges made against DDT were completely true—or even true at all—Carson can be credited with creating environmental awareness. Although Carson insisted that she was never for a complete ban on DDT, these newly created environmentalists celebrated a huge victory on June 14, 1972, when the Environmental Protection Agency banned the use of DDT after 7 months of hearings.

Unfortunately, Carson did not live to see this accomplishment, but her efforts have not been forgotten. She was named one of the 100 "people of the century" by *Life* magazine in 1990 and was given the same honor by *Time* magazine in 1999. In 1992, *Silent Spring* was selected as the most influential book of the past 50 years and was a motivating factor in the creation of Earth Day in 1970. Despite her accolades and supporters, Rachel Carson and her book still remain controversial so many decades later. There are those that still argue that DDT's link to cancer has not yet been established. Since the ban on DDT, there has been a significant rise in malaria cases in poorer countries, which has caused many anti-DDT activists to concede that there are benefits to the use of pesticides. Long after the release of *Silent Spring* the question remains: Is the disruption of nature beneficial to man, or will it come back to haunt him? This question is truly the legacy of Rachel Carson.

See Also: Agriculture; Environmental Movement; Environmental Protection Agency.

Further Readings

Bailey, Ronald. "*Silent Spring* at 40." *Reason Online* (June 12, 2002). http://www.reason.com/news/show/34823.html (Accessed February 2009).

Beyl, Caula A. "History of the Organic Movement." http://www.hort.purdue.edu/newcrop/history/lecture31/r_31-3.html (Accessed January 2009).

Carson, Rachel. *Silent Spring*. Boston: Houghton Mifflin, 1962.

Davis, Kenneth S. "The Deadly Dust: The Unhappy History of DDT." *American Heritage Magazine*, 22/2 (February 1971).

Lytle, Mark Hamilton. *The Gentle Subversive: Rachel Carson,* Silent Spring, *and the Rise of the Environmental Movement*. New York: Oxford University Press, 2007.

Milne, Lorus and Margery Milne. "There's Poison All Around Us Now." *New York Times* (September 23, 1962) http://www.nytimes.com/books/97/10/05/reviews/carson-spring.html (Accessed February 2009).

National Resources Defense Council. "The Story of *Silent Spring*." http://www.nrdc.org/health/pesticides/hcarson.asp (Accessed April 1997).

Tierney, John. "Fateful Voice of a Generation Still Drowns Out Real Science." *New York Times* (June 5, 2007) http://www.nytimes.com/2007/06/05/science/earth/05tier.html?n=Top/Reference/Times%20Topics/People/T/Tierney,%20John (Accessed February 2009).

Arthur Mathew Holst
Widener University

Skeptical Environmentalism

Skeptical environmentalism is an umbrella term used, sometimes pejoratively, to describe dissent from mainstream views on environmental problems. Skeptical environmentalists tend to dispute the authenticity of specific environmental problems, especially anthropogenic climate change, but also the more general idea that the human enterprise is in a dangerous state of ecological overshoot. Historically, skeptical environmentalists have been known to take an antiregulatory stance, arguing that the environmental movement intentionally distorts scientific data to support its own political agenda, that environmental threats are exaggerated in the public debate, and that most environmental protection policies are based on "junk science." Despite this alleged concern with scientific standards, it is striking to note that almost no work by skeptical environmentalists has been published in refereed academic journals or by publishers who subject their books to peer review. Instead, skeptical environmentalism is primarily published as monographs, often by conservative or libertarian think tanks such as the Cato Institute, the Heritage Foundation, and the Hudson Institute. These connections have recently been studied in a large literature survey that concluded that of 141 English-language environmental skeptical books published between 1972 and 2005, 92 percent were linked to different conservative think tanks.

Witnessing the global warming controversy, some have argued that skeptical environmentalism may be more about creating sufficient "noise" and conflict within epistemic communities than about settling a sincere debate between claims. Yet if skeptical environmentalism may be mistaken in its dismissal of scientific evidence on empirical conditions, it does not follow that there should not be legitimate room for debate on the politics of the environment, in particular, on the role of technology in securing sustainability. Although many authors in the green field believe that environmental sustainability will require radical decentralization and an end to consumer capitalism, skeptical environmentalists have argued that exactly the opposite; namely, sustained economic growth and accelerated technological innovation, may offer a more promising path to long-term sustainability, especially when considering the needs for a dramatic improvement of living standards in the developing world.

Much of the debate on these issues has been focused around the existence of an Environmental Kuznets Curve. The curve suggests that the pattern of environmental degradation follows an inverted parabolic curve, meaning that environmental degradation increases during the early stages of industrial development but then begins to decrease as the economy matures. Empirical studies have shown that there is indeed such a U-shaped pattern for certain specific pollutants and that, for instance, air quality has improved dramatically in many large cities over the last century. Yet green critics have argued that much of the improvement in environmental performance can be attributed to the migration of polluting industries to the developing world (pollution haven hypothesis) and that studies employing ecological footprint analysis have shown that the increased eco-efficiency of mature economies remains insufficient to compensate for their larger productive capacity and consumption rates. At the same time, skeptical environmentalists may argue that with rising affluence comes an accelerated innovation rate and that the real substantial improvements in environmental performance may still be ahead of us as breakthrough technologies (such as nuclear fusion, geoengineering, or space industrialization) are made available. All these questions seem to offer ample room for reasonable disagreement and political debate. To understand why such debates nonetheless often remain fruitless, it is important to realize that for skeptical environmentalists and radical greens alike, this is not merely a debate about specific policy measures but, rather, a debate about modernity as such. Although radical greens may advance a kind of "future primitivism," skeptical environmentalists like to cast themselves as defenders of civilization and of the very idea of scientific progress. With such an underlying polarization, it is only natural that learning from each others' positions becomes frustratingly difficult.

To some, the emerging school of bright-green environmentalism may offer a bridge across this fractured political landscape, as it accepts the empirical reality of the sustainability crisis yet combines it with a normative belief in human ingenuity otherwise characteristic of skeptical environmentalism. Instead of green frugality and a focus on resisting capitalism, bright-green environmentalism takes a stance in favor of science and social innovation, arguing that to win sufficient democratic support, green visions have to project a future of prosperity and improved quality of life.

Lomborg and Controversy

Returning to the work of individual skeptical environmentalists, the Danish political scientist Bjørn Lomborg seems to warrant a special note. Lomborg became internationally famous in 2001 when his book on environmental economics, *The Skeptical Environmentalist*, was published by Cambridge University Press after an extensive peer-review process, in itself a rare exception for the genre. Not only did Lomborg's book popularize the notion of skeptical environmentalism but it also sparked one of the most heated debates ever in environmental politics. In the book, Lomborg makes the claim that over the last centuries, most indicators of human welfare have steadily been improving and that, serious as specific environmental problems may be, the overall prospect for humanity's future appears encouraging. Contrary to what he refers to as the "litany of our ever deteriorating environment," Lomborg presents hundreds of statistical tables and figures to support his case, repeatedly suggesting that the environmental movement has either unintentionally or maliciously fabricated or at least grossly exaggerated different global problems.

Unsurprisingly, these charges made by Lomborg did not go unanswered for long. Members of the Danish and international scientific community accused Lomborg of scientific dishonesty, and in 2002 a case was brought against Lomborg to the Danish Committees on Scientific Dishonesty. In its ruling the following year, the Danish Committees on Scientific Dishonesty concluded that the book was indeed scientifically dishonest even as Lomborg himself was found not guilty, arguing that he lacked scientific expertise in many of the fields he was writing about.

Though the ruling of the Danish Committees on Scientific Dishonesty was in itself disputed later on, any critical reading of *The Skeptical Environmentalist* reveals that Lomborg makes many of the mistakes he accuses the environmental movement of, such as selective citation, deliberate misinterpretation of scientific results, and incorrect application of statistical methods. That said, most critics of the book also seem to have failed to see the more insightful side of Lomborg's concluding argument; namely, that in a world filled by worthy causes (among which environmental protection certainly should be one) we are forced into making hard decisions and prioritizing. Skeptical environmentalists may then be correct when suggesting that many green thinkers care too much about potential suffering in the future and too little about actual suffering today. Yet, given our imperfect understanding of the sciences and processes involved, it can be that failure to take immediate action on, for instance, global climate change that may put all our survival at risk, effectively invalidating the assumptions behind the kind of cost-benefit analyses recently championed by Lomborg as part of the so called "Copenhagen Consensus."

Moving toward a conclusion, it seems clear that although much of the literature on skeptical environmentalism can rightfully be reduced to simple denial or seen as the biased views of narrow industrial interests, we are still left with a core set of crucial big-picture questions concerning the future of humanity that will require our best political judgment. Although skeptical environmentalism traditionally has spent almost all its energy on falsifying the science behind environmental problems, it is likely that the coming years will see a move toward a more policy-oriented position in which the question will not be so much whether the environmental problems are real or important but, rather, what we are to do politically about them.

See Also: Ecological Modernization; Kuznets Curve; Wise Use Movement.

Further Readings

Beckerman, W. *Small Is Stupid: Blowing the Whistle on the Greens*. London: Duckworth, 1995.

Brockington, Dan. "Myths of Skeptical Environmentalism." *Environmental Science & Policy*, 6/6:543–46 (2003).

Jacques, P. "The Rearguard of Modernity: Environmental Skepticism as a Struggle of Citizenship." *Global Environmental Politics*, 6/1:76–101 (2006).

Jacques, P. J., et al. "The Organisation of Denial: Conservative Think Tanks and Environmental Scepticism." *Environmental Politics*, 17/3:349–85 (2008).

Lomborg, B. *The Skeptical Environmentalist: Measuring the Real State of the World*. Cambridge, MA: Cambridge University Press, 2001.

Simon, J. *The Ultimate Resource 2*. Princeton, NJ: Princeton University Press, 1996.

Rasmus Karlsson
Lund University

Social Ecology

Social ecology stresses the link between the domination of humans and the domination of nature, envisioning the creation of a nonhierarchical society as the solution to both contemporary ecological and social crises. Social ecology's ideal society mirrors the integrative and communitarian order of natural ecology—characterized by dynamic unity in diversity. For social ecology, theory and activism are inevitably linked, as expressed in the programs of the Social Ecology Institute.

Social ecology, as developed by Murray Bookchin, has as its central premise the idea that the domination of other humans occurs in concert with the domination of natural systems. Both stem from hierarchical social arrangements that set men over women, rich over poor, race over race, humans over nature, and mind over matter. In such stratified societies, ideologies of objectification and instrumentalization develop in concert with a market system that prices everything—including human and natural life. Domination alienates humans from their true nature and potential for life in community, as well as from their essential freedom.

Technology flowing from capitalism (the supreme example of a dominating system) produces the toxic results laid out in *Our Synthetic Environment,* which Murray Bookchin published a few months before Rachel Carson's *Silent Spring*. In 1964, Bookchin predicted the greenhouse effect in his analysis of capitalism's "grow or die" imperative that turns water and airways into "sewers."

Social ecology's method is that of an "integrative science," consciously bridging ideas rather than ranking and separating them in the manner of a hierarchical paradigm. Thus Bookchin conceives of social ecology as a science in dynamic interaction with imagination, a rational search for truth in dynamic interaction with concrete history, and an evolving theory in dynamic interaction with activism.

The aspect of dynamism here is a key one. Both ecologically sound social systems and the systems of thought that nurture them must continually evolve through dialogue and criticism, reflecting the dynamism of living systems rather than the stasis of final answers and stagnant institutions produced under paradigms of domination and control.

Social ecology's central integration is its joining of social critique with an ecological model, thus giving ecology a "revolutionary edge" and socialism a focus on the major contradiction of our time: that between capitalism and natural systems. In combining socialist with ecological perspectives, it also differs from each of these. Though social ecology concurs with Marx's standard, "from each according to their ability, and to each according to their need," it broadens the socialist focus on class oppression to encompass the oppression flowing from all hierarchical arrangements. It specifically rejects the instrumentalization of nature expressed in the socialist analysis of human labor as adding primary value to natural resources. It emphasizes not only the ways that humans work on nature but also the ways that nature works on humans.

Further, Bookchin stresses the distinction between social ecology and any environmentalism caught up in emotion and lacking rigorous assessment of the historical roots of ecological crises.

According to social ecology, the only way to counter the intertwined social and ecological crises of our day is to abolish all hierarchy. This is both necessary and intentionally utopian. As Bookchin puts it, those who cannot imagine the impossible will have to live with the unthinkable.

But social ecology also insists that its vision is realistic—as it is grounded in both nature and history. In nature, the ecological order expresses mutualism and reciprocity, as illustrated by the circulation of food through ecological systems. Here no species can properly be considered higher or lower in that they all feed and are fed by others. In parallel fashion, natural evolution creates both increasing systemic complexity and potential freedom through its proliferation of particulars in the ordering of unity in diversity. Thus nature "on its own terms" (rather than a transcendent being or principle) provides the model of communitarian ethics necessary to an ecologically sound and just society.

In turn, Bookchin notes humans lived in a state of harmony with nature in the Neolithic period, in nonstratified "organic societies." According to his analysis, however, equality and freedom were so naturally embedded in these societies that their members never consciously chose them. Here Bookchin locates the justification for historical evolution away from such organic societies to the state societies of today.

He stipulates that human freedom and equality had to be lost to be understood as concepts to value and protect. Thus the "dialectic of freedom" is historically intertwined with the "dialectic of domination"—leading to the possibility of consciously choosing freedom and justice today. If humans fail to choose these, however, hierarchical capitalism will continue to move toward both totalitarianism and the destruction of ecological systems.

Though it is grounded in natural and historical example, social ecology's ideal society must be consciously designed, not merely inherited from peeling away the destructive detritus of social institutions, as the anarchists Bookchin broke with later in his life believed. Rationality must guide the choice of how and when to act in creating such a society. In this context, critical thinking and education are essential forms of activism.

Modern science has an essential part to play in social ecology's goals. Though capitalism currently misuses technology, the scientific knowledge that produces it has brought humans to a place in history where "almost anything is possible." Appropriate technology can release human imagination as it alleviates the need for physical drudgery and moves societies beyond the scarcity that creates competition.

Thus social ecology's sustainable society is not one of depravation, but one rich in quality of life and occasions for expressing creativity. In making choices that actualize human potential by implementing such a society, Bookchin posits that human nature ("second nature") expands the potential of the "first (ecological) nature" in which it is inevitably embedded.

Social ecology also avoids the "indulgent individualism" of both anarchy and modern consumerism. In contrast, it proposes a communitarian model of interdependent individuals living in small-scale communities, confederated into a network of participatory democracies in which all those affected by decisions help make them. In this scenario, leadership and responsibility voluntarily circulate among different individuals.

This network of direct democracies is grounded in "ethical economics" geared to the well-being and sustainability of ecological systems and human life, rather than amassing wealth for a few.

Social ecology's integration of theory and action is expressed in the Social Ecology Institute Bookchin cofounded in 1974. The institute, affiliated with various colleges during its career, has sponsored degrees, conferences, and projects focusing on direct democracy, ethical economics, resistance to biotechnology and nuclear technologies and the development of alternative technologies, permaculture, the shaping of the U.S. Greens, ecofeminism, and support for Native American self-determination.

See Also: Anarchism; Capitalism; Domination of Nature; Ecofeminism; *Silent Spring*.

Further Readings

Bookchin, Murray. "The Communalist Project." *Harbinger,* 3:1 (2003).
Bookchin, Murray. *The Ecology of Freedom.* Palo Alto, CA: Cheshire Books, 1982.

Bookchin, Murray. *Post Scarcity Anarchism.* Montreal: Black Rose Books, 1986.
Bookchin, Murray. "Reflections, An Overview of the Roots of Social Ecology." *Harbinger,* 3:1 (2003).
Bookchin, Murray and Eirik Eigland. *Social Ecology and Communalism.* Oakland, CA: AK Press, 2007.
Staudenmeier, Peter. "Economics in Social-Ecological Society." *Harbinger,* 3:1 (2003).
Tokar, Brian. "Social Ecology and Social Movements." *Harbinger*, 3:1 (2003).

Madronna Holden
Oregon State University

Steady State Economy

The dominant model of modern economics and politics relies on continuous economic growth, regardless of the consequences for the environment. Many economists and politicians argue that economic growth is good and necessary for the environment, as we need to grow to a certain level to be able to afford to clean up the costs of pollution. This permanent growth in number and types of goods and services is countered by the steady state economy concept, which focuses on the relative stabilization of population and consumption over time. The steady state economy is a central element of ecological economics, though it was first developed by conventional economists to refer to a condition in which output and capital per worker are balanced over time.

Many economists and other proponents of unlimited economic growth hold the view that resources are relatively unlimited, and thus a steady state economy is not necessary. Indeed, humanity is in the midst of an era of unprecedented growth and affluence. However, one of the critical underlying principles of steady state economics is the second law of thermodynamics, commonly known as the law of entropy. This law simply means that energy, in the form of heat, is lost when activity occurs—a fact that modern technological innovation has not altered. Thus, ecological economists and other environmentalists argue that ultimately, the realities of increasing resource usage and ecological degradation will run up against the ecological limits to growth, necessitating the development of a steady state economy. The ecological limitations appear to be manifesting themselves in the form of significant threats to the supplies of most of the major resources humanity relies on for survival—water, fisheries, oil, natural gas, and many other basic resources are on the decline, some at a fairly rapid rate. In addition, mounting threats to plant and animal biodiversity, and the intensification of pollution in the water, in the soil, and in the air, all contribute further support for the position of proponents of a steady state economy.

The three key principles of a steady state economy include

- sustainable scale,
- fair distribution of wealth, and
- efficient allocation of scarce resources.

The first principle, development on a sustainable scale, prioritizes a balance between the size of the economy and the capacity of the planet's ecosystems. This fundamentally alters the dominant economic thinking, which relies on continuous growth of the economy, disregarding the stock of available natural resources as well as any damage to those resources caused by production activities. This has profound consequences in the industrial era, as human economic activity has increased exponentially, resulting in ecological damage and dwindling resources. A steady state economy seeks to develop and maintain harmony between environment and ecology.

A fair distribution of wealth is the second principle of a steady state economy. Poverty has remained an endemic condition worldwide during the centuries of sustained economic growth. A steady state economy would obviously lead to reduced levels of economic growth, likely contributing to the growth of worldwide poverty, assuming continued patterns of social and economic inequality. Thus, more equal distribution of economic resources is critical, as those who are poor are less able to concern themselves with sustainability when their very survival is at stake.

Finally, efficient allocation of scarce resources is the third element that contributes to the development of a steady state economy. Simply put, minimizing the influence of external forces and maximizing the ability of actors to participate in economic exchange allows for the most efficient allocation of resources in the marketplace. Ecological economists would contend that it is only necessary to consider efficient allocation after achieving sustainable scale and fair distribution, as it means little in an unjust, unsustainable economic system (i.e., one that is out of balance or not in a steady state).

Societies have not moved in the direction of a steady state economy, though some contend that policies like ecosystem or endangered species protections are the types of efforts that are necessary to move toward a steady state economy. However, continued growth ideologies and policies are dominant, leading adherents of the steady state economy to argue for a rapid shift in economic and social activity toward sustainability. One concern today is that politicians who champion "green" growth are simply using buzzwords to curry favor among their constituencies while offering policies that provide little or no progress toward sustainability or a steady state economy.

Development of a steady state economy offers humanity a different path for the future, one that would lead to a radically different relationship to nature. Achieving the consensus that such a balance is necessary after centuries of growth will be difficult to achieve, short of ecological catastrophe or the development of an ethic of sustainability among a large portion of the Earth's population, particularly those in power in the advanced industrial economies responsible for most of the ecological degradation and resource depletion to date.

See Also: Ecological Economics; Limits to Growth; Malthusianism; Sustainable Development.

Further Readings

Attarian, John. "The Steady-State Economy: What It Is, Why We Need It." http://www.npg.org/forum_series/steadystate.html (Accessed January 2009).

Center for the Advancement of the Steady State Economy. http://www.steadystate.org (Accessed January 2009).

Daly, Herman. *Steady-State Economics: Second Edition With New Essays*. Washington, D.C.: Island Press, 1991.

Todd L. Matthews
University of West Georgia

Stockholm Convention

The Stockholm Convention on Persistent Organic Pollutants is a binding global treaty that protects human health and the environment from a group of dangerous man-made chemicals known as persistent organic pollutants (POPs). The convention bans or severely limits the production, use, and trade of 12 particularly toxic POPs, often referred to as the "dirty dozen." These are the pesticides aldrin, chlordane, DDT (dichlorodiphenyltrichloroethane), dieldrin, endrin, heptachlor, mirex, and toxaphene; two industrial chemicals, polychlorinated biphenyls and hexachlorobenzene (which is also a pesticide); and two byproducts of certain industrial processes and combustion, dioxins and furans. The convention also establishes criteria and procedures for placing controls on additional POPs, requires efforts to reduce emissions from other existing POPs, and seeks to prevent the development and commercial introduction of new POPs.

POPs possess four key characteristics: toxicity, persistence, bioaccumulation, and long-range environmental transport. POPs are among the most toxic pollutants that humans release into the environment. Although extensive variations occur across substances, species, and exposure variables, the effects of POPs on humans and wildlife can include cancers, reproductive disorders including birth defects and infertility, developmental impairment, disruption of the immune system, damage to nervous systems, allergies, hypersensitivity, endocrine disruption, and in large doses, acute and lethal toxic poisoning.

POPs are also very stable, persistent compounds. Their relatively long half-lives allow them to remain in the environment for extended periods before breaking down into less harmful substances. POPs also bioaccumulate within individuals or food chains, magnifying the effect of individual exposures. Once ingested, POPs tend to remain in the fatty tissue of living organisms, meaning that even very small exposures can lead to larger effects within particular individuals. Indeed, concentrations in animals and people have been found at levels up to 70,000 times the background levels found in the surrounding local environments. Mammals and humans can then pass these chemicals to their offspring through breast milk. In addition, fish, predatory birds and mammals, and humans can absorb high concentrations of POPs quickly if they eat animals in which POPs have already accumulated.

Finally, POPs can travel long distances from their emission source via air currents, waterways, species migrations, and the "grasshopper effect," in which they are transported through an often seasonal process of evaporation and redeposit through rainfall. This ability to engage in "long-range transport" across international borders means that POPs can be found in ecosystems, waterways, animals, and people thousands of miles from the nearest location of their production, use, or release.

The Stockholm Convention is the most recent in a series of treaties adopted by the international community to create global policy to protect human health and the environment from hazardous chemicals, concern about which started in the 1960s. It follows in the footsteps of the 1989 Basel Convention on Hazardous Wastes, the 1998 Rotterdam Convention on Prior Informed Consent, and the 1999 Aarhus POPs Protocol to the Convention on Long-Range Transboundary Air Pollution. It also builds on the work done over the last four decades by a number of international organizations active on issues related to particular aspects of global chemicals policy, including the International Forum on Chemical Safety, United Nations Environment Programme (UNEP), United Nations Food and Agricultural Organization, and World Health Organization.

Efforts to initiate global action on POPs began during preparations for the 1992 Earth Summit but accelerated significantly in June 1996, when an international scientific assessment led the Intergovernmental Forum on Chemical Safety to conclude that the risks posed by the dirty dozen POPs warranted international action, including a global legally binding agreement. These recommendations were forwarded to the governing bodies of UNEP and World Health Organization.

In February 1997, at the Governing Council of UNEP, the world's governments agreed that a global legally binding treaty was needed to reduce the risks to human health and the environment posed by the dirty dozen and requested that the UNEP Secretariat convene formal negotiations. These began in Montreal on June 29, 1998, with the first session of the Intergovernmental Negotiating Committee for an International Legally Binding Instrument for Implementing International Action of Certain Persistent Organic Pollutants. Three years of negotiations ensued. Individual sessions included five official week-long meetings of the committee; two meetings of a Criteria Expert Group, which focused on criteria and procedures for identifying and potentially adding new chemicals; a separate set of consultations on financial assistance issues; numerous formal contact groups; countless informal consultations; and a final Conference of Plenipotentiaries.

On May 23–24, 2001, more than 110 countries and the European Commission unanimously adopted the final text of the convention during the Conference of Plenipotentiaries in Stockholm. The convention officially entered into force in May 2004, following its 50th ratification. As of March 2009, more than 160 countries have ratified the convention. The United States and Russia are the only major countries that are nonparties. The convention requires all parties to

- immediately ban the production and use (with certain time-limited exceptions) of aldrin, chlordane, dieldrin, endrin, hexachlorobenzene, heptachlor, mirex, and toxaphene;
- severely restrict the production and use of DDT; the only permitted use is for disease vector control, especially against malaria mosquitoes, and this is to be reduced over time as alternatives become less expensive;
- take measures to minimize and, where possible, eliminate releases of dioxins and furans;
- ban the import or export of POPs controlled under the convention except for narrowly defined purposes or environmentally sound disposal;
- promote the use of best available technologies and practices for replacing existing POPs, reducing emissions of POPs, and managing and disposing of POPs and materials containing POPs; and
- make efforts to prevent the development and commercial introduction of new POPs.

To ensure that the treaty continues to grow in response to new information on threats to human health and the environment, the convention established clear, scientifically based criteria and a specific step-by-step procedure for identifying, evaluating, and adding POPs to the treaty's control regime. A POPs Review Committee made up of technical experts meets on a regular basis to consider nominated chemicals in detail and develop risk profiles and risk management evaluations of candidate POPs. The Conference of the Parties, the supreme decision-making body of the convention (composed of governments that have ratified the treaty) holds final decision-making authority. The review committee has met four times as of March 2009, and it recommended that the Conference of the Parties consider placing controls on nine more chemicals, including chlodecone, hexachlorocyclohexane, lindane, and pentachlorobenzene. The Conference of the Parties considered these recommendations at its fourth official meeting in May 2009.

Another critical feature of the convention is the financial mechanism that assists developing countries in fulfilling their treaty obligations. Developed countries provide these resources, which are administered, at least initially, by the Global Environment Facility. Technical assistance is also provided to developing countries, most notably through 12 Stockholm Convention Regional Centers. The centers cooperate with implementation activities for the other global chemical and hazardous waste treaties, creating synergies and reducing the overall costs of implementing the treaties. The convention also includes very important provisions on reporting, monitoring, research, information exchange, and public education.

The Stockholm Convention is the first global treaty that seeks to eliminate substances directly toxic to human health and the environment. It is the centerpiece of global policy to control toxic chemicals, and its success or failure could prove critical to mankind's efforts to safeguard the environment and human health from the effect of such substances.

See Also: Anti-Toxics Movement; Basel Convention; Bhopal; Endocrine Disrupters; PCBs; Toxics Release Inventory.

Further Readings

Colborn, Theo. *Our Stolen Future: Are We Threatening Our Fertility, Intelligence, and Survival? A Scientific Detective Story*. New York: Dutton, 1996.

Downie, David and Terry Fenge, eds. *Northern Lights Against POPs: Combating Global Toxic Threats at the Top of the World*. Montreal: McGill-Queens University Press, 2003.

Stockholm Convention Secretariat. http://www.pops.int/ (Accessed February 2009).

Stockholm Convention Secretariat. *Ridding the World of POPs: A Guide to the Stockholm Convention on POPs*. Geneva: United Nations Environment Programme (UNEP), 2005.

David Downie
Fairfield University

Structural Adjustment

Beginning with the Third World debt crisis in the 1970s and 1980s, the World Bank and the International Monetary Fund (IMF) have made loans contingent on debtor nations deploying sweeping structural adjustment programs (SAPs). Enforcement of structural adjustment as a condition for receiving IMF and World Bank loans serves to integrate debtor nations into the world system in ways that typically advantage transnational corporate and financial interests in the First World at the expense of economic and social interests of Third World citizens. Structural adjustment is premised on the idea that a debtor nation's economic problems derive entirely from structural deficiencies within its own economy. The national economy must therefore adjust to the world economy. In reality, the economic crises Third World debtor nations face result largely from legacies of colonialism combined with external pressures deriving from the ability of First World nations and transnational corporations to secure for themselves positions of relative economic advantage. Structural adjustment programs usually entail all or most of the following: fiscal austerity, privatization of public industries and services, elimination of barriers to trade, liberalization of capital markets, and currency devaluation. Strings attached to loans can also be political in nature and designed to foster close integration of a nation with the neoliberal economic project. Neoliberals aim to reduce the size and reach of government and unleash the purported ability of a self-regulating market to serve effectively as the ultimate arbiter of economies and all social life. SAPs undermine the political sovereignty of debtor nations while creating conditions for the exploitation of people and environments to fuel profits for powerful interests in the global economy.

SAP conditions often undermine democratic decision making in Third World nations, thereby interfering with a nation's ability to advance self-determined domestic and international social and economic policies. The depth and breadth of prescriptive SAP policies alone reveal them as antidemocratic. Client government officials are called on to force the will of more powerful economic and political interests onto their nations, whereas citizens are barred from participation in loan negotiations. The IMF and World Bank may even refuse a government permission to inform its citizens what a given loan agreement entails. Some agreements have even stipulated which laws a country's government must pass to meet loan agreement requirements or economic targets. In negotiating loans, the IMF and World Bank are in a position of considerable advantage with regard to debtor nations seeking loans in the midst of a crisis.

Who Really Benefits?

Furthermore, these antidemocratic policies generally have not benefited debtor nations economically. Fiscal austerity can include all or some of the following: reduced government spending, increased taxation, requirements to reduce national and/or international debt, and requirements to raise interest rates. Although these strategies may help to reverse a currency crisis in the short term, they create a poor foundation for social and economic development. Reduced government spending typically yields such outcomes as near-complete disappearance of the social safety net, reduced investment in education, reduced ability to address environmental problems, and reductions in the government's ability to stimulate domestic small business development. These impacts undermine a nation's ability to prepare its citizens for occupations other than low-skill manufacturing and service jobs. This lack of social investment virtually condemns a nation to dependence on wealthy and powerful entities within the world system that supply high-tech products and services. Furthermore, governments lose taxes that could be generated from higher percentages of their populations engaging in skilled work, and these negative effects on the tax base reinforce governments' inability to implement social programs. Increased taxation, especially when combined with reduced social services, places further stress on people living under marginal circumstances, thereby reducing their chances to contribute to economic development. Paying down national and international debt also diverts money toward creditors and away from potential social investment. Raising interest rates slows the economy and makes it more difficult for consumers to purchase durable goods and homes or start or expand businesses. Economic contraction caused by rising interest rates reduces the tax base and increases the need for social services at a time when these services are being cut. All of these strategies applied simultaneously in a fiscal austerity package are more likely to cause economic recession than to stimulate development. In most nations implementing SAPs, these programs have resulted in widespread unemployment, declining real incomes, economically damaging levels of inflation, capital flight, persistent trade deficits, rising levels of external debt, the destruction of the social safety net, and deindustrialization.

Removing barriers to free trade has also become a standard condition for obtaining a loan from the IMF or the World Bank. Free trade is purported to level the playing field in the global market through the removal of tariffs and subsidies. As economic protections are removed, weak economies are exposed to competitive forces they often cannot withstand, especially in market areas where economies of scale are achieved by applying capital-intensive production methods. Free trade regimes also further undermine the ability of national and local powers to regulate industries in the areas of social justice and environmental protection because government regulations are conceptualized as barriers to trade. Intense competition among large corporations globally, coupled with competition among debt-ridden countries seeking opportunities to earn foreign exchange, further constrains possibilities for regulation. Free trade also paves the way for further concentration of wealth in the hands of First World corporate entities that can more easily achieve economies of scale and undercut their competitors with low prices. Similarly, capital market liberalization, another free trade–oriented condition for loans, supports the interests of the global financial sector at the expense of communities and nations as foreign banks capture markets and profits that had been the province of the domestic economy while also reducing opportunities for small-scale businesses and citizens of modest means to obtain loans.

Exports Encouraged

Loan agreements also encourage countries to maximize exports to earn foreign currencies necessary for repaying loans. Third World nations typically export raw materials, agricultural products, or relatively simple commodities on a large scale. As countries sell their raw materials and other products in the competitive global market, prices fall to match those of the lowest cost providers, so that economic pressure increases to secure production cost advantages by engaging in socially and environmentally damaging practices. Devaluation of domestic currencies compounds problems associated with export orientation of economies. Devaluation advantages producers and consumers in the First World because prices of debtor nation exports fall while prices of imports rise. In export-oriented debtor nation economies, imports typically include food and other basics of life.

The IMF and World Bank have also repeatedly required privatization of public utilities, services, and infrastructure as a condition for loan approvals, thereby creating opportunities for which transnational corporations may be uniquely positioned as a result of their ability to obtain credit and, in some cases, their technological advancement and prior experience running large-scale, technically complex industries and services. Privatization of basic services such as water reinforces the power and reach of global corporate interests by handing them both new business opportunities and captive markets. The continual privatization of the commons also raises the question of how those without money will fulfill their basic needs and who will speak on the behalf of nature.

Debtor nations are also encouraged by the IMF and the World Bank to lure transnational corporate entities to locate production facilities within their borders. Efforts to attract foreign direct investment typically undermine environmental and social protections that are perceived as barriers to trade. Economic gains from foreign investments are also lower than might be expected, as a high proportion of the profits earned by transnationals operating in Third World nations is repatriated to First World economies. Transnationals may therefore contribute relatively little to a debtor nation's efforts to earn foreign exchange, especially as many of the factories they operate in Third World nations reside in tax-free zones.

Poorly conceived loan conditions can deepen an economic crisis, and bank-instigated policy initiatives can prove politically unsustainable because citizens of debtor nations tend to perceive them as intrusions by a colonial power. Although some economic growth success stories of IMF policies exist, one must question whether these success stories have produced socioecologically sustainable and economically resilient societies. Neoliberal economic development policies stipulated in SAPs have also resulted in countries trading environmental health for economic growth. By requiring SAPs as a condition for loans, the IMF and the World Bank have enforced the dependency of many Third World nations within the world system.

See Also: Capitalism; Globalization; North–South Issues.

Further Readings

Kaplinsky, Raphael. *Globalization, Poverty and Inequality: Between a Rock and a Hard Place*. Malden, MA: Polity, 2005.

Stiglitz, Joseph E. *Globalization and Its Discontents*. New York: W. W. Norton, 2002.

Tina Lynn Evans
Fort Lewis College

SUBURBAN SPRAWL

The phrase *suburban sprawl* is a pejorative term that has not only become part of the popular lexicon of cities but is also shorthand for all that is ugly, inefficient, crass, and dysfunctional in metropolitan areas. Although commonly discussed in a North American context, sprawling development patterns may also be found elsewhere, notably in Europe, Australia, and more recently, China. Despite its widespread use, there often is lacking in the literature proper empirical support for or sufficient definition of the term, resulting in some confusion among conditions, causes, and impacts. However, whatever imprecision may plague its definition, the land use patterns characterized as suburban sprawl are recognized as being the result of a combination of such political factors as land economics, tax policies, zoning regulations, master planning processes, and—according to some critics—exclusionary lending and insurance practices motivated by race and class.

Given that suburbanization has for so long been associated with the "American Dream" and all that implies, it is unsurprising that a number of enduring debates have emerged over suburban sprawl, ranging from the validity of the term itself to the very validity of the planning function.

William H. Whyte was the first to use the term *urban sprawl* in his 1958 essay of the same name, which set the tone for much of the discourse to follow by disparaging much of newer postwar America as "a mess." Since then, there has been an enormous body of literature criticizing suburban America on economic, aesthetic, and social equity grounds, but several reviews of this literature have revealed little in the way of common definitions or attempts to empirically operationalize the term *suburban sprawl*. Generally, however, suburban sprawl is seen as a combination of low residential densities; the segregation of

Suburban developments have encroached on farmland all over the United States. These houses in Forsyth County, North Carolina, abut the edge of a farmer's vegetable field.

Source: U.S. Department of Agriculture, Natural Resources Conservation Service

land uses through zoning, so that homes are separated from jobs, retail, and services; little in the way of identifiable town "centers"; and a car-dependent street network with poor connectivity. All of these factors make for developments in which public transportation is uneconomical and people become separated by economic class.

In a landmark paper, G. Galster et al. determined that sprawl should be seen as a pattern of land use that exhibits low levels of a combination of the following empirically verifiable eight dimensions: density (number of residents within a developable square mile in an urban area), continuity (extent to which such developable land has been built upon in an unbroken manner), concentration (degree to which most development occurs within a few square miles in an urban area), clustering (minimizing of land used in development), centrality (concentration of development close to the Central Business District), nuclearity (concentration of developments within a single center or in multiple ones), mixed uses (multiple land found throughout an urban area), and proximity (these mixed uses are close together).

As verifiable as these dimensions may be, it is important to stress that none of these can be seen only as the result of urban design decisions: They are also the manifestations of a decades-long process of economic restructuring and globalization, in which American cities lost manufacturing capacity to low-wage centers in the developing world and the economy became characterized instead by highly paid financial, communication, and information service professions as well as low-wage retail and service sector employment. As a part of this restructuring, urban development has emphasized downtown convention centers, entertainment and sports complexes, and "edge city" office and retail developments and suburban enclaves. Older inner-city neighborhoods and obsolete industrial infrastructure, meanwhile, are left to deteriorate.

The consequences and effects for which sprawl is frequently criticized are economic, political, social, ecological, and aesthetic, and each of these provides the basis for a protracted debate.

Extending development ever farther away from existing centers requires municipal and suburban governments to fund utilities, services, and the construction and maintenance of public infrastructure such as roads, schools, and fire stations. These costs can only be borne by the sale and development of even more greenfield developments or by trimming budgets elsewhere.

Sprawling developments contribute to regional disparities and inequities. Because public services such as schools are funded by local property taxes, thriving suburban neighborhoods can afford to fund excellent schools, whereas those in inner cities remain impoverished. This contributes to ever more severe concentrations of poverty, which become entrenched and discourage investment.

Some argue that such social, economic, and racial segregation is not an unintended consequence of sprawl-friendly policies but, rather, is integrated with land use processes and embedded in regional power structures. Blacks, Latinos, and other marginalized populations become isolated through a combination of "gatekeeping" on the part of banks and real estate interests and development policies favoring high-end residential development over affordable housing.

With these disparities arise political implications as metropolitan areas become increasingly fragmented politically. Suburbs end up competing not only with each other but also with central cities, thereby exacerbating regional economic, social, and environmental challenges.

The ecological impacts of sprawl are widely noted. Expanding cities consume irreplaceable farmland—a real concern when so much of North America's most productive croplands are near growing cities. Lower residential densities also bring with them an expanded need for travel, mostly by private vehicle, with concomitant increases in paving, air pollution, and contaminated runoff. Such developments are also highly energy intensive, requiring greater levels of generated heating, cooling, and transport-related fuels.

Automobile-dependent sprawl often lacks public gathering places, and so has been targeted by social critics for engendering alienation, particularly among young people who have little to do locally and must rely on parents to drive them to almost all their activities. For these and other reasons, suburban sprawl has been the focus of long-standing debates in the literature and in the political arena.

"Antisprawl" activists see in present and developing subdivisions a host of grave environmental and social problems and a degradation of the overall quality of life. In contrast, "free market," "property rights," or "wise use" advocates see in present arrangements a normal exercise of consumer preferences in the fulfillment of personal and quality-of-life aspirations. Arguments in support of maintaining conventional suburban development patterns are generally associated with social conservatives and libertarians, who favor market-based development decisions and oppose the imposition of taxes to fund new public transit or government regulations on house size and location. Libertarians in particular have been quite vocal in countering what they see as a violation of property owners' rights and have used ballot-box initiatives to combat Smart Growth initiatives.

Others argue that there is some common ground in these debates: Some fiscal conservatives, for example, oppose suburban sprawl on economic grounds, whereas some conservatives reject sprawl's ugliness and the ruination of natural places. Similarly, left-leaning antisprawl activists can be just as opposed to zoning laws as are libertarians.

These "sprawl debates" are thus grounded in several potent normative political ideologies oriented to rights, liberties, and distributive justice. Addressing suburban sprawl then becomes a matter not only of understanding and incentivizing land markets and reinventing policies but also engaging in political debates over the very nature of the good society.

In the 1990s, several planning and design movements that were designed to address many of the perceived shortcomings of standard suburban developments gained prominence. Smart Growth, probably the best-known of these movements, sets out guidelines for compact neighborhood design; growth controls; higher densities; pedestrian-, cycling-, and transit-friendly environments; and a strong role for planning. The Congress for the New Urbanism also promotes these principles but adds strong aesthetic and "placemaking" elements, evoking in particular American urbanism of the 1920s. On a larger scale, regional approaches to governance are becoming a means of implementing stronger measures of joint authority and resource sharing between jurisdictions so that growth controls and environmental objectives can be more easily funded and managed.

However, none of these measures may be more effective in curtailing the seemingly inexorable spread of suburbia than external economic conditions. The collapse of the American housing and credit markets in 2007–08, combined with highly volatile gas prices, has resulted in abandoned housing developments and recurring speculation in the media over "the end of suburbia." Future debates over the fate of suburban sprawl may therefore be quite different than those of the past.

See Also: Green Discourse; Political Ideology; Urban Planning; Wise Use Movement.

Further Readings

Burchell, Robert W., et al. *Sprawl Costs: Economic Impacts of Unchecked Development.* Washington, D.C.: Island Press, 2005.
Ewing, Reid, et al. *Measuring Sprawl and Its Impact, Volume I.* Washington, D.C.: Smart Growth America, 2002.
Galster, G., et al. "Wresting Sprawl to the Ground: Defining and Measuring an Elusive Concept." *Housing Policy Debate,* 12/4 (2001).
Orfield, Myron. *Metropolitics: A Regional Agenda for Community and Stability.* Washington, D.C.: Brookings Institution, 1998.
Squires, Gregory D., ed. *Urban Sprawl: Causes, Consequences and Policy Responses.* Washington, D.C.: Urban Institute, 2002.
Whyte, W. "Urban Sprawl." In *The Exploding Metropolis*, edited by W. Whyte. Berkeley: University of California Press, 1958.

Michael Quinn Dudley
University of Winnipeg

Sustainable Development

Sustainable development is a term that has come to define human and environmental interactions in the 21st century. Sustainable development is the promotion of developmental patterns that will enable current human generations to meet their needs without compromising the ability of future generations to meet their own needs. The most popular definition of the term was coined by the United Nations Commission on Environment and Development in 1987.

The report of the commission, *Our Common Future,* indicated that sustainable development was development that meets the needs of the present without compromising the ability of future generations to meet their own needs. The concept embodies the consequences that the continued increase in human population and the uneven and inefficient use of natural resources have on the planet's environment. Sustainable development has become increasingly important in the last three decades in light of growing scientific acceptance that humanity's current developmental activities are having a detrimental effect on the global environment.

Examples of these environmental problems include global climate change, rising sea levels, pollution, desertification, deforestation, loss of biodiversity, and more. Although the term initially represented environmental concerns, it has evolved into a concept that incorporates all facets of human and environmental interaction. The terms *sustainable development* and *sustainability* are often used interchangeably; however, the two terms have different meanings. Sustainability indicates only the desire to perpetuate something. By combining this term with the idea of development, a significant set of values, normative assumptions, and power dynamics are involved.

The way that sustainable development is interpreted will affect the debates surrounding how the concept is to be promoted and implemented. A useful way to understand the concept is to identify sustainable development on a linear scale that is strong on one end and weak on the other. Strong sustainable development indicates that there would need to be a significant reordering of current sociopolitical and economic structures. Such interpretations of the term are linked to ideas that are ecocentric in nature, where the environment is put first. In this interpretation the environment has an intrinsic value and should be protected at all costs.

Desertification, exemplified by this drying lakebed, is just one of the serious threats to the environment that have resulted from unsustainable development practices.

Source: iStockphoto.com

The stronger form of sustainable development has been linked to the broader debates that suggest a reflexive modernity. Reflexive modernity suggests that current developmental patterns are so detrimental to the environment that a significant reordering of political and social processes should occur. In this interpretation sustainable development can only be achieved by altering the central tenets of industrial society. These include science and technology and the nation-state.

From the opposite end of the spectrum, a weak view of sustainable development maintains that the solutions to problems inherent within human environmental interaction can be found in the existing sociopolitical framework. A weak interpretation of sustainable development focuses on the ability of science and technology combined with fiscal, legislative, and educational mechanisms to encourage a sustainable form of development. The weak form of sustainable development can be said to be anthropocentric or human orientated in nature and linked to the more overarching idea of ecological modernization.

Ecological modernization foresees a future based within the present system of capital development, redefining the relationship between economy and environment in such a way that economic growth and environmental protection are seen as mutually reinforcing objectives. Ecological modernization relies on the underlying assumption that environmental crisis will necessitate innovation and technical development, providing the necessary tools to abate an environmental catastrophe.

Perspectives on sustainable development can exist in varying degrees on the scale between strong and weak interpretations. There is a general consensus that the pervasive nature of neoclassical economics has also come to permeate throughout thinking on sustainable development with a broad acceptance that intragenerational equity and intergenerational equity can only be achieved within the confines of economic growth. Intergenerational equity is concerned with equity between current and future generations. Intragenerational equity is concerned with equity between people of the same generation. Whether interpretations of sustainable development are strong or weak, one of the simplest and most popular ways of understanding sustainable development is through the metaphor of the three pillars of sustainable development.

The three pillars are society, economy, and the environment and are intended to represent the holistic nature of the concept. Although useful as a starting point for understanding sustainable development, the three pillars approach and the strong and weak interpretations are overly simplistic and reductionist. Sustainable development includes concepts such as democracy, justice, burden sharing, care, access, and more. Josef Jabareen provides a useful way of understanding the diversity of sustainable development by outlining seven areas to which the concept applies. The following outlines these areas.

First is the ethical paradox dimension. This represents the ethical dilemma that sustainable development produces; that is, the continued debate about the inconsistency between the terms *sustainable* and *development*. Second is the area of natural capital stock, which refers to the material dimension of sustainable development. In essence, this refers to the quantifiable natural assets of the Earth's biosphere on which development is based—a position that is frequently used in the natural sciences. Third is the idea of fairness, which refers to the social dimension of sustainable development. This includes issues such as social equity, equal rights for development, democracy, public participation, and empowerment. Fourth, the idea of eco-form or spatial design refers to the integration of sustainable development into planning frameworks. It focuses on human settlement in the built environment and includes both rural and urban dimensions. Fifth is integrative management, which refers to the overall management of sustainable development.

This form of management is representative of the three pillars approach to sustainable development. Sixth is the idea of global discourse, which highlights the political dimension of sustainable development. As a starting point, this can refer to the unifying discourse of the need to alter developmental patterns, as well as a vision of a single Earth and one world. However, it can also refer to the fractured and disjointed global discourses that are inherent in political discussions of sustainable development. Finally, there is the idea of the Utopian vision of the achievement of sustainable development. More than actually achieving an end state, with society in harmony with its environment, this dimension of sustainable development highlights the need to promote altered developmental patterns. This element can also reflect the possibility of a dystopian society, in which appropriate developmental patterns are not achieved.

By outlining the weak and strong interpretations of sustainable development, introducing the three pillars of sustainable development, and then discussing the various metaphors of sustainable development, a picture has developed of the complexity of the issues involved. These perspectives are, of course, not exhaustive, with many complex interpretations of sustainable development existing.

Because of the complex set of issues that sustainable development embodies, including environmental, social, and economic factors, new forms of governance structures are developing based around the ideas of sustainable development. Global environmental issues such as global warming and transnational global pollution mean that sustainable development cannot be achieved by a single nation-state. Instead, what is needed is multinational cooperation on a global level, as well as active involvement at the regional, local, and individual levels.

Considering the significant inequalities in the global political system, substantial barriers exist. With this said, a significant network of international regimes has been established over the past three decades that has provided a framework from which sustainable development can address the various environmental phenomena. This section explores sustainable development at the international and national scales. At the international scale, the United Nations has been pivotal in establishing the concept on the world stage, and therefore there is a focus on the main United Nations events that have promoted sustainable development.

The concept of sustainable development in its modern guise can be said to have developed and been galvanized through four main events. The first event of considerable significance is the United Nations Conference on the Human Environment at Stockholm in 1972. The Stockholm Conference became a key symbol of political acknowledgement for the growing worldwide awareness of the environment. Although this conference produced little in the way of state policy, it was pivotal in raising awareness on environmental and developmental issues. The Stockholm Conference also resulted in the commissioning of the United Nations Environment Programme.

The second event toward the advancement of sustainable development was the World Commission on Environment and Development, popularly known as the Brundtland Commission, in 1987. The commission provided a common and easily identifiable definition to a previously faceless concept. The report *Our Common Future* established sustainable development on the political stage and raised awareness of the concept on a global scale.

More than raising awareness, however, the report arguably represents a convergence between the processes of modernity and the effect that these processes have on the environment. The commission elaborated on this initial concept by maintaining that sustainable development is a process whereby the exploitation of natural resources combined with the way financial investment is directed should affect technical development and institutional change that should apply to both current and future generations. The World Commission was ratified in 1983 by the United Nations General Assembly and was designed to exist in a supranational political space beyond the control of nations.

The third event, and the most publicized, was the United Nations Conference on Environment and Development, held in Rio de Janeiro in 1992 and popularly named the Earth Summit. The summit was seen as a milestone in political history for creating a political space in which the world's political leaders were able to focus on global environmental issues. The summit represented a common recognition by many nations and relevant organizations of the world that issues of development needed to be tempered with effective environmental measures.

Substantively, a number of agreements emerged from the conference. The Framework Convention on Climate Change set out international targets for reducing the anthropogenic causes of climate change. The Biodiversity Convention established broad aims to conserve international biodiversity, with the aim of making use of the components of biodiversity in a suitable manner and enabling an equitable distribution of the benefits of using these resources.

The Desertification Convention was designed to create localized action frameworks to address the degradation of dryland environments. Most notable was the production of

Agenda 21, which mapped out a blueprint for sustainable development by outlining the main issues implicated in global environmental change and how they might be tackled. The rhetoric contained within this document has had far-reaching implications for the advancement of sustainable development discourse in a multitude of ensuing negotiations and meetings.

Agenda 21 is a legally nonbinding program of action for sustainable development. Adopted at the United Nations Conference on Environment and Development, it is a document comprising 40 chapters and is intended to guide the actions of governments, aid agencies, local government, and other actors on environment and development issues. Agenda 21 covers four broad areas for the promotion of sustainable development. The first is social and economic dimensions to development. This includes issues such as production and consumption, health, human settlement, and the promotion of an integrated decision-making process. Second is the conservation and management of natural resources.

This includes the atmosphere, oceans and seas, land, forests, mountains, biological diversity, ecosystems, biotechnology, freshwater resources, toxic chemicals, and hazardous radioactive and solid wastes. Third, Agenda 21 aims to strengthen the role of groups and actors involved in promoting sustainable development. This includes youth groups, women, indigenous populations, nongovernmental organizations, local and regional authorities, trade unions, business groups at all levels, scientific and technical communities, and farms. Fourth, Agenda 21 attempts to outline areas for implementing sustainable development.

Areas identified include finance, technology transfer, information coordination and public awareness, capacity building, education, legal instruments, and institutional frameworks. In essence, Agenda 21 establishes a framework, or package, of long-term goals. It has been seen as the most comprehensive document negotiated between governments that considers the interaction among economic, social, and environmental trends at every level of human activity.

Essentially, it establishes a framework within which a number of long-term goals are presented. This said, however, Agenda 21 has been criticized for promoting a vision of sustainable development that does little more than perpetuate the Enlightenment goals of progress through economic growth and industrialization at all costs. Moreover, it has been suggested that Agenda 21 advances the neoliberal goals of the wealthy Western societies. Despite the criticisms, Agenda 21 is a pivotal document that deals with a diverse range of sustainable development issues. Since the publication of Agenda 21 there has been a substantial movement in the way that sustainable development is perceived. This is best represented by examining the role of the World Summit on Sustainable Development in this process.

The World Summit on Sustainable Development is the most recent event to represent the importance of sustainable development on the world political stage. The resultant document, the "Plan of Implementation," specified a number of commitments to the nations of the world. There were commitments to reduce the loss of biological diversity by 2010, halve the proportion of people without access to drinking water and sanitation by 2025, restore world fish stocks by 2015, and promote the production of chemicals that are harmless to human health and the environment by 2020.

There was also a reiteration of the important role that sustainable consumption plays in the future of sustainable development. However, many commentators have noted the real absence of any substantial international agreements, particularly from the United States. Importantly, however, eight core action themes for achieving sustainable development at the national level were identified. These included poverty eradication, sustainable production and consumption, protection of the natural resource base of economic and social development, globalization, health and sustainable development, small island developing states, and Africa and other regional initiatives. Overall, the United Nations and the above-mentioned events have established sustainable development on the political stage.

Sustainable development has become a central element of European Union policy. Although this remains somewhat fragmented and contested, it nonetheless is a concept that, as with the United Nations, has infiltrated the mechanisms of the European Union. Evidence of sustainable development at the European level can be found in commission white papers, European Action Plans, and legal and legislative reform documents. In 1998, the Cardiff process was initiated. The Cardiff process requires the council of ministers on a cross-sector basis to integrate sustainable development into their policy areas. At the Helsinki Summit in 1999, a commitment to sustainable development was reinforced, and the European Commission outlined its priorities toward promoting sustainable development.

In 2001, the sixth environmental action plan was released, which set out substantive action plans in areas such as climate change and environmental and human health. Four priority areas were identified. These included climate change, nature and biodiversity, environment and health and quality of life, and natural resources and waste. The main avenues for action identified in the Sixth Programme include the effective implementation and enforcement of environmental legislation necessary to set a common baseline for all European Union countries; the integration of environmental concerns, in which environmental problems have to be tackled where their source is—frequently in other policies; the use of a blend of instruments that offer the best choice of efficiency and effectiveness; and the stimulation and participation of all actors from business to citizens, nongovernmental organizations, and social partners through better and more accessible information and joint work on solutions. All of these are closely tied to the framework outlined in Agenda 21 for the achievement of sustainable development.

Sustainable development at the national level has been integrated into state practices to varying degrees. Agenda 21 called for the development of national sustainable development strategies in line with international recognition that sustainable development requires a realignment of governance processes. There is a move away from a model that focuses on central planning to one that focuses on creating enabling conditions that should be based around making or improving strategic connections between existing strategic planning frameworks. These connections should be made at all levels of government and on a cross-sector basis, as well as incorporating the needs of other actors that are active in sustainable development governance.

At a very fundamental level, sustainable development is considered to be a contradiction in terms, with sustainable and development representing opposing ideals and hence being unobtainable. It has been argued that there are so many different interpretations of the term that its use in any effective way to address contemporary social economic and environmental issues will result in a waste of time and resources. Conversely, it has been suggested that the very quality that forces a criticism of sustainable development—ambiguity—is also a strength, as it enables the concept to be versatile and to respond to multiple perspectives in multiple arenas.

Sustainable development and the issues that it encompasses have had a profound effect on the way that research agendas and academic work have developed. The interconnected nature of the issues has involved both the natural and the social sciences. This article has provided a brief introduction to the concept of sustainable development and outlined the multiple interpretations of the concept while providing a tangible guide as to how it is being translated into governance structures.

The defining characteristics of sustainable development are that the term is ambiguous, vague, and often contested. There is continued debate not only on how best to integrate sustainable development into dynamic, nonlinear, and complex social and environmental systems but also how the term should be understood in the first instance. To understand such a multidimensional term there has been significant collaborative work among the social and natural sciences with the establishment of stronger cross-disciplinary networks, as well as the establishment of multidisciplinary research centers. *Sustainable development* is a term that is constantly evolving.

See Also: Agenda 21; Future Generations; Globalization; Kyoto Protocol; Precautionary Principle; UN Conference on Environment and Development.

Further Readings

Baker, Susan. *Sustainable Development.* London: Routledge, 2006.

Borne, Gregory. *Sustainable Development: The Reflexive Governance of Risk.* Lampeter, UK: Edwin Mellen, 2009.

Division of Sustainable Development, United Nations Department of Social and Economic Affairs. http://www.un.org/esa/dsd/index.shtml (Accessed October 2009).

Jabareen, Yosef. "A Knowledge Map for Describing Variegated and Conflict Domains of Sustainable Development." *Journal of Environmental Planning and Management,* 47/4:623–42 (2004).

Lafferty, William, ed. *Governance for Sustainable Development: The Challenge of Adapting Form to Function.* Cheltenham, UK: Edward Elgar, 2004.

Gregory Borne
University of Plymouth

Technology

For many people who are apprehensive about anthropogenic environmental impairment in the world, thoughts about the role of technology and technological change in that phenomenon are never far away. Technology, according to them, is often either the cause or the solution for environmental degradation. The way modern technologies are designed, some would argue, introduces intrinsic properties that can either irreparably damage or, conversely, eliminate all damage to the environment. More recently, scholars have stressed the importance of viewing technology as inherently ambivalent in its properties. Drawing on empirical studies from recent social science scholarship, they argue that technology can be used by humans to be a force either of good or of destruction, but that choice can be made eventually only by humans.

Some key questions about technology and the environment are as follows:

- What do we mean when we speak of technology in the context of environmental change?
- Historically, how has the relationship between technology and the environment evolved?
- What have been some of the early conceptions of the relationship between technology and environmental destruction?
- What are some of the contemporary issues surrounding environmentally benign technological development?

Technology is first and foremost a human creation that is designed and constructed to serve often narrowly defined utilitarian goals of the user(s). Two points are especially noteworthy in this definition—that technology is always a designed human creation and that it serves utilitarian goals. Technology is thus anything that has been consciously created by humans with a purpose in mind. That purpose, according to this definition, should always serve the goal of maximizing the well-being of the individual or the collective. But by stressing the importance of goals, this definition overlooks the significance of means in achieving those goals. The significance of means is especially heightened in the environment–technology relationship because, as we shall see, much of the controversy in this relationship is because of its inattention to means to attain goals.

In 2009, this scientist at the U.S. Department of Energy's Sandia National Laboratory in New Mexico discovered a new type of membrane for hydrogen fuel cells that may help make environmentally friendly hydrogen cars a reality.

Source: U.S. Department of Energy, Sandia National Laboratories

What Counts as Technology?

The definition of technology is also notable because it overlooks another aspect that is relevant in understanding the technology–environment relationship. First, it does not define what counts as technology. The definition does not specify whether a lever, a water supply system, or even music is a technology. Indeed, historians and other theorists of technology have understood technology as a multifaceted and unstable concept because of the multiple meanings it has acquired in different settings. However, for the purposes of this article, we understand technology here to encompass at least four connotations: technology as tool, technology as techniques, technology as process, and technology as system. Technology as tool or object is, for most of us, the most recognizable aspect. It refers to technology as tool, machine, or appliance that performs a function. Technology as a technique refers to the skills that humans acquire in using tools or machines. Technology as process refers to specific operational procedures followed in assembly lines, for example, that proceed from beginning to end in an organized and regulated fashion. Finally, technology as system denotes a complex arrangement of machinery and institutions that perform a specialized task. The electric power system is one such example of technology as a system.

The historical articulation of the relationship between technology and the environment can be enlightening. To start, anthropological research into existing hunter-gatherer societies around the world reveals some interesting clues about the use of technology by Neolithic-age humans in their relations with the environment. Two aspects are especially significant: the need to live within the carrying capacity of their region and the need to maintain autonomy of the social unit. Maintaining carrying capacity was essential for hunter-gatherers because they were inextricably reliant on their surroundings for their daily sustenance.

Maintaining autonomy of the social unit from surrounding tribes or households was equally important to ensure that the household could harvest an uninterrupted supply of natural resources from the environment. Technology was largely limited to the development of tools and techniques that facilitated hunter-gatherer lifestyles. The use of technology acquired an entirely different connotation with the increase in social complexity, the founding of empires, and the creation of "hydraulic civilizations" in the fertile valleys of the old world. With the rise of population and a settled agricultural mode of existence, humans required technology to play a role in harnessing environmental resources—land, forests, and water. Massive systems of irrigation were developed by states for farming that required the creation of a managerial organization, as well as an array of technologies such as dams, levees, and canals. The creation of states with a class of governing elites was accompanied by the institutionalization of redistributive mechanisms such as taxation that provided for the elites. The surplus of the agriculture or trade product was siphoned from the commoners into the hands of the few who governed the state. Basic transportation and information infrastructure had to be created to oversee the mechanisms for redistributing natural resources.

Complexity Grows With the Scientific Revolution

The industrial and scientific revolution in the 18th and 19th centuries added to the complexity of the relationship between technology and the environment in at least two ways. At a fundamental level, the scientific revolution altered the conception of nature and the environment. From an image of nature as a nurturing mother, the scientific revolution sought to mechanize and rationalize the dominant Western worldview about the environment. Second, technological advancement was understood explicitly as a project for bringing the vast resources of nature under the domain of human manipulation and control. These beliefs in a scientific worldview and technological advancement came to be encapsulated within the human condition of modernity. To be considered modern, humans were expected to at the very least believe in a rationalist explanation of the workings of the environment and in the ability of technology to manage that environment for human needs. It was in societies of Western Europe and North America (often referred to collectively as "the West") that modernity became the widespread worldview by the mid-20th century.

The widespread nature of technology as tool and as technique is visible in our daily lives by the rapid turnover in electronic gadgets and by the sprouting of new specialists in technology. The prevalence of expansive technological systems is another defining facet of modern Western existence. All services essential for human survival such as water, waste, sewerage, energy, and transportation are provided through established technological systems. The extensive use of technology in regulated processes such as manufacturing assembly lines to achieve specific aims is yet another facet of technology that has become ubiquitous in modern times.

The ubiquity of modern technology and the predominant rationalistic worldview in the West have given rise to much intellectual discourse on the role of technology in human society and in the world in general. Several thinkers from the late 19th and early 20th centuries were dismayed and disillusioned by the drudgery and dreariness of industrial modernity, especially in Western cities. These critics blamed Western civilization's total subservience to machines and technics as the primary reason for the unenviable condition of human lives. For them a complete break from our overly technicized lives with a retreat to a pastoral world was the only way out. Referred to as luddites or dystopians for their often-extreme views, these critics nevertheless engendered a healthy sense of skepticism about technology and its abilities. In contrast, there was no dearth of those who were enthusiastic that technology could usher in a "second creation"—re-create an Eden-like paradise for humans. Technology, these enthusiasts argued, being a product of human ingenuity was as remarkable as nature's creations and, as such, capable of evoking the same sense of sublime that nature stirred in men.

The Rise of Environmentalism

With the rise of environmentalism and the green movement in the 1970s in the West, discourses about human intervention into the natural world were constructed on existing claims about technology's role in modern lives. While the Prometheans emerged as the counterpart of technology enthusiasts, radical ecologists developed their claims on the tradition of technological dystopians. Prometheans are enthusiastic about the ability of human ingenuity to surmount obstacles. To the Prometheans, natural resources on the Earth are infinite because these resources are themselves products of the human ability to develop technologies. It is the constant development of new technologies that converts matter in the world to new resources that will counter any existing scarcity. In polar opposition to the Prometheans, the radical ecologists, especially of the deep ecology and ecofeminist persuasion, are convinced that it is humans, and especially the modern technological outlook of humanity, that are responsible for the destruction of the environment. Humans are a cancer on the planet that is rapidly consuming it.

Whereas the Prometheans uncritically celebrate modern technology, the radical ecologists view it with unbridled hostility. Appropriate technology emerged from the environmental movement as a technology-oriented experiment to reconcile these opposing views. This effort attempted to design technologies that were appropriate to the environmental, social, and ethical concerns of the people using it. By the late 20th century, the appropriate technology movement had spawned at least three efforts directed at the development of sustainable technologies: intermediate technology, green technopole, and green localism. Intermediate technology refers to the creation of implements and devices for developing country contexts that are a significant improvement over traditional methods but at the same time are not accompanied by the social and environmental dislocations that technologies transferred from the West entail. A key aspect about intermediate technologies is that they can be operated and serviced by unskilled users with the help of locally available materials. Green technopole emerged as an effort that sought to incorporate environmentally benign technologies into modern industrial manufacturing enterprises—a process that has also been referred to as ecological modernization. Arising out of a need for corporations to inject an environmental orientation into their production, a range of technological innovations has come to be characterized as ecological modernization. Green production strategies have transitioned from incremental, end-of-pipe clean technologies to sustainable production techniques such as biomimicry, cradle-to-cradle design, green chemistry, and industrial metabolism. Green local innovations, in contrast, have evolved from a grassroots vision of environmentally sustainable lifestyles. Spearheaded by small civil society groups such as small businesses, nonprofits, and public sector organizations, green localism actions have focused their attention on sustainable agriculture, reuse/resale, home power, and green transportation. Although lacking the economic ability of the green technopole to drive technological innovation, green localism has evolved an extensive network of like-minded organizations and the public that assist in generating the momentum for progressive innovation strategies.

The study of technology as a factor in social and environmental transformation was for a long time the object of excessive pessimism or optimism, but the entrenchment of the environmental movement has injected a fresh realism into the process of environmental technology innovation.

See Also: Appropriate Technology; Domination of Nature; Environmental Justice; Innovation, Environmental.

Further Readings

Dryzek, John S. *The Politics of the Earth: Environmental Discourses*. New York: Oxford University Press, 1997.

Hess, David J. *Alternate Pathways in Science and Industry: Activism, Innovation and the Environment in an Era of Globalization*. Cambridge, MA: MIT Press, 2007.
Hess, David J. "The Green Technopole and Green Localism: Ecological Modernization, the Treadmill of Production, and Regional Development." Paper Prepared for Presentation at the Symposium on the Treadmill of Production, University of Wisconsin–Madison, October 2003.
Hughes, Thomas P. *Human-Built World: How to Think About Technology and Culture*. Chicago: University of Chicago Press, 2005.
Merchant, Carolyn. *The Death of Nature: Women, Ecology and the Scientific Revolution*. New York: Harper & Row, 1980.
Nye, David E. *Technology Matters: Questions to Live With*. Cambridge, MA: MIT Press, 2006.
Schumacher, E. F. *Small Is Beautiful: A Study of Economics as If People Mattered*. London: Abacus, 1974.
Winner, Langdon. *Autonomous Technology: Technics-Out-of-Control as a Theme in Political Thought*. Cambridge, MA: MIT Press, 1977.

Govind Gopakumar
Rensselaer Polytechnic Institute

Toxics Release Inventory

The U.S. Emergency Planning and Community Right-to-Know Act, 1986, requires owners and operators of covered facilities to self-report specific data including aggregate and trend data that are to be available to the public. Based on the principle of community right to know, this publicly accessible record is referred to as the Toxics Release Inventory (TRI). The data to be reported are the release into the environment of substances and the threshold quantities listed in the regulations. Since its introduction, the TRI has been amended by the Pollution Prevention Act (1990), which requires data on enterprise waste management and source reduction to also be reported in the TRI. Similar registers are maintained in other countries. In Canada, the registry is known as the National Pollutant Release Inventory, first published by authority of the Canadian Environmental Protection Act in 1995 containing data reported for 1993. In Australia, the National Pollutant Inventory was officially available in 2000. Because the registers now include release and transfer data, they may accurately be termed pollutant release and transfer registers. Complementary reporting requirements may also exist in some local jurisdictions.

The legal obligation for an enterprise to self-report is not unlike the obligation to file income tax returns. Standard forms are annually available for the enterprise to complete and return to the responsible agency, which compiles the data and makes them available to the public as the TRI. Access to such records is now available on the Internet.

The TRI alone cannot be used to determine whether or not the specified substances may be legally sourced, used, released, and transferred in the course of business operations. Unless otherwise regulated, to be in compliance with the TRI, an enterprise must first determine whether a facility is subject to the reporting requirement. This determination is made by identifying whether any substances at a facility are identified in the regulations and, if so, in what quantities they occur. Once that determination is made, the second part is to calculate the quantity by which a substance has exceeded the threshold quantity, if any, and report such data in the required form to the responsible agency. As noted, the expanded concept of the TRI includes the transfer of regulated substances beyond the property of the enterprise, however that boundary may occur. As a result, releases into the air, the water, and land—whether intentional or not—are all reportable. The term *transfer* recognizes that regulated substances may be intentionally removed from the enterprise property either as waste or by agreement, whether for off-site recycling, treatment, or disposal. With the inclusion of reporting sources, as well as the transfer of the regulated substances, it is possible to systematically track their movement and enable means of intentionally implementing actions at multiple levels and for various reasons.

Knowledge is power, and its disclosure is of interest to all parties. Reporting enterprises' business interests must be balanced with the public's right to know. The requirement to maintain records can therefore be considered from the perspective of the reporting enterprise, as well as of the public interest. Public access to the TRI allows a broad range of stakeholders and a corresponding complex regime for accountability in the matter of the listed substances. Where perception is reality, the TRI database would be a correspondingly significant descriptor of reality.

The availability of the collected data could influence internal management decision making. Some enterprises would likely collect the data for internal management reasons regardless. For other enterprises, such regulations provide incentives necessary to design and institute management controls. Internally, therefore, data required by TRI can be a means to raise corporate awareness of operations. Workers, visitors, and any persons on the premises or vicinity can also access the reports to determine their exposures to the specified substances and take appropriate actions. Enterprises and governments can use the data to demonstrate compliance with local and international reporting obligations.

The primary incentive intended by the TRI publication of data is to elicit public interest, and thus public pressure, on industry to improve conduct with respect to particular substances. Such pressure can arise from neighbors whose persons and property are directly or indirectly affected by accidental or intentional releases of the substances into the environment. Service providers, related businesses, researchers, consultants, and regulators can use the same data to advance their interests and improve the condition of societies. With the institution of registries in multiple jurisdictions, it is possible to extend the scope of the actions to a larger—and possibly global—scale.

In a dynamic and increasingly complex world, identifying substances to be listed is a rather laborious task. Although the TRI now includes well over 600 substances for reporting, there are difficulties in identifying which substances are reportable. For instance, lead and compounds including lead are separately identified in the TRI. Industrial processes could involve elemental lead, lead compounds, alloys, and mixtures for legally permitted reasons. It is not always easy for the reporting enterprise to determine which category applies at the reportable stage of source, release, or transfer.

Where threshold quantities above zero apply to each TRI-regulated category, it is conceivable that a facility may actually source, release, and transfer a large quantity of regulated substances and not be subject to reporting requirements if thresholds in each individual category are not exceeded. In this sense, setting out reporting thresholds can also operate to represent a permitted maximum release or transfer of each substance for a facility. Over the long term, releases could continue below reporting threshold quantities and could bioaccumulate to toxic levels without triggering any TRI provisions.

Lowering threshold quantities could result in greater transparency. However, the same transparency could compromise the confidentiality of business information. In extreme cases, this level of transparency could affect the existence of an enterprise. Even service providers, investors, insurers, and creditors who typically have an interest in access to internal business information as part of their due diligence inquiries may not want the same information publicly accessible.

A systemic issue with self-reporting regimes is that accuracy of data can be hard to verify, at least in the short term. For the reasons given earlier and personal disposition, resistance to reporting information will exist. Such difficulties can alleviate over time with cross-referenced databases, familiarity of the systems, and evolving cradle-to-grave information on the substances. Mitigating steps including auditing the reports can be done as the example in Canada suggests. Appropriate disclosure, compliance with accessibility requirements, and use of information can continue to be issues.

TRI or, rather, pollutant release and transfer register–type requirements implemented globally could lead to more comprehensive tracking of substances of interest. Evolving formulations and usages of substances would operate to continue current difficulties in ongoing reporting requirements.

See Also: Agenda 21; Bhopal; Corporate Responsibility; Kyoto Protocol; Regulatory Approaches.

Further Readings

Basset, Susan. "Lead Emissions." *Pollution Engineering*, 33/10 (2001).

Brehm, John and James T. Hamilton. "Noncompliance in Environmental Reporting: Are Violators Ignorant, or Evasive, of the Law?" *American Journal of Political Science*, 40/2 (1996).

Howes, Michael. "What's Your Poison? The Australian National Pollutant Inventory Versus the US Toxics Release Inventory." *Australian Journal of Political Science*, 36/3 (2001).

Marchi, Scott and James Hamilton. "Assessing the Accuracy of Self-Reported Data: An Evaluation of the Toxics Release Inventory." *Journal of Risk & Uncertainty*, 32/1 (2006).

U.S. Emergency Planning and Community Right-To-Know Act. http://www.epa.gov/oecaagct/lcra.html (Accessed February 2009).

U.S. Pollution Prevention Act. http://www.epa.gov/oppt/p2home/pubs/p2policy/act1990.htm (Accessed February 2009).

Lester de Souza
Independent Scholar

Tragedy of the Commons

The "tragedy of the commons" describes a social dilemma in which rational human behavior in using a resource as an unmanaged commons leads to overuse and, consequentially, to the destruction of that resource. It has proven a useful concept for understanding how resources are ruined not by some outside forces but by the inherent behavior that can be expected whenever a number of individuals own a scarce resource in common. A famous article in the journal *Science* by Garrett Hardin (1968) introduced the concept to common parlance.

Central to the tragedy of the commons is a metaphor: Imagine a common parcel of land (the commons) used by farmers. As a rational being, each herder will try to keep as many

The concept of the tragedy of the commons can be applied to the problem of overuse in national parks like Yosemite, where large numbers of visitors like these can threaten the survival of the resource.

Source: iStockphoto.com

cows as possible on the land, even if the consequence of this behavior would be the destruction of the commons. Each farmer tries to maximize his gain. The utility of adding one more animal on the land has one negative and one positive effect:

- Positive effect: The farmer receives all the benefits from each additional animal.
- Negative effect: The land is slightly damaged by each additional animal.

Because of the unequal division of the costs and benefits, each rationally behaving farmer concludes that the highest utility to him is to add another animal on the commons. And another, and another, and so on. He gains all the benefits, but the disadvantages are shared among all farmers using the commons. The benefits are privatized, and the costs are socialized. This is no big problem as long as the carrying capacity of the commons is not exceeded. As soon as the number of farmers surmounts a certain threshold, however, the tragedy of the commons takes effect: The carrying capacity is exceeded, but each farmer still seeks to pursue his short-term interest. Because all farmers reach the same conclusion, the long-term consequence is that everyone loses their share in the collective resource.

The tragedy herein lies in the paradox that individually rational strategies lead to collectively irrational outcomes. The rational behavior of each individual leads—in a finite world—inevitably to the ruin of the commons. In a nutshell, the central cause of any tragedy of the commons is that individuals do not have to bear the entire costs of their behavior when using a public good. The consequences are externalized. Individuals gain by internalizing benefits and externalizing costs.

Central to the tragedy of the commons is the perception that humans are rational self-interested actors. Actors behave purposefully in maximizing their utility. They assess the costs and benefits associated with possible options and choose the option that they expect will be the most beneficiary to them at no great cost, regardless of the costs to others. This assumption, that humans are rational self-interested actors, serves as the ontological basis for the factors causing a tragedy of the commons to arise.

Although one could think that almost all environmental problems involve a tragedy of the commons, this is not the case. Many problems involve problem structures that are more complex. For example, water pollution of transboundary rivers involves a so-called upstream–downstream problem structure. In this case, the relationship between the victims and perpetrators is asymmetric, with "downstream" states being the victims of and "upstream" states being the perpetrators of pollution, whereas in the case of a tragedy of the commons problem, such as overfishing, all states are simultaneously victims and perpetrators.

The term became popular when ecologist Garrett Hardin used it in his 1968 *Science* essay "The Tragedy of the Commons." However, Hardin's tragedy was not a new concept: Related theories can be traced back to authors such as Aristotle, as well as Thomas Hobbes and his *Leviathan*. More recently, around 180 years ago, William Foster Lloyd outlined a theory of the commons. Lloyd identified some of the problems resulting from property used in common. If the only "commons" of importance were a few grazing areas, this would be of little general interest, but that is not the case. In addition to using the tragedy as a metaphor for the general problem of overpopulation, Hardin developed some examples of "commons" problems such as the oceans (overfishing), national parks (overuse), and air and water pollution. Hardin realized that this concept is applicable to many modern problems. As a consequence, the tragedy of the commons has been used to describe such diverse problems as traffic jams in major cities, the rise of resistant diseases because of the careless use of antibiotics, the problem of acid rain, the problem of international cooperation, and the relationship between the public and the private sector in modern economies, to name only a few.

Is the Tragedy Inevitable?

When analyzing commons problems, past research constituted that at least two fundamental conditions must exist before a tragedy of the commons can emerge.

First, the resource needs to be subtractable (one person's use reduces the quantity or quality available to others) and must inherit the ability to be overused. This relates to the nature of the resource itself. A common distinction is made in this regard between a public good and a common-pool resource (CPR), a synonym for the term *commons*. The use of a public good cannot lead to a tragedy of the commons because it is nonconsumptive. For instance, one person's use of the weather forecast does not reduce its availability for another. However, the use of a CPR is by definition high in subtractability, such as in the use of a grazing field, where the grass is eaten by the animals of herder A reduces the use of that grass for the animals of herder B. Nevertheless, this does not lead to a tragedy as long as the carrying capacity is not exceeded; that is, the commons is very large, and there are only very few grazing animals on it.

Second, access to the resource needs to be open. Not every use of a subtractable and able-to-be-overused resource will inevitably lead to its ruin. Past research in this regard found that a tragedy is more likely in a situation in which the exclusion of beneficiaries from these resources is especially difficult. Potential individuals will be strongly tempted to free-ride on the efforts of others if they are not hindered.

Solutions to Tragedy of the Commons Problems

Even knowing the conditions that lead to a tragedy, avoiding it is not easily ensured. Because the nature of a resource is fixed, it is not possible to make a subtractable resource nonsubtractable; solutions have to concentrate on the second condition: managing access to resources. In this regard, Hardin already proposed two solutions to tragedy of the commons problems: the transfer of the resources either to government control or to private property. In addition to these two solutions, empirical studies of sustainable resources provided us with more approaches so that one can finally distinguish between four basic ideal types of solutions:

- Accessibility: Absence of enforced property rights; the resource is open to all
- Private property: Rights to the resource held by individuals (or firms) who may exclude others

- Communal/Group property: Rights to the resource held by a group of users who may exclude others
- State property: Rights to the resource held by the government that can regulate access to and use of the resource

In the decades after Hardin's article, the extensive theoretical as well as empirical research has shown that tragedies of the commons exist, but that they are not unavoidable. Unfortunately, there is no panacea for dealing with CPRs. All approaches are themselves subject to failure in some instances. In particular, both solutions Hardin suggested—privatization and state property—have propelled intensive research. However, the idea that privatization could solve CPR problems has been vigorously contested. Many critics point to empirical evidence where private owners exploit and destroy resources in an unprecedented way. In regard to the latter, Hardin's idea of mutual coercion, generally managed by outside agents, found support by many scholars and public officials. Nevertheless, much of the evidence found in instances when turning administrative responsibilities for managing resources over to centralized agencies points to counterproductive effects. In these cases, resource deterioration has frequently been accelerated by massive problems of inefficiency and corruption.

Despite these little-encouraging results, the most promising approach for creating robust, durable CPR regimes comes from Elinor Ostrom (1990). One of the most promising findings of this research is the capability of individuals facing tragedy-of-the-commons problems to solve these without the need of external intervention. In her work, Ostrom documents examples from grazing and forest institutions in Switzerland and Japan that have worked well and lasted for centuries, in which individuals succeeded in averting the tragedy of the commons by self-government and self-organization.

Moreover, Ostrom (1990) identified eight design principles of successful CPR regimes that were extracted from the analysis of CPR case studies. No single set of conditions seems to be sufficient. Nevertheless, most successful, robust, durable CPR regimes include six or more of these principles, whereas failed and fragile CPR regimes do not use more than a few of them.

These design principles were reformulated as follows for this article, with the aim of giving guidance to the creators of effective CPR regimes:

1. Define boundaries clearly: Defining the boundaries and the rules of who has the right to use a resource ensures that users can identify the geographic extent of the resource itself, as well as identify undoubtedly anyone who does not have the rights to prevent him or her from further use of the resource.
2. Enhance proportional equivalence between benefits and costs: The division between the rules that assign costs and the rules that assign benefits should be congruent. This way, participants consider the rules to be fair and the matching of both types of rules to local conditions should be ensured.
3. Allow collective-choice arrangements: Most users should be involved in the process of modifying the rules over time. This is important to take the information about the costs and benefits as perceived by different participants more fully into account.
4. Ensure effective monitoring: Supervisors should be accountable to appropriators or are the appropriators themselves.
5. Include graduated sanctions: It is important that action of an appropriator who breaks the rules is called attention to in the process and is assessed with graduated sanctions (depending on the seriousness and context of the offense).

6. Include conflict-resolution mechanisms: Ensure the existence of fair and low-cost conflict resolution arenas to resolve conflicts.
7. Ensure minimal recognition of rights to organize: Appropriators should have the right to devise their own institutions and have these rights recognized by international, national, regional, and local governments.
8. Enable nested enterprises: Larger resources with many participants' nested enterprises, ranging in size from small to large, are needed.

Altogether, tragedies of the commons have shown to be real but not unavoidable. Even though tragedies of the commons can be the outcome of many problems like overfishing, air and water pollution, traffic jams, and the rise of resistant diseases, they are not inevitable. Learning from successful efforts as well as from failures seems to be even more important when considering that most of the commons we looked at were on a local or regional scale. Although the number and importance of commons problems on these scales will not decline, for the global commons that crosses national boundaries, the need for effective approaches that are to be developed and deployed quickly enough will certainly arise. If not, the large-scale tragedies promise to become ever more serious as the world's population continues to grow and global problems increase.

See Also: Common Property Theory; Cost-Benefit Analysis; Deforestation; Water Politics.

Further Readings

Cox, Buck and Susan Jane. "No Tragedy of the Commons." *Environmental Ethics*, 7/1 (1985).

Hardin, Garrett. "The Tragedy of the Commons." *Science*, 162/3859 (1968).

Hardin, Garrett and John Baden, eds. *Managing the Commons*. San Francisco: W. H. Freeman & Co., 1977.

Lloyd, William Foster. *Two Lectures on the Checks to Population*. Oxford: Oxford University Press. Reprinted in part in *Managing the Commons*, edited by Garrett Hardin and John Baden. San Francisco: W. H. Freeman & Co., 1977, 1833.

Ostrom, Elinor. *Governing the Commons: The Evolution of Institutions for Collective Action*. Cambridge, MA: Cambridge University Press, 1990.

Ostrom, Elinor, et al. "Revisiting the Commons: Local Lessons, Global Challenges." *Science*, 284 (1999).

Martin Köppel
Independent Scholar

Transnational Advocacy Organizations

The end of World War II marked a major global political change. Many of the former colonial nations became free, and the world was divided into capitalism and communism. Another significant development during the same period was the increase in the number of nongovernment organizations (NGOs). NGOs are defined as "self-governing, private, not-for-profit organizations that are geared toward improving the quality of life of disadvantaged people." These NGOs were formed to work in communities neglected by the government because of their deep connection to traditional lifestyle.

As a result, many organizations targeted communities in developing nations, where the government failed to meet the expectations of the people and practiced favored politics. Advocacy organizations were primarily involved in advocating for policy changes at the national level. Because these organizations were operating across borders, they became transnational. Regardless of their type, the majority of NGOs work in areas like environment, health, human rights, children, poverty, and indigenous peoples' rights. Recently, NGOs working in particular issues have been forming "networks" and making civil society stronger than before. One such network is the Transnational Advocacy Network, formed by advocacy organizations working independent of the borders.

In general, there are two types of NGOs, service providing and advocacy. Service-providing NGOs work directly in the communities affected and thereby influence local well-being (e.g., World Bank and United Nations). In contrast, advocacy NGOs advocate for change in policies at the national level (e.g., Oxfam, CARE, Action Aid, etc.). Because the NGOs are private organizations working to solve global problems, they are different from national governments. Several features differentiate NGOs from national governments including the following: focusing on practical issues (e.g., education, trafficking and development); relying on media for public support; receiving donations and volunteer works as a result of popularity and legitimacy; actively promoting democracy and improving political management in societies; engaging rather than criticizing the private sector; and rejecting the existing global order, particularly with regards to the environment and the social justice sectors.

During the 19th and early 20th centuries, there were few international organizations working mostly on antislavery and relief works during wars. However, many experts consider the 1950s as the period of the NGOs' boom. NGO formation occurred in three phases.

The first-generation NGOs formed in the 1950s were primarily relief organizations like the Red Cross. The second-generation NGOs formed in the late 1960s promoted the integration of local organization with the work of international NGOs. The third-generation NGOs created after the 1980s (end of the Cold War) made political statements on behalf of the communities marginalized by government policies. The third-generation NGOs' use of advocacy strategy was directed toward facilitating sustainable changes through international advocacy, which means less direct involvement at the grassroots level but greater need for maintaining strong institutional links with partners at the local level. These NGOs have been able to provide advocacy because of their attachments to local communities and their links to the global network.

CARE is an excellent example of an organization that transformed from a relief organization to an international advocacy organization fighting poverty and working for women's and marginalized communities' rights. CARE was founded in 1945 in Atlanta, Georgia, to support World War II survivors, mostly in Europe and America. Later, CARE expanded its work in Asia and Africa and to other developing nations. Today, CARE works in 66 nations and employs more than 12,000 people. Recently, CARE has focused on women at the community level to improve basic education, prevent the spread of HIV, increase access to clean water and sanitation, expand economic opportunity, and protect natural resources. Similarly, Human Rights Watch is another international organization, formed in 1978. Human Rights Watch operates on the "name and shame" policy by reporting accurate information on human rights abuses globally. Because of its global influence, Human Rights Watch meets with governments, the United Nations, regional groups like the African Union and the European Union, financial institutions, and corporations to press for changes in policies and practices that promote human rights and justice around the world. These two examples give an idea of what advocacy organizations do and how they operate to influence policies at all levels. Both organizations work toward influencing policies at the national level to promote human rights at all levels.

What Transnational Advocacy Organizations Do

Transnational advocacy organizations (TAOs) are defined as "self-organized advocacy groups undertaking voluntary actions across state borders in pursuit of what they deem the wider public interest." Advocacy organizations are known by different names: nonstate actors, NGOs, transnational advocacy networks, and transnational or global civil society. Some of the earliest works by TAOs were against female circumcision in Kenya and foot binding in China, and for environmental management in the Himalayas and human rights in Latin America. These organizations are complex, multilayered, and large, and they operate across borders. Transnational organizations work through international campaigns to influence policies and are led by activists. In developing nations, domestic NGOs use international allies to put pressure on their government on issues like human rights and sustainable development. They also tend to connect microlevel grassroots organizations with international advocacy networks of NGOs. Therefore, NGOs have to balance their relationship with different stakeholders, from local to international. As a result, NGOs use different methods to legitimize their work. They

- seek to influence policy by pointing to practical experience on the ground,
- promote a particular value recognized in the society as being crucial,
- act as experts on an issue and help national experts on issues of importance,
- work in transnational networks with grassroots and other organizations,
- work with alliances and networks,
- promote notions of universal human dignity or global justice,
- function within international legal norms and regulations (human rights),
- are politically independent with independent sources of funding, and
- give voice to or represent marginalized groups.

Until 1960, advocacy organizations were mostly based in the country of their origin. However, with globalization, most of the organizations became international. Since then, the number of TAOs has significantly increased, with environmental organizations having the biggest increase. The 1972 Conference on Human and Environment paved the way for the creation of environmental organizations. Many of the large advocacy organizations—Natural Resources Defense Council, World Wildlife Fund, World Vision, Action Aid, Rainforest Alliance, the Mountain Institute, and so on—were created in the last 30 years. At this time, there are more than 30,000 international organizations working on various issues from household livelihoods to global security. Some estimate the nonprofit sector to be worth over $1 trillion a year. Cheap air travel and communications in the last two decades has further helped in making these organizations transnational in operations.

Although the role of TAOs is to bring about institutional change toward creating policies that empower marginalized communities, the act of advocacy is not just to empower marginalized communities. The underlying function of an advocacy organization is often to enhance the self-respect of weaker communities, to improve their self-confidence, to constitute integrity, and to promote mutual trust, which are the essential ingredients to developing a healthy community.

In addition, TAOs have to fight against cynicism and despair, which powerless communities tend to fall victim to in the face of massive political and practical obstacles impairing them from improving their lot. The advocacy organizations are challenged by political responsibility in communities where they work. Some actions taken to overcome this challenge are dividing political arenas, agenda setting and strategy building, raising and allocating financial resources, information flow, information frequency and format, information translation into useful forms, and the formalization of relationships with different stakeholders.

How TAOs Have Changed

In recent years, international organizations have become more visible, better organized, and more sophisticated in dealing with different political institutions and the media. In addition, many organizations are creating networks to create scope and common ground for a particular issue. For example, the global alliance to protect tropical forests, the common fight against poverty, and the fight against HIV/AIDS. Transnational advocacy networks are very difficult to form, but the formation of networks on a particular subject gives high value. The density and strength of networks come from identity as defined by principles, goals, and targets of the member organizations. The voice of a network is not the sum of the component voices, but it is the product of interaction of voices. Likewise, the various organizations that form the network are not egalitarian. Power is exercised within the networks, as stronger actors tend to dominate weaker ones, and the weaker and smaller organizations depend on stronger and/or larger organizations for funds and operation.

Despite their global significance and achievements, transnational organizations are not free from criticism. According to the World Polity Theory, international NGOs and networks act as conveyor belts carrying Western liberal norms to developing nations. Some authors claim NGOs operate under the neoliberalism framework and therefore follow Western ideologies in their work. Sometimes, to become successful, international NGOs concentrate their efforts on issues not directly related to their main issue. For example, many organizations reframe issues like poverty into environmental problems, when it is connected to economic and political policies.

Sometimes the efforts of international organizations do not reach the grassroots level and, in the process, only elites and the educated benefit from their activities. The other major criticism of NGOs is their transparency, accountability, legitimacy, and representation. There are also increasing differences between NGOs in the North and the South. TANs in the South are different and do not share the same beliefs as those of Northern NGOs. Even though Third World activists may oppose the policies of their own governments, they have no reason to believe that international actors would do better, and they suspect the contrary. The issue of sovereignty for Third World activists is deeply embedded in the issue of structural inequality, and therefore success of transnational organizations is limited.

Despite their criticisms, NGOs are influential at the global level in directly engaging with communities and influencing policies at all levels and by participating in global networks and associations across borders. NGOs are also better than aid agencies, because aid agencies are weak and transient at the grassroots level and have little influence on decision makers at national and international levels. Therefore, in the future we can expect more robust networks of advocacy organizations working for global change. One of the issues in which TAOs can play a significant role in the future is climate change. Advocacy organizations can help developing and even developed nations to frame policies toward mitigating global warming and investing in the future.

See Also: Environmental Nongovernment Organizations; North–South Issues; Organizations.

Further Readings

Castells, Manuel. "Global Governance and Global Politics." *PS: Political Science and Politics* (January 2005).
Chapman, Jennifer and Thomas Fisher. "The Effectiveness of NGO Campaigning: Lessons From Practice." *Development in Practice,* 10/2 (2000).
Collingwood, Vivien. "Nongovernmental Organizations, Power and Legitimacy in International Society." *Review of International Studies,* 32 (2006).
Evans, Peter. "Fighting Marginalization With Transnational Networks: Counter-Hegemonic Globalization." *Contemporary Sociology,* 29 (2000).
Hudson, Alan. "NGOs Transnational Advocacy Networks: From 'Legitimacy' to 'Political Responsibility'?" *Global Networks,* 1/4 (2001).
Keck, Margaret. E. and Kathryn Sikkink. *Activists Beyond Borders: Advocacy Networks in International Politics.* Ithaca, NY: Cornell University Press, 1998.
Lerche, Jens. "Transnational Advocacy Networks and Affirmative Actions for Dalits in India." *Development and Change,* 39/2 (2008).
Madon, S. "International NGOs: Networking, Information Flows and Learning." *Journal of Strategic Information Systems,* 8 (1999).
Price, Richard. "Transnational Civil Society and Advocacy in World Politics." *World Politics,* 55 (2003).
Van Tuijl, Peter and Lisa Jordan. "Political Responsibility in Transnational/NGO Advocacy." Washington, D.C.: Bank Information Center (1999). http://www.bicusa.org/en/Article.138 .aspx (Accessed January 2009).

Krishna Roka
Penn State University

Transnational Capitalist Class

The transnational capitalist class may also be referred to by a number of other different terms such as *superclass* or *global ruling class.* The transnational capital class is described as a segment of the international bourgeoisie that stands for transnational capital—the owners of transnational corporations, which in turn own the means of production globally, and the organizations, institutions and individuals who support this structure.

Many sources acknowledge that since the 1970s the global capitalist system or "world capitalism" has undergone a period of restructuring toward a more internationally oriented model in a process that has come to be known as "globalization." Some would argue that the roots of globalization are much older and that international trade has always formed a key component of the economic system; however, it can be seen that this process of international trade has been radically accelerated by the growth and availability of information and communication technologies made possible by the electronics revolution, as well as by the explosion in international travel that has been enabled by the rapid growth of the aviation industry and development of the jet airliner. This has quickened the pace of internationalization.

The transnational capitalist class is likely to be seen as increasingly important actors in the sustainability debate, as the discourse of globalization posits that increasing power is being transferred from the state to private corporations. This "hollowing out of the state" means that old state-centrist views of governance are replaced with global notions of power that transcend nation-state boundaries and in which the transnational capitalist class play an increasingly important role.

"Global System Theory" is an attempt to understand capitalist globalization. Protagonists of Global System Theory advance that it possesses three main components:

- the transnational capitalist class,
- the discourse of globalization embodied in transnational corporations, and
- the "culture-ideology" of consumerism and consumption.

Leslie Sklair, a sociologist at the London School of Economics, characterizes the transnational capitalist class as consisting of four "fractions," or subsets of people.

1. The owners and controllers of transnational corporations (this could also include agents acting on their behalf locally as affiliates, representatives, or franchisees)
2. The bureaucrats and politicians who advance the discourse of globalization
3. Professionals and business people who advance the discourse of globalization through their work
4. The elites who advance the "culture-ideology" of consumerism, whether this be the traders of goods and owners of brands or the media and advertising professionals

He states that there is a temporal and a spatial dimension to the distribution of these four fractions and the institutions that give them their agency within the system. Sklair highlights that there are differences and tensions between the worldviews of these four fractions and that they are "not always entirely united on every issue."

The transnational capitalist class exhibits the following identifying characteristics:

- Members of the transnational capitalist class have economic interests with global linkages rather than being exclusively embedded within their local and national business communities.
- Members of the transnational capitalist class may have a portfolio of property, assets, and shares that are global in their scope. This is aided by the fact that enabling communications technologies and international political economy allow money and assets to change hands with ease across nation-state boundaries in an increasingly globalized economy.
- Members of the transnational capitalist class advance the notion that unless local workers are prepared to accept inferior pay and benefit packages to compete with the international labor force, jobs and competitiveness will be lost as foreign competition can surpass local efforts. This is characterized as "the race to the bottom."

- Members of the transnational capitalist class tend to have expansive outward global orientations toward problems of economy, politics, culture, and ideology, rather than introspective, local-looking views.
- Members of the transnational capitalist class tend to share similar lifestyles whose patterns of consumption transcend national boundaries: partaking of luxury goods and services; patronizing brands and chains that transcend what is locally available; and sharing similar patterns of higher education, often involving "business school" or "International MBA" experience.
- Members of the transnational capitalist class seek to advance the notion that they are "citizens of the world," in addition to identifying with their own historical connections with places of birth or residence. They may change their nationality or take a second passport to advance their business interests, and they are likely to seek residence in a country that advances their economic interests, rather than choosing their country of birth.

The transnational capitalist class is opposed by those who reject capitalism as a way of life or as an economic system that they wish to embrace. However, it should also be noted that there is a portion of capitalists who reject the discourse of globalization and the effects it entails. Most local businesses cannot prosper when faced with competition from transnational corporations, and so here we can see an example of where the proprietors of such businesses would consider themselves broadly "capitalists" and yet would be opposed to the transnational capitalist class in ideology.

William Robinson and Jerry Harris argue that the transnational capitalist class "is a ruling class because it controls the levers of an emergent trans-national state apparatus and of global decision making." In particular, the transnational capitalist class is accused of constructing a new hegemony based around a new globalized structure of consumption, production to satiate this consumption, and accumulation of wealth. This structure is embodied in various political and economic forces that conspire to advance the notion of globalization.

The discourse of globalization has profound effects on the sustainability of business and trade. Providing goods and services that are shipped over long distances is viewed by many as inherently unsustainable; however, the transnational capitalist class is responsible for advancing and extending this notion, fueling the demand for luxury goods and services and encouraging this behavior in others who would seek to emulate the lifestyles of the transnational capitalist class.

Sklair also advances the notion that the transnational capitalist class have "used the discourses of national competitiveness and sustainable development to further the interests of global capital." He argues that sustainable development has become "a major industry" in which the goal of sustainability is viewed by transnational corporations as "a prize," and the advantage of being seen in a leading position is that the "winner[s] . . . get to redefine the concept." To this end, the transnational capitalist class can be viewed as complicit in this commoditization of sustainability.

See Also: Capitalism; Globalization; Sustainable Development.

Further Readings

Robinson, William I. *A Theory of Global Capitalism: Production, Class, and State in a Transnational World (Themes in Global Social Change)*. Baltimore, MD: Johns Hopkins University Press, 2004.

Robinson, William I. and Jerry Harris. "Towards a Global Ruling Class? Globalization and the Transnational Capitalist Class." *Science and Society*, 64/1 (Spring 2000).
Sassen, Saskia. *A Sociology of Globalization* (Contemporary Societies Series). New York: Norton, 2007.
Sklair, Leslie. "The Transnational Capitalist Class and the Discourse of Globalization." *Cambridge Review of International Affairs*, 14/1:67–85 (Autumn 2000).

Gavin D. J. Harper
Cardiff University

Transportation

Developing better alternatives to current transportation systems and technologies is critical to any sustainable development strategy. Globally, mobility services are responsible for more than 60 percent of world oil consumption. The International Energy Agency predicts that global oil consumption will increase from its 2007 level of about 83 million barrels per day to more than 106 million barrels per day in 2030. In addition, despite all of the talk in recent years about alternative fuels, petroleum-based products remain dominant, constituting 98 percent of all transport energy consumption in 2008.

Given the growing demand in carbon-based transportation services, the following questions will be critical to meeting the challenges of sustainability. What are the principal modes of transportation? How do environmental externalities vary among modes? What are some common factors that influence the economic, social, and environmental characteristics of transportation systems? Also, what strategies can be employed to develop more efficient and environmentally sustainable methods to move people, goods, and services from point A to point B?

There are many available alternative transportation choices, including:

- Human-powered: walking, biking
- Private vehicle: rented or owned
- For hire, single party: taxi or limousine
- Shared ride: carpool, vanpool
- Rubber-tired: public transit bus or trolley bus
- Rail transit: street cars/light rail, monorail, heavy rail, commuter rail
- Airborne: airplane, helicopter
- Waterborne: ferry or personal boat

Each mode varies widely in cost. Environmental economists use an inclusive concept called triple bottom line accounting to measure cost. Triple bottom line accounting integrates social and environmental externalities associated with economic activity into overall cost calculations. According to this standard, air travel and private vehicles are generally more costly, and high-occupancy ground and waterborne public transit are usually less costly. For instance, carbon dioxide emissions per person-kilometer traveled by subway can be up to 90 percent lower than those generated by private vehicle use. However, despite their negative attributes, private vehicles continue to be the most sought-after mode of transport. This is evidenced by the fact that global vehicle production hovers around 70 million units per year and continues to grow.

Interestingly, China is poised to overtake the United States as the biggest national producer of vehicles. It is also the world's leader in use of the most energy-efficient mode of human-powered transportation (as measured by distance traveled per calorie used)—the trusty bicycle.

The term *transportation system* typically refers to how various transportation modes are deployed and connected to one another to provide efficient and safe movement of goods and people. What agents affect the evolution of transportation systems, and particularly motorized vehicle use? Common factors include

- economic structure and growth;
- spatial and demographic characteristics;
- technological change, resource availability, and public policy;
- consumption culture; and
- urban planning culture.

In general, economically wealthy countries exhibit higher per capita mobility requirements (kilometers traveled/year) than less wealthy ones. Wealthy economies generate more economic activity, which generates more demand for mobility. There is an especially strong correlation between per capita wealth and demand for private modes of transport. Economic wealth also generates social and cultural change, including higher citizen interest in freedom of movement and engagement in mobility-intensive leisure and social activities.

Spatial development characteristics heavily influence the evolution of transportation systems. Countries with more geographically dispersed, lower-density urban systems such as the United States, Canada, and Australia generate more per capita transport demand, and especially private vehicle use. Density also increases the financial cost of building and maintaining public transit systems relative to private automobile transportation (though public transport is almost always cheaper if environmental externalities are factored into the calculation). On the other side of the density spectrum, in Asian megacities such as Hong Kong or Tokyo, mass transit is the preferred choice for most residents.

New public transportation like this monorail in Kuala Lumpur, Malaysia, which opened in 2003, is crucial to reducing fossil fuel use. Worldwide, transportation accounts for over 60 percent of world oil consumption.

Source: iStockphoto.com

Demographic factors are also a major driver behind transportation demand. For example, mobility demand growth in populous developing countries like India and China greatly exceeds Organisation for Economic Co-operation and Development averages. China has a relatively low population growth rate, but a high economic growth rate. In contrast, India has experienced steady but slower lower economic growth but faster population growth. Among these cases, economic factors seem to be the primary driver behind growth, as mobility demand in China has increased more quickly than in India.

Technological advances have greatly expanded the menu of mobility choices available to societies. Most notable are innovations leading to the manufacture and mass consumption of private vehicles. Why do societies favor private vehicle use over more efficient and environmentally friendly alternatives? The answer to this question is partly related to how governments influence prices for transport goods and services. Price signals have a major effect on which modes of transportation consumers choose to use. Governments can potentially use triple bottom accounting methods to calculate externalities associated with transportation choices and then employ a combination of voluntary (e.g., rebates to encourage alternative fuel use) and/or mandatory (taxes, fees, emissions caps, etc.) incentives and disincentives to influence consumer behavior. Unfortunately, governments rarely make full use of their regulatory powers to fully price in costs paid by society to support the high-carbon transportation choices of private actors.

Political competition plays an important role in mediating government involvement in energy and transportation markets. Auto and energy producers use lobbyists and political contributions to keep their elected allies in power. Public entities, in turn, provide direct subsidies to petroleum producers, including investment in extraction and refining infrastructure and technology. They also reduce producer and user monetary expenditures by not requiring auto firms to meet higher mileage standards and using public funds to provide road infrastructure for private vehicles. The scope of these incentives and investments is so immense that they create a self-perpetuating cycle of dependency on high-carbon industries that seems difficult to overcome. For instance, auto executives commonly justify their actions by insisting that tightening mileage or other regulatory standards will result in layoffs and financial losses.

However, it is possible to break this dependency. Governments can choose to use regulation to internalize the true costs of high carbon transportation. This can actually stimulate growth of new high-profit, high-employment transport firms. In fact, auto producers in countries where governments have used regulatory levers to shift the flow of investment toward more energy efficient vehicles now have a distinct competitive advantage in the global marketplace.

Many of the same scientific tools that got us into our current predicament also provide hope for the future. The potential for alternative/renewable fuels and vehicles is bright. Biodiesel, compressed natural gas, and liquefied petroleum gas (propane) offer less-damaging alternatives to gasoline. Electric vehicles may also provide low-carbon mobility. However, environmental costs will vary depending on the fuel mix used to produce electricity (e.g., hydro versus coal).

Soon, solar energy may power the electric vehicles of the future. Hydrogen also offers promise if a way can be found to produce it cheaply and cleanly. Governments can play a critical role in incentivizing private investment in low carbon fuels by providing rebates and subsidies to alternative fuel producers and users to help them reach economy-of-scale efficiencies.

Consumer transportation choices are driven as much by culture as price. Our global love affair with the automobile continues unabated. This is especially true in the United States, where few possessions have been more symbolic of individual freedom. For teens, getting that first drivers license ranks among the most important events in their lives.

Many people are so tied to their cars that they cannot imagine living without them. In a keeping-up-with-the-Joneses culture, the type of vehicles people own or the number of frequent flyer miles they amass directly affects their social status. However, this may be changing. In many communities, green living is the new "cool," where owning an electric vehicle, brewing biodiesel in the garage, or living without a car garners praise.

Many governments actively encourage urban planning models that favor auto use while penalizing other modes of transport. Low-density suburban communities rarely provide good bus service, rail links, or dense mixed-use developments. Ubiquitous shopping malls stick out like islands in seas of concrete.

The challenge to develop more sustainable transportation alternatives in the midst of suburban sprawl is immense. The poster child for global car culture is Los Angeles, California, where around half of available land surface has been paved over to accommodate motorized vehicles. It was not always this way. During the first few decades of the 20th century, streetcars crisscrossed the city (almost 2,000 kilometers of lines in 1915). But by the 1950s, superhighways linking widely dispersed suburbs became the norm. This same planning model has been replicated all across the United States.

However, many communities are learning from past mistakes. Even in car-loving Los Angeles, commuter rail is making a comeback. Hundreds of cities and towns are building new mass transit infrastructure to serve "built out" neighborhoods while also developing more comprehensive approaches to planning future development. Various schools of thought have evolved in recent decades (new urbanism, smart growth, etc.) to guide this process. Common urban planning recommendations include the following:

- Design communities to conserve energy
- Provide infrastructure for and access to public transportation networks (e.g., train stations, commuter parking, bike storage, etc.)
- Build higher-density "compact"' communities
- Limit expansion of outer suburbs/urban fringe settlements
- Encourage development of mixed housing/retail/commercial complexes that provide both services and employment to local residents
- Expand infrastructure for bike/pedestrian traffic
- Deepen multimodal transport linkages
- Channel development to already disturbed areas and limit use of greenfields
- Build infrastructure to support new transport technologies such as electric vehicles
- Require developers to provide more social and environmental amenities (e.g., schools, parks, affordable housing)
- Preserve undisturbed and limited use ecological zones bordering built out areas

As humanity lurches toward a carbon-constrained future, the need for sustainable transportation alternatives will grow. However, already-developed tools and strategies, if fully implemented, offer a path toward a low-carbon world. If communities can muster the will to act, achieving real sustainable development is within their grasp.

See Also: Ecological Economics; Ecotax; Petro-Capitalism; Regulatory Approaches; Suburban Sprawl; Sustainable Development.

Further Readings

American Public Transportation Association. *Public Transportation Reduces Greenhouse Gases and Conserves Energy.* Washington, D.C.: American Public Transportation Association, 2008. http://www.apta.com/ (Accessed December 2008).

American Public Transportation Association. *Transit Vision 2050*. Washington, D.C.: American Public Transportation Association, 2008. http://www.apta.com/ (Accessed January 2009).

Energy Information Administration. *Annual Energy Review 2007*. Washington, D.C.: Energy Information Administration (DOE/EIA-0384, 2007).

Environmental Protection Administration. *Fuel Economy Guide 2009*. Washington, D.C.: Environmental Protection Agency, 2008. http://www.fueleconomy.gov/ (Accessed January 2009).

Fotsch, Paul. *Watching the Traffic Go By: Transportation and Isolation in Urban America*. Austin: University of Texas Press, 2006.

Goilias, Konstadinos, ed. *Transportation Systems Planning: Methods and Applications*. Boca Raton, FL: CRC Press, 2002.

Grava, Sigurd. *Urban Transportation Systems: Choices for Communities*. New York: McGraw-Hill, 2003.

Harrington, Jonathan. *The Climate Diet: How You Can Cut Carbon, Cut Costs and Save the Planet*. London: Earthscan, 2008.

International Energy Agency. *Key World Energy Statistics: 2007*. Paris: International Energy Agency, 2007.

Lindstrom, Matthew, et al., eds. *Suburban Sprawl: Culture, Theory and Politics*. Lanham, MD: Rowman & Littlefield, 2003.

Szold, Terry, et al., eds. *Smart Growth: Form and Consequences*. Toronto: Lincoln Institute of Land Policy, 2002.

Jonathan Harrington
Troy University

UNCERTAINTY

Uncertainty can be described as a state of knowledge—or lack thereof—in which the probability of any adverse effect or the effects themselves cannot be reliably assessed. Because of the complex and ambiguous nature of humanity's interaction with the environment, uncertainty is inherent in any analysis of this relationship. The notion of uncertainty is often contrary to the positivist perspective on science, which observes cause-and-effect relationships between phenomena with the aim of resolving, avoiding, or negating adverse effects of environmental problems. With this in mind, it is often the linear relationship in the absence of uncertainty that policymakers adhere to when deciding on appropriate courses of action.

It is increasingly recognized, however, that the cause-and-effect relationship in the absence of uncertainty is overly reductionist. This applies to lots of different fields of study, from quantum mechanics to the use of computer models, for predicting what effects human activity will have on the environment. For example, atmospheric, oceanic, and hydrologic systems more broadly interact in such a complex manner that any form of prediction or certainty is unrealistic.

These issues are brought into sharp relief within the varying debates surrounding global climate change. Although scientific evidence is increasing on the effect that human populations are having on the environment, there remains significant uncertainty over precise cause-and-effect relationships. At the most fundamental levels, questions are still raised over the causes of global climate change. For example, to what degree does humanity have an effect on climate? What other factors are involved? To what extent do such issues as climate cycles, sun spots, and volcanic activity contribute to climate fluctuations?

Uncertainty may be said to fall into two categories. The first is risk, which is an event with a known probability. Risk can usually be applied to localized systems in which the variables involved are limited. Second, there is true uncertainty—this is an event with an unknown probability. Drawing on climate change again, both risk and true uncertainty can be applied to the release of carbon dioxide. For example, scientific observation can ascertain the degree to which the release of carbon dioxide from exhaust fumes are able to affect the climate in local areas, producing smog and leading to a number of respiratory disorders in local populations. From a boarder perspective, the release of carbon dioxide has uncertain effects on the global environment.

Uncertainty poses significant challenges for the development of effective policies that encourage adaptive and mitigative mechanisms designed to address environmental issues. To accommodate risk uncertainty, a number of mechanisms have been developed; for example, sustainability assessment, environmental impact assessments, and general risk assessments. Accommodating true uncertainty is not as straightforward. Two principles that have dominated with regard to uncertainty over the past decade are the polluter-pays principle and the precautionary principle. The polluter-pays principle simply states that the perpetrator of the pollution incident must compensate all those parties who have been detrimentally affected by this event. In reality, this principle can only appropriately be applied to the pollution that has occurred where there is a certain degree of linearity between the cause-and-effect relationships or where the uncertainty is at a relatively low level.

There is an ongoing debate over the effectiveness of the use of the precautionary principle in the face of scientific uncertainty. For example, it has been argued that the use of the precautionary principle with the logic that it is better to be safe than sorry could be seen as an excuse to prevent any activity that may be perceived as having a negative side effect, the possible consequences of which would be to inhibit innovation and technological advancement. It has also been argued that uncertainty can be used by powerful interest groups for their own ends. In contrast, uncertainty as represented in the precautionary principle does not suggest that there should be a ban on all activities that cannot be quantified or the outcomes absolutely predicted. Instead, what is argued is that there is a measured phase-out of risky and uncertain activities until more information and experience are available.

Extending this debate, there are distinctions between strong and weak forms of the precautionary principle. Strong precaution suggests that regulation is required whenever there is a possible risk to health, safety, or the environment, even if the supporting evidence is speculative and even if the economic costs of regulation are high. A strong form of the precautionary principle, for example, would call for measures to lessen the risk of environmental harm by climate change by incorporating a focus on reducing or preventing the emission of greenhouse gases. Over time, there has been a gradual transformation of the precautionary principle from what appears in the Rio Declaration to a stronger form that arguably acts as a restraint on development in the absence of firm evidence that it will do no harm. Weak precaution indicates that lack of scientific evidence does not preclude that action of damage would otherwise be serious and irreversible. The weak version of the precautionary principle is seen as the least restrictive and allows preventive measures to be taken in the face of uncertainty but does not necessarily require them. Examples of this can be witnessed in the Rio Declaration and the United Nations Framework Convention on Climate Change.

See Also: Global Climate Change; Globalization Precautionary Principle: Sustainable Development.

Further Readings

Borne, Gregory. *Sustainable Development: The Reflexive Governance of Risk.* Lampeter: Edwin Mellen, 2009.

Renn, O. *Risk Governance: Coping With Uncertainty in an Uncertain World.* London: Earthscan, 2009.

Gregory Borne
University of Plymouth

UN Conference on Environment and Development

The United Nations Conference on Environment and Development (UNCED), also known as the Earth Summit, was held in Rio de Janeiro June 3–14, 1992, ushering in a new era of global environmental politics. The Earth Summit, held in part to mark the 20th anniversary of the UN Conference on the Human Environment (UNCHE), surpassed expectations in terms of both attendance and influence. With representatives from 178 governments, including 108 heads of state and over 2,400 representatives from nongovernmental organizations (NGOs), the Earth Summit was the largest event of its kind to date. The Earth Summit was also novel in that over 17,000 people attended the parallel NGO Forum, something that made headlines around the world, as 8,000 journalists provided daily coverage of both the official conference and the parallel forum. The attendance of so many heads of state, and the extensive press coverage in turn, raised the profile of the agreements signed at the Earth Summit, especially the Framework Convention on Climate Change, the Convention on Biological Diversity, and the Rio Declaration on Environment and Development. The Earth Summit also further legitimated the participation of NGOs in international negotiations and "sustainable development" as the organizing principle for national development strategies and multilateral treaties. Finally, principles and funding mechanisms formulated in the expansive Agenda 21 have had far-reaching effects on the domestic politics of most countries, as well as on international environmental politics.

UNCED History

The foundations for the UNCED were established 20 years earlier at the UNCHE held in Stockholm in 1972. Preparations and then negotiations at UNCHE were marked by a split between industrialized countries and developing countries, many of whom had recently gained independence from colonial powers. Industrialized countries pushed for regulations on transboundary air and water pollution, arguing that the origins of environmental problems were the same regardless of economic system or level of industrialization, and that developing countries stood to gain the most from measures to protect the environment. Developing countries countered that air and water pollution were primarily caused by industrialized countries, which should thus bear the economic burdens associated with pollution mitigation. Moreover, developing countries were very concerned that an environmental logic would be used to restrict their development, and that assistance for environmental protection would take the place of development aid. Negotiations at UNCHE were not able to overcome these fundamental conflicts, though they did succeed in putting global environmental issues on the international political agenda, as well as leading to the formation of the United Nations Environment Programme, the first United Nations secretariat with headquarters located in a developing country.

The conflict between environmental protection and economic development continued throughout the 1970s. In the early 1980s, the United Nations General Assembly established the World Commission on Environment and Development to formulate a long-term agenda. The resulting Brundtland Commission produced the widely influential report *Our Common Future* in 1987, solidifying the call for linking environmental protection efforts with social and economic development. In particular, the Brundtland report focused on intergenerational equity, stating that development was sustainable only if the needs of future generations are not compromised in meeting the needs of the present.

The years leading up to the Earth Summit were marked by the preparation of national reports and policy proposals by governments, international institutions, and NGOs, as well as significant negotiations on the documents signed in Rio. The buildup to the summit was especially intense, as the same tensions between industrialized and developing countries and environmental protection and development marked many of the negotiations. Moreover, coming at the end of the Cold War and after waves of democratization in Latin America and east Asia, preliminary negotiations and the summit itself were the first time that many NGOs came into contact with like-minded groups from other countries.

UNCED Official Documents

The UNCED produced five documents: the Rio Declaration, the Framework Convention on Climate Change, the Convention on Biodiversity, Agenda 21, and the Declaration on the Forest.

The Rio Declaration comprised 27 principles, including the sovereign right of every state to exploit its own resources pursuant to its own environmental and development policies, as long as those activities within its jurisdiction do not damage the environment of other states or areas beyond its national jurisdiction. This emphasis on national sovereignty and the right to development was taken almost verbatim from the Stockholm Conference. The Rio Declaration also states that to achieve sustainable development, appropriate demographic policies and the reduction of unsustainable production and consumption patterns is needed. The Rio Declaration also promotes participation by women and indigenous peoples in decision-making processes. Finally, both the precautionary and the polluter-pays principles were included in the draft, which was unanimously approved.

The Convention on Biological Diversity is a legally binding treaty whose stated goals are the conservation of biodiversity, the sustainable use of the components of biodiversity, and sharing the benefits arising from the commercial and other use of genetic resources in a fair and equitable way. Debates around the Convention on Biological Diversity were often contentious, especially with regard to rules governing genetic engineering and intellectual property rights, financing of the convention, and instituting mechanisms for profit sharing. By the end of the Earth Summit, 153 states had signed the convention, with the notable exception of the United States.

The text for the Framework Convention on Climate Change, which was signed by 153 states in Rio, was drafted in 1988 by the Intergovernmental Panel on Climate Change and then revised by the International Negotiating Committee. Although the former included significant input from the scientific community, the latter did not, leading to a less ambitious final document. The most contentious of the five Rio documents, the Framework Convention on Climate Change negotiations were marked not only by divisions between the industrialized North and the developing South but also between European countries and the United States, which refused to accept any agreement that included a target of reducing carbon dioxide emissions to their 1990 level by 2000. Developing countries, for their part, also refused any limits on their right to develop. The Framework Convention on Climate Change did, however, include the principle that developing countries should be refunded the full incremental costs associated with measures taken under its rubric and calls for technological transfers and financial assistance to assist them in developing in a less carbon-intensive way.

Agenda 21 is an immense document of policy recommendations for achieving the various goals encompassed in the concept of sustainable development. Drafted through a broad participatory process, Agenda 21 is written in a style very different from the other UNCED documents. Rather than being a legally binding agreement, Agenda 21 presents the political consensus of that time on the causes of environmental degradation, as well as a contemporary listing of best practices in regard to solving specific problems. As such, Agenda 21 deals with a range of issues, including urban and rural poverty, health, excessive consumption in developed countries, and education, especially for girls.

The first eight chapters of Agenda 21 deal with social and economic development issues such as combating poverty, changing consumption patterns, the role of population in hindering sustainable development, and the promotion of health and sustainable human settlements. This first section of Agenda 21 also outlines specific measures for the planning and implementation of "local Agenda 21s" by local authorities and supports the increased use of market-based instruments for achieving environmental and developmental goals. The second section of Agenda 21 covers the conservation and management of resources for development, dealing specifically with the atmosphere, forests, deserts, mountains, oceans, freshwater ecosystems, sustainable agriculture, and biodiversity.

The third section of Agenda 21 focuses on strengthening the role of major groups, specifically calling for greater participation in decision-making processes by women, children and youth, indigenous communities, NGOs, local authorities, workers, transnational corporations and entrepreneurs, the scientific and technological community, and farmers. The fourth section of Agenda 21 covers the means of implementation, including financial resources and mechanisms, science and technology transfer, education, increased public awareness and capacity building, and international institutional arrangements and legal instruments.

The Statement of Forest Principles represents both success and failure, as it is the first global agreement dealing with forest management but is not a legally binding convention. Along with finding economic and social substitutes for forestry, signatories are expected to implement forest conservation and reforestation programs.

Despite its limitations, the Earth Summit was a landmark event not only for sustainable development but also for the expanding participation in the United Nations policymaking process.

See Also: Agenda 21; Brundtland Commission; Environmental Nongovernmental Organizations; Sustainable Development; UN Framework Convention on Climate Change.

Further Readings

Grubb, Michael, et al. *The Earth Summit Agreements: A Guide and Assessment.* London: Earthscan, 1993.

Hass, P. M., et al. "Appraising the Earth Summit: How Should We Judge UNCED's Success." *Environment*, 34/8 (1992).

UN Convention on Biodiversity. http://www.cbd.int (Accessed July 2009).

UN Environment Programme. "Agenda 21." http://www.unep.org/Documents.Multilingual/Default.asp?documentID=52 (Accessed July 2009).

UN Framework Convention on Climate Change. "Documentation." http://unfccc.int/documentation/items/2643.php (Accessed July 2009).

W. Chad Futrell
Cornell University

UN Framework Convention on Climate Change

The United Nations Framework Convention on Climate Change (UNFCCC) is an environmental treaty that entered into force on March 21, 1994, and has been updated a number of times since. The UNFCCC itself is a declaration of intent, set up to allow for the creation of specific mandatory emission limits through the adoption of subsequent updates; thus, signing the UNFCCC does not bind the signatory nation to specific actions the way signing and ratifying those updates does. The most recent update, the Kyoto Protocol, has eclipsed the UNFCCC in fame and recognition in part because of Presidents Bill Clinton and George W. Bush's decisions not to forward it to the U.S. Senate to be ratified (thus retaining the United States as a signatory of the protocol, but one that is not bound by it) because of economic impact concerns. The UNFCCC is primarily concerned with greenhouse gas levels in the atmosphere and the effects on the climate and the world's population.

The UNFCCC was negotiated in 1992 at the United Nations Conference on Environment and Development, better known as the Earth Summit, and was one of the two legally binding agreements drafted at the conference. (The other was the Convention on Biological Diversity, which was also signed but not ratified by the United States.) The operations of the UNFCCC are overseen by a United Nations secretariat, the executive secretary of which is Yvo de Boer (b. 1954), the Austrian-born son of a Dutch diplomat. De Boer has been criticized throughout the green community for his pro-U.S. statements, despite the United States' effective nonparticipation in the UNFCCC.

There are three categories of membership in the UNFCCC: the developed country groups of Annex I and Annex II, and the developing countries. Annex I countries are obligated to reduce greenhouse gas emissions below specific levels set by the protocols and can exceed those levels only by purchasing emission allowances or otherwise offsetting them through means agreed to by the UNFCCC. Annex II countries are a subset of the Annex I and are expected to help pay the costs of developing countries. Developing countries are expected to work toward emissions reductions as is feasible without interfering with their development and may voluntarily join Annex I, but no requirements are placed on them—a fact that some critics oppose, as long-term environmental reform requires all countries to reduce their greenhouse gas emissions. At the moment, however, the mechanism of reducing emissions does have an undeniable economic impact—perhaps short term, perhaps not so short—a justification for avoiding requiring these emissions used both by the UNFCCC with respect to developing nations and by former president George W. Bush with respect to the highly developed United States.

Purchasing emissions allowances under the authority of the UNFCCC works much as it does within a country (or political union, as with the European Union Emission Trading Scheme): A particular entity that has been assigned a cap on its emission levels (a business, when the overseeing body is a government; a country as represented by its government, when the overseeing body is the UNFCCC) wishes to exceed that cap, and so purchases allowances from other entities working under the same system. The total amount of allowances—the total maximum level of emissions—thus remains the same. It is somewhat analogous to sports teams' trading their draft picks; there is a system that determines the initial allotment, but once allowances have been allotted, they may be traded within the system. In theory this provides the potential for a significant emissions allowance market, if reducing emissions is cheaper in one country than in another, such that both profit from the trade. This market would not have significantly long-term potential, as the long-term goal is for all countries to reduce emissions substantially, which means reducing the number of allowances, eventually to such a small level that too few entities will have "slack" available for trade. (By extension, it seems reasonable to assume that if no significant progress is made toward that long-term goal as of some medium-term point in time, the allowances system will be abandoned as ineffectual and the market for them will cease to exist.) Overly active emissions allowance trade may indicate an overabundance of allowances.

The other major method by which UNFCCC members may offset their emissions excesses is through carbon projects (sinks) that, when approved by or developed with the Clean Development Mechanism and Joint Implementation projects created by the Kyoto Protocol, generate allowance credits that may be used or traded, thus increasing the number of available allowances in the pool. Potential projects include carbon capture and storage (such as scrubbing a certain amount of carbon dioxide from the atmosphere through geo-engineering), reforestation (planting trees), and development of renewable energy projects. Some believe that carbon projects have better long-term potential than emissions allowance trading. In both cases, there is for most countries a financial motive to remain within their initial allotment of allowances, and even those countries for whom it is initially cheaper or more practical to exceed that initial allotment should find it more practical in the long run to make plans to reduce emissions.

The sum of all these emissions and the sinks that offset them is called the greenhouse gas inventory. Emission inventories are regularly used by the scientific community and by governments to track changes in pollution and greenhouse gases, and they generally note the source of the emissions and their type, within a specified geographic area over a certain period of time (usually a year). Governments use this information to inform public policy and, when applicable, to confirm that no quotas have been exceeded; members of the UNFCCC are required to file annual greenhouse gas inventory reports. Scientific bodies use emission inventories to model air quality, climate change, and other types of information. The greenhouse gas inventory lists specific emissions to consider, including carbon dioxide, methane, nitrous oxide, and many fluorinated compounds (such as hydrofluorocarbons), but it has been criticized for not being inclusive enough. Various methodologies can be used in constructing the inventory, which ideally are made transparent in the final report.

Conferences

There have been 14 UNFCCC conferences—called Conferences of the Parties (COPs)—since the drafting of the UNFCCC.

COP-1 1995 Berlin: The first COP was concerned principally with the projected difficulty some countries would have meeting the promised climate change commitments. The resulting Berlin Mandate established the Analytical and Assessment Phase, a 2-year period during which countries could pick from a wide range of options to fit their economic and environmental needs. The Berlin Mandate has been since criticized for exempting from further obligation the countries outside Annex I despite the fact that the countries that were industrializing most rapidly—and who were therefore most rapidly increasing their emissions—were not Annex I countries.

COP-2 1996 Geneva: COP-2 discussed binding midterm targets and scientific findings pertaining to climate change and followed the recommendation of the United States in favoring flexible requirements for different countries rather than a uniform emissions policy to be followed globally, or even among all Annex I or II nations.

COP-3 1997 Kyoto: COP-3 established the Kyoto Protocol, which set legally binding emissions levels limits (varying nation to nation, but in the case of industrialized nations, averaging about a 7 percent reduction from 1990 levels) that must be achieved by signatories during 2008–12, the first emissions budget period of UNFCCC. The Clinton administration never formally rejected this but declined to send it to the Senate to be ratified, postponing ratification indefinitely. The subsequent Bush administration explicitly refused to send it to the Senate. As a result, the United States is a signatory of the UNFCCC but is not legally bound by the Kyoto Protocol requirements, which has brought international criticism on the nation.

COP-4 1998 Buenos Aires: Intended to resolve issues left over from COP-3, COP-4 made insufficient progress and established a two-year plan for implementing the Kyoto Protocol.

COP-5 1999 Bonn: A technical meeting addressing various minor concerns.

COP-6 2000 The Hague: Much of COP-6 discussion centered around the U.S. proposal to allow carbon projects to generate credit that would permit countries to satisfy emissions reductions requirements by "buying them down," rather than actually by reducing emissions. Many opponents felt this was against the spirit of the UNFCCC and was an unsustainable solution. Other topics of discussion were the appropriate punishments for signatory and bound countries that failed to meet their targets. Disagreement between the United States and the European Union led to the talks collapsing, and COP-6 was suspended and resumed eight months later in Bonn, in July 2001.

In the intervening time, George W. Bush was inaugurated as president of the United States, having been elected in the weeks before COP-6, and explicitly rejected the ratification of the Kyoto Protocol two months into his presidency. The United States opted not to continue negotiations in COP-6, and its representatives attended only as observers. Despite the lack of behind-the-scenes progress in those eight months, the lack of U.S. involvement made negotiations much easier and in fact, the original U.S. proposal for carbon projects was approved and the Joint Implementation and Clean Development Mechanism projects was created. Significantly, no limit was placed on how many emissions allowance credits could be generated for a country by its carbon projects. As long as the carbon projects were carried out domestically; a heavily industrialized nation could theoretically afford to increase emissions so long as it funded enough carbon projects to offset them. The terms carbon footprint and carbon neutral became part of the general public's lexicon in part because of discussions of the Kyoto Protocol and its carbon projects. Ultimately, COP-6 did not resolve the question of how to deal with unmet requirements but did stipulate that repercussions could include a binding "compliance action plan" (much like being on probation) and the suspension of the right to sell allowances.

COP-7 2001 Marrakech: COP-7 finally ironed out the details for implementing the Kyoto Protocol through the Marrakech Accords. Once more, the U.S. representatives excused themselves from negotiations. Once more, negotiations proceeded more easily without them. Major decisions included formulating the rules for international emissions trading and the specifics of the carbon projects programs.

COP-8 2002 New Delhi: Representatives discussed the commitments of developing countries, which signed the UNFCCC but were not actually bound to any specific actions. Guidelines were prepared for the technical review of greenhouse gas inventories in Annex I countries, and preparations were made for the Special Climate Change Fund and Least Developed Countries Fund, which would provide financial help (funded by Annex II) to countries trying to reduce their emissions levels.

COP-9 2003 Milan: The meeting primarily addressed procedures for reforestation carbon projects and various technical and administrative concerns.

COP-10 2004 Buenos Aires: The 10th-anniversary COP continued to develop the discussions already in progress while evaluating the state of the UNFCCC to date.

COP-11 2005 Montreal: Continued the dialogue and discussed emissions trading programs and the Clean Development Mechanism project.

COP-12 2006 Nairobi: Although several Kyoto Protocol administrative issues were discussed, the lack of extensive accomplishments led to some delegates being designated *climate tourists*, a derisive term referring to delegates attending the meeting for more the exotic locale rather than climate change initiatives.

COP-13 2007 Bali: The primary goal here was to formulate a timeline for negotiation on the critical post-2012 period that will succeed the Kyoto Protocol, with COP-14 and COP-15 being set aside for such negotiations.

COP-14 2008 Poznan: The first of two initial negotiations on the post-2012 period were held with the primary focus being on forest conservation and the financial effect of climate change on poor developing nations.

COP-15 2009 Copenhagen: COP-15 took place in December 2009, with the goal of establishing the global climate change agreement for the post-2012 period.

Membership

The current list of UNFCCC signatories is as follows:

Afghanistan
Albania
Algeria
Angola
Antigua
Barbuda
Argentina
Armenia
Australia**
Austria**
Azerbaijan
Bahamas
Bahrain
Bangladesh
Barbados
Belarus*
Belgium**
Belize
Benin
Bhutan
Bolivia
Bosnia and Herzegovina
Botswana
Brazil
Brunei
Bulgaria*
Burkina Faso
Myanmar (Burma)
Burundi
Cambodia
Cameroon
Canada**
Cape Verde
Central African Republic
Chad
Chile
China
Colombia
Comoros
Democratic Republic of the Congo
Republic of the Congo
Cook Islands
Costa Rica
Côte d'Ivoire
Croatia*
Cuba
Cyprus
Czech Republic*
Denmark**
Djibouti
Dominica
Dominican Republic
Ecuador
Egypt
El Salvador
Equatorial Guinea
Eritrea
Estonia*
Ethiopia
European Union
Fiji
Finland**
France**
Gabon
Gambia
Georgia
Germany**
Ghana
Greece**
Grenada
Guatemala
Guinea
Guinea-Bissau
Guyana

(Continued)

(Continued)

Haiti
Honduras
Hungary*
Iceland**
India
Indonesia
Iran
Ireland**
Israel
Italy**
Jamaica
Japan**
Jordan
Kazakhstan
Kenya
Kiribati
North Korea
South Korea
Kuwait
Kyrgyzstan
Laos
Latvia*
Lebanon
Lesotho
Liberia
Libya
Liechtenstein*
Lithuania*
Luxembourg**
Republic of Macedonia
Madagascar
Malawi
Malaysia
Maldives
Mali
Malta
Marshall Islands
Mauritania
Mauritius
Mexico
Federated States of Micronesia
Moldova
Monaco*
Mongolia
Montenegro
Morocco
Mozambique
Namibia
Nauru
Nepal
Netherlands**
New Zealand**
Nicaragua
Niger
Nigeria
Niue
Norway**
Oman
Pakistan
Palau
Panama
Papua New Guinea
Paraguay
Peru
Philippines
Poland*
Portugal**
Qatar
Romania*
Russia*
Rwanda
Saint Kitts and Nevis
Saint Lucia
Saint Vincent and the Grenadines
Samoa
San Marino
São Tomé and Principe
Saudi Arabia
Senegal
Serbia
Seychelles
Sierra Leone
Singapore
Slovakia*
Slovenia*
Solomon Islands
South Africa
Spain**
Sri Lanka
Sudan
Suriname
Swaziland
Sweden**
Switzerland**
Syria
Tajikistan
Tanzania
Thailand
Timor-Leste
Togo
Tonga
Trinidad and Tobago
Tunisia
Turkey*
Turkmenistan
Tuvalu
Uganda
Ukraine*
United Arab Emirates
United Kingdom**
United States**
Uruguay
Uzbekistan
Vanuatu
Venezuela
Vietnam
Yemen
Zambia
Zimbabwe.

Asterisked countries are Annex I; double-asterisked are Annex II (and by extension also Annex I).

See Also: Global Climate Change; Kyoto Protocol; UN Conference on Environment and Development.

Further Readings

Aldy, Joseph E. and Robert N. Stavins, eds. *Architectures for Agreement: Addressing Global Climate Change in a Post-Kyoto World*. New York: Cambridge University Press, 2007.
Dessler, Andrew E. and Edward A. Parson. *The Science and Politics of Global Climate Change: A Guide to the Debate*. New York: Cambridge University Press, 2006.
Stewart, Richard B. *Reconstructing Climate Policy: Beyond Kyoto*. Washington, D.C.: AEI, 2003.
The United Nations Framework Convention on Climate Change. http://unfccc.int/essential_background/convention/background/items/1353.php (Accessed July 2009).

Bill Kte'pi
Independent Scholar

Urban Planning

Although cities have always to some extent been planned, the modern profession of urban planning emerged during the 20th century in response to deteriorating conditions of contemporary industrial cities and to the dramatic transformations brought about by rapid urbanization. Planning encompasses a wide range of city-making processes and functions, including land use decision making, policy guidance, economic development, and neighborhood revitalization. It seeks to plan cities in such a way as to reconcile the goals of economic development, social justice, and environmental protection.

However, as it evolved from its early design orientation as architecture writ large to its later conceptualization as a generic methodology of rational public decision making, urban planning has become increasingly controversial, both within and without the profession. Critical planning theorists recognize that all planning is inherently distributive and therefore inherently political as well.

The planner's role as an "expert" has been questioned by many, especially in regard to achieving a proper balance between expertise and citizen input. The justifications, assumptions, methods, and outcomes of planning have all come under scrutiny and criticism, leading to new paradigms seen as more appropriate to addressing the needs and concerns of a pluralistic, globalized society.

Three basic eras characterize the arc of modern planning history: the formative decades in which the problems of the rapidly growing cities in Europe and North America inspired not just movements of reform but utopian alternatives (1800–1910); the period of institutionalization, professionalization, and self-recognition, as well as national and regional planning efforts (1910–45); and the postwar era, in which planning became standardized with systems and rational approaches, only to enter a time of crisis, diversification, and reevaluation.

As the northern industrial city expanded to unprecedented size and density in the 19th century, reformist and visionary efforts sought to address overcrowding and to separate dangerous industries from housing. Such ideas were famously realized in the so-called White City of the 1893 Columbian Exposition, which in turn inspired the American "City Beautiful" movement and its somewhat deterministic notions of inspiring civic virtue through monumentality.

In the early decades of the 20th century, informal associations concerned with urban and regional development became formalized into planning commissions, which produced smaller-scale area plans and later, metropolitan comprehensive plans. Planning at this point was seen as an extension of architecture and was more concerned with the rational layout of cities through zoning rather than explicit social policy goals.

Throughout the Depression and through the war years, urban development in North America was slowed by the Depression, as well as the exigencies of wartime production. However, New Deal–era home financing, the exploding demand for housing, and the need to rebuild European and Japanese cities after the war contributed to a tremendous demand for planning expertise in urban development and rebuilding. In the face of these challenges, planners adopted instrumentally rational decision making to implement large-scale suburbanization, urban renewal, and public works projects, largely through the use of Euclidean zoning separating residential from commercial, industrial, and institutional uses.

In the mid-20th century, planning adhered to what is referred to as its "modernist project," wherein cities were to be transformed by technological innovation and decision making informed by scientific knowledge, which was to be applied to all aspects of society through rational planning. Much like Le Corbusier, whose shocking 1925 Plan Voisin was intended to replace much of Paris' Right Bank district with identical and massive towers, modernist planners felt free to ignore historic precedent as a source of urban design. Formalism—or the notion that "form follows function"—dominated planning.

The justification for such modernist planning depended first of all on the conception of "the public good"—that is, a unitary public interest for whom the goals of planning were self-evident and shared universally—and that decision making for "the public" could be accomplished through applied comprehensive rationality.

This Rational Comprehensive Planning Model dominated planning through most of the 20th century and is still commonly used as a de facto planning model. It relies on a positivist approach to the use and analysis of data, with the underlying assumption that such rational decision making can be objective and apolitical. Having taken into account all relevant facts and sources of information, planning can be confidently made in the public interest. Rational planning allows the planner to identify a problem, identify a goal to address it, collect background data and information, identify and assess alternative scenarios, select the preferred alternative, implement the plan, and finally, monitor, evaluate, and revise the plan.

Rational Comprehensive Planning became subject to a long string of critiques starting in the 1960s and 1970s, fueled not only by public dissatisfaction with large-scale planning fiascos like the infamous Pruitt-Igoe public housing development and destructive freeway projects (to which Jane Jacobs's popular manifesto *The Death and Life of Great American Cities* contributed significantly) but also by theorizing within the discipline itself. Planners began to question almost every assumption underlying their profession, from the possibility of objectivity to the existence of a unitary public interest. Planning problems themselves were viewed as "wicked"—that is, unique and symptomatic of still further problems—that any claim to comprehensiveness in addressing them was insupportable.

In the face of these criticisms, new planning paradigms emerged. Incrementalism rejects the broad scope and unrealistic goals of comprehensiveness, opting instead to consider only a limited range of policy options, and then only in terms of a few potential outcomes. Likewise, "mixed scanning" takes exception to comprehensiveness but also rejects the minimalist aspirations of incrementalism. Instead, mixed scanning offers a compromise between both that would set reasonable limits on possible considerations included in decision making, while allowing for longer-term goal setting.

Primary among the critiques of planning is the role of the planner him or herself. Seen as an "expert" operating within a top-down hierarchy, the planner is capable of making decisions for communities. Advocacy planners argue against this model, preferring to see the planner work alongside members of the community and act on their behalf. Critics have pointed out, however, that advocacy also maintains the stance of planner as "expert," and that by claiming to speak for communities it can be disempowering.

This gets to the heart of the critiques offered by Marxist and radical planners, who question mainstream planning for its "ameliorative" stance. Politically, planners are seen as embedded within the very power structures that reinforce social and economic disadvantage and as therefore unable to address issues of equity and injustice. Instead of working only for incremental change to the status quo, planners should be more attuned to radical or transformation solutions.

Feminists have taken exception to much of what constitutes planning's "heroic narrative," dominated as it is by "founding fathers" and middle-class values. The rational, zoning-based land use planning that has dominated much of the North American landscape is seen to have been particularly detrimental to the interests of women, isolating stay-at-home mothers from all the everyday services their families need.

As a function primarily associated with state actors, planning has come under attack by libertarians, who see urban growth controls and other forms of land use regulation as an unwarranted infringement on private property rights.

These critiques and others have led to the conception of a "postmodern" planning that values pluralism, rejects solely positivist approaches and embraces reflexivity, and sees planners as fully embedded within the communities with which they plan. Rejecting as it does all metanarratives, postmodernism holds that any totalizing knowledge is impossible, so claims to comprehensive rationality are untenable. For the postmodern planner, there can only be plural "discourses" and a valorization of local "knowledges."

Informed by decades of such critiques, some planning theorists have reconceptualized planning as a communicative practice, based more on a social learning model in which planners and citizens learn from one another. Following Jurgen Habermas's notions of communicative action, based on social actors engaging qualitative interpretation of various forms of communication, "transactive" planners recognize their own embeddedness in communities and their own role in the ongoing social construction of information. Transactive planning retains a role for the planner as a valued participant—but only as one social actor among many.

For all these critiques, it must be acknowledged, however, that urban planners generally lack real political power. Actual planning and land use decisions are generally made by powerful economic and political actors, with planners acting in merely an advisory role. It remains to be seen to what extent planning's postmodern, transactive stance will serve it as environmental, social, and economic conditions grow more complex, demanding, and urgent.

See Also: People, Parks, Poverty; Political Ideology; Suburban Sprawl.

Further Readings

Allmendinger, Philip. *Planning Theory*. New York: Palgrave, 2002.

Dear, M. J. "Postmodernism and Planning." *Environment and Planning D: Society and Space*, 4/3:367–84 (1986).

Mandelbaum, Seymour J., et al., eds. *Explorations in Planning Theory*. New Brunswick, NJ: Center for Urban Policy Research, 1996.

Sandercock, Leonie. *Towards Cosmopolis: Planning for Multicultural Cities*. Toronto: J. Wiley, 1998.
Stein, Jay M., ed. *Classic Readings in Urban Planning*, 2nd ed. Chicago: American Planning Association, 2004.

Michael Quinn Dudley
University of Winnipeg

UTILITARIANISM

Moral and ethical decisions relating to environmental issues are made on the basis of ethical principles that tend to either privilege individuals or the good of the society or consider all the systems of life on the planet including those nonhuman goods or plants and animals. Utilitarianism tends to reflect the influence of democratic principles that determine policy based on majority rule and to value the greatest good of the society. Principles of utility are also reflected in economic considerations of cost-benefit analysis and free market values.

Utilitarianism was initially defined as an ethical concept of valuing an action for its contribution to happiness or pleasure and later expanded to advocate for the "greatest good for the greatest number." The liberal philosophical perspective purported by Jeremy Bentham was refined by John Stuart Mill. Advocating for societal norms rather than for the self-interest that was advanced by John Locke, Thomas Hobbes, and Adam Smith, the utilitarian position was one that supported norms that would deem the good perceived by the majority as the best practice.

Utilitarianism is often cited in situations involving quantitative measurement, economic models, and pragmatic approaches to decisions involving humans. This form of ethical analysis is influential in the areas of economics, public policy, and government regulation. If a policy tends to maximize good consequences, it is often regarded as effective. The ethical theory most commonly associated with free market economic theory is frequently called preference utilitarianism. A frequently cited example would be the conservation movement supported by agencies like the U.S. Forest Service that would argue that the environment is an instrumental value that should serve the greatest good of human interests. Preservationists argue for the intrinsic and aesthetic value of the forest that serves the needs of thousands of wildlife species and want humans kept out of many wilderness areas so that they do not corrupt the natural ecological balance. Agencies like the Forest Service rely on experts who can measure and compare options to make decisions that maximize the overall good.

Often these decisions have economic consequences. Some examples might be decisions about the development of ski areas on wilderness lands, the mining of Hopi and Navaho lands by Peabody Coal that dramatically lowered the water table, or decisions about mountaintop removal in the process of mining the Appalachians. In these cases, the highest and best use was determined to be the one that produced the highest economic yield. However, using this methodology to determine the value of clean air, clean water, or species protection is problematic because it requires expressing all values in economic terms to perform a cost-benefit analysis on outcomes. Valuing one species over another or determining that one group is more entitled to clean water than another often brings up issues of environmental racism or anthropocentrism.

The type of utilitarianism that emphasizes pleasure, or the absence of pain, is termed hedonistic utilitarianism. Animal rights advocate Peter Singer expanded the definition to include the happiness and suffering of animals. He argued that all sentient beings, those that feel pain, should have the same consideration as human beings. Singer asserted that although both humans and animals deserve equal moral consideration, humans have sophisticated mental and emotional capacities that render them more sensitive to certain types of suffering and would not grant animals identical treatment or consideration.

Environmental ethics continues to consider whether ideas of justice should be based on equality or rights. Green politics frequently rejects utilitarian reasoning in favor of advocating a change in patterns of consumption in the interest of sustainability. The utilitarian position is rejected by those who favor principles of deep ecology that consider the intrinsic value of all life, including the other-than-human world. In recent years, deep ecologists have come under fire as ecofascists with comparisons drawn between their favoring the natural world over less fortunate humans who might, for example, cut down the rainforest to produce cash crops to support their communities. Critics of deep ecology claim that favoring a natural order connects ecological values with nationalism, as in the frequently cited example of German National Socialism equating the German soil with the German people, which in their view established a link between environmental purity and racial purity.

Although many green party ideologies initially leaned toward a critique of capitalism and a tendency toward anarchism, the center has shifted from a total rejection of industrialism to a reformist agenda that concedes the necessity of working with the capitalist system and a liberal democratic framework based on a broader social justice agenda. Although the limits to growth and finitude of resources are still central issues of debate, the potential for alternative technologies and carbon offsets allows for a more pragmatic adoption of green principles in a globalized economy. Issues of individual liberty and a personal sense of morality collide with the normative democratic sensibility of majority rule and coercive methods to achieve green goals. For example, the taxation of drivers based on mileage to lower the carbon footprint of a community or the phase-out of incandescent light bulbs in favor of compact fluorescent lightbulbs mandate specific goals without allowing a choice for those who want to opt out of such programs.

See Also: Cost-Benefit Analysis; Ecofascism; Green Parties; Limits to Growth; Pinchot, Gifford.

Further Readings

Carter, Neil. *The Politics of the Environment: Ideas, Activism, Policy.* Cambridge, MA: Cambridge University Press, 2001.

DesJardins, Joseph R. *Environmental Ethics.* Belmont, CA: Thomson Wadsworth, 2006.

Singer, Peter. *Animal Liberation: A New Ethics for Our Treatment of Animals.* New York: Avon Books, 1977.

Stephens, Piers H. G. "Green Liberalisms: Nature, Agency and the Good." *Environmental Politics*, 10/3:1–22 (Autumn 2001).

Zimmerman, Michael E. *Contesting Earth's Future: Radical Ecology and Postmodernity.* Berkeley: University of California Press, 1994.

Stephanie Yuhas
University of Denver

Water Politics

Water politics refers to the multifaceted relationship between the availability and use by humans of increasingly scarce water resources on the one hand, and political processes and power relationships on the other. The study of water politics (also known as hydropolitics) has traditionally been viewed within the prism of interstate conflict and cooperation over water resources that transcend international borders. The debate over potential "water wars" between riparian states, for instance, has loomed large in the literature, countered by analyses detailing institutional mechanisms and international legal principles that induce cooperation and win–win situations between states such as the Nile Basin Initiative and the Mekong River Commission. However, other water politics–related areas such as the environment, expanded conceptions of security, and society and culture have increasingly become important, as have the roles of nonstate players and the uneven distribution of and access to water within states. Indeed, the rise of human security and right-to-water discourses has challenged traditional state-centric perspectives and the construction of particular remedies embedded in mainstream debates about water politics.

Three factors contribute to the urgency of water politics as a field of concern. First, water is a limited natural resource that is essential for life and that is indispensable for human health, human well-being, and modern society and the economy. Water drives both agricultural and industrial activity, maintains ecological assets, generates electricity, enables development and the growth of cities, alleviates poverty, and raises living standards. However, although 70 percent of the world's surface is covered with water, only about 1 percent of it is currently available for human use as freshwater; this water resource is recharged through rain and stored in rivers and underground aquifers. Most of this freshwater is used for irrigation in agriculture, which accounts for about 70 percent of global water withdrawals and increased fivefold over the past century. Industry and domestic use account for the rest.

Second, water has become increasingly scarce as a result of unsustainable rises in total global population (expected to reach 7.2 billion by 2015), water consumption patterns (which grew sixfold during the 20th century—more than twice the rate of population growth), pollution (which renders available freshwater supply unusable), and climate change (which shifts precipitation patterns and has increased desertification in vulnerable areas). As a result, surface water and, more recently, groundwater resources globally are being depleted faster than they can be replenished. This in turn has lead to alarming levels of water stress in arid and semiarid areas, as well as to the severe degradation of global freshwater ecosystems such as wetlands. The United Nations estimates that by 2025 over 5.5 billion (two-thirds of the world's population) will live in areas facing moderate to severe water stress compared with 1.76 billion people out of 5.7 billion total world population who lived under severe water stress conditions in 1995.

Finally, the distribution of and access to these scarce water resources, and the services derived from them, are both uneven and subject to power considerations internationally and within states. The areas most affected by such water shortages are in Africa, the Middle East, and south Asia, where chronic water stress and drought is common. Over 1.1 billion poor people (18 percent of the world's population) lack access to safe water, 2.6 billion (42 percent) lack access to basic sanitation, and over 2 million die annually because of water-related diseases. Moreover, industrialized-country water consumption runs as high as 380 liters/capita/day (such as in the United States), whereas many sub-Saharan countries must make due with only 20–30 liters per capita per day.

This combination of decreasing water resources and the disparities in access to, and use of, such water has led some analysts to link water scarcity with interstate conflict and to hypothesize about the increasing potential of "water wars" between states. Former World Bank vice president Ismail Serageldin famously predicted that although many of the wars of the 20th century were based on oil, wars of the 21st century would be over water. The Middle East and the surrounding region provides the setting for the most likely of such water-based conflicts, with the World Bank estimating that half the region's people already live under water stress and seven countries already use more water than is available, largely as a result of overpumping aquifers. Indeed, it is the world's driest region, with water availability per person around 1,200 m^3 per year compared with an average of about 7,000 m^3 per year globally. The Jordan River basin, for instance, contains five riparian players mediated via the larger Arab-Israeli war: Jordan, the Occupied Palestinian Territory, Israel, Syria, and Lebanon.

The conceptualization of water politics has grown beyond a narrow focus on interstate conflict and cooperation mechanisms. Instead, water politics often signifies the exacerbation of social tension between competing groups and sectors within societies as a result of both globalization and uneven domestic power structures. The Indian activist Vandana Shiva, for instance, is a leading voice within the anti-neoliberal globalization movement that has raised the issue of privatization of water supplies and their concentration among large transnational companies. Increasingly, a discourse centered on human right to water is challenging a state's authority to allocate water resources as it sees fit within traditional security and developmental paradigms.

In either case, control over such scarce water resources by particular states or groups within states confers power and potential wealth, whereas lack of such control produces insecurity and dependence.

See Also: Agriculture; Globalization; Groundwater; Power.

Further Readings

Allen, John A. *The Middle East Water Question: Hydropolitics and the Global Economy.* London: I. B. Tauris, 2002.

Elhance, Arun P. *Hydropolitics in the 3rd World: Conflict and Cooperation in International River Basins*. Washington, D.C.: U.S. Institute of Peace, 1999.
Gleick, Peter. *The World's Water 2008–2009: The Biennial Report on Freshwater Resources*. Washington, D.C.: Island Press, 2008.
Homer-Dixon, Thomas F. *Environment, Scarcity and Violence*. Princeton, NJ: Princeton University Press, 1999.
Lowi, Miriam R. *Water and Power: The Politics of a Scarce Resource in the Jordan River Basin*. Cambridge, MA: Cambridge University Press, 1995.
Postel, Sandra. *Last Oasis: Facing Water Scarcity*, 2nd ed. New York: W. W. Norton, 1997.
Shiva, Vandana. *Water Wars: Privatization, Pollution and Profit*. Boston: South End Press, 2002.
Turton, Anthony and Roland Hedwood, eds. *Hydropolitics in the Developing World: A Southern African Perspective*. Pretoria, South Africa: Center for International Political Studies, 2002.
United Nations. "Water for Life: International Decade for Action." http://www.un.org/waterforlifedecade/factsheet.html (Accessed May 2009).
Zeitoun, Mark. *Power and Water in the Middle East: The Hidden Politics of the Palestinian-Israeli Water Conflict*. London: I. B. Tauris, 2008.

Karim Makdisi
American University of Beirut

WETLANDS

Wetlands are distributed throughout the world except Antarctica. A key facet of a wetland is that water is the primary factor controlling the environment and its associated biota. Several definitions of wetlands have been described. The Ramsar definition of wetlands, which is widely used, is a broad one, encompassing not just marshes and lakes but also coral reefs, peat forests, temporary pools, riparian systems, underground caves, and other systems found everywhere from the mountains to the sea, including man-made habitats such as aquaculture ponds and irrigated agricultural lands. According to the U.S. Environmental Protection Agency, they are areas where water covers the soil or is present either at or near the surface of the soil all year or for varying periods of time during the year.

The Environmental Protection Agency defines two broad categories of wetlands: coastal or tidal wetlands and inland or nontidal wetlands. Coastal wetlands are distributed throughout almost the entire coastal zone, where saline water and the influence of tides are the defining factors. The salinity of saltwater wetlands is not a fixed value but, rather, changes with distance from the ocean, water depth, freshwater input (e.g., precipitation, runoff, ground water), and season. Nontidal or inland wetlands are most common on floodplains along rivers and streams—small to large portions of water surrounded by dry land, along the margins of lakes and ponds, and in other low-lying areas where the groundwater intercepts the soil surface or where precipitation sufficiently saturates the soil. Inland wetlands include marshes and wet meadows dominated by herbaceous plants, swamps dominated by shrubs, and wooded swamps dominated by trees.

About 6 percent of the world's surface is made up of a variety of wetlands, including swamps like this one at the Cypress Creek National Wildlife Refuge in southern Illinois.

Source: U.S. Fish & Wildlife Service

Wetlands and their functions are inextricably linked to the watershed and its surroundings, encompassing a vast and diverse range of ecosystems and, as such, representing a wide array of relationships and services to both people and biodiversity. Wetlands are valuable and among the most productive ecosystems on Earth, occupying about 6 percent of the world's land surface. They include a wide spectrum of habitats ranging from extensive peat bogs in northern latitudes to tropical mangrove forests, from seasonal ponds and marshes to floodplains and permanent riparian swamps, from freshwater shallow lakes and margins of large reservoirs to the salt lakes, brackish lagoons, estuaries, and coastal salt marshes. Extensive sea grass beds along coasts and coral reefs are also wetlands. Thus, wetlands show great differences in their habitat characteristics, hydrological regimes, water quality and soils, and the nature and diversity of their biota.

Wetlands are key to the life cycles of waterfowl and other wildlife and provide a variety of irreplaceable ecosystem services; filter our water and help provide clean, secure water sources; provide environmental and societal value by moderating the effects of droughts, floods, climate change, and erosion; maintain soil fertility; decompose and detoxify wastes; recycle essential nutrients; and provide recreation and learning opportunities for people of all ages. They are essential habitat for many species of wildlife, have the potential to remove and store greenhouse gases from the Earth's atmosphere, and are a key component of watershed management planning. One wetland acre is seven times more valuable than an acre of tropical forest in terms of ecosystem benefits. Wetlands provide numerous ecosystem goods and services not only to the local people living around them but also to communities living outside wetland areas. Mangroves are considered a natural barrier protecting the lives and property of coastal communities from storms and cyclones, flooding, and soil erosion. Values attributed to this service have been calculated by different authors to range from $3,700 per hectare to $7,700 per square kilometer.

In spite of these facts, wetlands are often given too little weight in policy decisions—not understood and appreciated—compromising the sustainability of human activities. Hence, we have to protect, enhance, and restore ecosystem goods and services of wetlands on multiple scales.

Use and User Conflicts

As in most natural resource use systems, in wetlands, too, conflicts do arise within and between the different natural resource user groups. The drivers of this type of conflict could fall into any of the four major groups of attributes associated with wetlands:

- biophysical and ecological (e.g., perceived competition for the same shrimp resource),
- social, human, cultural, and political (e.g., competition between traditional leaders and local government),

- economic (e.g., livelihood options eroded), and
- policy, institutional, and legal (e.g., lack of conflict resolution mechanisms between different fisheries gear users or between farmers and herders, or between farmers and fishermen).

Sustainable Management of Wetlands

Wetlands are dynamic areas, influenced by both natural and human factors. To maintain their biological diversity and productivity, and to permit the wise use of their resources, there is an urgent need for their conservation and sustainable use through well-focused management actions. Wetlands were the first major ecosystem to be protected by an international treaty. Much of the international effort has been directed toward supporting the concept of wise use of wetlands advocated under the Ramsar Wetlands Convention (entered into force in 1975) and supported by Wetlands International and its partners. Yet, until recently, wetlands have traditionally been perceived by policymakers as "wastelands" with no value unless drained.

The Millennium Ecosystem Assessment emphasizes that achieving the United Nations Millennium Development Goals such as the eradication of poverty, therefore, partly depends on maintaining or enhancing wetland ecosystem services, and that to do so, a cross-sectoral focus is urgently needed that emphasizes securing wetland ecosystems and their services in the context of sustainable development and improving human well-being. The following are the root causes for the mismanagement of wetlands:

- Poor understanding of local context within the framework of broader ecological, political, and economic landscapes and drivers leading to changes to natural hydrological, chemical, and physical regimes—ruthless "reclamation" for "productive uses," dumping of solid wastes and untreated sewage, unregulated upstream modifications leading to deterioration in water quality, flood damage, increased bank erosion, and reduced dry season flows
- Failure to incorporate the dynamic nature of ecological and social processes
- Inappropriate and unrealistic project formats and time frames
- Insufficient or too much devolution or empowerment and local resource stewardship institutions
- Poor monitoring and evaluation mechanisms
- Future strategies for ensuring wetland ecosystem sustainability may include, but are not limited to, the following:
    - Identify, describe, and document the wetland resources and wetland values (mapping and/or inventory/assessment of wetlands)
    - Identify, address, and document key threats and/information gaps (e.g., wetlands inventory data and mapping)
    - Construct or update a comprehensive interdisciplinary spatiotemporal database—biophysical and ecological—using geomatics (GIS, RS, GPS, etc.)
    - Define and prioritize targets for wetlands conservation and management in the region
    - Collect multidisciplinary scientific information for the legitimacy of a zoning system, linking closed seasons with a suitable system of allocating natural resources equitably during open seasons
    - Identify and take account of national and state planning and policy frameworks to enhance conservation and management (including restoration) of wetlands
    - Facilitate transboundary cooperation, strategic frameworks, and guidelines along with specific laws and institutions for ensuring conservation/sustainable use (e.g., Sunderbans)
    - Disseminate information about the importance, value, and functions of wetlands and about best-practice conservation and management of wetlands
    - Describe how wetlands condition will be monitored and reported

- Ensure suitable compensatory mechanisms/market services to encourage participation by the poor (e.g., integration of conservation with sustainable livelihoods) through livelihood enhancement and diversification aided by microfinance/credit mechanisms to support alternate income-generating activities
- Employ a combination of Sustainable Livelihoods—a "people-centered" approach, using the Sustainable Use analytical framework (the sustainability of uses of wild living natural resources)—ecological or natural resources perspective; and the institutional analysis and development framework
- Include stakeholders to avoid alienation and resistance and to accurately understand complex relationships: community-based comanagement
- Empower the community through training/capacity building for greater ownership of their new situation after project completion
- Maintain and enhance wetlands functions and values by strengthening control measures on pollution/habitat degradation/loss
- Take corrective actions for fisheries, tourism, industrial pollution, urban and construction and biodiversity use by enforcing the relevant legislation and standards governing them
- Promote pro-poor community-based ecotourism wherever feasible for integration of conservation with livelihoods
- Promote the communication, education and public awareness approach with the initial focus on the values and interests of the stakeholders, rather than exclusively on the biodiversity values

See Also: Agenda 21; Biodiversity; Sustainable Development; Urban Planning; Wilderness.

Further Readings

Ramsar Convention Secretariat. "Wise Use of Wetlands: A Conceptual Framework for the Wise Use of Wetlands." *Ramsar Handbooks for the Wise Use of Wetlands*, Vol. 1, 3rd ed. Gland, Switzerland: Ramsar Convention Secretariat, 2007.

Sellamuttu, Senaratna, et al. *Good Practices and Lessons Learnt in Integrating Ecosystem Conservation and Poverty Reduction Objectives in Wetlands*. Sri Lanka: Wetlands International, 2008.

Smardon, Richard. *Sustaining the World's Wetlands Setting Policy and Resolving Conflicts*. New York: Springer, 2009.

Van der Duim, R. and R. Henkens. *Wetlands, Poverty Reduction and Sustainable Tourism Development, Opportunities and Constraints*. Wageningen, Netherlands: Wetlands International, 2007.

Gopalsamy Poyyamoli
Kumaraswamy Ilangovan
Pondicherry University

Wilderness

Wilderness preservation historically is one of the central concerns of the environmental movement. Wilderness also is a central feature in conversations that go back for millennia about what nonhuman nature is, why it has value, and how people are related to it. Large sections of the Earth are now protected as wilderness areas. There are, however, many criticisms of wilderness and its preservation, and a debate now ensues about what wilderness is and how and why it should or should not be preserved or protected.

Some of the problems with wilderness preservation arise from controversy over what the term *wilderness* means. The term stems from the Old English term *wildeornes*, which can be traced back to *wil(d)deor*—meaning wild deer or wild animals—or *wildeorern*—wild in the sense of untamed or uncontrolled. The central etymological meaning of *wildeornes* is thus uncultivated or wild land, land that is uninhabited by people, and/or land inhabited only by wild animals. In its contemporary usage, many people think of wilderness as land that is empty of humans and human impacts. For many environmentalists, wilderness is quintessential nonhuman nature.

Some wilderness scholars argue that distinguishing cultivated from wild lands is a hallmark development that marked the transition from Paleolithic hunter-gatherer peoples to Neolithic semipermanent agricultural and herder communities approximately 10,000 to 12,000 years ago. A defining mark of wilderness was a lack of human control. As humans achieved more economic, political, scientific, and technological mastery over nature in places such as Europe, romantics in the 18th and 19th centuries championed a return to a wilder and less cultivated nature. This led to the roots of the wilderness movement in the United States through writers such as Henry David Thoreau and John Muir.

Wilderness Preservation in the United States

Wilderness preservation in the United States is often linked to the creation of Yellowstone National Park in 1872, but the self-identified wilderness movement that emerged in the 1930s with the Wilderness Society was in part a reaction against national parks. Wilderness proponents such as Robert Marshall and Aldo Leopold championed roadless wilderness areas as a foil to mechanized, commercial tourism that was prevalent in many of the national parks. Wilderness preservation found a home in the U.S. Forest Service, which administratively designated wilderness areas in 1924, 1929, and 1939. Concern that U.S. Forest Service administrative designation was inadequate led groups such as the Wilderness Society and the Sierra Club to lobby for a National Wilderness Preservation System (NWPS). After eight years of legal maneuvering and 65 rewrites, the Wilderness Act of 1964 established 54 federally designated wilderness areas totaling 9.1 million acres, all within national forests.

Section 2c of the Wilderness Act famously defined wilderness as "an area where the earth and its community of life are untrammeled by man, where man himself is a visitor who does not remain" and contrasted this with "those areas where man and his works dominate the landscape." This was qualified in a legal sense as "undeveloped Federal land retaining its primeval character and influence, without permanent improvements or human habitation, which is protected and managed so as to preserve its natural conditions and which . . . generally appears to have been affected primarily by the forces of nature, with the imprint of man's work substantially unnoticeable." Critics and proponents of wilderness preservation claim that this definition cemented what has become known as the received wilderness idea, which informs many wilderness preservation efforts in the United States and elsewhere.

Understanding wilderness areas to be natural areas, there were three important dimensions of this definition. First, the area must be undeveloped and have no permanent improvements or human inhabitants—this legally qualifies the untrammeled and primeval character. Second, the area must be roadless, as roads make human developments and inhabitation possible. Third, there is a subjective quality to wilderness naturalness, such that human impacts and trammels are substantially unnoticeable. In the 10 years following passage of the 1964 Wilderness Act, there was debate within the U.S. Forest Service about whether lands had to be pure of all prior human effects to be legal candidates for possible inclusion within the NWPS—assuming that "human" meant Euro-American and largely ignoring the effects of indigenous peoples of North America. This debate was settled by the U.S. Congress in 1975 when it rejected such a purity definition of wilderness and designated 16 new wilderness areas in the eastern United States, all of which had been substantially affected by people in the past. This implied that naturalness was not synchronic but diachronic and extended into the future as human effects disappeared across time and naturalness washed back into an area.

Since 1964, the U.S. Congress has passed over 100 other wilderness bills that have expanded the NWPS. There are, as of 2008, 704 federal wilderness areas in 44 states that collectively total 107,361,680 acres, or 4.7 percent of the land mass of the United States. There are 74 state-designated wilderness areas (2,668,903 acres) in seven states that have wilderness systems resembling the NWPS. The Wild and Scenic Rivers Act of 1968 now mandates protection of 11,409 miles of rivers in the United States, most of which are in "wild" and "scenic" categories that offer wilderness-style protection for free-flowing sections of rivers and adjacent riparian areas. In addition, there are a variety of other land classifications beyond federal and state protection schemes in the United States that classify public lands as wilderness and/or offer wilderness-style protection within counties and cities and on some Native American reservations. Finally, a number of private land efforts offer wilderness-style protection, the largest being the Nature Conservancy, which has bought and protected more than 12 million acres of land since the 1950s.

International Growth of Wilderness Movements

Concurrent with the wilderness movement in the United States there were wilderness movements in Australia, Canada, New Zealand, and South Africa, all of which established designated wilderness areas. Other countries that now have protected areas designated as wilderness include Finland, Italy, Mexico, Namibia, the Philippines, Russia, Sri Lanka, Ukraine, and Zimbabwe. Championing wilderness preservation around the world, the Wild Foundation was created in 1974 and has sponsored World Wilderness Congresses in South Africa (1977, 2001), Australia (1980), Scotland (1983), the United States (1987, 2005), Norway (1983), and India (1998), along with the upcoming 2009 Congress in Mexico.

In 1992, the World Commission on Protected Areas (International Union for the Conservation of Nature) added wilderness as a new category of protection, now spearheaded by the World Commission's Wilderness Task Force. The Wild Foundation defines wilderness as areas that are mostly biologically intact, legally protected to remain wild and free of industrial infrastructure, and open to traditional indigenous use and/or low-impact recreation. Conservation International defines wilderness areas as distinct biogeographic units or a series of ecoregions that are at least 70 percent covered by natural vegetation, have approximately 0.5 or less inhabitants per square kilometer (after subtracting urban areas), and have at least 0.5 percent of the world's 300,000 vascular plant species that are endemic to the region or contain at least 1,500 endemic vascular plant species; using these criteria, Conservation International has identified 37 major wilderness areas. By some estimates approximately 36 percent of the Earth is human dominated, 37 percent is partially disturbed by humans, and 27 percent is undisturbed. Many wilderness proponents would like to see as much land as possible in these last two categories protected as wilderness.

Criticism of Wilderness Preservation

Although support for wilderness preservation seems to be growing, wilderness has come under attack for a number of different reasons. First, there is considerable debate over what the term *wilderness* means. If *Homo sapiens* is a naturally evolved species and humans are a part of nature like many people proclaim, then it seems that everything is natural. Preserving wilderness areas because they are natural might seem nonsensical. In a contrary sense, many people claim that nature is a social construct or is something that is socially constructed or produced. If wilderness as a subset of nature also is socially constructed, preserving wilderness areas because they are natural also might seem nonsensical. Although it is not always clear what it means to say that something is socially constructed, wilderness proponents have been challenged to develop some kind of middle-ground concept of wilderness that lies between the naturalist and the social constructivist poles to escape this definitional problem.

Second, if wilderness is defined largely as an absence of people and their effects, then—beyond Antarctica—there has been no wilderness for millennia because of the presence of indigenous and native peoples. Their inhabitation, developments, and effects typically were ignored or rendered invisible by colonizing Europeans, who found North America to be in a "wilderness condition," Australia to be a vacant continent (the English doctrine of *terra nullius*), and other "new worlds." To address the problem of ethnocentrism, indigenous peoples increasingly have entered the debate over rethinking what wilderness is and how it should be managed. New work in history and the social sciences is providing more accurate accounts of past human impacts and how these impacts did and did not control and trammel landscapes.

Third, many environmental justice proponents have directed much-needed attention toward the problem of how some instances of wilderness preservation have involved the displacement and removal of indigenous and native peoples from their home landscapes, leading to dispossession and exclusion. In light of these past injustices, should wilderness areas be decommissioned, or can wilderness preservation be reconciled with social justice for marginalized peoples? This is a subset of the larger problem of nature protection versus environmental justice.

Fourth, the relationships between wilderness preservation and the sciences of ecology, conservation biology, and restoration ecology are less than clear. Some critics claim that wilderness preservation is built on the idea of preserving or freeze-framing a balance of nature, and that this flies in the face of the new nonequilibrium paradigm of ecology: In the past, natural stability was thought to be the norm in the wake of exceptional disturbances, whereas today, disturbances are thought to be the norm in the wake of exceptional stability. To be fair, many disturbance and landscape ecologists are not opposed to wilderness preservation, but their work suggests a need to rethink what it is that wilderness preservation preserves. Many people now champion biodiversity conservation, arguing that nature protection schemes such as wilderness preservation should protect biodiversity at genetic, species, and ecosystem/landscape levels. However, given the scale and scope of human impacts, biodiversity protection might necessitate significant, long-term human control and manipulation of natural entities and processes, something anathema to preserving wilderness as natural areas that are not controlled or dominated by people. Still more human control and domination might be necessary for ecological restoration efforts to repair the past damages of human impacts.

There is much to consider when thinking about the future of wilderness. Global climate change already has modified the biotic and abiotic compositions of many wilderness areas. Market forces and the treadmill of economic growth continue to turn wilderness into resources for human use and profit. An increasingly urban population of people has less contact with the natural world and more contact with virtual wilderness in theme parks and on the Internet. A rethinking of what wilderness is and what its values consist of will likely reshape a rethinking of how and why wilderness should or should not be preserved.

See Also: Conservation Enclosures; Domination of Nature; Environmental Management; Environmental Movement; Forest Service; Indigenous Peoples; People, Parks, Poverty.

Further Readings

Callicott, J. Baird and Michael P. Nelson, eds. *The Great New Wilderness Debate*. Athens: University of Georgia Press, 1998.

Kormos, Cyril F. *A Handbook on International Wilderness Law and Policy*. Golden, CO: Fulcrum, 2008.

Lewis, Michael, ed. *American Wilderness: A New History*. New York: Oxford University Press, 2007.

Nash, Roderick Frazier. *Wilderness and the American Mind*, 4th ed. New Haven, CT: Yale University Press, 2001.

Nelson, Michael P. and J. Baird Callicott, eds. *The Wilderness Debate Rages On: Continuing the Great New Wilderness Debate*. Athens: University of Georgia Press, 2008.

Oelschlaeger, Max. *The Idea of Wilderness: From Prehistory to the Age of Ecology*. New Haven, CT: Yale University Press, 1991.

Mark Woods
University of San Diego

Wise Use Movement

The term *wise use* does not describe an organization or a formal alliance but, rather, is an umbrella term for a worldview more or less shared by hundreds of grassroots and business organizations and think tanks. Despite appealing to a broad base of rural landowners and other sincere adherents, the wise use movement has faced criticism and suspicion from environmentalists, who counter that its populist positions are little more than a cynical front for the resource extraction industries that fund them.

Hostile equally to environmental regulations and environmentalists themselves, the wise use movement has since the late 1980s sought to curtail or roll back conservation measures by changing the terms of debate over the environment from protecting wilderness to protecting private property rights. Offering as its foundation a questionable interpretation of the Fifth Amendment of the U.S. Constitution protecting private property, the movement has challenged regulations aimed at preserving wilderness from resource extraction. Philosophically, the movement counters the "limits to growth" consensus on the part of environmentalists with a cornucopian outlook that posits nature's ability to provide for humanity's needs far into the future, granting our ability to exercise our ingenuity and freedom.

The rhetoric of wise use makes clear not so much that it represents an alternate viewpoint from that of mainstream green politics but, rather, that it is the enemy of environmentalism, which is viewed as irrational, pagan, antihumanist, socialist, and contrary to American values.

The fundamental philosophy of wise use can be seen to have emerged from Enlightenment classical liberal values. Most notably we can see its origins in the writings of John Locke, who argued for rights of the individual, the right to property, the right to use unclaimed nature for one's private benefit, and the limitations of governments to alienate such property from its owners without just compensation.

The term itself derives from its use by Gifford Pinchot, the first head of the U.S. Forest Service, who in 1910 called for the "wise use" of national forests and minerals. Unlike his contemporary and Sierra Club founder John Muir, who sought to preserve nature for its own sake, Pinchot believed in setting a framework for fee-based resource extraction by private interests that sought to moderate between economic and ecological priorities.

However, there is nothing moderate about its present meaning. In response to decades of surging interest in environmental protection and success in implementing corresponding governmental regulations, the movement's founder Ron Arnold sought to fire a broadside at all environmental organizations, employing the rhetoric of "war" and with the often-stated intention of essentially destroying them.

The political foundation for wise use was laid in the so-called Sagebrush Rebellion in the late 1970s, in which a coalition of cattle farmers, miners, loggers, and developers lobbied against federal control over public lands, which constitute some 60 percent of the western states. They argued that federal control and regulations—in particular the Federal Land Policy and Management Act—were adversely affecting their livelihoods and violated the principle of states' rights. This movement was successful to the extent that its language and values were adopted by the administration of President Ronald Reagan, who declared himself their ally and who appointed James Watt as Interior Secretary, following Watt's tenure as director of the Mountain States Legal Foundation, which was sympathetic to the rebellion.

The movement coalesced following the 1988 Multiple-Use Strategy Conference organized by Ron Arnold, Allan Gottlieb, and their Center for the Defense of Free Enterprise. The following year, Gottlieb and Arnold released their book *The Wise Use Agenda*, which argued for the elimination of all governmental restrictions on the use of private property, as well as for unrestricted access to public lands.

Arnold, believing that environmentalists are antihuman, terroristic, and dedicated to the destruction of modern industrialized economies, decided that an activist movement was the only thing that would be capable of defeating another activist movement.

Environmental groups have sought to counter the wise use agenda for years, even as its tenets became widely accepted in successive Republican administrations. Following President Reagan, George W. Bush chose as his Interior Secretary Gale Norton, who like Watt was associated with the Mountain States Legal Foundation. Also, Bush's Agriculture Secretary Ann Veneman once represented various wise use–related groups in California.

As articulated by Arnold, the main principles of wise use hold that humans must use the resources of the Earth to survive, and that far from being fragile, the Earth is resilient and enduring. So too are human beings, whose ingenuity and capacity to adapt can overcome the limits of the natural world so as to make its carrying capacity essentially infinite. In this, wise use is heavily cornucopian in outlook.

To operationalize this philosophy—at least in the United States—wise use advocates have turned to a controversial interpretation of the Fifth Amendment to the U.S. Constitution, which states in part that "private property [shall not] be taken for public use, without just compensation." Environmental regulations are seen under this interpretation as a "taking" for which property owners must be compensated. As well, federal control of public lands must be removed and these lands opened to the resource extraction industries, as well as to a full range of recreational uses, including all-terrain vehicles.

As documented by David Helvarg in his 1997 book *The War Against the Greens* (which he updated in 2004), the wise use movement is not so much a grassroots effort as it is a so-called astroturf movement—or front—for the mining, logging, and energy industries, as well as being tied to right-wing militia groups.

He points out that the movement has corrupted the political arena concerning environmental issues through a systematic abuse of language that sees industry-friendly objectives camouflaged in ostensibly green language: For example, a public relations coalition for the coal industry called themselves the Greening Earth Society. This creative approach to semantics was to find its way into the environmental initiatives of President George W. Bush; namely, the Clear Skies and Healthy Forests programs—the former eliminated air pollution controls, and the latter opened forests to commercial logging under the guise of fire prevention.

The other significant element of the wise use agenda that would become de facto policy under George W. Bush would be the denial of climate change and the deliberate promotion of doubt regarding climate change science.

Helvarg's accusations that the movement is an industry-backed front have been rejected not just by wise use proponents but also by independent journalists, who point out that although corporate dollars certainly are involved in funding major wise use groups, individual citizens and organizations—feeling their way of life threatened by environmental regulations—have responded sincerely to wise use arguments.

With the administration of President Barack Obama, and his declared intention of pursuing climate-friendly and green energy policies, it would appear that wise use has suffered a political setback. However, its influence has been of such endurance and significance that it will for the foreseeable future continue to influence American environmental politics.

See Also: Conservation Movement; Forest Service; Political Ideology; Sagebrush Rebellion.

Further Readings

Brick, Philip D. and R. McGreggor Cawley, eds. *A Wolf in the Garden: The Land Rights Movement and the New Environmental Debate*. Lanham, MD: Rowman & Littlefield, 1996.

Echeverria, John D. and Raymond Booth Eby. *Let the People Judge: Wise Use and the Private Property Rights Movement*. Washington, D.C.: Island Press, 1995.

Gottlieb, Alan. *The Wise Use Agenda*. Bellevue, WA: Free Enterprise, 1989.

Helvarg, David. *The War Against the Greens*. Neenah, WI: Big Earth, 2004.

Michael Quinn Dudley
University of Winnipeg

World Trade Organization

The World Trade Organization (WTO) is a central player in the process of contemporary globalization. Although its core function is to facilitate trade liberalization (the reduction or removal of tariffs and other trade practices that inhibit the free flow of goods and services) among its member states, the WTO also plays an important role in certain aspects of environmental governance. For example, the WTO is currently discussing how to liberalize trade in environmental goods and services and also plays an important role in adjudicating disputes relating to exemptions to trade rules for environmental protection purposes. This article introduces the organization's history, core principles, and institutional structure before summarizing environmental politics at the WTO.

The photograph shows a dolphin caught in a controversial type of fish seine in the Pacific in 1966. The World Trade Organization's involvement in the dolphin-safe nets dispute led to changes within the organization in the early 1990s.

Source: National Oceanic and Atmospheric Administration

The two primary functions of the WTO are to provide a forum for trading partners to negotiate legally binding rules to govern international trade liberalization, and to settle disputes among members related to the implementation and interpretation of those rules. The WTO was established in 1995 as the organization responsible for housing and servicing not only a revamped version of its predecessor, the 1947 General Agreement on Tariffs and Trade (GATT), but also a host of new international trade-related agreements. These new agreements, which emerged from the Uruguay Round of GATT negotiations, cover such areas as trade in goods, services, textiles, and agriculture, as well as rules for dispute settlement and intellectual property protection, among many trade-related topics.

The core principle of the WTO is nondiscrimination. This means that all WTO member governments must treat each other equally with respect to international trade. In this context, most-favored-nation treatment means WTO members cannot offer special terms of trade, such as lower tariffs, to one trading partner over another. Similarly, national treatment means that WTO members cannot discriminate between their own products and those imported from other member states. Although there are some exceptions to the principles of nondiscrimination—for example, to facilitate developing-country market access—the WTO agreements only permit these exceptions under strict conditions.

Institutional Structure

The core of the organization is its membership. The WTO currently has 153 members who negotiate and implement the policies that govern the bulk of international trade. The WTO's chief policymaking body is the Ministerial Conference, which consists of one representative from each member state and meets approximately every two years to make decisions on any matter related to any of the WTO agreements. Just below the Ministerial Conference hierarchically is the General Council, which is also made up of one representative from each member state; it is responsible for the operation of the WTO between Ministerial Conferences and makes all decisions by consensus. The General Council operates as the Dispute Settlement Body, administers the Trade Policy Review Mechanism, and oversees operation of all councils and committees.

The General Council delegates responsibility for the day-to-day operation of the organization to the Councils on Goods, Services, and Trade-Related Aspects of Intellectual Property Rights. The councils in turn establish various committees to discuss the various aspects of their work. In 2001, the General Council established the Trade Negotiation Committee and various additional subcommittees in "special session," such as the Committee on Trade and Environment in Special Session, to negotiate the various mandates contained in the Doha Development Agenda.

The WTO secretariat, located in Geneva, is the bureaucratic body responsible for servicing the needs of its members, including organizing meetings, taking minutes, and providing technical and professional support. The staff is primarily made up of economists, trade lawyers, and others with specialization in international trade. It currently has 625 regular staff members, representing over 70 nationalities, and is headed by a director-general, currently Pascal Lamy, who is chosen by consensus by WTO members to serve a four-year term of office. The WTO website contains more detailed explanations of all of these bodies and their work.

Environmental Politics at the WTO

WTO members discussed the relationship between trade liberalization and environmental protection/degradation long before the contemporary Doha Round of negotiations began in 2001. Although negotiations on environmental issues did not begin in earnest until 2001, discussions about the relationship between trade and environment were actually initiated in 1971, when the GATT secretariat (the WTO's precursor) produced a paper that examined the relationship between domestic environmental protection measures and international trade. The paper prompted GATT contracting parties to call for a mechanism under the GATT to examine the trade–environment relationship more closely.

Specifically, shortly after the 1971 paper was released, GATT contracting parties established the GATT Group on Environmental Measures and International Trade to discuss issues such as eco-labels, the relationship between certain multilateral environmental agreements and the WTO, and transparency of national environmental regulations. However, the Group on Environmental Measures and International Trade did not convene to discuss these issues until 20 years later in 1992, on the request of the European Free Trade Association. Environment-related discussions were ongoing in the interim, however. For example, during the Tokyo Round (1973–79), WTO members discussed the extent to which environmental regulations and standards could form barriers to trade.

Attention was further honed on trade–environment issues in the 1990s. In 1991, the GATT dispute settlement system adjudicated the well-known tuna–dolphin dispute, in which Mexico challenged U.S. regulations pertaining to dolphin-safe tuna fishing. The GATT's tuna–dolphin case hinged on the interpretation of GATT Article XX, which has become an increasingly important site of environmental governance at the WTO over time. In part, Article XX sets out conditions for exemptions to WTO rules for the purpose of environmental protection. It says that WTO members can implement nondiscriminatory policies that otherwise conflict with GATT rules if those policies are either necessary for the protection of human, animal, or plant life or health or if they relate to the conservation of a natural resource.

In 1994, three years after the tuna-dolphin dispute, the WTO membership established the Committee on Trade and Environment (CTE) as part of the Marrakech Ministerial Decision on Trade and Environment. The CTE has been operating according to its original terms of reference since it was established, with the primary aim to "build a constructive relationship between trade and environment concerns." It reports regularly to the Ministerial Conference on its work, and although the CTE has provided a forum for discussion of environmental issues in the WTO, it has yet to make any recommendations regarding substantive rule changes for environmental protection purposes.

In 2001, WTO's Trade Negotiations Committee established the CTE in Special Session to negotiate a specific set of issues delineated in the Doha Development Agenda (contained in paragraph 31). The CTE in Special Session negotiations have focused on defining the relationship between the WTO and specific trade obligations set forth in multilateral environmental agreements, procedures for information exchange with multilateral environmental agreement secretariats and for observer status, and the reduction or, as appropriate, elimination of tariff and nontariff barriers to environmental goods and services. As of the time of writing, these negotiations are ongoing.

Alongside discussions and negotiations within the CTE and CTESS, the WTO's dispute settlement system has remained the most publicly visible and contentious site of environmental politics at the WTO. The international trade regime has adjudicated numerous disputes under GATT Article XX that raised concerns about the relationship between international trade liberalization and domestic environmental protection. These concerns made the international newspapers in 1999 when activists dressed in turtle costumes targeted the Seattle Ministerial Conference to protest WTO decisions that they believed challenged the sovereign right to domestic environmental protection.

Over the years, the WTO's Panel and Appellate Body have considered the trade law compatibility of environmentally justified import restrictions and conditions on such products as shrimp, retreaded tires, gasoline, and products containing asbestos. Although some argue that the WTO has made progress over the years, making some pro-environment rulings (such as certain aspects of the shrimp–turtle case in 1998 and the asbestos case in 2001), the organization has also received a great deal of criticism from the environmental community related to its Article XX rulings. The full text of these decisions is available at the WTO's website.

Although it is common to criticize the WTO for its environmental shortcomings, it is important to note that the WTO is just an international organization—it is a forum for its member governments to discuss trade issues. It is these individual member governments, not the organization itself, that make the decisions that guide the WTO's role in global governance, including its approach to environmental politics. As the WTO continues to address environmental issues in its dispute settlement system and negotiations, it is these member governments that will make the decisions that dictate the WTO's future role in environmental politics.

See Also: Globalization; North American Free Trade Agreement; Organizations.

Further Readings

DeSombre, Elizabeth and Samuel Barkin. "Turtles and Trade: The WTO's Acceptance of Environmental Trade Restrictions." *Global Environmental Politics*, 2/1 (2002).

Esty, Daniel. *Greening the GATT*. Washington, D.C.: Institute for International Economics, 1994.

Steinberg, Richard, ed. *The Greening of Trade Law: International Trade Organizations and Environmental Issues*. Lanham, MD: Rowman & Littlefield, 2002.

World Trade Organization. "Dispute Settlement Gateway." http://www.wto.org (Accessed January 2009).

Sikina Jinnah
Brown University

Green Politics Glossary

A

Accident Site: The location of an unexpected occurrence, failure, or loss, either at a plant or along a transportation route, resulting in a release of hazardous materials.

Adulterants: Chemical impurities or substances that by law do not belong in a food, or pesticide.

Advisory: A nonregulatory document that communicates risk information to those who may have to make risk management decisions.

Affected Public: 1. The people who live and/or work near a hazardous waste site. 2. The human population adversely affected following exposure to a toxic pollutant in food, water, air, or soil.

Agent: Any physical, chemical, or biological entity that can be harmful to an organism (synonymous with stressors).

Air Quality Standards: The level of pollutants prescribed by regulations that are not be exceeded during a given time in a defined area.

Alternative Fuels: Substitutes for traditional liquid, oil-derived motor vehicle fuels like gasoline and diesel. Includes mixtures of alcohol-based fuels with gasoline, methanol, ethanol, compressed natural gas, and others.

Appliance Standards: Standards established by the U.S. Congress for energy-consuming appliances in the National Appliance Energy Conservation Act of 1987.

Asbestos: A mineral fiber that can pollute air or water and cause cancer or asbestosis when inhaled. The U.S. Environmental Protection Agency (EPA) has banned or severely restricted its use in manufacturing and construction.

B

Basalt: Consistent year-round energy use of a facility; also refers to the minimum amount of electricity supplied continually to a facility.

Best Available Control Measures (BACM): A term used to refer to the most effective measures (according to EPA guidance) for controlling small or dispersed particulates and other emissions from sources such as roadway dust and soot and ash from woodstoves and open burning of rush, timber, grasslands, or trash.

Biodegradable: Capable of decomposing under natural conditions.

Bottle Bill: Proposed or enacted legislation that requires a returnable deposit on beer or soda containers and provides for retail store or other redemption. Such legislation is designed to discourage use of throwaway containers.

Brownfields: Abandoned, idled, or underused industrial and commercial facilities/sites in which expansion or redevelopment is complicated by real or perceived environmental contamination. These sites can be in urban, suburban, or rural areas. The EPA's Brownfields initiative helps communities mitigate potential health risks and restore the economic viability of such areas or properties.

C

Carcinogen: Any substance that can cause or aggravate cancer.

Chlorinated Hydrocarbons: 1. Chemicals containing only chlorine, carbon, and hydrogen. These include a class of persistent, broad-spectrum insecticides that linger in the environment and accumulate in the food chain. Among them are dichlorodiphenyltrichloroethane (DDT), aldrin, dieldrin, heptachlor, chlordane, lindane, endrin, mirex, hexachloride, and toxaphene. Other examples include trichloroethylene, used as an industrial solvent. 2. Any chlorinated organic compounds including chlorinated solvents such as dichloromethane, trichloromethylene, and chloroform.

Clean Fuels: Blends or substitutes for gasoline fuels, including compressed natural gas, methanol, ethanol, and liquefied petroleum gas.

Comparative Risk Assessment: Process that generally uses the judgment of experts to predict effects and set priorities among a wide range of environmental problems.

Conservation: Preserving and renewing, when possible, human and natural resources. The use, protection, and improvement of natural resources according to principles that will ensure their highest economic or social benefits.

Contaminant: Any physical, chemical, biological, or radiological substance or matter that has an adverse effect on air, water, or soil.

Cost-Benefit Analysis: An economic method for assessing the costs and benefits of pursuing public policy.

Cost Recovery: A legal process by which potentially responsible parties who contributed to contamination at a Superfund site can be required to reimburse the Trust Fund for money spent during any cleanup actions by the federal government.

D

DDT: The first chlorinated hydrocarbon insecticide; chemical name dichlorodiphenyltrichloroethane. It has a half-life of 15 years and can collect in the fatty tissues of certain animals. EPA banned registration and interstate sale of DDT for virtually all but emergency uses in the United States in 1972 because of its persistence in the environment and accumulation in the food chain.

Decontamination: Removal of harmful substances such as noxious chemicals, harmful bacteria or other organisms, or radioactive material from exposed individuals, rooms and furnishings in buildings, or the exterior environment.

Disposal Facilities: Repositories for solid waste, including landfills and combustors intended for permanent containment or destruction of waste materials.

Dump: A site used to dispose of solid waste without environmental controls.

E

Ecological Entity: In ecological risk assessment, a general term referring to a species, a group of species, an ecosystem function or characteristic, or a specific habitat or biome.

Ecological Risk Assessment: The application of a formal framework, analytical process, or model to estimate the effects of human actions(s) on a natural resource and to interpret the significance of those effects in light of the uncertainties identified in each component of the assessment process.

Emergency and Hazardous Chemical Inventory: An annual report by facilities having one or more extremely hazardous substances or hazardous chemicals above certain weight limits.

Emission Cap: A limit designed to prevent projected growth in emissions from existing and future stationary sources from eroding any mandated reductions.

Endangered Species: Animals, birds, fish, plants, or other living organisms threatened with extinction by anthropogenic (man-caused) or other natural changes in their environment. Requirements for declaring a species endangered are contained in the Endangered Species Act.

Environmental Equity/Justice: Equal protection from environmental hazards for individuals, groups, or communities regardless of race, ethnicity, or economic status. This applies to the development, implementation, and enforcement of environmental laws, regulations, and policies and implies that no population of people should be forced to shoulder a disproportionate share of the negative environmental effects of pollution or environmental hazards because of a lack of political or economic strength levels.

Ethanol: An alternative automotive fuel derived from grain and corn; usually blended with gasoline to form gasohol.

F

Fossil Fuel: Fuel derived from ancient organic remains; for example, peat, coal, crude oil, and natural gas.

Free Trade Associations/Zones: Groups of nations or regions in which tariff and quota barriers are reduced or eliminated to spur increased economic activity.

Fugitive Emissions: Emissions not caught by a capture system.

G

Global Warming: An increase in the near surface temperature of the Earth. Global warming has occurred in the distant past as the result of natural influences, but the term is most often used to refer to the warming predicted to occur as a result of increased emissions of greenhouse gases. Scientists generally agree that the Earth's surface has warmed by about 1 degree Fahrenheit in the past 140 years.

Greenhouse Effect: The warming of the Earth's atmosphere attributed to a buildup of carbon dioxide or other gases; some scientists think that this build-up allows the sun's rays to heat the Earth while making the infrared radiation atmosphere opaque to infrared radiation, thereby preventing a counterbalancing loss of heat.

H

Hazard: A natural or human product or process that poses a potential or actual threat to people.

Heavy Metals: Metallic elements with high atomic weights; for example, mercury, chromium, cadmium, arsenic, and lead. They can damage living things at low concentrations and tend to accumulate in the food chain.

High Seas: Portions of the ocean beyond the limits of national jurisdictions as defined by the Third United Nations Convention on the Law of the Sea.

I

Industrial Waste: Unwanted materials from an industrial operation; may be liquid, sludge, solid, or hazardous waste.

Institution: Norms through which groups self-organize for collective gain. Distinct from "organizations," the term "institution" refers to habits, rules, and laws governing behavior.

Interstate Commerce Clause: A clause of the U.S. Constitution that reserves to the federal government the right to regulate the conduct of business across state lines.

Interstate Waters: Waters that flow across or form part of state or international boundaries; for example, the Great Lakes, the Mississippi River, or coastal waters.

L

Lead (Pb): A heavy metal that is hazardous to the health if breathed or swallowed. Its use in gasoline, paints, and plumbing compounds has been sharply restricted or eliminated by federal laws and regulations.

Lifetime Exposure: Total amount of exposure to a substance that a human would receive in a lifetime (usually assumed to be 70 years).

Litter: The highly visible portion of solid waste carelessly discarded outside the regular garbage and trash collection and disposal system.

M

Media: Specific environments—air, water, soil—that are the subject of regulatory concern and activities.

Montreal Protocol: Treaty, signed in 1987, governing stratospheric ozone protection and research, and the production and use of ozone-depleting substances.

Municipal Solid Waste: Common garbage or trash generated by industries, businesses, institutions, and homes.

N

Netting: A concept in which all emissions sources in the same area that are owned or controlled by a single company are treated as one large source, thereby allowing flexibility in controlling individual sources to meet a single emissions standard.

NIMBY: An acronym for "Not in My Backyard" that identifies the tendency for individuals and communities to oppose the siting of noxious or hazardous materials and activities in their vicinity. It implies a limited or parochial political vision of environmental justice.

O

Ozone Depletion: Destruction of the stratospheric ozone layer, which shields the Earth from ultraviolet radiation harmful to life. This destruction of ozone is caused by the breakdown of certain chlorine- and/or bromine-containing compounds (chlorofluorocarbons or halons), which break down when they reach the stratosphere and then catalytically destroy ozone molecules.

P

Permit: An authorization, license, or equivalent control document issued by the EPA or an approved state agency to implement the requirements of an environmental regulation; for example, a permit to operate a wastewater treatment plant or to operate a facility that may generate harmful emissions.

Point Source: A stationary location or fixed facility from which pollutants are discharged; any single identifiable source of pollution; for example, a pipe, ditch, ship, ore pit, or factory smokestack.

Political Ecology: a field of research concerned with the relationship of systems of social and economic power to environmental conditions, natural resources, and conservation.

Pollution: In general, the presence of a substance in the environment that because of its chemical composition or quantity prevents the functioning of natural processes and produces undesirable environmental and health effects. Under the Clean Water Act, for example, the term has been defined as the man-made or man-induced alteration of the physical, biological, chemical, and radiological integrity of water and other media.

Polychlorinated Biphenyls: A group of toxic, persistent chemicals used in electrical transformers and capacitors for insulating purposes and in gas pipeline systems as lubricant. The sale and new use of these chemicals, also known as PCBs, were banned by law in 1979.

Prior Appropriation: A doctrine of water law that allocates the rights to use water on a first-come, first-served basis.

Public Water System: A system that provides piped water for human consumption to at least 15 service connections, or regularly serves 25 individuals.

R

Radiation: Transmission of energy though space or any medium. Also known as radiant energy.

Radioactive Waste: Any waste that emits energy as rays, waves, streams, or energetic particles. Radioactive materials are often mixed with hazardous waste from nuclear reactors, research institutions, or hospitals.

Recycle/Reuse: Minimizing waste generation by recovering and reprocessing usable products that might otherwise become waste (e.g., recycling aluminum cans, paper, bottles, etc.).

Riparian Rights: Entitlement of a land owner to certain uses of water on or bordering the property, including the right to prevent diversion or misuse of upstream waters. In general, a matter of state law.

Risk: A measure of the probability that damage to life, health, property, and/or the environment will occur as a result of a given hazard.

Risk Assessment: Qualitative and quantitative evaluation of the risk posed to human health and/or the environment by the actual or potential presence and/or use of specific pollutants.

S

Smog: Air pollution typically associated with oxidants.

Solid Waste: Nonliquid, nonsoluble materials ranging from municipal garbage to industrial wastes that contain complex and sometimes hazardous substances.

Stakeholder: Any organization, governmental entity, or individual that has a stake in or may be affected by a given approach to environmental regulation, pollution prevention, energy conservation, and so on.

Structural Adjustment: A set of policies, typically imposed by multilateral lending agencies like the World Bank and the International Monetary Fund during a national financial crisis, that imposes restrictions on government trade regulations, subsidies, and labor/environmental standards.

T

Tailpipe Standards: Emissions limitations applicable to mobile source engine exhausts.

Toxicity: The degree to which a substance or mixture of substances can harm humans or animals.

Trust Fund (CERCLA): A fund set up under the Comprehensive Environmental Response, Compensation and Liability Act (CERCLA) to help pay for cleanup of hazardous waste sites and for legal action to force those responsible for the sites to clean them up.

U

Ultraviolet Rays: Radiation from the sun that can be useful or potentially harmful. Ultraviolet, or UV, rays from one part of the spectrum (UV-A) enhance plant life. UV rays from other parts of the spectrum (UV-B) can cause skin cancer or other tissue damage. The ozone layer in the atmosphere partly shields us from ultraviolet rays reaching the Earth's surface.

W

Waste Treatment Plant: A facility containing a series of tanks, screens, filters, and other processes by which pollutants are removed from water.

Watershed Approach: A coordinated framework for environmental management that focuses public and private efforts on the highest priority problems within hydrologically defined geographic areas, taking into consideration both ground and surface water flow.

Wildlife Refuge: An area designated for the protection of wild animals, within which hunting and fishing are either prohibited or strictly controlled.

Source: U.S. Environmental Protection Agency (http://www.epa.gov/OCEPAterms)

Green Politics Resource Guide

Books

Barrow, C. J. *Environmental Management and Development.* London: Routledge, 2005.
Bryant, Bunyan, ed. *Environmental Justice: Issues, Policies, and Solutions.* Washington, DC: Island, 1995.
Cerrell Associates, Inc. *Political Difficulties Facing Waste-to-Energy Conversion Plant Siting.* Prepared for the California Waste Management Board. Sacramento, CA: Cerrell Associates, Inc, 1984.
Chatterjee, Pratap and Mathias Finger. *The Earth Brokers: Power, Politics and World Development.* London: Routledge, 1994.
Coggin, Terrance P. and John M. Seidl. *Politics American Style: Race, Environment and Central Cities.* New York: Prentice-Hall, 1972.
Commission for Racial Justice, United Church of Christ. *Toxic Wastes and Race in the United States: A National Report on the Racial and Socioeconomic Characteristics of Communities With Hazardous Wastes Sites.* New York: Public Data Access Inc., 1987.
Conca, Ken and Geoffrey Dabelko, eds. *Environmental Peacemaking.* Washington, DC: Woodrow Wilson Center, 2002.
Diehl, Paul and Nils Gleditsch Petter, eds. *Environmental Conflict.* Boulder, CO: Westview, 2001.
Downie, David Leonard. "Global POPs Policy: The 2001 Stockholm Convention on Persistent Organic Pollutants," in David Leonard Downie and Terry Fenge, eds., *Northern Lights Against POPs: Combatting Toxic Threats in the Arctic.* Montreal: McGill-Queen's University Press, 2003.
Dryzek, John. *The Politics of the Earth: Environmental Discourses.* New York: Oxford University Press, 1997.
Durett, Dan. *Environmental Justice: Breaking New Ground.* Washington, DC: Committee of the National Institute for the Environment, 1993.
Gadgil, Madhav and Ramachandra Guha. *Ecology and Equity.* London: Routledge, 1995.
Gerlak, A. and L. Parisi. "An Umbrella of International Policy: The Global Environment Facility at Work," in Dennis L. Soden and Brent S. Steel, eds., *Handbook of Global Environmental Policy and Administration.* New York: Marcel Dekker, 1999.

Gerrard, Michael B. *Whose Backyard, Whose Risk: Fear and Fairness in Toxic and Nuclear Waste Siting*. Cambridge, MA: MIT Press, 1994.
Gillroy, J. M., ed. *Environmental Risk, Environmental Values, and Political Choices: Beyond Efficiency Trade-Offs in Public Policy Analysis*. Boulder, CO: Westview, 1993.
Global Environment Facility. *Producing Results for the Global Environment*. New York: Global Environment Facility, 2005.
Gould, J. M. *Quality of Life in American Neighborhoods: Levels of Affluence, Toxic Waste, and Cancer Mortality in Residential Zip Code Areas*. Boulder, CO: Westview, 1986.
Griffiths, T. and L. Robin, eds. *Ecology and Empire*. Seattle: University of Washington Press, 1997.
Harrison, D., Jr. *Who Pays for Clean Air: The Cost and Benefit Distribution of Automobile Emission Standards*. Cambridge, MA: Ballinger, 1975.
Held, David, et al. *Global Transformations: Politics, Economics and Culture*. Stanford, CA: Stanford University Press, 1999.
Hellawell, John M. *Biological Indicators of Freshwater Pollution and Environmental Management*. London: Elsevier Applied Science Publishers, 1986.
Hill, David, ed. *The Quality of Life in America; Pollution, Poverty, Power, and Fear*. New York: Holt, Rinehart and Winston, 1973.
Hofrichter, Richard, ed. *Toxic Struggles: The Theory and Practice of Environmental Justice*. Philadelphia, PA: New Society Publishers, 1993.
Horta, Korinna, et al. *The Global Environment Facility: The First Ten Years—Growing Pains or Inherent Flaws?* Washington, DC; Halifax, Canada: Environmental Defense and Halifax Initiative, 2002.
Mandelker, D. R. *Environment and Equity: A Regulatory Challenge*. New York: McGraw Hill, 1981.
Manwaring, Max, ed. *Environmental Security and Global Stability*. Lanham, MD: Lexington Books, 2002.
Marsh, George. *Man and Nature*. Cambridge, MA: Harvard University Press, 1965.
Meffe, G. K. and C. R. Carroll. *Principles of Conservation Biology*. Sunderland, MA: Sinauer Associates, 1994.
Merchant, Carolyn, ed. *Major Problems in American Environmental History*, 2nd edition. Boston, MA: Houghton Mifflin Co., 2005.
Neuman, Roderick. *Making Political Ecology*. New York: Hodder Arnold, 2003.
Perrin, Constance. *Everything in Its Place: Social Order and Land Use in America*. Princeton, NJ: Princeton University Press, 1992.
Sachs, Aaron. *Eco-Justice: Linking Human Rights and the Environment*. Worldwatch Paper 127. Washington, DC: Worldwatch Institute, 1995.
Schwab, Jim. *Deeper Shades of Green: The Rise of Blue-Collar and Minority Environmentalism in America*. San Francisco: Sierra Club, 1994.
Snow, Donald. *Inside the Environmental Movement: Meeting the Leadership Challenge*. Washington, DC: Island, 1992.
Sweet, William. *Kicking the Carbon Habit*. New York: Columbia University Press, 2006.
Turner, B. L., II, et al., eds. *The Earth as Transformed by Human Action: Global and Regional Changes in the Biosphere Over the Past 300 Years*. Cambridge, UK: Cambridge University Press, 2003.
U.S. Environmental Protection Agency. *Environmental Equity: Reducing Risk for All Communities*, Volumes 1 and 2. Office of Policy, Planning, and Evaluation,

EPA230-R-92-008 and EPA230-R-92-008A. Washington, DC: Government Printing Office, 1992.

U.S. Environmental Protection Agency. *Environmental Justice 1994 Annual Report: Focusing on Environmental Protection for All People*. Office of Administration and Resources Management (3103), EPA/200-R-95-003. Washington, DC: Government Printing Office, 1995.

U.S. Environmental Protection Agency. *Environmental Justice Strategy: Executive Order 12898*. Office of Administration and Resources Management (3103), EPA/200-R-95-002. Washington, DC: Government Printing Office, 1995.

U.S. Environmental Protection Agency. *Toxics in the Community: National and Local Perspectives*. Office of Pesticides and Toxic Substances (TS-779), EPA 560/4-91-014. Washington, DC: Government Printing Office, 1991.

U.S. General Accounting Office. *Hazardous and Nonhazardous Waste: Demographics of People Living Near Waste Facilities*. GAO/RCED-95-84. Washington, DC: Government Printing Office, 1995.

Vallette, J. *The International Trade in Wastes: A Greenpeace Inventory*. Washington, DC: Greenpeace, 1989.

Waxman, Henry A. *The Real Story Behind EPA's Environmental Equity Report: An Evaluation of Internal EPA Memoranda*. Washington, DC: US House of Representatives, 1992.

Wenz, Peter S. *Environmental Justice*. New York: State University of New York Press, 1988.

World Commission on Environment and Development. *Our Common Future*. New York: Oxford University Press, 1987.

Journals

Alternatives (Alternatives Inc.)
American Naturalist (Thomson Corporation)
Amicus Journal (National Resources Defense Council)

Biodiversity and Conservation (Chapman and Hall)
Biological Conservation (Elsevier Science)
BioScience (American Institute and Biological Sciences)
Boston College Environmental Affairs Law Review (Boston College)

Capitalism, Nature, Socialism (Routledge)
Conservation Biology (Blackwell Publishing)
Critical Reviews in Environmental Science and Technology (Taylor & Francis)

Ecological Economics (International Ecological Economics)
Energy and Environment (Multi-Science Publishing)
Environment (Voyage Publications)
Environmental Action (American Chemical Society)
Environmental Ethics (Center for Environmental Philosophy)
Environmental Law (Oxford University Press)
Environmental Management (Academic Press)
Environmental Politics (Frank Cass)
Environmental Science and Technology (Center for Environment and Energy Research and Studies)

Environment and Behavior (SAGE Publications)
EPA Journal (Environmental Protection Agency)

Global Environment Politic (MIT Press)

Human and Ecological Risk Assessment (Taylor & Francis)
Human Ecology (Springer Science and Business Media)

International Journal of Sustainable Development & World Ecology (Taylor & Francis)

Journal of Environmental Economics and Management (Academic Press)
Journal of Environmental Management (Academic Press)
Journal of Environment and Development (SAGE Publications)
Journal of Policy Analysis and Management (Wiley InterScience)
Journal of Risk: Issues in Health and Safety (Franklin Pierce Law Center)

Nature (Palgrave Macmillan)
New Scientist (Reed Business Information)

Planning (Oxford University Press)
Policy Studies Journal (Blackwell Publishing)
Population and Environment (Center for Environment and Population)
Progressive (Progressive)

Sierra (Sierra Club)
Society and Natural Resources (Routledge)

Trends in Ecology and Evolution (Oxford University Press)

Waste Age (Prism Business Media)
Whole Earth Review (Point Foundation)

Websites

Alliance for Energy and Economic Growth
www.yourenergyfuture.org

Alternative Technology Association
www.ata.org.au

Building Green, LLC
www.buildinggreen.com

Climate Ethics
www.climateethics.org

Ecology and Society
www.ecologyandsociety.org

Global Green Grants Fund
www.greengrants.org

Green Energy Council
http://greenenergycouncil.com

Green Peace
www.greenpeace.org

International Green Party Index
www.greens.org

National Energy Education Department (NEED)
www.need.org

National Renewable Energy Laboratory
www.nrel.gov

The New York Times Blog: DotEarth
www.dotearth.blogs.nytimes.com

U.S. Green Party
www.gp.org

The Wall Street Journal Blog: Environmental Capital
www.blogs.wsj.com/environmentalcapital

Worldwatch Institute
www.worldwatch.org

Green Politics Appendix

Convention on Biological Diversity

http://www.cbd.int

The purposes of the Convention on Biological Diversity, an international treaty that came into force in December 1993, are to promote the conservation, sustainable use, and fair and equitable sharing of biodiversity. This website is a resource for information about biodiversity in general as well as the convention. It includes a brief history of the convention and its text in Arabic, Chinese, English, French, Russian, and Spanish, a list of parties to the convention, with information about national focal points, strategies, reports, status, and profiles. It also contains information about programs and plans to promote the goals of the convention, a history of decision and meetings since the convention was first passed, and news items, press releases, and official statements relevant to the convention. Special sections present information about biodiversity relevant to particular groups including business, local authorities, parliamentarians, universities and the scientific community, children and youth, and nongovernmental organizations. The website also has information sections about the 2010 Biodiversity Target and the International Day for Biological Diversity.

Endangered Species Program

http://www.fws.gov/Endangered

This website, created and maintained by the U.S. Fish & Wildlife Service, includes news about conservation of endangered species in the United States, including the text of the Endangered Species Act (ESA) as well as an overview of the act, relevant policies and regulations, and a glossary of relevant terms. Indexes by species and by state list the species included, the states where they are listed, and other information such as experimental populations and listings by year. Many publications are downloadable from this site including information about the ESA, conservation partnerships with states, communities, and landowners, consultancy programs, habitat conservation planning, and the Endangered Species Recovery Program. A section of the website is devoted to the Candidate Program, which identifies species in need of protection, including a state-by-state listing of candidate species. A "Kid's Corner" provides educator resources and information for students.

Global Greens

http://www.globalgreens.org

This is the website of the Global Green Network, created in 2001 as a network of representatives of national Green Parties from all over the world. The purpose of the Global Green Network is to facilitate communication and increase understanding among Green Parties and promote the Global Green Charter which commits the parties to the guiding principles of ecological wisdom, social justice, participatory democracy, nonviolence, sustainability and respect for diversity. The Charter is available for download from the website in English, German, Spanish, Esperanto, French, Portuguese, and Swedish and the website also documents the history of the drafting process for this Charter and previous Global Greens statements. The website also includes information about Global Green Congresses, a calendar of events, press releases and Global Greens statements, information about elections and office holders, and a listing (with internet links if possible) of national Green Parties belonging to the network.

Greenpeace International

http://www.greenpeace.org/international

Greenpeace International is an independent global campaigning organization founded in 1971 that aims to promote peace and preserve the environment. The website contains information about the organization and its history as well as current causes which include climate change, preservation of the forest and marine environments, prohibition of genetically engineered food, reduction or elimination of toxic chemicals in manufacturing as well as greater regulation of their disposal, elimination of nuclear power and nuclear weapons, and encouragement of sustainable trade. Many Greenpeace reports and press releases are available from the website while many photographs and videos for press use may be previewed on the website and ordered by email. The website also includes links to national Greenpeace organizations, webcam links to the Greenpeace ships, and several blogs covering current events.

International Work Group for Indigenous Affairs

http://www.iwgia.org/sw617.asp

The International Work Group for Indigenous Affairs (IWGIA) is an international membership funded primarily by the Nordic Ministries of Foreign Affairs and the European Union that supports the human rights of indigenous peoples including self-determination, cultural integrity, right to territory and control of land and resources, and the right to development. A primary purpose of the IWGIA is the documentation and dissemination of information about indigenous peoples, and many of their publications are available for download from this website (primarily in English and Spanish, with a few in other languages). The website also includes a calendar of international meetings and a news archive. One section of the website is devoted to basic information about indigenous issues with links to further information: Topics covered include identification of indigenous peoples, climate change, sustainable development, land rights, self-determination, racism, international and national policies, political participation, and intellectual property rights. Country profiles are also available for download for countries in Africa, Asia, Latin America, and the Arctic.

Kyoto Protocol

http://unfccc.int/kyoto_protocol/items/2830.php

This website, part of the United Nations Frame Work Convention on Climate Change website, includes basic information about the history and purpose of the Kyoto Protocol as well as the downloadable text of the protocol in six languages: Arabic, Chinese, English, Spanish, French, and Russian as well as a reference manual about emissions accounting. A database provides country-by-country information about ratification and implementation status. A hypertext document includes links to further information about important terms and concepts related to the Kyoto Protocol, including emissions trading, clean development mechanism, joint implementation, registry systems, reporting, compliance, and adaptation. The site also includes links to other aspects of the Climate Change Convention, including the conference program, documents, speeches, calls for action, data, links to scientific information, press briefings, and links to information about the December 2009 conference in Copenhagen, Denmark.

Union of Concerned Scientists: Citizens and Scientists for Environmental Solutions

http://www.ucsusa.org

The Union of Concerned Scientists is a U.S.-based nonprofit organization whose purpose is to promote responsible changes in government policy, corporate practices, and consumer choice through a combination of scientific research and citizen action. The website is organized into sections on scientific integrity, global warming, clean vehicles, clean energy, nuclear power, nuclear weapons and global security, food and agriculture, and invasive species. A separate section provides information about issues the union is currently trying to influence ("action alerts"), with suggestions on how individuals can support the union's position including instructions on writing effective letters to the editor, contacting legislators, and so on. There is also a news section about different issues of concern to the union, an index to actions in different parts of the United States, and a section of fact sheets, position papers, and background information intended for legislators and other policymakers. Many union reports and newsletters are downloadable from this website as well.

Sarah Boslaugh
Washington University in St. Louis

Index

Article titles and their page numbers are in **bold**.